Principles of Zoology

Principles of Zoology

Edited by **Carlos Wyatt**

Callisto Reference
New York

Published by Callisto Reference,
106 Park Avenue, Suite 200,
New York, NY 10016, USA
www.callistoreference.com

Principles of Zoology
Edited by Carlos Wyatt

International Standard Book Number: 978-1-63239-694-5 (Hardback)

Printed in China.

Contents

Preface

This book was inspired by the evolution of our times; to answer the curiosity of inquisitive minds. Many developments have occurred across the globe in the recent past which has transformed the progress in the field.

Zoology refers to that branch of biology which deals with the study of animals. It incorporates the study of the structure, embryology, evolution, classification, habits and distribution of all species of animals. It also examines their relationship with the environment. This book traces the progress of this field and highlights some of its key discoveries. It talks in detail about the new concepts and innovations related to zoology. The topics included in it are of utmost significance and bound to provide incredible insights about all the different branches of zoology for example, zoography, animal physiology, behavioural ecology, etc. This book will prove to be a beneficial source of information for zoologists, geneticists, ecologists, professionals, students and researchers associated with the field of zoology.

This book was developed from a mere concept to drafts to chapters and finally compiled together as a complete text to benefit the readers across all nations. To ensure the quality of the content we instilled two significant steps in our procedure. The first was to appoint an editorial team that would verify the data and statistics provided in the book and also select the most appropriate and valuable contributions from the plentiful contributions we received from authors worldwide. The next step was to appoint an expert of the topic as the Editor-in-Chief, who would head the project and finally make the necessary amendments and modifications to make the text reader-friendly. I was then commissioned to examine all the material to present the topics in the most comprehensible and productive format.

I would like to take this opportunity to thank all the contributing authors who were supportive enough to contribute their time and knowledge to this project. I also wish to convey my regards to my family who have been extremely supportive during the entire project.

Editor

1

Population genetics of the fiddler crab *Uca sindensis* (Alcock, 1900) (Crustacea: Brachyura: Ocypodidae) from the Arabian Sea

Hsi-Te Shih[1], Noor Us Saher[2], Ehsan Kamrani[3], Peter KL Ng[4], Yu-Ching Lai[1] and Min-Yun Liu[5*]

Abstract

Background: The fiddler crab *Uca sindensis* (Alcock, 1900) (Crustacea: Brachyura: Ocypodidae) is distributed in the northern coasts of the Arabian Sea (Pakistan, Iran, Iraq, and Kuwait). Its typical habitat is on high intertidal areas with higher salinity, which might restrict its distribution, especially within the Persian Gulf. The purpose of the present phylogeographic study is to understand whether the Strait of Hormuz acts as a barrier to the gene flow of this species.

Results: The genetic analyses of the mitochondrial 16S rRNA, cytochrome oxidase subunit I (COI), and control region (CR) of specimens from various localities showed that there was no genetic differentiation between the populations inside and outside of the Persian Gulf.

Conclusions: We conclude that the narrow Strait of Hormuz does not form a barrier for the larval dispersal in this species. Its restricted distribution in the northern Arabian Sea may instead be associated with its preference for higher salinity sediments present in the coasts of this region.

Keywords: *Uca sindensis*; Phylogeography; Arabian Sea; Persian Gulf; 16S rRNA; COI; Control region

Background

Uca sindensis (Alcock, 1900) (Ocypodidae) is a species of fiddler crab with a relatively restricted distribution in the northern Arabian Sea. Described from Karachi, Pakistan (Alcock 1900), it has since been reported from Iran, Iraq, and Kuwait (Collins et al. 1984; Yamaguchi 1994; Apel and Türkay 1999; Naser et al. 2010; Mokhlesi et al. 2011; Naderloo and Türkay 2012). Its ecology and behavior have been studied in Kuwait (Collins et al. 1984), Iran (Mokhlesi et al. 2011), and Pakistan (Saher and Qureshi 2012).

With regard to the taxonomy of *U. sindensis*, Alcock (1900) originally considered it as a variety of *Uca inversa* (Hoffmann, 1874), a decision followed by Crane (1975). She treated it as a subspecies of *U. inversa* and established a subgenus *Amphiuca* (= *Paraleptuca* Bott, 1973) for the two taxa that also included *Uca chlorophthalmus* (H. Milne Edwards, 1837) and *Uca crassipes* (White, 1847). However, Beinlich and von Hagen (2006) established a new subgenus, *Cranuca*, exclusively for *U. inversa* on the basis of its unusual pleonal structure and form of the dactylar tooth of the major chela. They also transferred the other taxa in Crane's (1975) Indo-West Pacific *Celuca* (= *Austruca* Bott, 1973) to the subgenus *Paraleptuca*, an action later followed by Ng et al. (2008). However, Beinlich and von Hagen's (2006) definition of *Paraleptuca* seems to be paraphyletic (Naderloo et al. 2010). Shih et al. (2013a,b) studied the phylogeny of the Indo-Pacific broad-fronted group and the results support the concept of the subgenera *Austruca* (= Crane's *Celuca* for Indo-West Pacific taxa), *Parathelphuca*, and *Cranuca*, with *U. sindensis* included in *Austruca*. The morphological similarity of *U. sindensis* and *U. inversa* (cf. Alcock 1900; Crane 1975) thus appears to be superficial. In fact, Alcock (1900) had mentioned that *U. sindensis* resembles *Uca annulipes*, and this has been supported genetically (Shih et al. 2013b).

Most phylogenetic and phylogeographic studies of marine crabs use the mitochondrial 16S rRNA and cytochrome oxidase subunit I (COI) genetic markers (e.g.,

* Correspondence: mylalex@tori.narl.org.tw
[5]Taiwan Ocean Research Institute, National Applied Research Laboratories, Qieding, Kaohsiung City 85243, Taiwan
Full list of author information is available at the end of the article

Levinton et al. 1996; Sturmbauer et al. 1996; Shih and Suzuki 2008; Shih et al. 2009, 2010, 2012; Davie et al. 2010; Wong et al. 2010, 2011, 2012). In addition, the mitochondrial control region marker, with highly variable and fast divergence rates (Avise 2000), has also recently been used for phylogeographic studies of crabs, e.g., *Portunus trituberculatus* (Miers, 1876) (Portunidae) (Cho et al. 2009; Guo et al. 2012), *U. crassipes* (Aoki and Wada 2013), *Uca boninensis* (Ocypodidae) (Shih et al. 2013a), *Grapsus grapsus* (Linnaeus, 1758) (Grapsidae) (Ströher et al. 2011), *Episesarma versicolor* (Tweedie, 1940) (Sesarmidae) (Supmee et al. 2012), *Ucides cordatus* (Linnaeus, 1763) (Oliveira-Neto et al. 2007) (Ucididae), *Cardisoma guanhumi* (Latreille, in Latreille, Le Peletier, Serville & Guérin, 1828) (Oliveira-Neto et al. 2008), and *Discoplax celeste* Ng and Davie, 2012 (Gecarcinidae) (Turner et al. 2013).

The Persian Gulf is a shallow semi-enclosed basin with average depth of only 35 m. It has a low-salinity current flowing towards the northwest from the Gulf of Oman and along the coast of Iran, forming a series of eddies around the Central Trough of the Gulf. It also has a high-salinity current flowing towards the southeast, mainly along the coast of Arabia and out through the Strait of Hormuz (Thoppil and Hogan 2010). The narrow (approximate width 50 km) outlet of the Strait of Hormuz has a major influence on the circulation and salinity inside the Gulf, making it quite different from conditions outside (Johns et al. 2003). As such, it may act as a barrier to the gene flow of marine organisms, resulting in different genetic structures inside and outside the Gulf. Examples studied showing this pattern include corals (Eghtesadi-Araghi 2011), barnacles (Tsang et al. 2012), sea cucumber (Claereboudt and Al-Rashdi 2011), and reef fishes (Burt et al. 2011). However, the Persian Gulf was completely dried in glacial periods due to its shallow depth, with marine organisms reestablishing themselves in the Gulf after that. This suggests that historic factors are less important in determining geographical distribution of marine organisms (Tsang et al. 2012).

Compared with the Persian Gulf, another semi-enclosed basin, the Red Sea, has a very different geography and geological history, with a prominent sill to constrain the outflow and the formation of land bridge across the Arabian Sea during glaciation events (Johns et al. 2003; Sirocko 2003); resulting in a higher endemism (Tsang et al. 2012; DiBattista et al. 2013).

To understand whether the larval dispersal of *U. sindensis* is blocked by the narrow Strait of Hormuz, we examine the genetic variation of 16S, COI, and control region (CR) from different localities of its range. The possible biogeographic mechanisms for the limited distribution of this species are discussed.

Methods

The specimens of *U. sindensis* have been collected from Pakistan, Iran, and Iraq (Table 1; Figure 1), preserved in 70% to 95% ethanol and deposited in Zoological Collections of the Department of Life Science, National Chung Hsing University, Taichung, Taiwan (NCHUZOOL) and Zoological Reference Collection, Raffles Museum of Biodiversity Research, National University of Singapore (ZRC).

Genomic DNA was isolated from the muscle tissue of the legs by using the GeneMark tissue and cell genomic DNA purification kit (Taichung, Taiwan). A region of approximately 550 basepairs (bp) of the 5′-end of the 16S gene was selected for amplification with polymerase chain reaction (PCR) using the primers 1471 and 1472 (Crandall and Fitzpatrick 1996). A portion of the COI gene was amplified with PCR using the primers LCO1490, HCO2198 (Folmer et al. 1994), and COH6 (Schubart and Huber 2006). The PCR conditions for the above primers were denaturation for 50 s at 94°C, annealing for 70 s at 45°C to 47°C, and extension for 60 s at 72°C (40 cycles), followed by extension for 10 min at 72°C. The primers for a region of approximately 960 bp of CR gene are the recently designed CR-F4 (Shih et al. 2013a) and ILEUCAR3 (Pie et al. 2008), with the annealing temperature of 45°C to 50°C in PCR condition. Sequences were obtained by automated sequencing (Applied Biosystems 3730, Applied Biosystems Inc., Foster City, CA, USA) and were aligned manually, with the aid of ClustalW v 1.4 (Thompson et al. 1994), after verification with the complementary strand. According to Shih et al. (2013b), *U. sindensis* belongs to the monophyletic *Austruca*, so other consubgeneric species from Indian Ocean are used as outgroups, viz. *Uca albimana* (Kossmann, 1877), *U. annulipes* (H. Milne Edwards, 1837), *Uca* aff. *annulipes*, *Uca bengali* Crane, 1975, and *Uca iranica*, Pretzmann, 1971. Sequences of different haplotypes have been deposited in the DNA Data Bank of Japan (DDBJ), with other sequences which were published in earlier papers of author HTS (accession numbers see Table 1).

For a combined analysis of 16S and COI markers, phylogenetic congruence among the two dataset partitions was tested under the maximum parsimony (MP) criterion using the incongruent length difference (ILD) test (Farris et al. 1994) implemented in the PAUP* program v 4.0b10 (Swofford 2003) as the partition homogeneity test. The parameters included 1,000 reiterations of a heuristic search with 100 randomly added sequence replications and tree bisection-reconnection (TBR) branch-swapping, using Steepest Descent and the MULTREES option enabled, and total number of rearrangements for each search was limited to 500,000 to avoid excessive computation time. The topologies of the two datasets were congruent ($p = 0.901$) and as such, the sequences were combined.

Table 1 Haplotypes of 16S rRNA, COI, and CR genes

Species	Locality	Sample size	Haplotype of 16S	DDBJ accession number	Haplotype of COI	DDBJ accession number	Haplotype of CR	DDBJ accession number
U. sindensis	Pakistan: Sandspit, Karachi [1]	1	Usd-1	AB813652	Usd-C2	LC015060	Usd-CR1	LC015065
		1	Usd-1	AB813652	Usd-C3	LC015061	Usd-CR2	LC015066
		1	Usd-1	AB813652	Usd-C4	LC015062	Usd-CR3	LC015067
		1	Usd-2	LC015053	Usd-C1	AB813673	Usd-CR4	LC015068
		2	Usd-1	AB813652	Usd-C1	AB813673	Usd-CR5, 6	LC015069, LC015070
		2	Usd-1	AB813652	Usd-C2	LC015060	Usd-CR7, 8	LC015071, LC015072
		1	Usd-3	LC015054	Usd-C1	AB813673	Usd-CR9	LC015073
		1	Usd-4	LC015055	Usd-C1	AB813673	-	
	Pakistan: Sonmiani, Balochistan [2]	1	Usd-1	AB813652	Usd-C1	AB813673	Usd-CR10	LC015074
		1	Usd-1	AB813652	Usd-C2	LC015060	Usd-CR11	LC015075
		1	Usd-5	LC015056	Usd-C2	LC015060	Usd-CR12	LC015076
		1	Usd-6	LC015057	Usd-C2	LC015060	Usd-CR13	LC015077
		1	Usd-1	AB813652	Usd-C3	LC015061	-	
	Iran: Qeshm [3]	1	Usd-1	AB813652	Usd-C1	AB813673	Usd-CR14	LC015078
		1	Usd-1	AB813652	Usd-C2	LC015060	Usd-CR15	LC015079
		2	Usd-1	AB813652	Usd-C1	AB813673	Usd-CR16, 17	LC015080, LC015081
	Iran: Bandar Abbas [4]	1	-		Usd-C2	LC015060	Usd-CR18	LC015082
		1	Usd-7	LC015058	Usd-C1	AB813673	Usd-CR19	LC015083
		3	Usd-1	AB813652	Usd-C2	LC015060	Usd-CR20, 21, 23	LC015084, LC015085, LC015087
		1	Usd-1	AB813652	Usd-C5	LC015063	Usd-CR22	LC015086
	Iraq: Khur Al-Zubair [5]	1	Usd-8	LC015059	Usd-C2	LC015060	Usd-CR24	LC015088
		1	Usd-1	AB813652	Usd-C2	LC015060	Usd-CR24	LC015088
Outgroups								
U. bengali	Thailand: Ranong			AB813651		LC015064	-	
U. annulipes	India: Tamil Nadu, Porto Novo			AB471894		AB471907	-	
U. aff. *annulipes*	Madagascar			AB813648		AB813669	-	
U. iranica	Iran: Hormozgan, Qeshm			AB471897		AB471911	-	
U. albimana	Egypt: Sinai			AB471893		AB471906	-	

Haplotypes of 16S rRNA, COI, and control region (CR) genes for the specimens of *Uca sindensis* from the Arabian Sea and the outgroups. The numbers within brackets after the localities correspond to those in Figure 1.

Maximum likelihood (ML) analysis was conducted in RAxML v 7.2.6 (Stamatakis 2006) for the combined dataset. The model GTR + G (i.e., GTRGAMMA) was used for all subsets with 100 runs and found the best ML tree by comparing the likelihood scores. The robustness of the ML tree was evaluated by 1,000 bootstrap pseudoreplicates under the model GTRGAMMA. A consensus MP tree was constructed using PAUP* with 2,000 bootstrap replications of a simple heuristic search, TBR branch-swapping, and 100 random addition sequence replications. Gaps in MP tree construction were treated as missing. All characters were equally weighted. To examine the relationships of the combined 16S and COI haplotypes in detail, a gene genealogy was constructed using TCS v 1.20 (Clement et al. 2000), with the treatment of gaps as missing states.

Figure 1 Collection sites. Collection sites (red solid circles; nos. 1 to 4) for specimens of *Uca sindensis* used in this study. Red empty circle (no. 5) means the additional record of Kuwait (Collins et al. 1984).

To understand whether there is genetic differentiation between the populations inside and outside the Persian Gulf, we treated the specimens from Iran and Iraq (nos. 3 and 4; Figure 1) as one population, and another population is composed of specimens from Pakistan (nos. 1 and 2; Figure 1). The nucleotide diversity (π), number of haplotypes (nh), and haplotype diversity (h) for each population were estimated by using the program DnaSP v 5.10.01 (Librado and Rozas 2009). To determine whether patterns of sequence variation was consistent with prediction of the neutral model, Tajima's D (Tajima 1989) and Fu's F_S (Fu 1997) tests were calculated by Arlequin v 3.5.1.2 (Excoffier et al. 2005), with 1,000 permutations for the significance assessment of the neutrality test. To determine whether a population expansion is valid, the mismatch distribution, with the Harpending's raggedness index (Hri, Harpending 1994), was computed by Arlequin using the demographic expansion model for pairwise differences with 1,000 bootstrap replicates. If the Hri value is significant ($p < 0.05$), the sudden population expansion model should be rejected (Schneider and Excoffier 1999). To estimate the degree of differentiation between populations, pairwise F_{ST} between groups was calculated by Arlequin. Sometimes the value of F_{ST} is negative, which was discussed by several authors and we follow the usual treatment to round it to zero (Weir 1996; Roesti et al. 2012; Willing et al. 2012).

Results

A 548-bp segment of the 16S was amplified and aligned; six positions are variable and none are parsimoniously informative. Among the total number of sequences, eight different haplotypes are distinguished (Table 1). The studied segment of 16S sequences is AT rich (68.7%) (T, 36.0%; A, 32.7%; G, 19.9%; and C, 11.4%). For the COI gene, a 658-bp segment was compared, resulting in five different haplotypes (Table 1). The studied segment of the COI sequence is also AT rich (58.4%) (T, 32.2%; A, 27.2%; G, 17.8%; and C, 23.8%). In this gene, five positions are variable and two parsimoniously informative. A 963-bp segment of CR was compared, and 24 different haplotypes were obtained (Table 1). The segment of CR is AT rich (75.8%) (T, 35.1%; A, 40.7%; G, 10.1%; C, 14.1%), with 75 positions variable and 34 parsimony informative.

The genealogic tree of the combined 16S and COI (Figure 2) shows that there is no subdivision of *U. sindensis* specimens from various localities. The haplotype network of the combined dataset (Figure 3) shows the haplotype 'Usd-1 + Usd-C1' to be central, relative to all other haplotypes. As such, it is here regarded as the ancestral haplotype (cf. Clement et al. 2000). It is also clear that a lot of individuals, inside and outside the Gulf, share the same haplotypes. Although some haplotypes seem to exclusively appear in each locality, their difference from the common ones is only 1 or 2 bp.

The molecular diversity indices for 16S, COI, and CR are shown in Table 2. For the 16S of all populations, the nucleotide diversity per site (π) = 0.076 and the haplotype diversity (h) = 0.380; for COI, π = 0.13 and h = 0.63; and for CR, π = 1.35 and h = 0.996. The divergence of COI is about five times that of 16S and CR is about 3.5 times that of COI.

The pairwise F_{ST} value between populations is −0.0309 (and has rounded to zero) which implies that the gene flow between them is very high. As a result, the two populations are combined into a single population. Tajima's D is −1.7519 (p = 0.02) and Fu's F_S value is −27.9037 (p < 0.001). The significant negative values of the neutrality tests may imply that the population has experienced a recent expansion (Tajima 1989; Fu 1997). The Hri value is 0.08172 (p = 0.29), which is not significant and also implies that the population might experience sudden population expansion. In addition, the mismatch distribution is smooth and unimodal (Figure 4), with the simulated sum of squared deviation (SSD) being 0.00555 (p = 0.44), which also supports the recent population expansion.

Discussion

The results of the present study suggest that the Strait of Hormuz is not a barrier to gene flow between the populations of *U. sindensis* inside and outside the Gulf, at least on the basis of the low nucleotide diversities of 16S (0.08%), COI (0.13%), and CR (1.35%) (Table 2),

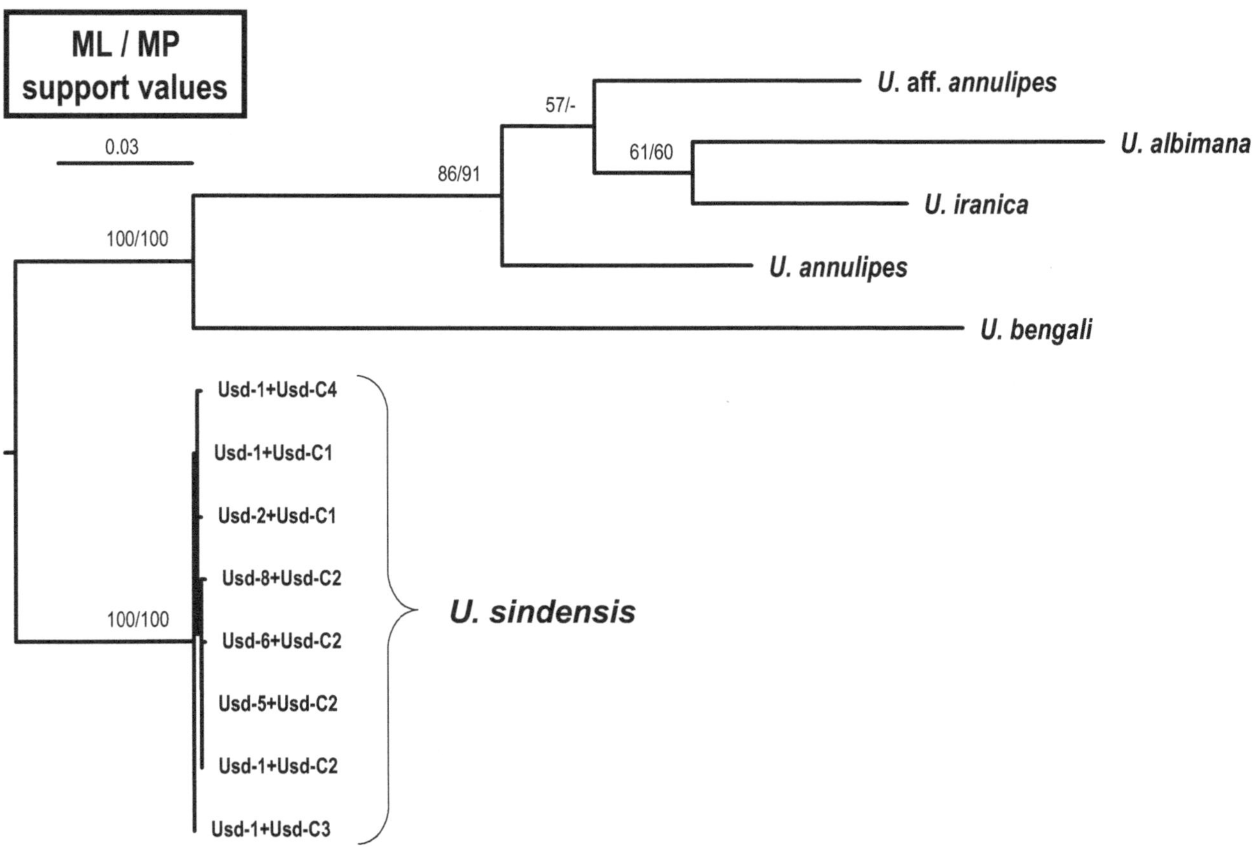

Figure 2 A maximum likelihood (ML) tree for *Uca sindensis* and outgroups. The maximum likelihood (ML) tree is based on the combined 16S and COI markers. Probability values at the nodes represent support values for ML and maximum parsimony (MP). For haplotype names, see Table 1.

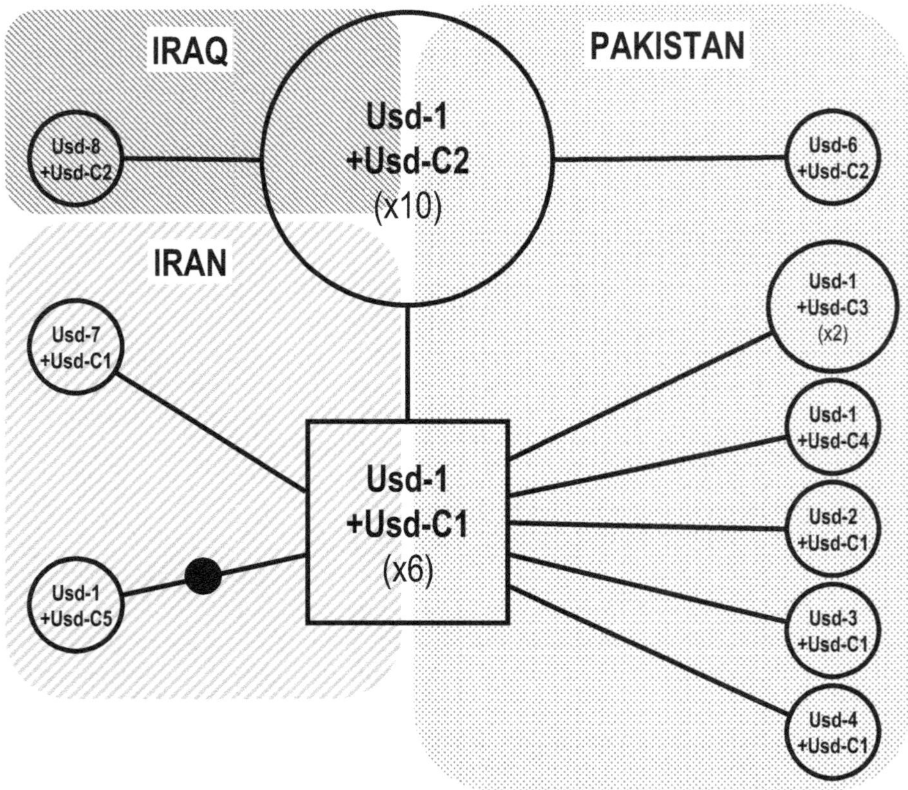

Figure 3 Genealogical network for the combined 16S and COI haplotypes observed from *Uca sindensis*. The specimens were taken from the Arabian Sea. The ancestral haplotype is indicated by a square. The unlabeled node indicates haplotype not found in the sampled populations. For haplotype names, see Table 1.

Table 2 The measures of 16S, COI, and CR

	n	nh	*π* (%)	*h*
16S				
Inside the Gulf	11	3	0.067	0.345
Outside the Gulf	13	4	0.084	0.423
Total	24	6	0.076	0.380
COI				
Inside the Gulf	11	3	0.138	0.618
Outside the Gulf	13	4	0.129	0.679
Total	24	5	0.130	0.630
CR				
Inside the Gulf	11	10	1.406	0.982
Outside the Gulf	13	13	1.290	1
Total	24	23	1.351	0.996

The measures of 16S, COI, and control region (CR) diversity for specimens of *Uca sindensis* from the Arabian Sea. Inside the Gulf includes Iraq and Iran; outside the Gulf includes only Pakistan. *n*, number of sampled individuals; nh, number of recovered haplotypes; *π*, nucleotide diversity; *h*, haplotype diversity.

genealogic tree (Figure 2), haplotype network (Figure 3), and the estimate of gene flow (F_{ST}).

Low genetic variability of COI has been reported in other *Uca* species, which is usually attributed to the high levels of larval dispersal, e.g., *U. annulipes* (= *U.* aff. *annulipes*, see Shih et al. 2013b) from East Africa (0.04%, Silva et al. 2010), *U. crassipes* from eastern Indian Ocean to Central Pacific Ocean (0.02%, Shih et al. 2012), and *Uca maracoani* from Brazilian coast (0.291%, Wieman et al. 2014). However, some *Uca* species and other intertidal crabs are known to show relatively high genetic variability, e.g., *Uca pugnax* (Smith, 1870) from the Atlantic coast of USA (0.61%, Sanford et al. 2006), *Uca splendida* (Stimpson, 1858) from Vietnam and East Asia (0.82%, Shih et al. 2012), and *Neohelice granulata* (Dana, 1851) (Varunidae) from Brazil (0.706%, Ituarte et al. 2012), which may be explained by past historical barriers to their larval dispersal that may have affected their behavior, physiology, etc. (McEdward 1995; Brodie et al. 2007; Anger et al. 2008; López-Duarte et al. 2011; discussed below).

The genetic variability of *U. sindensis*, with a nucleotide diversity of 1.35% of CR, is 3.5 times that of the COI (Table 2). This CR divergence is relatively low

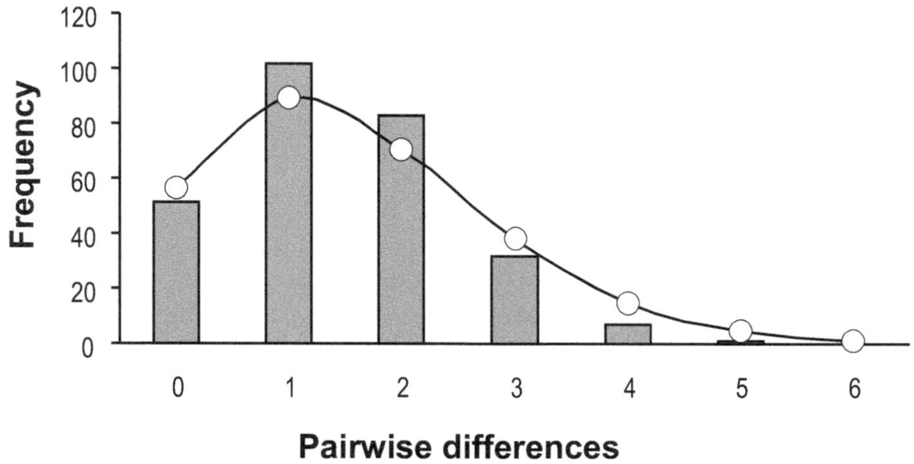

Figure 4 Mismatch distribution for the populations of *Uca sindensis* from the Arabian Sea. The observed pairwise differences are shown in bars, and the expected values under the sudden expansion model are in solid line.

compared with most crab species with wide distributions like *Portunus trituberculatus* (2.72%, Cho et al. 2009; 2.05%, Guo et al. 2012), *Cardisoma guanhumi* (3.0%, Oliveira-Neto et al. 2008; 3.1%, Pie et al. 2008), and *Ucides cordatus* (3.5%, Oliveira-Neto et al. 2007; 4.4%, Pie et al. 2008). Even a species with a restricted distribution like *U. boninensis* (known only from the Ogasawara (= Bonin) Islands, Japan) has a high nucleotide diversity (3.18%, Shih et al. 2013a). The pattern is not always easy to decipher; some species with a low diversity like *D. celeste* (1.0%, Turner et al. 2013), has a small range and is endemic to Christmas Island in the Indian Ocean, while *E. versicolor* (0.7%, Supmee et al. 2012) (studied along the Andaman Sea Coast of Thailand) has a wide Southeast Asian distribution. The highly variable CR marker may therefore provide another line of evidence to better understand the genetic structure and molecular diversity at the population and species level.

The neutrality tests and mismatch distribution (Figure 4) suggest that *U. sindensis* has undergone a recent expansion from a single source, with the most common haplotype 'Usd-1 + Usd-C1' likely to be the ancestral one (Figure 3) (Clement et al. 2000; Avise 2009). Rarer haplotypes, with only 1 or 2 bp difference from the most common haplotype (Figure 3), may have been derived relatively recently through separate mutations (Avise 2009).

Uca sindensis is restricted to the northern Arabian Sea (Pakistan, Alcock 1900) and the northern and northwestern coasts of the Persian Gulf (Kuwait, Collins et al. 1984; Iran, Apel and Türkay 1999; Iraq, Naser et al. 2010). An analysis of the fiddler crab fauna in this region may help understand the factors affecting the distribution of *U. sindensis*. In Pakistan, *U. sindensis*, *Uca urvillei*, *U. annulipes*, and *U. iranica* have been reported (Alcock 1900; Crane 1975; Saher and Qureshi 2012; Saher et al. 2014); Iran has *U. sindensis*, *U. iranica*, and *U. inversa* (Pretzmann 1971; Apel and Türkay 1999; Shih et al. 2009; Naderloo and Türkay 2012), while Kuwait has *U. sindensis* and *U. iranica* (Collins et al. 1984; Apel and Türkay 1999). Interestingly, no *Uca* species has been reported from the southern coasts of the Persian Gulf, and Apel and Türkay (1999) speculated that this may be due to the very high salinity of the area. An exception is the sea in United Arab Emirates (UAE) which has relatively lower salinity, and here, five species are present: *U. iranica*, *U. albimana*, *U. inversa*, *Uca hesperiae*, and *Uca tetragonon* (Apel and Türkay 1999; see below). It is interesting that *U. albimana*, *U. inversa*, *U. hesperiae*, and *U. tetragonon* also occur in the western part of the Indian Ocean (i.e., in the Red Sea and along East Africa) and are believed to be widespread because their larvae are dispersed by currents along the African coast (Apel and Türkay 1999). Only *U. iranica* has a comparatively restricted range.

The importance of salinity as a limiting factor for both larvae and adults of *Uca* species has previously been emphasized (Crane 1975; Hedgecock 1986; Anger 2003; Spivak and Cuesta 2009; Hopkins and Thurman 2010; López-Duarte et al. 2011; Thurman et al. 2013). Salinity has also been proposed as a major factor controlling the crab distribution in the Persian Gulf, especially for *Uca* species by Apel and Türkay (1999). They hypothesized that areas in the Persian Gulf without *Uca* are because of the very high salinities present (>40‰), with the crabs preferring habitats with salinities <39‰. It is likely that the distribution of *U. sindensis*, from Pakistan to the northwestern Persian Gulf, is also influenced by salinity.

Although *U. sindensis* is sometimes sympatric with *U. iranica*, its preferred habitat is in higher intertidal areas or even in sheltered shores far from the sea (Crane 1975; Collins et al. 1984; Mokhlesi et al. 2011). These are generally habitats with higher salinities than open waters. Within the Persian Gulf and the Gulf of Oman, *U. sindensis* (and other *Uca* spp.) occur in the coast of Iran, Iraq, and Kuwait, areas that have salinities of between 36‰ to 37‰ at least in summer (Apel and Türkay 1999; Johns et al. 2003; Naderloo and Türkay 2012). In the habitats in Pakistan (Sandspit and Korangi, Karachi), the salinity was always above 36.0‰ (Saher 2008). From the Gulf of Oman to Pakistan, the salinity is relatively lower (36.5‰ to 36‰) (Chodankar et al. 2005), but still high enough for *U. sindensis* to survive. Once we reach the western Indian subcontinent, the salinities become even lower (34‰ to 36‰) (Subrahmanyam et al. 2011), and interestingly, there are no records of *U. sindensis* here. From the known range at least, the preferred salinity for *U. sindensis* appears to be between 36‰ to 39‰, and according to Thurman et al. (2013), it should be classified as a hypersaline species since this value exceed 35‰. Other hypersaline species are *Uca burgersi* Holthuis, 1967, *Uca panacea* Novak and Salmon, 1974, *Uca rapax* (Smith, 1870), and *Uca subcylindrica* (Stimpson, 1859) from America (Thurman 2002; Thurman et al. 2013), *U. inversa* (Hoffmann, 1874) from the Indian Ocean (Spaargaren 1977), and *Uca formosensis* Rathbun, 1921 from Taiwan (Shih et al. 1999, 2005; Shih 2008).

López-Duarte et al. (2011) found that the larvae of estuarine *Uca* species have a rhythmic vertical swimming behavior following hatching an action that promotes seaward transport (and dispersal) due to the physiological stresses to low salinities in estuaries. Coastal species with euhaline habitats, on the other hand, exhibit only weakly rhythmic or arrhythmic behavior. The habitats and the adjacent coastal areas of *U. sindensis* are highly saline (see above) which is presumably suitable for the larval development, there is no need to move too far, and as such, the larvae only have weak or no vertical migration behavior (López-Duarte et al. 2011). Surface currents therefore have a greater influence, with larvae transported primarily by advection from adult habitats (Queiroga and Blanton 2005; López-Duarte et al. 2011). As such, populations may become localized due to the direction of surface currents and the hypersaline habits of the species; explaining the low genetic differentiation observed between localities. This habitat salinity hypothesis could explain the low genetic variability of *U. sindensis* in this study as well as that of other coastal and oceanic species like *U. crassipes* (Shih et al. 2012). This is in contrast to the high genetic variability present in estuarine *U. splendida* (Shih et al. 2012), *U. pugnax* (Sanford et al. 2006; López-Duarte et al. 2011), and *N. granulata* (Ituarte et al. 2012). The current understanding is that the distribution of many marine organisms in the hypersaline Red Sea have been influenced by salinity (e.g., the barnacles, Tsang et al. 2012; reef fishes, DiBattista et al. 2013). This evidence suggests that salinity plays an important role in the distribution of intertidal and shallow subtidal organisms which face extreme environment stresses frequently.

Further studies on the larval development and behavior, as well as the salinity and temperature tolerance for larvae and adults, of *U. sindensis* will be important to better understand how these aspects affect their distribution.

Conclusions

Based on the low nucleotide diversities of mitochondrial 16S, COI, and control region (CR), genealogic tree, haplotype network and the estimate of gene flow, which suggest high levels of larval dispersal, we conclude there was no genetic differentiation between the populations of *Uca sindensis* inside and outside of the Persian Gulf; and the narrow Strait of Hormuz does not form a barrier for the larval dispersal in this species. Its restricted distribution in the northern Arabian Sea may instead be associated with its preference for sediments with higher salinity (36‰–39‰) present in the coasts of this region and *U. sindensis* is best classified as a hypersaline species. The neutrality tests and mismatch distribution also suggest that this species has undergone a recent expansion from a single source, with some rarer haplotypes the result of relatively recent but separate mutations.

Competing interests

The authors declare that they have no competing interests.

Authors' contributions

HTS conceived this study, performed the molecular genetic analysis, and drafted the manuscript. NUS and EK collected and processed the samples and drafted the manuscript. PKLN participated in the discussion and drafted the manuscript. YCL carried out the molecular work and helped draft the manuscript. MYL conceived this study and carried out the molecular work. All authors read and approved the final manuscript.

Acknowledgements

This study was supported by grants from the National Science Council (NSC 98-2621-B-005-001-MY3, 101-2621-B-005-001-MY3), Executive Yuan, Taiwan, to HTS. Thanks are also due to the members of the HTS's laboratory for their help in molecular work. We acknowledge two anonymous referees who greatly improved this manuscript.

Author details

[1]Department of Life Science, National Chung Hsing University, Taichung 40227, Taiwan. [2]University of Karachi, Karachi 75270, Pakistan. [3]Marine Biology Department, Hormozgan University, Bandar Abbas 79145, Iran. [4]Department of Biological Sciences, National University of Singapore, Singapore 119260, Republic of Singapore. [5]Taiwan Ocean Research Institute, National Applied Research Laboratories, Qieding, Kaohsiung City 85243, Taiwan.

References

Alcock A (1900) Materials for a carcinological fauna of India. No. 6. The Brachyura Catometopa or Grapsoidea. J Asia Soc Bengal 69:279–456

Anger K (2003) Salinity as a key parameter in the larval biology of decapod crustaceans. Invertebr Reprod Develop 43:29–45

Anger K, Spivak E, Luppi T, Bas C, Ismael D (2008) Larval salinity tolerance of the South American salt marsh crab, *Neohelice* (*Chasmagnathus*) *granulata*: physiological constraints to estuarine retention, export and reimmigration. Helgol Mar Res 62:93–102

Aoki M, Wada K (2013) Genetic structure of the wide-ranging fiddler crab *Uca crassipes* in the west Pacific region. J Mar Biol Ass UK 93:789–795

Apel M, Türkay M (1999) Taxonomic composition, distribution and zoogeographic relationships of the grapsid and ocypodid crab fauna of intertidal soft bottoms in the Arabian Gulf. Estuar Coast Shelf Sci 49(Suppl A):131–142

Avise JC (2000) Phylogeography: the history and formation of species. Harvard University Press, Cambridge

Avise JC (2009) Phylogeography: retrospect and prospect. J Biogeogr 36:3–15

Beinlich B, von Hagen HO (2006) Materials for a more stable subdivision of the genus *Uca* Leach. Zool Meded Leiden 80:9–32

Brodie RJ, Styles R, Borgianini S, Godley J, Butler K (2007) Larval mortality during export to the sea in the fiddler crab *Uca minax*. Mar Biol 152:1283–1291

Burt JA, Feary DA, Bauman AG, Usseglio P, Cavalcante GH, Sale PF (2011) Biogeographic patterns of reef fish community structure in the northeastern Arabian Peninsula. ICES J Mar Sci 68:1–9

Cho EM, Min GS, Kanwal S, Hyun YS, Park SW, Chung KW (2009) Phylogenetic analysis of mitochondrial DNA control region in the swimming crab, *Portunus trituberculatus*. Anim Cells Syst 13:305–314

Chodankar AR, Banakar VK, Oba T (2005) Past 100 ky surface salinity gradient response in the Eastern Arabian Sea to the summer monsoon variation recorded by δ18 O of *G. sacculifer*. Glob Planet Change 47:135–142

Claereboudt MR, Al-Rashdi KM (2011) Shallow-water sea cucumber inventory in the Sultanate of Oman. SPC Beche-de-mer Inform Bull 31:25–29

Clement M, Posada D, Crandall KA (2000) TCS: a computer program to estimate gene genealogies. Mol Ecol 9:1657–1659

Collins MJ, Jones DA, Clayton DA (1984) Redescription of *Uca sindensis* (Alcock, 1900) (Brachyura: Ocypodidae) with notes on the ecology of a population from Kuwait. J Crustacean Biol 4:318–328

Crandall KA, Fitzpatrick JFJ (1996) Crayfish molecular systematics: using a combination of procedures to estimate phylogeny. Syst Biol 45:1–26

Crane J (1975) Fiddler crabs of the world (Ocypodidae: genus *Uca*). Princeton University Press, Princeton

Davie PJF, Shih HT, Chan BKK (2010) A new species of *Mictyris* (Decapoda, Brachyura, Mictyridae) from the Ryukyu Islands, Japan. Crustaceana Monogr 11:83–105

DiBattista JD, Berumen ML, Gaither MR, Rocha LA, Eble JA, Choat JH, Craig MT, Skillings DJ, Bowen BW (2013) After continents divide: comparative phylogeography of reef fishes from the Red Sea and Indian Ocean. J Biogeogr 40:1170–1181

Eghtesadi-Araghi P (2011) Coral reefs in the Persian Gulf and Oman Sea: an integrated perspective on some important stressors. J Fish Aquat Sci 6:48–56

Excoffier L, Laval G, Schneider S (2005) Arlequin ver. 3.0: an integrated software package for population genetics data analysis. Evol Bioinf Online 1:47–50

Farris JS, Källersjö M, Kluge AG, Bult C (1994) Testing significance of incongruence. Cladistics 10:315–319

Folmer O, Black M, Hoeh W, Lutz R, Vrijenhoek R (1994) DNA primers for amplification of mitochondrial cytochrome c oxidase subunit I from diverse metazoan invertebrates. Mol Mar Biol Biotechnol 3:294–299

Fu YX (1997) Statistical tests of neutrality of mutations against population growth, hitchhiking and background selection. Genetics 147:915–925

Guo EM, Liu Y, Cui ZX, Li XL, Cheng YX, Wu XG (2012) Genetic variation and population structure of swimming crab (*Portunus trituberculatus*) inferred from mitochondrial control region. Mol Biol Rep 39:1453–1463

Harpending HC (1994) Signature of ancient population growth in a low resolution mitochondrial DNA mismatch distribution. Hum Biol 66:591–600

Hedgecock D (1986) Is gene flow from pelagic larval dispersal important in the adaptation and evolution of marine invertebrates? Bull Mar Sci 39:550–564

Hopkins MJ, Thurman CL (2010) The geographic structure of morphological variation in eight species of fiddler crabs (Ocypodidae: genus *Uca*) from the eastern United States and Mexico. Biol J Linn Soc 100:248–270

Ituarte RB, D'Anatro A, Luppi TA, Ribeiro PD, Spivak ED, Iribarne OO, Lessa EP (2012) Population structure of the SW Atlantic estuarine crab *Neohelice granulata* throughout its range: a genetic and morphometric study. Estuar Coast 35:1249–1260

Johns WE, Yao F, Olson DB, Josey SA, Grist JP, Smeed DA (2003) Observations of seasonal exchange through the Straits of Hormuz and the inferred heat and freshwater budgets of the Persian Gulf. J Geophys Res 108(C12):3391

Levinton JS, Sturmbauer C, Christy J (1996) Molecular data and biogeography: resolution of a controversy over evolutionary history of a pan-tropical group of invertebrates. J Exp Mar Biol Ecol 203:117–131

Librado P, Rozas J (2009) DnaSP v5: a software for comprehensive analysis of DNA polymorphism data. Bioinformatics 25:1451–1452

López-Duarte PC, Christy JH, Tankersley RA (2011) A behavioral mechanism for dispersal in fiddler crab larvae (genus *Uca*) varies with adult habitat, not phylogeny. Limnol Oceanogr 56:1879–1892

McEdward L (1995) Ecology of marine invertebrate larvae. CRC Press, Boca Raton

Mokhlesi A, Kamrani E, Backwell P, Sajjadi M (2011) Study on the behaviour of two fiddler crabs, *Uca sindensis* and *Uca annulipes* (Decapoda: Ocypodidae), in Bandar Abbas, Iran. J Mar Biol Ass UK 91:245–249

Naderloo R, Türkay M (2012) Decapod crustaceans of the littoral and shallow sublittoral Iranian coast of the Persian Gulf: faunistics, biodiversity and zoogeography. Zootaxa 3374:1–67

Naderloo R, Türkay M, Chen HL (2010) Taxonomic revision of the wide front fiddler crabs of the *Uca lactea* group (Crustacea: Decapoda: Brachyura: Ocypodidae) in the Indo-West Pacific. Zootaxa 2500:1–38

Naser MD, Ali MH, Yasser AG (2010) New record of the fiddler crab *Uca* (*Paraleptuca*) *sindensis* (Crustacea: Brachyura: Ocypodidae) from Khor Al Zubair, Basrah, Iraq. Mar Biodivers Rec 3:e87

Ng PKL, Guinot D, Davie PJF (2008) Systema Brachyurorum: Part I. An annotated checklist of extant brachyuran crabs of the world. Raffles Bull Zool Suppl 17:1–296

Oliveira-Neto JF, Pie MR, Boeger WA, Ostrensky A, Baggio RA (2007) Population genetics and evolutionary demography of *Ucides cordatus* (Decapoda: Ocypodidae). Mar Ecol 28:460–469

Oliveira-Neto JF, Pie MR, Chammas MA, Ostrensky A, Boeger WA (2008) Phylogeography of the blue land crab, *Cardisoma guanhumi* (Decapoda: Gecarcinidae) along the Brazilian coast. J Mar Biol Ass UK 88:1417–1423

Pie MR, Oliveira-Neto JF, Boeger WA, Ostrensky A, Baggio RA (2008) The organization of the mitochondrial control region in 2 brachyuran crustaceans: *Ucides cordatus* (Ocypodidae) and *Cardisoma guanhumi* (Gecarcinidae). J Hered 99:432–437

Pretzmann G (1971) Ergebnisse einiger Sammelreisen nach Vorderasien 2. Teil: marine Brachyura. Ann Naturhist Mus Wien 75:477–487

Queiroga H, Blanton J (2005) Interactions between behaviour and physical forcing in the control of horizontal transport of decapod crustacean larvae. Adv Mar Biol 47:107–214

Roesti M, Salzburger W, Berner D (2012) Uninformative polymorphisms bias genome scans for signatures of selection. BMC Evol Biol 12:94

Saher NU (2008) Population dynamics and biology of fiddler crabs in the mangrove areas of Karachi coast. Ph.D. Dissertation, University of Karachi, Karachi, Pakistan

Saher NU, Qureshi NA (2012) Spatial distribution of *Uca sindensis* (Crustacea, Ocypodidae) along the coast of Pakistan. Egypt Acad J Biol Sci Zool 4:119–129

Saher NU, Sahir O, Shih HT, Kamal M, Qureshi NA (2014) On a new record of the genus *Uca* (Decapoda, Brachyura, Ocypodidae) found along the coast of Pakistan. Crustaceana 87:666–673

Sanford E, Holzman SB, Haney RA, Rand DM, Bertness MD (2006) Larval tolerance, gene flow, and the northern geographic range limit of fiddler crabs. Ecology 87:2882–2894

Schneider S, Excoffier L (1999) Estimation of past demographic parameters from the distribution of pairwise differences when the mutation rates very among sites: application to human mitochondrial DNA. Genetics 152:1079–1089

Schubart CD, Huber MGJ (2006) Genetic comparisons of German populations of the stone crayfish, *Austropotamobius torrentium* (Crustacea: Astacidae). Bull Fr Peche Piscic 380–381:1019–1028

Shih HT (2008) A study on the fiddler crab, *Uca formosensis*, and its habitat. Hsinchu City Government, Hsinchu (in Chinese)

Shih HT, Suzuki H (2008) Taxonomy, phylogeny, and biogeography of the endemic mudflat crab *Helice/Chasmagnathus* complex (Crustacea: Brachyura: Varunidae) from East Asia. Zool Stud 47:114–125

Shih HT, Mok HK, Chang HW, Lee SC (1999) Morphology of *Uca formosensis* Rathbun, 1921 (Crustacea: Decapoda: Ocypodidae), an endemic fiddler crab from Taiwan, with notes on its ecology. Zool Stud 38:164–177

Shih HT, Mok HK, Chang HW (2005) Chimney building by male *Uca formosensis* Rathbun, 1921 (Crustacea, Decapoda, Ocypodidae) after pairing: a new hypothesis for chimney function. Zool Stud 44:242–251

Shih HT, Kamrani E, Davie PJF, Liu MY (2009) Genetic evidence for the recognition of two fiddler crabs, *Uca iranica* and *U. albimana* (Crustacea: Brachyura: Ocypodidae), from the northwestern Indian Ocean, with notes on the *U. lactea* species complex. Hydrobiologia 635:373–382

Shih HT, Naruse T, Ng PKL (2010) *Uca jocelynae* sp. nov., a new species of fiddler crab (Crustacea: Brachyura: Ocypodidae) from the Western Pacific. Zootaxa 2337:47–62

Shih HT, Ng PKL, Wong KJH, Chan BKK (2012) *Gelasimus splendidus* Stimpson, 1858 (Crustacea: Brachyura: Ocypodidae), a valid species of fiddler crab from the northern South China Sea and Taiwan Strait. Zootaxa 3490:30–47

Shih HT, Komai T, Liu MY (2013a) A new species of fiddler crab from the Ogasawara (Bonin) Islands, Japan, separated from the widely-distributed sister species *Uca (Paraleptuca) crassipes* (White, 1847) (Crustacea: Decapoda: Brachyura: Ocypodidae). Zootaxa 3746:175–193

Shih HT, Ng PKL, Liu MY (2013b) Systematics of the Indo-West Pacific broad-fronted fiddler crabs (Crustacea: Ocypodidae: genus *Uca*). Raffles Bull Zool 61:641–649

Silva IC, Mesquita N, Paula J (2010) Lack of population structure in the fiddler crab *Uca annulipes* along an East African latitudinal gradient: genetic and morphometric evidence. Mar Biol 157:1113–1126

Sirocko F (2003) Ups and downs in the Red Sea. Nature 423:813–814

Spaargaren DH (1977) On the water and salt economy of some decapod crustaceans from the Gulf of Aqaba (Red Sea). Neth J Sea Res 11:99–106

Spivak ED, Cuesta J (2009) The effect of salinity on larval development of *Uca tangeri* (Eydoux, 1835) (Brachyura: Ocypodidae) and new findings of the zoeal morphology. Sci Mar 73:297–305

Stamatakis A (2006) RAxML-VI-HPC: maximum likelihood-based phylogenetic analyses with thousands of taxa and mixed models. Bioinformatics 22:2688–2690

Ströher PR, Firkowski CR, Freire AS, Pie MR (2011) A molecular method for the detection of sally lightfoot crab larvae (*Grapsus grapsus*, Brachyura, Grapsidae) in plankton samples. Genet Mol Biol 34:529–531

Sturmbauer C, Levinton JS, Christy J (1996) Molecular phylogeny analysis of fiddler crabs: test of the hypothesis of increasing behavioral complexity in evolution. Proc Natl Acad Sci U S A 93:10855–10857

Subrahmanyam B, Murty VSN, Heffner DM (2011) Sea surface salinity variability in the tropical Indian Ocean. Remote Sens Environ 115:944–956

Supmee V, Ngernsiri L, Sriboonlert A, Wonnapinij P, Sangthong P (2012) Population genetics of the violet vinegar crab (*Episesarma versicolor*) along the Andaman Sea coast of Thailand. Zool Stud 51:1040–1050

Swofford DL (2003) PAUP*: Phylogenetic Analysis Using Parsimony (*and other methods), v 4. Sinauer, Sunderland

Tajima F (1989) Statistical method for testing the neutral mutation hypothesis by DNA polymorphism. Genetics 123:585–595

Thompson JD, Higgins DG, Gibson TJ (1994) CLUSTAL W: improving the sensitivity of progressive multiple sequence alignment through sequence weighting, position-specific gap penalties and weight matrix choice. Nucleic Acids Res 22:4673–4680

Thoppil PG, Hogan PJ (2010) A modeling study of circulation and eddies in the Persian Gulf. J Phys Oceanogr 40:2122–2134

Thurman CL (2002) Osmoregulation in six sympatric fiddler crabs (genus *Uca*) from the northwestern Gulf of Mexico. Mar Ecol 23:269–284

Thurman CL, Faria SC, McNamara JC (2013) The distribution of fiddler crabs (*Uca*) along the coast of Brazil: implications for biogeography of the western Atlantic Ocean. Mar Biodivers Rec 6:e1

Tsang LM, Achituv Y, Chu KH, Chan BKK (2012) Zoogeography of intertidal communities in the West Indian Ocean as determined by ocean circulation systems: patterns from the *Tetraclita* barnacles. PLoS One 7:e45120

Turner LM, Hallas JP, Smith MJ, Morris S (2013) Phylogeography of the Christmas Island blue crab, *Discoplax celeste* (Decapoda: Gecarcinidae) on Christmas Island, Indian Ocean. J Mar Biol Ass UK 93:703–714

Weir BS (1996) Genetic data analysis II: methods for discrete population genetic data. Sinauer, Sunderland

Wieman AC, Berendzen PB, Hampton KR, Jang J, Hopkins MJ, Jurgenson J, McNamara JC, Thurman CL (2014) A panmictic fiddler crab from the coast of Brazil? Impact of divergent ocean currents and larval dispersal potential on genetic and morphological variation in *Uca maracoani*. Mar Biol 161:173–185

Willing EM, Dreyer C, van Oosterhout C (2012) Estimates of genetic differentiation measured by F_{ST} do not necessarily require large sample sizes when using many SNP markers. PLoS One 7(8):e42649

Wong KJH, Chan BKK, Shih HT (2010) Taxonomy of the sand bubbler crabs *Scopimera globosa* De Haan, 1835, and *S. tuberculata* Stimpson, 1858 (Crustacea: Decapoda: Dotillidae) in East Asia, with description of a new species from the Ryukyus, Japan. Zootaxa 2345:43–59

Wong KJH, Shih HT, Chan BKK (2011) Two new species of sand-bubbler crabs, *Scopimera*, from North China and the Philippines (Crustacea: Decapoda: Dotillidae). Zootaxa 2962:21–35

Wong KJH, Shih HT, Chan BKK (2012) The ghost crab *Ocypode mortoni* George, 1982 (Crustacea: Decapoda: Ocypodidae): redescription, distribution at its type locality, and the phylogeny of East Asian *Ocypode* species. Zootaxa 3550:71–87

Yamaguchi T (1994) Fiddler crabs of the genus *Uca* in the collections of three natural history museums in Europe. 1. The specimens held by the Nationaal Natuurhistorisch Museum, Leiden and the Natural History Museum, London. Calanus 11:151–189

2

Thicker filaments of *Aphanizomenon gracile* are more harmful to *Daphnia* than thinner *Cylindrospermopsis raciborskii*

Lukasz Wejnerowski*, Slawek Cerbin and Marcin Krzysztof Dziuba

Abstract

Background: Filamentous cyanobacteria are known to negatively affect the life history of planktonic herbivores through mechanical interference with filtering apparatus. Here, we hypothesise that not only the length but also the thickness of cyanobacterial filaments is an important factor shaping the life history of *Daphnia*.

Results: To test our hypothesis, we cultured *Daphnia magna* with non-toxin-producing strains of either *Aphanizomenon gracile* or *Cylindrospermopsis raciborskii*. The former possesses wide filaments, whereas the latter has thinner filaments. The strain of *A. gracile* has two morphological forms differing in filament widths. The exposure to the thicker *A. gracile* filaments caused a stronger body-length reduction in females at maturity and a greater decrease in offspring number than exposure to the thinner *C. raciborskii* filaments. The width of filaments, however, did not significantly affect the length of newborns. The analysis of mixed thick and thin *A. gracile* filament width distribution revealed that *D. magna* reduces the number of thinner filaments, while the proportion of thicker ones increases. Also, the effects of cyanobacterial exudates alone were examined to determine whether the changes in *D. magna* life history were indeed caused directly by the physical presence of morphologically different filaments and not by confounding effects from metabolite exudation. This experiment demonstrated no negative effects of both *A. gracile* and *C. raciborskii* exudates.

Conclusions: To our knowledge, this is the first study that demonstrates that the thickness of a cyanobacterial filament might be an important factor in shaping *D. magna*'s life history. At a given biomass, thicker filaments of *A. gracile* were more detrimental to *D. magna* than thinner ones of *C. raciborskii*. There was also a strong interaction between species of the cyanobacterium and filament biomass, where species with thicker filaments and at higher biomass had the strongest negative impact on *D. magna* life history.

Keywords: Filter-feeding; Cyanobacterial exudates; Mechanical interference; *Daphnia magna* filamentous cyanobacteria; *Aphanizomenon gracile*; *Cylindrospermopsis raciborskii*; Filament morphology

Background

In the presented work, an attempt was made to determine whether the width of cyanobacterial filaments is a factor responsible for significant changes in the life histories of large-bodied cladocerans such as *Daphnia magna*.

Filamentous cyanobacteria are recognised to negatively affect the performance of filter-feeding cladocerans because of a lack of polyunsaturated fatty acids (for more details see: Von Elert and Wolffrom 2001; Von Elert et al. 2003, Oren 2010) and the toxicity of some species (Carmichael 1994; DeMott 1999). Moreover, filaments clog filtration apparatuses and impair their functioning (Gliwicz and Siedlar 1980). Daphnids are forced to frequently clean their filtering apparatus from filaments by post-abdominal rejection movements, thereby losing a majority of earlier-acquired food particles (Gliwicz and Siedlar 1980), or invest in detoxification mechanisms if toxic filaments are ingested (Pflugmacher et al. 1998). It is well known that filamentous cyanobacteria can induce various changes in the life histories of cladocerans, e.g. decreased growth rate (Gliwicz 1990a) and delayed reproduction and the production of fewer offspring

* Correspondence: wejner@amu.edu.pl
Department of Hydrobiology, Faculty of Biology, Adam Mickiewicz University, Umultowska 89, 61-614 Poznań, Poland

(Repka 1997). Claska and Gilbert (1998) demonstrated that *Anabaena affinis* and *Anabaena flos-aquae* cause an increase in the interclutch interval of *Daphnia pulex*. These authors also observed an increase in the number of aborted eggs and dead neonates in daphnids that fed on cyanobacterial filaments. However, filamentous cyanobacteria are frequently consumed and can support the growth of *Daphnia*; for example, a non-toxic strain of *Aphanizomenon flexuosum* improved the growth of *Daphnia galeata* (Kurmayer 2001). Moreover, Von Elert et al. (2003) observed that the growth rate of *D. galeata* that were fed filaments enriched with sterols was nearly equal to the growth rate of daphnids that were fed *Scenedesmus obliquus*. These contradictory data have been at least partially explained by Gliwicz (1990a), who found that the growth and reproduction of daphnids are strongly suppressed above a critical concentration of filaments. Another supporting example is the study by Soares et al. (2009) that demonstrated that only a high biovolume of *Cylindrospermopsis raciborskii* can reduce the somatic growth of *D. magna*.

The degree of the negative effects of the morphology of cyanobacterial filaments on daphnids is usually explained by the density of filaments and only one morphological parameter, such as length. Several authors have reported that *Daphnia* prefer short filaments (Holm and Shapiro 1984; Oberhaus et al. 2007) because they are easier to ingest (Dawidowicz 1990). Panosso and Lürling (2010) studied the role of filament length of *C. raciborskii* on the grazing of *D. magna*. These authors did not support hypothesis that long filaments affect more negatively on *Daphnia* clearance rate than short filaments. In their view, only filaments longer than 200 µm should reveal the importance of filament length for *Daphnia*. Recently, Bednarska et al. (2014) demonstrated that filaments of *C. raciborskii* longer than 50 µm caused reduction in growth and fecundity of several clones of *Daphnia*. However, the effect of the thickness of filaments has been overlooked, and there is a lack of knowledge concerning the importance of filament thickness to the life history of *Daphnia*. Various species of filamentous cyanobacteria can have substantially different filament widths. For example, morphological analyses of planktonic cyanobacteria from Belgium and Luxemburg carried out by Willame et al. (2006) revealed clear differences in cell widths among filamentous cyanobacteria such as *Aphanizomenon issatschenkoi* (mean ± SD: 2.8 ± 0.3 µm), *Aphanizomenon gracile* (3.6 ± 0.3 µm), *Planktothrix agardhii* (5.5 ± 0.5 µm) and *Lyngbya sp.* (8.5 ± 0.7 µm). The results of Nadin-Hurley and Duncan (1976) suggest that filament thickness, rather than length, appears to be a limiting factor for *Daphnia* grazing. These authors observed long but narrow and pliant filaments of the yellow-green alga *Tribonema sp.* in *Daphnia*'s guts, while wider filaments became stuck and crowded closely together in the mouth. Moreover, Cerbin et al. (2013) observed a widening of *A. gracile*'s filaments in the presence of *Daphnia*, which was interpreted as a defence mechanism against grazing. These results suggest that filament width can shape the interactions between *Daphnia* and filamentous cyanobacteria. Considering the above-mentioned reports, we hypothesised that, when the biomasses of two non-toxic cyanobacteria are similar, the species with thicker filaments will have a stronger negative effect on the life history of *Daphnia* than the species with thinner filaments.

To test our hypothesis, we conducted two experiments using two species of filamentous cyanobacteria that differ in the thickness of their filaments. These species were *C. raciborskii*, a species with thin and pliant filaments, and *A. gracile*, which has thicker filaments. Moreover, the strain of *A. gracile* used in experiments expresses two morphological forms that differ markedly in thickness of filaments. This allowed us to reduce the interspecies variability of the filament's properties and see how *Daphnia*'s grazing influences the width distribution. There should be no changes in the proportion of thin to thick filament numbers after exposure to *Daphnia* grazing, if the width is not the important factor influencing *Daphnia*'s life history. In the second experiment, we checked a life history response of *D. magna* to the presence of cyanobacterial exudates alone. This allowed us to exclude the physical effects of filaments from the possible negative effects of cyanobacterial exudates.

Methods

Experimental organisms and conditions

The following species of filamentous cyanobacteria were used in this study: *A. gracile* Lemmermann 1907, strain SAG 31.79, and *C. raciborskii* (Woloszyńska) Seenayya and Subba Raju 1972, strain SAG 1.97. These strains were obtained from the culture collection of algae (Sammlung von Algenkulturen (SAG)) at the University of Göttingen, Germany. Both strains are not known for toxin production, and no cylindrospermopsin was detected when concentrations were measured using an ELISA test (96 well-plate, ABRAXIS, Warminster, PA, USA). These two species differed in filament thickness; *A. gracile* had thicker filaments than *C. raciborskii* (Table 1). Additionally, the strain of *A. gracile* has been in culture since 1962, and for this reason, it is a mixture of two morphological forms distinctly differing in width. However, both cyanobacteria species had similar filament lengths; the mean length of *A. gracile* filaments was 254.1 ± 146.1 µm (mean ± SD), while *C. raciborskii* filaments had a mean length of 280.7 ± 170.7 µm. Additionally, there were no differences in length between thick and thin filaments of *A. gracile*. In addition to these two cyanobacteria, we used

Table 1 Filament thicknesses (for each treatment, $n = 210$ measurements) and biomasses of *Aphanizomenon* and *Cylindrospermopsis* used for the experiments

Treatment	Thickness of filaments (μm)	Biomass of filaments (mg chl a/L)
Controls without cyanobacteria		
No-Aph	-	0
No-Cyl	-	0
Aphanizomenon gracile		
Low-Aph	2.12 ± 0.44	0.07 ± 0.01
High-Aph	2.24 ± 0.53	0.27 ± 0.03
Cylindrospermopsis raciborskii		
Low-Cyl	1.62 ± 0.13	0.08 ± 0.01
High-Cyl	1.57 ± 0.14	0.26 ± 0.003

Values are given as the means ± standard deviations.

the green algae *S. obliquus* (Turpin) Kützing 1834 strain SAG276-3a and *Chlorella vulgaris* Beyerinck [Beijerinck] (1890) strain SAG211-11b as high-quality food sources for daphnids. The stocks of phytoplankton were cultivated in a walk-in phytotron (Conviron; Winniped, Canada) in a 2-L chemostat on WC medium (Guillard and Lorenzen 1972). The cyanobacterial trichomes and algal cells of each strain were harvested and then separated from their media by centrifugation for 30 min at 3,000 rpm (Rotina 380R centrifuge, Hettich, Kirchlengern, Germany). The pellets were resuspended in filtered lake water prior to their use in experiments.

A long-established laboratory clone of *D. magna* Straus 1820 was maintained in an incubator in 1-L glass beakers that were filled with 0.45 μm filtered and conditioned lake water (Lake Kierskie Duże, western Poland). Before starting the experiments, the daphnids were cultured using a defined food source (a mixture of *S. obliquus* and *Ch. vulgaris*, chlorophyll a = 0.019 mg/L which is far above an incipient limiting level, refreshed every second day), temperature (20°C) and light intensity (PAR: 44 μmol quanta m^{-2} s^{-1}, 16:8 h light-dark cycle) for several generations. The neonates from the third clutch, born on the same day and less than 12 h old, were used in the experiments to minimise the carryover of maternal effects into the experiment.

Cyanobacterial filament width experiment

In the first experiment, the effects of *A. gracile* and *C. raciborskii* trichomes on the life history of *Daphnia* were studied. We used equal biomasses of both species of cyanobacteria. The biomasses were estimated based on spectrophotometric measurements (Metertech, SP-830 Plus, Taipei, Taiwan) of the concentrations of chlorophyll a (Table 1). The treatments were as follows: the low biomass of *A. gracile* (Low-Aph) and *C. raciborskii* (Low-Cyl) filaments, and the high biomass of *A. gracile* (High-Aph) and *C. raciborskii* (High-Cyl) (Table 1). The low concentrations of chlorophyll a correspond to those found in Polish hypertrophic lakes dominated by filamentous cyanobacteria (e.g. Kowalczewska-Madura and Gołdyn 2006), and high concentrations to those found in extremely hypetrophic lowland dam reservoirs (Gorniak et al. 2003). All the treatments with cyanobacterial filaments also contained a mixture of *S. obliquus* and *Ch. vulgaris* (1:1) in concentrations far above the incipient limiting level (in total 0.019 mg/L of chlorophyll a). Daphnids fed with only green algae constituted control treatments (No-Aph, No-Cyl). For each treatment, seven newborns of *D. magna* were individually placed in 150 mL tubes. Daphnids were transferred every second day to clean tubes with freshly prepared algae and cyanobacteria suspensions. Cyanobacteria were sampled during the transfer before placing daphnids into the tubes and after 2 days of grazing. This procedure was undertaken twice, in the beginning and at the end of the experiment (56 cyanobacteria samples of both species, in total). These samples were used for morphological measurements. The thickness of 30 and length of 150 filaments was measured from each sample of each species. Additionally, the clear difference in thickness of *A. gracile* filaments allowed to divide them into two groups: thin (<2.5 μm) and thick (>2.5 μm). This procedure allowed to test whether there were any changes in the ratio of thin to thick filaments due to *Daphnia* grazing. Since the thickness of those filaments sampled at the first and last transfer of animals did not differ significantly in the corresponding groups, they were pulled together to rise the number of measurements per tube (60). Also, several life history traits were recorded for the daphnids: the body lengths and ages of mothers at maturity as well as the numbers and body lengths of neonates. Body length was defined as the distance from the most posterior point on the head to the base of the junction of the tail spine with the carapace.

Cyanobacterial exudates experiment

In the second experiment, we tested the influence of any possible exudates of *A. gracile* and *C. raciborskii* on the life history of *D. magna*. These exudates could be unknown toxins or metabolites that are detrimental to daphnids. The cyanobacterial biomasses used were the same as described for the previous experiment. The treatments with cyanobacteria culture water were set up similarly to the first experiment (No-Aph, No-Cyl, Low-Aph, Low-Cyl, High-Aph, High-Cyl). The water with cyanobacterial exudates was obtained from 24-h incubations of filaments in lake water. After this time, the water was purified of trichomes by centrifugation followed by filtration through GF-C Whatman filters before being used in experiments. For a control, we used clean, filtered lake water that had no subsequent contact with cyanobacterial filaments (No-Aph, No-Cyl). All the treatments contained equal concentrations of green algal food (*S. obliquus* and *Ch. Vulgaris*) that was far above the incipient limiting level. The cyanobacterial water and green algal food were refreshed every day. Similarly to the first experiment, the body lengths and age of mothers at maturity as well as the numbers and body lengths of neonates were recorded.

Statistical analyses

Differences among treatments for each life history parameter were determined using two-way ANOVA tests. Because there were missing replicates for some of the treatments, we used the Type III sums of squares. All the assumptions were tested and met. When the ANOVAs detected significant differences, Tukey's *post hoc* tests were applied to perform pairwise comparisons. The only exception was the age at the time of the first reproduction; in this case, all the daphnids matured at the same age except for a few individuals in one treatment. In order to test the significance of influence of grazing (time) on the number of *Aphanizomenon* filaments differing in thickness, the rmANOVA was applied. All the statistical analyses and plots were performed and produced, respectively, using the R statistical software version 3.0.2 (R Core Team, 2013), which is under general public licence.

Results

Cyanobacterial filament experiment

In the first experiment, the one in which daphnids had direct contact with cyanobacterial filaments, the lengths of mothers at maturity (LM) depended on the biomass of cyanobacteria (expressed as the concentration of chlorophyll a) regardless of the species (biomass, two-way ANOVA: $F_{2,29} = 42.53$, $P < 0.0001$) (Figure 1a). The daphnids cultured in the low or high cyanobacteria biomass treatments were both significantly smaller than those grown without filaments (Tukey's HSD test: $P < 0.0001$; $P < 0.0001$, respectively) (Figure 1a). Moreover, the LM was significantly reduced for the high biomass in comparison to the low biomass (Tukey's HSD test: $P = 0.004$). However, the LM parameter was also dependent on the species of cyanobacteria (ANOVA: $F_{1,29} = 13.18$, $P = 0.001$). Mothers fed *A. gracile* were much smaller at maturity than those in *C. raciborskii* treatments, regardless of the biomass (Figure 1b). The interaction between

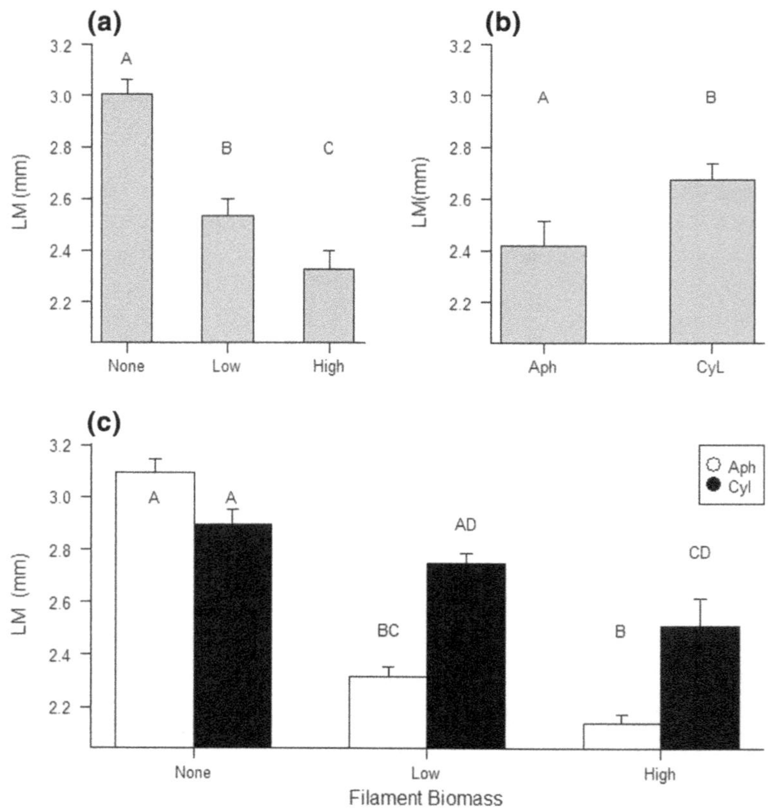

Figure 1 The influences of cyanobacteria. The influences of the cyanobacterial biomass only **(a)**, species only **(b)** and the interaction of species and biomass **(c)** on the length of mothers at maturity (LM). Different letters indicate significant differences among groups as indicated by Tukey's HSD tests.

biomass and species was also significant (ANOVA: $F_{2,29}$ = 10.56, $P = 0.0003$) (Figure 1c). Most treatments with filaments (Low-Aph, High-Aph and High-Cyl) were significantly smaller than those in controls (No-Aph and No-Cyl) (Tukey's HSD test: $P < 0.0001$, $P < 0.0001$, P = 0.01, respectively), but there was no difference in LM between the No-Cyl and Low-Cyl treatments (Tukey's HSD test: $P = 0.75$), indicating a much weaker biomass effect of *C. raciborskii* than *A. gracile* (Figure 1c). Likewise, mothers grown at a high biomass of *A. gracile* (High-Aph) had smaller sizes in comparison to those grown in Low-Cyl and High-Cyl conditions (Tukey's HSD test: $P < 0.0001$, $P = 0.001$, respectively). A similar effect was found for daphnids from the Low-Aph treatment, which were smaller than the females grown in Low-Cyl (Tukey's HSD test: $P = 0.0001$). However, no differences were detected between the Low-Aph treatment and either the High-Cyl or High-Aph conditions (Tukey's HSD test: $P = 0.1$, $P = 0.3$, respectively).

Referring to the age of mothers at the first reproduction (AFE), the females in all of the treatments laid their eggs at the same time, except some of the High-Aph and High-Cyl females delayed their clutch for one extra day. However, the amount of cyanobacterial biomass had a significant effect on the number of newborns (NB) (Figure 2). The NB was strongly dependent on the biomass of cyanobacteria, regardless of the species (ANOVA: $F_{2,29} = 19.12$, $P < 0.00001$) (Figure 2a). A high biomass of cyanobacterial filaments caused a stronger reduction in offspring number than a low biomass of filaments (Tukey's HSD test: $P = 0.005$), and both biomasses of filaments (Low and High) significantly reduced the NB in comparison to the controls without filaments (Tukey's HSD test: $P = 0.004$, $P < 0.0001$, respectively). Furthermore, when the effect of species was tested independently of the filament biomass, differences in NB were also detected (ANOVA: $F_{1,29} = 15.61$, $P = 0.0004$) (Figure 2b). The presence of *A. gracile*

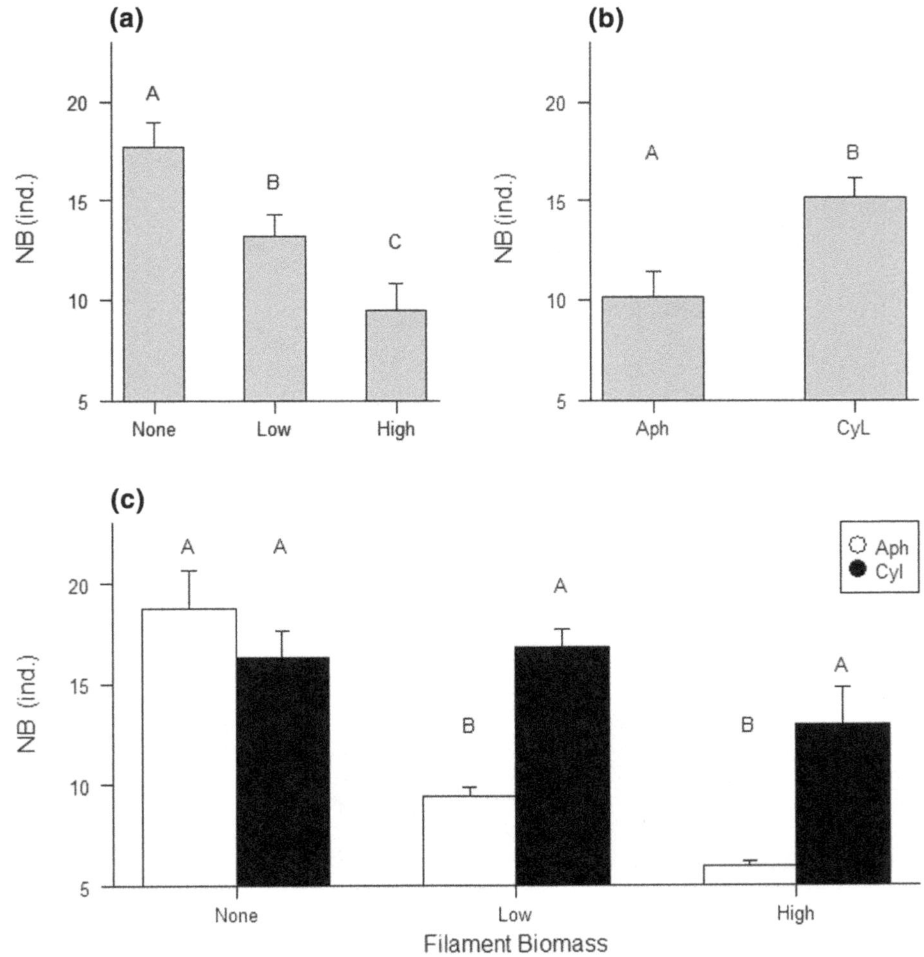

Figure 2 The influences of the cyanobacteria on the number of newborns (NB) in the first clutch. The influences of the cyanobacterial biomass only **(a)**, species only **(b)** and the interaction of species and biomass **(c)** on the number of newborns (NB) in the first clutch. Different indicate significant differences among groups as indicated by Tukey's HSD tests.

filaments caused a greater reduction in the clutch size of daphnids than the filaments of *C. raciborskii*. The interaction between both factors, the biomass of filaments and the effect of species, was also significant for offspring number (ANOVA: $F_{2,29} = 8.00$, $P = 0.001$) (Figure 2c). The clutch size in the presence of *A. gracile* (Low-Aph and High-Aph) was significantly reduced in comparison to the No-Aph treatment (Tukey's HSD test: $P = 0.0004$, $P < 0.0001$, respectively). However, no differences were found between the No-Cyl control and treatments with *C. raciborskii* (Low-Cyl and High-Cyl; $P = 0.9$, $P = 0.5$, respectively), again indicating a stronger effect of *A. gracile*. Furthermore, the females cultured at a low biomass of *C. raciborskii* produced more offspring in comparison to those fed on *A. gracile* at either the Low-Aph or High-Aph biomass ($P = 0.0004$, $P < 0.0001$, respectively). More neonates were also produced by females from the High-Cyl treatment in comparison to those from the High-Aph treatment ($P = 0.0009$).

Although females produced varying offspring numbers, as outlined above, when considering the neonate lengths (LNB), the effects of the filament biomass (ANOVA: $F_{2,29} = 1.22$, $P = 0.3$) (Figure 3a), the cyanobacteria species ($F_{1,29} = 0.31$, $P = 0.5$) (Figure 2b) and the interaction between these factors ($F_{2,29} = 0.58$, $P = 0.5$, respectively) (Figure 3c) were all negligible.

The numbers of thin and thick *A. gracile* filaments have changed within 2 days during the experiment (Figure 4). The number of thin filaments decreased in a sample while the number of thick filaments increased, showing that *D. magna* handles thinner filaments more effectively. There was a clear interaction between time and filament thickness (rmANOVA: $F(1, 26) = 138,89$, $P < 0.0001$) (Figure 5).

Cyanobacterial exudates experiment

The experiment with cyanobacterial exudates alone revealed a significant effect of the exudate concentration on the LM (ANOVA: $F_{2,29} = 4.43$, $P = 0.02$) (Figure 6a).

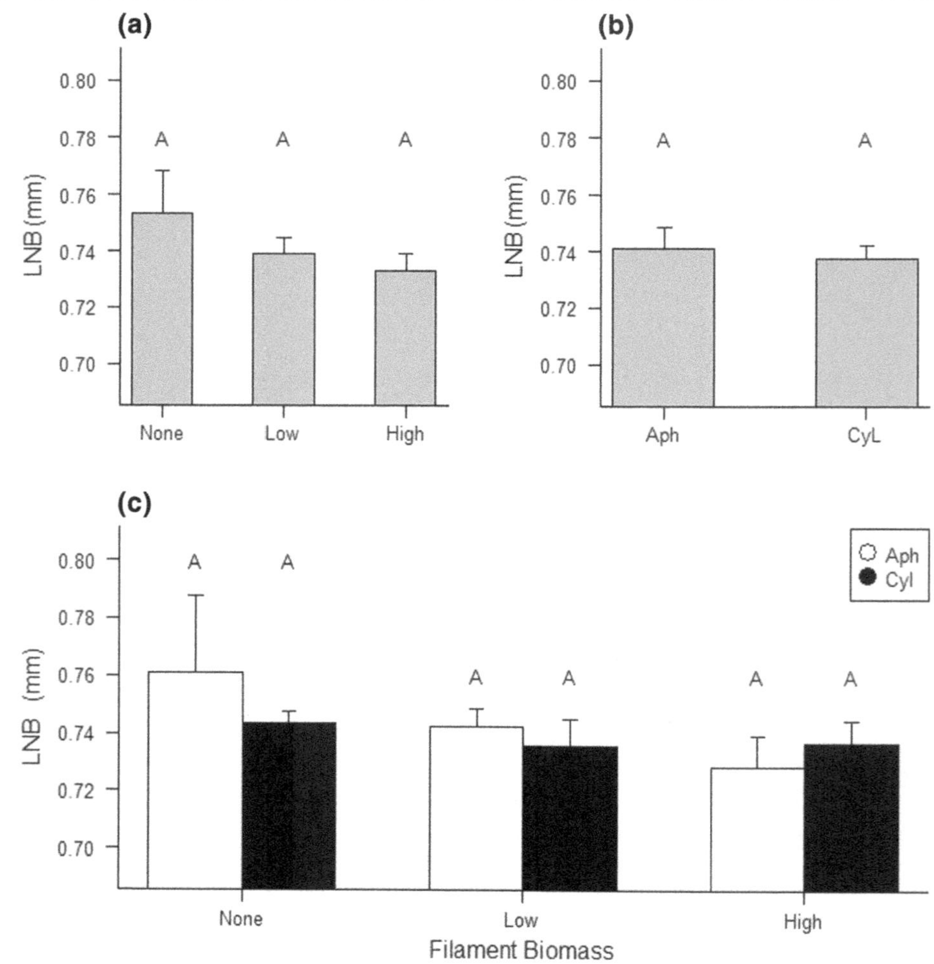

Figure 3 The influences of the cyanobacteria on the length of first newborns (LNB). The influences of the cyanobacterial biomass only **(a)**, species only **(b)** and the interaction of species and biomass **(c)** on the length of first newborns (LNB). Different letters indicate significant differences among groups as indicated by Tukey's HSD tests.

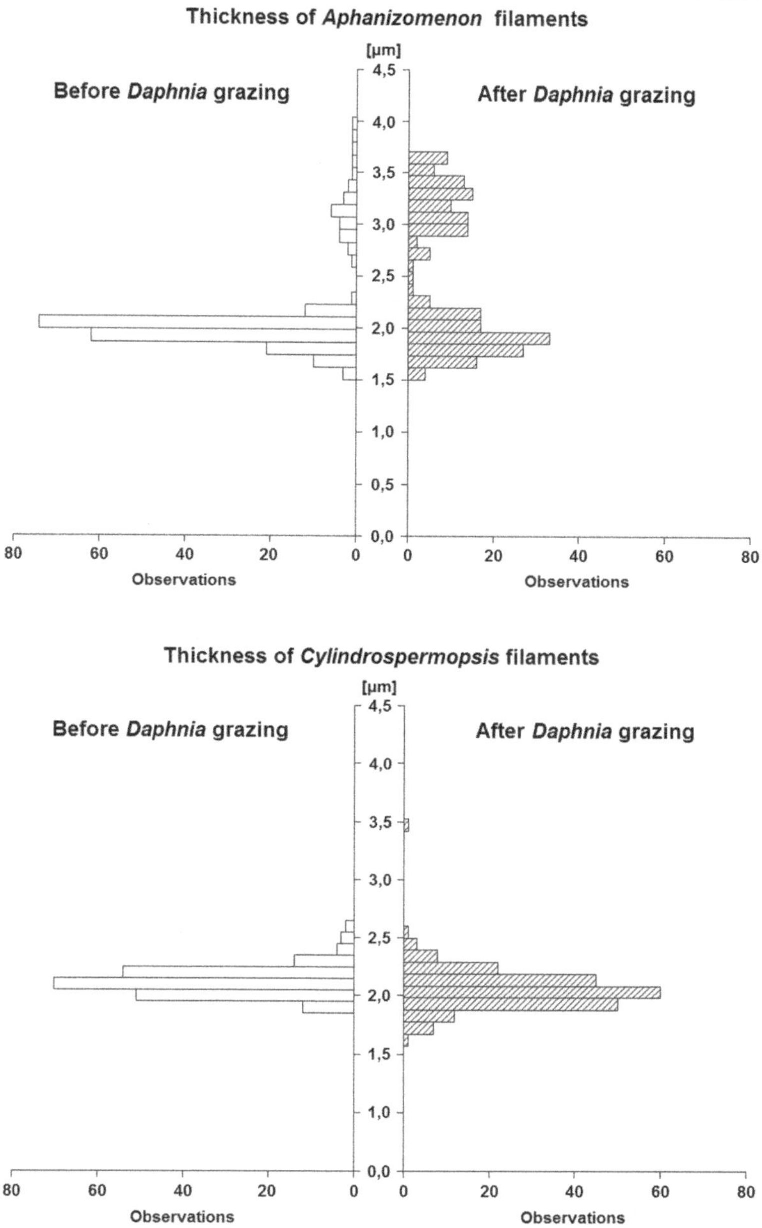

Figure 4 The histogram presenting distribution of measurements of widths of filaments during the filament experiment.

Daphnids grown at a high concentration of exudates were significantly larger at the first eggs in comparison to those exposed to low exudate levels (Tukey's HSD test: $P = 0.02$). However, the LMs for the two concentrations of exudates did not differ with respect to the controls. There was a significant influence of the species of cyanobacteria on the LM parameter (ANOVA: $F_{1,29} = 5.27$, $P = 0.02$) (Figure 6b). *D. magna* females exposed to *A. gracile* exudates were larger at the first eggs than those that were cultured with the exudates of *C. raciborskii*. However, the interaction between concentration and species was not significant for LM (Figure 6c). The age at the first eggs in all experimental treatments, including controls, was the same: all mothers produced eggs at the age of 7 days. Similarly, there were no significant effects of the filament density (ANOVA: $F_{2,28} = 0.3$, $P = 0.7$) (Figure 7a), the species of the cyanobacteria ($F_{1,28} = 0.06$, $P = 0.8$) (Figure 7b) or the interaction between these factors ($F_{2,28} = 0.2$, $P = 0.7$, respectively) (Figure 7c) for the NB parameter. There was a significant effect of the species of cyanobacteria on the LNB (ANOVA: $F_{1,28} = 4.24$, $P = 0.04$) (Figure 8b), but the effects of the exudate

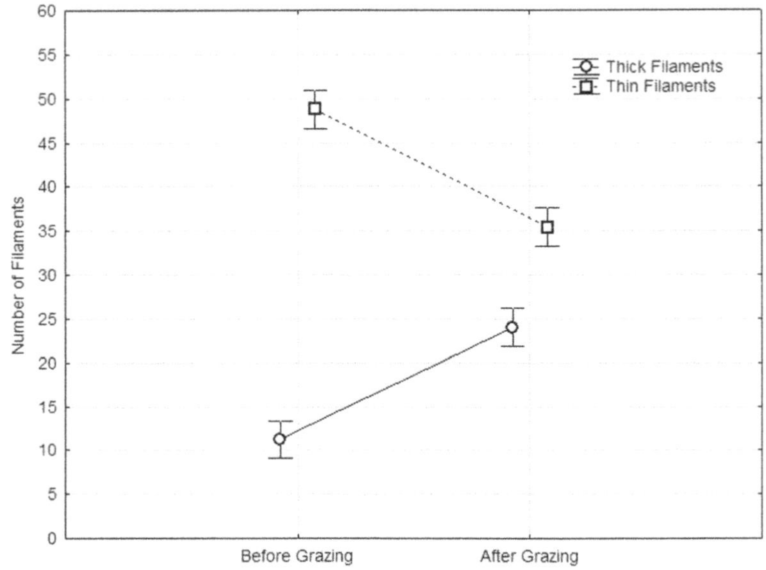

Figure 5 The interaction plot of time and filament thickness. The interaction plot of time (before and after exposure to grazing) and filament thickness on the number of thin (<2.5 μm) and thick (>2.5 μm) *Aphanizomenon* filaments.

concentration and the interaction between species and concentration were both insignificant (Figure 8a,c). Females exposed to *A. gracile* exudates produced newborns of a larger size than females grown with exudates of *C. raciborskii* ($P = 0.02$).

Discussion

The hypothesis that thicker cyanobacterial filaments have a stronger negative impact on the life history traits of *D. magna* seems to be plausible. The presence of thick *A. gracile* filaments suppressed daphnids more than the presence of thin *C. raciborskii* filaments. This effect was most pronounced in the length at maturity. The daphnids that fed on *A. gracile* were significantly smaller in comparison to those from tubes containing *C. raciborskii* (Figure 1b). Moreover, the females grown with the thicker filaments of *A. gracile* produced smaller clutches in comparison to those cultured in the presence of the thinner *C. raciborskii* (Figure 2b). The negative influence of filaments has been shown to be positively correlated with their density (Gliwicz 1990a), and this correlation was also observed in our experiment, where the high biomass treatments had a more severe effect on the size of females at maturity than the low biomass treatments. However, the filament thickness interacted with biomass. Thicker *A. gracile* filaments at a high biomass caused more severe effects than *C. raciborskii*. The importance of filament diameter is also supported by the data obtained for the specific strain of *A. gracile* which possesses two types of filaments: thick and thin (Figure 4). The number of thick filaments increased after 2 days of exposure to *D. magna*, and simultaneously, the number of thin ones decreased (Figure 4). This shows that *D. magna* handles thinner filaments more effectively. The increased number of thick *A. gracile* filaments suggest that they are more difficult to handle by *D. magna*, and that can be attributed to their thickness. Thicker filaments require *Daphnia* to spend more energy to break them or remove from the filtering chamber, and this energy cannot be spent on reproduction. The negative effects of filaments in our experiment was most likely due to the mechanical interference of filaments with *Daphnia*'s filtering apparatus, a phenomenon described first by Gliwicz and Siedlar (1980). The mechanical interference from filaments impairs the ability of daphnids to collect food and triggers frequent cleaning movements of the post-abdominal claws (Gliwicz and Siedlar, 1980). Bednarska et al. (2014) observed that filaments of *C. raciborskii* longer than 50 μm reduced growth and fecundity of *Daphnia*. However, the inhibition may depend not only on the length but also on the flexibility and toughness of a filament. It is possible that different reactions of daphnids to the presence of *A. gracile* or *C. raciborskii* were due to varying elasticity of filaments between these two species. In order to avoid any effects related to filament flexibility, future experiments should be based on different strains belonging to a single species of the cyanobacterium.

The significance of the physical presence of filaments is highlighted by the results of the second experiment conducted with cyanobacterial exudates alone. For these results, there were no clear negative effects of exudates on any of the measured parameters. This finding

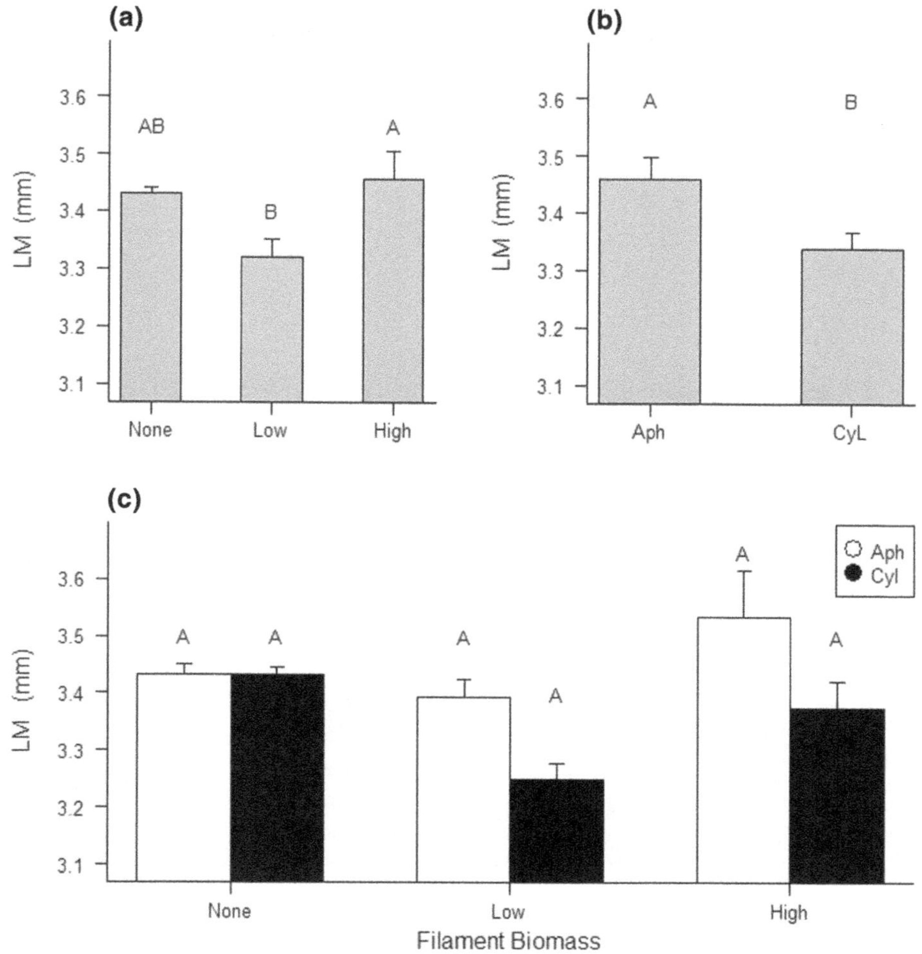

Figure 6 The influence of cyanobacterial exudates on the length of mothers at maturity. **(a)** Concentration of exudates only, **(b)** species only and **(c)** interaction of species and concentration. Different letters indicate significant differences among groups as indicated by Tukey's HSD tests.

excludes any negative influence of the chemicals possibly released by cyanobacteria.

The females that were exposed to filaments appear to invest more in the quality of their offspring. We observed no differences in the sizes of neonates in the presence of filaments (Figure 3c), and the females matured at almost the same time in all the treatments, but the filament treatments did reduce the sizes of mothers at first reproduction and the numbers of offspring. Moreover, daphnids grown with cyanobacterial exudates alone produced similar numbers of offspring and grew to sizes comparable to the controls (Figure 8c). We did, however, find that females exposed to *A. gracile* exudates produced significantly larger newborns compared to those cultured with the exudates of *C. raciborskii.* There are two possible explanations for this result. First, these differences in length could be interpreted as a maternal investment in the quality of offspring that will face unfavourable conditions. Because the presence of *A. gracile* filaments caused a stronger negative effect on *D. magna,* the production of larger newborns in the presence of *A. gracile* exudates could be a result of a greater investment in the quality of newborns. In this case, exudates should be considered as a signal of unfavourable conditions. An investment in the quality of offspring in the presence of *Cylindrospermopsis* was found by Gliwicz and Lampert (1990). Also Reinikainen et al. (1999) noted that an increased proportion of available resources were allocated to reproduction in the presence of toxic cyanobacteria. A second explanation could be the availability of additional food resources, such as bacteria growing on exudates. This explanation is also supported by the larger sizes of females at the first eggs. In this second case, exudates should be considered as a growth medium for bacteria. For example, Maszczyk and Bartosiewicz (2012) demonstrated that chemicals originating from fish, such as kairomones, also fertilise the environment and enhance the growth of zooplankton

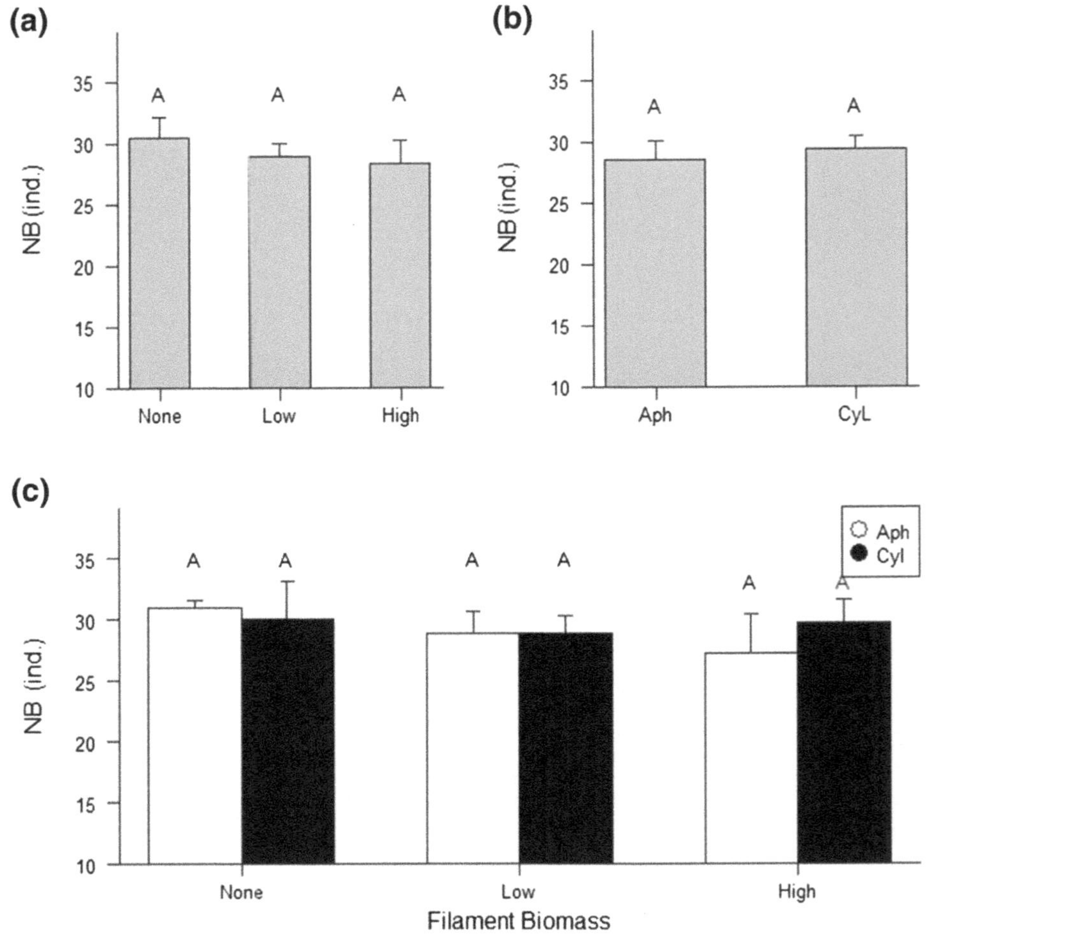

Figure 7 The influence of cyanobacterial exudates on the number of newborns in the first clutch. (a) Concentration of exudates only, **(b)** species only and **(c)** interaction of species and concentration. Different letters indicate significant differences among groups as indicated by Tukey's HSD tests.

by increasing bacterial food availability. This scenario could also be true for our cyanobacterial exudate experiment. Unfortunately, we have no data concerning bacterial growth and development during our experiments.

The mechanisms underlying the influence of filament thickness on the filtering efficiency and life history of filter-feeding plankton remain unknown. Smetacek (2001) argues that planktonic evolution is governed by protection and that the shapes of plankton reflect defence responses to specific attack systems. Cerbin et al. (2013) suggested that the widening of the filaments of *A. gracile* is a defence mechanism against grazing. It is plausible that the wider filaments used in our experiment somehow impaired in a greater extent of the feeding process. Thicker filaments could reduce the effectiveness of such filtering processes and cause higher energetic costs for daphnids. The functioning mechanism of *Daphnia*'s filtration apparatus (mechanical sieving or direct interception of food particles) has been studied and discussed widely (Rubenstein and Koehl 1977; Gophen and Geller 1984; Gerritsen et al. 1988; Hartmann and Kunkel 1991; Koehl 1996). However, recent studies have shown that both mechanisms can occur together. This co-occurrence is closely related to the Reynolds number (Re) of the setulaes (for more detail, see Abrusan 2004). At a low Re, viscous forces dominate the flow. The filtering appendages act as paddles (Abrusan 2004) and the majority of water with suspended particles (including cyanobacterial filaments) flows tangentially to the surface of the filter screens (Gerritsen et al. 1988). Under such conditions, food gathering is predominantly based on direct interception (adhesion) rather than filtration (Rubenstein and Koehl 1977; Koehl 1996). Bednarska and Dawidowicz (2007) observed a reduction in the intersetal and intersetular distances on the third pair of the thoracic legs of *Daphnia* in the presence of *C. raciborskii*. According to these authors, the morphological responses of daphnids to cyanobacterial filaments seem to be adaptive for reducing the mechanical interference of cyanobacterial filaments during food collection. *D. magna* has one of the smallest meshes among the Anomopoda (Geller

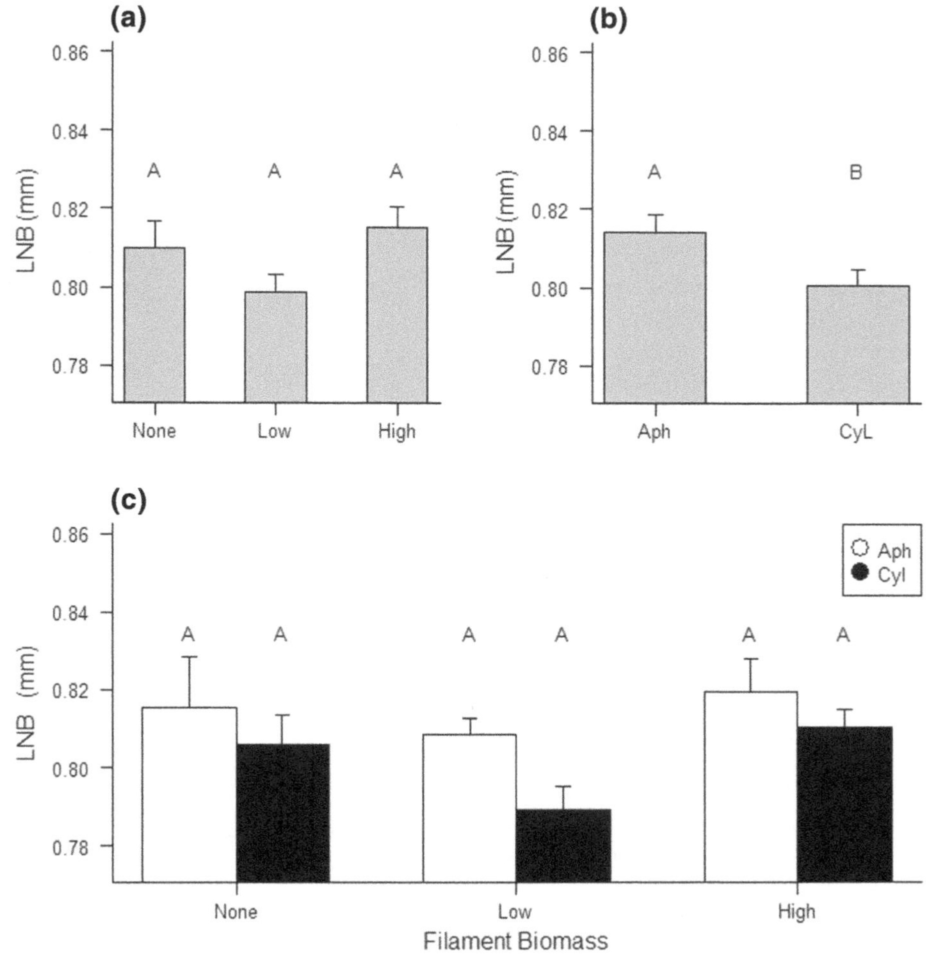

Figure 8 The influence of cyanobacterial exudates on the length of newborns. (a) Concentration of exudates only, **(b)** species only and **(c)** interaction of species and concentration. Different letters indicate significant differences among groups as indicated by Tukey's HSD tests.

and Müller, 1981), and it seems unlikely that filaments clog the meshes in the filter combs. However, Nadin-Hurley and Duncan (1976) reported problems with swallowing wider algal filaments and observed that narrower filaments were transferred further into the guts, while wider filaments became stuck and crowded closely together in the mouth. We can speculate that thin and pliant filaments interfere with the filtering apparatus to a lesser extent. These filaments may be broken with less effort and thus ingested at lower energetic costs. This suggestion is supported by the results of Soares et al. (2009), who found that there were no differences between animals reared on feeds mixed with *C. raciborskii* and those grown on a diet of pure *S. obliquus*.

Our results provide further insight into unravelling the controversies over the manipulation of fish stocks to reduce the top-down effects on zooplankton that can result in harmful cyanobacterial blooms in freshwater ecosystems. Some authors have questioned the usefulness of this technique when filamentous cyanobacteria dominate the phytoplankton community (McQueen 1990; Gliwicz 1990b; De Bernardi and Giussani 1990) because these filaments are not readily consumed by herbivorous zooplankton (Holm et al. 1983; Lampert 1987). A contradictory opinion has been expressed by Carpenter et al. (1995), who suggested that phytoplankton biomass can be successfully controlled by herbivores because the ingestion of cyanobacterial filaments by these organisms is possible (Sterner 1989; Epp 1996). Considering our results, we suspect that the ability of daphnids to graze on filamentous cyanobacteria and, therefore, also the efficiency of biomanipulations can also depend on the characteristic thicknesses of cyanobacterial filaments. If non-toxic species of filamentous cyanobacteria with thick filaments (e.g., those from the genus *Planktothrix* or *Oscillatoria*) predominate within the phytoplankton community, a biomanipulation may be unsuccessful. Simply, *Daphnia* can be too weak to break efficiently thicker filaments. That hypothesis is, to some extent, supported by the fact that the share of thinner filaments of *A. gracile* was decreasing while that of thicker filaments was increasing (Figure 5).

It seems that the thickness of filaments, in addition to length, is a significant factor influencing *Daphnia*'s life history patterns. Based on the differential responses of daphnids to morphologically different filamentous cyanobacteria, thicker filaments of *A. gracile* negatively affect some life history traits of *D. magna* (body growth and offspring number) more strongly than thinner filaments of *C. raciborskii*. This information can provide a reasonable explanation for the dominance structure within cyanobacterial communities and is important for determining top-down effects in freshwater ecosystems, as well as for informing strategies for biomanipulations and water quality control. If our predictions are correct, they will add a new perspective to the consequences of the presence of filamentous cyanobacteria for daphnids and extend the predictions of Gliwicz and Lampert (1990). It is possible that filament thickness is negatively correlated with ingestion processes and adds up to the negative mechanical interference with filtering appendages. In order to give a stronger evidence, further investigation should include more than two taxa of filamentous cyanobacteria of different diameters.

Conclusions

In conclusion, thicker filaments of *A. gracile* caused a stronger reduction of some life history parameter of *D. magna* (length of females at first maturity and offspring number) than *C. raciborskii* that possesses thinner filaments. The analysis of *A. gracile* filaments' width distribution revealed that this strain has two types of filaments, thin and thick. *D. magna* reduced significantly the number of thin filaments. In turn, the share of thick filaments increased. Cyanobacterial exudates had no negative effects on life history traits of *D. magna*. The results indicate that not only the length but also the thickness of cyanobacterial filaments can shape the life history of *D. magna*. It is suspected that thicker filaments were more resistant, thus they caused more harm to the animals.

Abbreviations

ANOVA: analysis of variance; ELISA: enzyme-linked immunosorbent assay; High-Aph: high concentration of *A. gracile* filaments or exudates; High-Cyl: high concentration of *C. raciborskii* filaments or exudates; LFE: length of mothers at maturity; LNB: length of newborns; Low-Aph: low concentration of *A. gracile* filaments or exudates; Low-Cyl: low concentration of *C. raciborskii* filaments or exudates; NB: number of newborns; No-Aph: control without *A. gracile* filaments or exudates; No-Cyl: control without *C. raciborskii* filaments or exudates; Re: Reynolds number; rmANOVA: repeated measures analysis of variance; SAG: Sammlung von Algenkulturen; WC: artificial medium used in cultivation of algae and filamentous cyanoacteria.

Competing interests

The authors declare that they have no competing interests.

Authors' contributions

LW carried out the experiments and drafted the manuscript. LW analysed samples. LW and SC designed the study and analysed data. SC revised and finalised the manuscript. MKD participated in sample collection. All authors read and approved the final manuscript.

Acknowledgements

We thank Alena Gsell, Maciej Bartosiewicz and two anonymous reviewers for their valuable comments which helped to improve the manuscript. This study was supported by the National Science Center in Poland, in part, by grant No. NN 304 014 539 (the influence of cyanobacterial filaments on daphnids) and grant No. UMO-2012/07/N/NZ8/00166 (the influence of cyanobacterial exudates on daphnids).

References

Abrusan G (2004) Filamentous cyanobacteria, temperature and *Daphnia* growth: the role of fluid mechanics. Oecol 141:395–401

Bednarska A, Dawidowicz P (2007) Change in filter-screen morphology and depth selection: uncoupled responses of *Daphnia* to the presence of filamentous cyanobacteria. Limnol Oceanogr 52:2358–2363

Bednarska A, Pietrzak B, Pijanowska J (2014) Effect of poor manageability and low nutritional value of cyanobacteria on *Daphnia magna* life history. J Plankton Res 36(3):838–847

Beyerinck [Beijerinck] MW (1890) Culturversuche mit Zoochlorellen, Lichenengonidien und anderen niederen Algen. Botanische Zeitung 47:725–739, 741–754, 757–768, 781–785

Carmichael WW (1994) The toxins of cyanobacteria. Sci Am 270:78–86

Carpenter SR, Christensen DL, Cole JJ, Cottingham KL, He X, Hodgson JR, Kitchell JF, Knight SE, Pace ML, Post DM, Schindler DE, Voichick N (1995) Biological control of eutrophication in lakes. Environ Sci Technol 29:784–786

Cerbin S, Wejnerowski L, Dziuba M (2013) *Aphanizomenon gracile* increases in width in the presence of *Daphnia*. A defence mechanism against grazing? J Limnol 72:505–511

Claska ME, Gilbert JJ (1998) The effect of temperature on the response of *Daphnia* to toxic cyanobacteria. Freshwater Biol 39:221–232

Dawidowicz P (1990) The effect of *Daphnia* on filament length of blue-green algae. Hydrobiologia 191:265–268

De Bernardi R, Giussani G (1990) Are blue-green algae a suitable food for zooplankton? An overview. Hydrobiologia 200(201):29–41

DeMott WR (1999) Foraging strategies and growth inhibition in five daphnids feeding on mixtures of a toxic cyanobacterium and a green alga. Freshwater Biol 42:263–274

Epp GT (1996) Grazing on filamentous cyanobacteria by *Daphnia pulicaria*. Limnol Oceanogr 41(3):560–567

Geller W, Müller H (1981) The filtration apparatus of Cladocera: filter mesh-sizes and their implications on food selectivity. Oecol 49:316–321

Gerritsen J, Porter KG, Stricker JR (1988) Not by sieving alone: observations of suspension feeding in *Daphnia*. Bull Mar Sci 43:366–376

Gliwicz ZM (1990a) *Daphnia* growth at different concentration of blue-green filaments. Arch Hydrobiol 120:51–65

Gliwicz ZM (1990b) Why do cladocerans fail to control algal blooms? Hydrobiologia 200(201):83–97

Gliwicz ZM, Lampert W (1990) Food thresholds in *Daphnia* species in the absence and presence of blue-green filaments. Ecology 71:691–702

Gliwicz ZM, Siedlar E (1980) Food size limitation and algae interfering with food collection in *Daphnia*. Arch Hydrobiol 88:155–177

Gophen M, Geller W (1984) Filter mesh size and food particle uptake by *Daphnia*. Oecol 64:408–412

Gorniak A, Grabowska M, Jekatierynczuk-Rudczyk E, Zieliński P, Suchowolec T (2003) Long-term variations of phytoplankton primary production in a shallow, polyhumic reservoir. Hydrobiologia 506–509:305–310

Guillard RRL, Lorenzen CJ (1972) Yellow-green algae with chlorophyllide c. J Phycol 8:10–14

Hartmann HJ, Kunkel DD (1991) Mechanisms of food selections in *Daphnia*. Hydrobiologia 225:129–154

Holm NP, Shapiro J (1984) An examination of lipid reserves and the nutritional status of *Daphnia pulex* fed *Aphanizomenon flos-aquae*. Limnol Oceanogr 29:1137–1140

Holm NP, Ganf GG, Shapiro J (1983) Feeding and assimilation rates of *Daphnia pulex* fed *Aphanizomenon flos-aquae*. Limnol Oceanogr 228:677–687
Koehl MAR (1996) When does morphology matter? Ann Rev Ecol Syst 27:501–542
Kowalczewska-Madura K, Gołdyn R (2006) Anthropogenic changes in water quality in the Swarzędzkie Lake (West Poland). Limnol Rev 6:147–154
Kurmayer R (2001) Competitive ability of *Daphnia* under dominance of non-toxic filamentous cyanobacteria. Hydrobiologia 442:279–289
Kützing FT (1834'1833') Synopsis diatomearum oder Versuch einer systematischen Zusammenstellung der Diatomeen. Linnea 8:529–620
Lampert W (1987) Laboratory studies on zooplankton-cyanobacteria interactions. New Zeal J Mar Fresh Res 21:483–490
Lemmermann E (1907) Algen I (Schizophyceen, Flagellaten, Peridineen). In: Band D (ed) Kryptogamenflora der Mark Branderburg und angrenzender Gebiete herausgegeben von dem Botanishcen Verein der Provinz Brandenburg. Verlag von Grebrüder Borntraeger, Leipzig, pp 1–304
Maszczyk P, Bartosiewicz M (2012) Threat or treat: the role of fish exudates in the growth and life history of *Daphnia*. Ecosphere 3(10):91
McQueen DJ (1990) Manipulating lake community structure: where do we go from here? Freshwater Biol 23:613–620
Nadin-Hurley CM, Duncan A (1976) A comparison of daphnid gut particles with the sestonic particles present in two Thames Valley reservoirs throughout 1970 and 1971. Freshwater Biol 6:109–123
Oberhaus L, Gélinas M, Pinel-Alloul B, Ghadouani A, Humbert J-F (2007) Grazing of two toxic *Planktothrix* species by *Daphnia pulicaria*: potential for bloom control and transfer of microcystins. J Plankton Res 29:827–838
Oren A (2010) Mats of filamentous and unicellular cyanobacteria in hypersaline environments. In: Seckbach J, Oren A (eds) Modern and Ancient Microorganisms in Strafied Systems, vol 14, Cellular Origin. Life in Extreme Habitats and Astrobiology. Springer, Heidelberg, pp 387–400
Panosso R, Lürling M (2010) *Daphnia magna* feeding on *Cylindrospermopsis raciborskii*: the role of food composition, filament length and body size. J Plankton Res 32(10):1393–1404
Pflugmacher S, Wiegand C, Oberemm A, Beattie KA, Krause E, Codd GA, Steinberg CE (1998) Identification of an enzymatically formed glutathione conjugate of the cyanobacterial hepatotoxin microcystin-LR: the first step of detoxification. Biochim Biophys Acta 1425:527–533
R Core Team (2013) R: A language and environment for statistical computing. R. Foundation for Statistical Computing, Vienna, Austria. ISBN 3-900051-07-0, URL http://www.r-project.org/
Reinikainen M, Hietala J, Walls M (1999) Reproductive allocation in *Daphnia* exposed to toxic cyanobacteria. J Plankton Res 21:1553–1564
Repka S (1997) Effects of food type on the life history of *Daphnia* clones from lakes differing in trophic state. I. *Daphnia galeata* feeding on *Scenedesmus* and *Oscillatoria*. Freshwater Biol 38:675–683
Rubenstein DI, Koehl MAR (1977) The mechanisms of filter feeding: some theoretical considerations. Amer Nat 111:981–994
Seenayya G, Subba Raju N (1972) On the ecology and systematic of the alga known as Anabaenopsis raciborskii (Wolosz.) Elenk. and a critical evaluation of the forms described under the genus Anabaenopsis. Desikachary TV (eds) In: Papers submitted to the First International Symposium on Taxonomy and Biology of blue-green algae. Madras. pp 52–57
Smetacek V (2001) A watery arms race. Nature 411:745
Soares MCS, Lürling M, Panosso R, Huszar V (2009) Effects of the cyanobacterium *Cylindrospermopsis raciborskii* on feeding and life-history characteristics of the grazer *Daphnia magna*. Ecotoxicol Environ Safety 72:1183–1189
Sterner RW (1989) Resource competition during seasonal succession toward dominance by cyanobacteria. Ecology 70(1):229–245
Von Elert E, Wolffrom T (2001) Supplementation of cyanobacterial food with polyunsaturated fatty acids does not improve growth of *Daphnia*. Limnol Oceanogr 46:1552–1558
Von Elert E, Martin-Creuzburg D, Le Coz JR (2003) Absence of sterols constrains carbon transfer between cyanobacteria and a freshwater herbivore (*Daphnia galeata*). Proc Biol Sci 207:1209–1214
Willame R, Boutte C, Grubisic S, Wilmotte A (2006) Morphological and molecular characterization of planctonic cyanobacteria from Belgium and Luxembourgh. J Phycol 42:1312–1332

3

Trophic structure of the pelagic food web in the East China Sea

Mei-Ling Bai[1,2*], Fan-Sian Lin[1], Yu-Ching Lee[1], Gwo-Ching Gong[3] and Chih-hao Hsieh[1,2,4]

Abstract

Background: Trophic structure and trophic transfer efficiency are among the most fundamental characteristics of an ecosystem. They characterize the transfer of nutrient and energy and are crucial in estimating the yield of harvestable biomass. In this study, we investigated the regulation of trophic structure (phytoplankton, zooplankton, and larval fish abundance) and biomass ratio of zooplankton to phytoplankton (as an indicator of transfer efficiency) in the East China Sea, one of the largest marginal seas in the world and an important fishing ground.

Results: The results showed that when sea surface temperature was below 25°C, temperature co-acted with resource availability (zooplankton for larval fish and phytoplankton for zooplankton) in determining the trophic structure. When sea surface temperature was above 25°C, resource availability dominated the regulation of trophic structure. Biomass ratio of zooplankton to phytoplankton decreased with increasing phosphate concentration.

Conclusions: Our study suggested that the trophic structure of the East China Sea might be controlled by bottom-up processes, and this control is mediated by temperature.

Keywords: Bottom-up control; Top-down control; Transfer efficiency; Phytoplankton; Zooplankton; Fish; Temperature; Nutrient status; East China Sea

Background

The trophic structure organizes the biological components of an ecosystem and channels the flow of nutrient and energy; thus, its regulation is a central issue in ecological research (Ware and Thomson 2005; Casini et al. 2008; Llope et al. 2012). In some marine ecosystems, the biomass of each trophic level is largely determined by available resources, suggesting a bottom-up control (e.g., Iverson 1990; Ware and Thomson 2005; Chassot et al. 2007), whereas in other marine ecosystems, the predator-prey interaction at higher trophic levels cascades down to the lower ones, demonstrating a top-down process (e.g., Frank et al. 2005; Daskalov et al. 2007; Casini et al. 2008). Actually, both processes could act simultaneously in an ecosystem, and the balance of bottom-up versus top-down control might depend on the ecosystem's state and integrity (Hunter and Price 1992; Gripenberg and Roslin 2007).

In addition to the direction of trophic interactions, the strength of these interactions, characterized by trophic transfer efficiency, is crucial for understanding many ecological processes (Pauly et al. 1998). Trophic transfer efficiency describes the proportion of prey production that is converted to predator production. It may influence trophic structure and food chain length (Hairston and Hairston 1993; Post 2002) and is necessary for estimating fishery production and the amount of carbon transfer (Pauly et al. 1998; Richardson and Jackson 2007). Therefore, great research efforts have been devoted to understand the regulation of trophic transfer efficiency, and the factors proposed include environmental conditions such as temperature, light or nutrients (Dickman et al. 2008; Davis et al. 2010; Hoekman 2010), the indirect effect from a third trophic level (bottom-up or top-down cascade) (Hairston and Hairston 1993; Dickman et al. 2008), and the composition, diversity and structure of the trophic levels in issue (Mallin and Paerl 1994; Ye et al. 2013).

Here, we investigated the trophic structure in the East China Sea (ECS), one of the largest marginal seas in the world. The ECS consists of a broad continental shelf

* Correspondence: meilingbai@ntu.edu.tw
[1]Institute of Oceanography, National Taiwan University, Taipei 10617, Taiwan
[2]Ocean Data Bank, Ministry of Science and Technology, Taipei 10617, Taiwan
Full list of author information is available at the end of the article

bordered by China Mainland, Taiwan Island, Japanese Archipelago, and Korean Peninsula and is characterized by a complex hydrographic system with highly seasonal dynamics (Ichikawa and Beardsley 2002; Takahashi and Morimoto 2013). Parallel to the eastern edge of the ECS, the Kuroshio Current flows northwards year round; along the China coast, the southward China Coastal Current dominates when the northeastern monsoon prevails, and the northward Taiwan Strait Warm Current enters from the south when the southwestern monsoon governs. Freshwater discharge from Changjiang at the northwest of the ECS and the Kuroshio upwelling at the southeast of the ECS are the main nutrient sources for the region (Liu et al. 1992; Gong et al. 2003). The ECS ever supported a productive fishery and rich biodiversity but has suffered heavy overfishing, eutrophication, and other disturbances in the past decades (Chen et al. 1997; Li et al. 2009).

Extensive studies have investigated the spatiotemporal variation in the abundance or composition of larval fish (Okazaki and Nakata 2007; Hsieh et al. 2011; Lefébure et al. 2013), zooplankton (e.g., Liao et al. 2006; Lan et al. 2008; Tseng et al. 2008), and phytoplankton (e.g., Gong et al. 2006; Liu et al. 2010; Gong et al. 2011) in the ECS. Many of these previous studies used the abundance of neighboring trophic levels to explain the variation of the focal study group, but the multi-level trophic interactions and the factors regulating them remained to be investigated. Furthermore, many studies focused on a specific season or a confined area in the ECS, and information about the overall pattern of this highly variable ecosystem was scarce. In this study, we analyzed the trophic interactions between larval fish, zooplankton, and phytoplankton using a dataset covering the entire ECS and comprised of data from different seasons. The main objective is to answer (1) whether the trophic structure of the ECS is controlled by a top-down or bottom-up process and how the processes are mediated by temperature, and (2) which factors may influence the trophic transfer efficiency from phytoplankton to zooplankton in the ECS.

Methods

Sample collection and data treatment

Biological samples and corresponding environmental data were collected in seven cruises between 2007 and 2010 over the ECS (Figure 1, Additional file 1: Table S1). Due to constraints of weather and ship time, not all stations were sampled during each cruise. In total, 120 sets of samples were used in the analyses. For collecting larval fishes and zooplankton, an ORI net (Omori 1965) with a mesh size of 330 μm and a mouth diameter of 160 cm was towed obliquely from near the bottom to the sea surface. The volume of filtered water per tow was obtained from readings of a flowmeter attached to the net. At each sampling station, water temperature

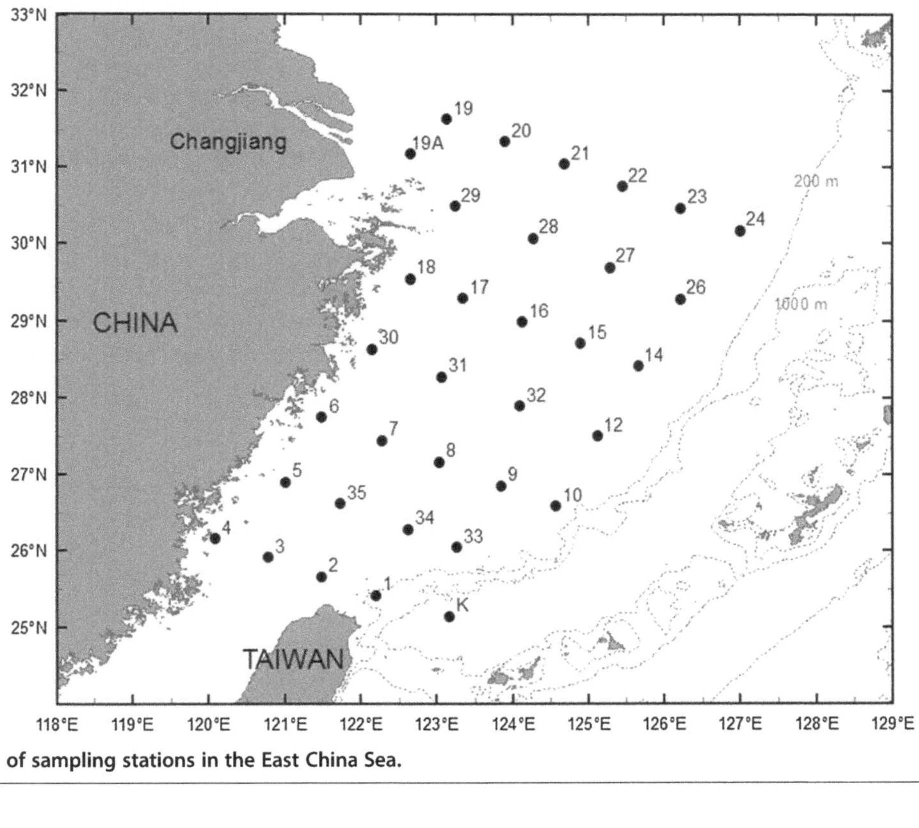

Figure 1 Location of sampling stations in the East China Sea.

and salinity along the vertical profile were recorded with a CTD profiler (SBE 9/11 plus, SeaBird Inc., Bellevue, WA, USA). Seawater samples were taken using Go-Flo bottles at four to six depths spanning the water column. From these samples, concentrations of chlorophyll-a, nitrate (NO_3), nitrite (NO_2), ammonium (NH_4), phosphate (PO_4), and silicate (SiO_3) were analyzed according to the standard methods (Gong et al. 2003).

Larval fishes were counted and standardized to per cubic meter. The majority of larval fishes sampled in plankton nets are in very early stages of their development; therefore, their density can be used as a rough proxy for the abundance of spawning adults (Hsieh et al. 2009). We did not convert larval fish density to carbon biomass, as we had no adequate information about the length distribution and the weight-length relationship of larval fish, which are necessary for the conversion. The zooplankton settled volume was determined using the displacement method (Ballón et al. 2011). To exclude carnivorous zooplankton from the analyses, the samples were digitized with the ZooScan integrated system, and the scanned images were manually classified to calculate the volume percentage of chaetognaths and jellyfishes (Ye et al. 2011; Chang et al. 2012; García-Comas et al. 2013). The displacement volume of each sample was then modified by multiplying a factor of (1 – percentage of carnivorous zooplankton) and then converted to carbon biomass based on the regression formula from the literature (Wiebe 1988; Sánchez and Olaso 2004). Chlorophyll-a concentration was averaged over the towing depth and converted to carbon biomass using a carbon: chlorophyll-a ratio of 30 (Geider 1987). The abundance of larval fish and the biomasses of zooplankton and phytoplankton were log-transformed prior to analyses. Concentrations of dissolved inorganic nitrogen (DIN, sum of NO_3, NO_2, and NH_4), PO_4 and SiO_3 were averaged over the towing depth to indicate the nutrient status of each station. Sea surface temperature (SST) and surface salinity (SSS) were adopted to indicate the physical features of the water.

The log-transformed ratio of zooplankton biomass to phytoplankton biomass (Z:P ratio) was calculated as an indicator of the trophic transfer efficiency from phytoplankton to zooplankton. The actual trophic transfer efficiency, defined as the production ratio of predator versus prey, is very difficult to derive in natural marine ecosystems. Therefore, the biomass ratio of adjacent trophic levels has been used to infer the trophic transfer efficiency (Gaedke and Straile 1994), despite that biomass ratios may differ from production ratios and biomass development of different trophic levels may not be perfectly synchronized. A recent work, using the shipboard incubation method to measure the production of zooplankton and phytoplankton in the ECS, has proved that there is indeed a very strong correlation between biomass ratio and production ratio for the two tropic levels (Su, in preparation).

Data analysis

We applied mixed-effects models to investigate which factors may affect the abundance of each trophic level and the Z:P ratio. For the larval fish density, we used the biomass of zooplankton (food for fish), SST and SSS as fixed-effect variables. For the biomass of zooplankton, we used both the larval fish density (proxy of predation on zooplankton) and the biomass of phytoplankton (food for zooplankton) as well as SST and SSS as fixed-effect variables. For the biomass of phytoplankton, we used the biomass of zooplankton (proxy of predation on phytoplankton), nutrient status (DIN, PO_4, and SiO_3, which represent resources for phytoplankton), as well as SST and SSS as fixed-effect variables. For the Z:P ratio, we used larval fish density, SST, SSS, and nutrient status as fixed-effect variables. We considered sampling cruise as a random effect on both intercept and slope in all analyses to account for variation among cruises, such as seasonal effects.

As ecological relationships might show an abrupt change which could indicate a change in underlying mechanisms, we used piecewise regression analysis to detect potential discontinuities in univariate models prior to multivariate analyses (Toms and Lesperance 2003; Casini et al. 2009). Both the mixed-effects piecewise regression and generalized linear mixed-effects model (GLMM) (Bolker et al. 2009) were applied to model the relationship between each response variable and each predictor. Fits of the two models were then compared based on Akaike's information criterion (AIC) (Akaike 1974). A reduction of 10 units or more in AIC values is generally considered as a significant improvement of the model fit (Bolker et al. 2009). Therefore, we treated a relationship as segmented when the piecewise model had an AIC value of at least 10 units lower than the linear model ($\Delta AIC \geq 10$), and conducted the following analyses for the specific response variable in groups delineated by the point of discontinuity estimated by the piecewise model with best fit.

We then used multivariate GLMMs with normal distribution and identity link function to identify the variable combinations that could best model the biomass of each trophic level and trophic efficiency. For each response variable, all corresponding fixed-effect variables were included in the initial model, and backward stepwise selection was applied. The best models were determined based on AIC. The piecewise regression analysis and GLMMs were performed with NLMIXED and GLIMMIX procedures, respectively, in SAS 9.3 (SAS Institute Inc., Cary, NC, USA).

We further examined how much variance of the abundance/biomass of each trophic level could be explained by the abundance/biomass of other trophic levels by calculating the marginal R^2_{GLMM} (Nakagawa and Schielzeth 2013). Although the AIC value and other information criteria are useful for comparing and selecting models, they provide no information about the absolute model fit and are not comparable between different datasets (Burnham et al. 2011). A recent statistical development offers a method for calculating the variance explained by the fixed effect (marginal R^2_{GLMM}) and comparing cross-datasets model fit for GLMMs (Nakagawa and Schielzeth 2013). We constructed GLMMs for the abundance/biomass of each trophic level using only the abundance of one of the other trophic levels as the fixed effect, and then calculated the marginal R^2_{GLMM} as an indicator of the strength of trophic interactions. The R^2_{GLMM} was calculated using the MuMIn package in R (R Core Team).

Results

In univariate analyses, the variation of larval fish density could be best described by the piecewise model of SST (Table 1). Larval fish density increased with increasing SST when SST was lower than 25.3°C, but decreased with increasing SST when SST was higher (Figure 2). For phytoplankton biomass, abrupt changes occurred in its relationships with SST (at SST = 24.8°C), SSS (at SSS = 32.0 psu), and DIN (at DIN = 10.4 μM), and the piecewise model based on SST had the best fit among all models. Therefore, we built multivariate GLMMs for larval fish density and phytoplankton biomass in the temperature ranges above and below 25°C, separately. However, for the zooplankton biomass and the Z:P ratio, no piecewise model performed significantly better than the corresponding linear model according to the criterion of ΔAIC ≥ 10. Thus, for the zooplankton biomass and the Z:P ratio, we used the complete dataset. Nevertheless, we additionally modeled zooplankton biomass in the two temperature ranges for comparison purposes.

The results of multivariate GLMMs showed that bottom-up forces played an important role in shaping the trophic structure of the ECS (Table 2). In both temperature ranges, the biomass of phytoplankton was determined only by physical and nutritional variables, the zooplankton biomass increased with increasing phytoplankton biomass, and the larval fish density increased with increasing zooplankton biomass. The larval fish density was positively correlated with both zooplankton biomass and SST when SST was below 25°C. When SST was above 25°C, only zooplankton biomass influenced larval fish density. Zooplankton biomass could be best explained by phytoplankton biomass when using the complete dataset, although, when SST was below 25°C, the model including the effect of SST in addition to phytoplankton biomass offered a better fit to zooplankton biomass than the model with phytoplankton biomass alone. For the Z:P ratio, only the negative effect of PO_4 concentration was included in the final model.

Table 1 Comparison of linear models and piecewise models

Response	Predictor	AIC linear model	AIC piecewise model	Cut point
Larval fish	SST[a]	452.1	440.5	25.3
	SSS	460.6	462.1	32.2
	Zooplankton biomass	446.6	447.0	−1.9
Zooplankton	SST	358.4	360.2	25.2
	SSS	363.9	362.9	32.8
	Phytoplankton biomass	354.2	358.0	8.9
Phytoplankton	SST[a]	216.0	192.6	24.8
	SSS[a]	216.6	204.7	32.0
	DIN[a]	211.4	198.9	10.4
	PO_4	219.4	223.4	0.2
	SiO_3	219.6	213.1	16.3
Zooplankton/phytoplankton ratio	SST	363.0	364.8	24.5
	SSS	362.4	365.4	28.3
	DIN	353.2	355.5	9.0
	PO_4	342.9	348.2	0.5
	SiO_3	353.8	356.8	4.7
	Larval fish density	356.2	360.2	−1.3

The relationships between the abundance/biomass of each trophic level and each predictor variable were analyzed using generalized linear mixed-effects models and mixed-effects piecewise regression models, and the AIC values of the two approaches were compared.
[a]The piecewise model had a much better fit compared to the linear model (ΔAIC ≥ 10), indicating a change of relationship at the cut point.

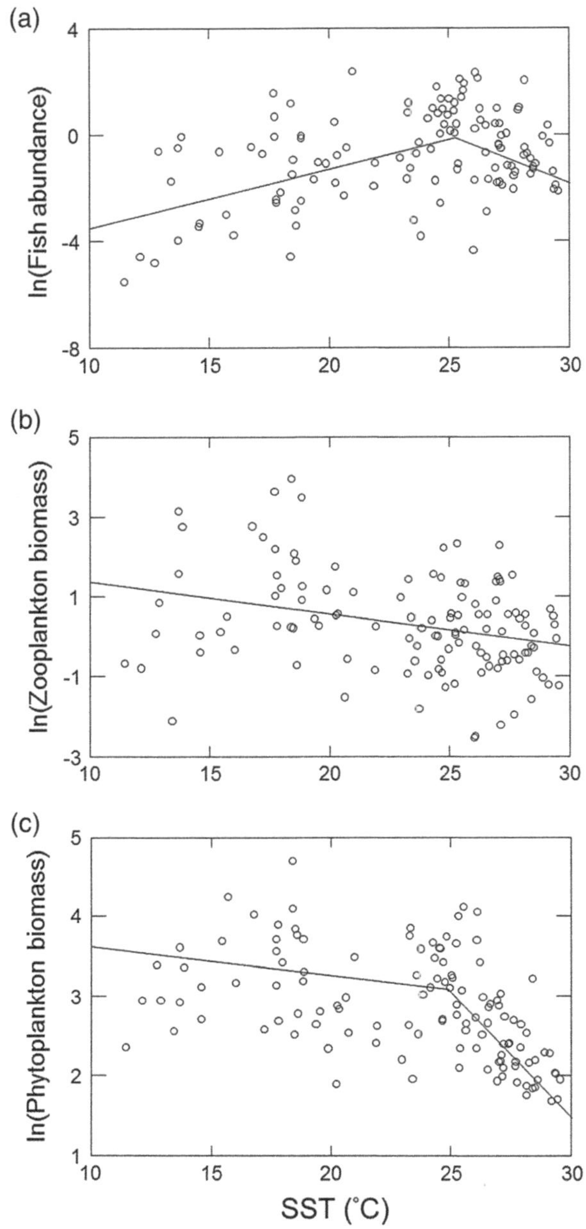

Figure 2 Relationships between the abundance/biomass of each trophic level and SST. (a) Larval fish abundance and SST, **(b)** zooplankton biomass and SST, and **(c)** phytoplankton biomass and SST. The trend lines were based on piecewise regression with random effects in **(a)** and **(c)**, and on generalized linear mixed models in **(b)**.

When SST was below 25°C, about 9% of the variation in larval fish density could be explained by zooplankton biomass, and about 5% of the variation in zooplankton biomass could be explained by phytoplankton biomass (Figure 3). When SST was above 25°C, the explained variation of larval fish density by zooplankton biomass and zooplankton biomass by phytoplankton biomass raised to 23% and 14%, respectively. Less than 1% of the variation in larval fish density could be explained by phytoplankton biomass when SST was below 25°C, while 21% could be explained by phytoplankton biomass when SST was higher.

Discussion

Our study suggested that the bottom-up control regulates the trophic structure in the East China Sea, and this control is mediated by temperature (Table 2, Figure 3). The results of multiple GLMMs showed that temperature co-acted with resource availability in determining the trophic structure when SST was below 25°C (Table 2). Correspondingly, phytoplankton biomass alone was a poor predictor for zooplankton biomass, and likewise, zooplankton biomass alone was a poor predictor for larval fish density (Figure 3). By contrast, while SST was above 25°C, the bottom-up process dominated the regulation of trophic structure; thus, the trophic levels were well coupled and the fish production could be better predicted by primary production.

It is worthy to clarify the effects of temperature on trophic structure in the ECS. We found that the relationship between phytoplankton biomass and temperature showed an abrupt change at SST of about 25°C (Table 1, Figure 2). Such a change may be related to the spatiotemporal variation of the temperature and nutrient conditions in the system. Specifically, the temperature of the coastal area affected by the Changjiang discharge (the main source of nutrient input into the ECS) is relatively low and changes seasonally (SST from 11°C to 26°C; Gong et al. 2003). On the contrary, the oligotrophic Kuroshio Current bordering the ECS is warm and relatively stable in temperature (SST from 20°C to 30°C). That is, the nutrient level is negatively correlated with water temperature in the ECS, and the decline is particularly sharp when the shelf water encounters the Kuroshio Current, on average at SST around 25°C. It is well known that SST is a reasonable proxy for nutrient status in the ECS (Gong et al. 2003). Thus, our finding of the relationship between phytoplankton biomass and temperature (Figure 2) basically signals the bottom-up effects of nutrient on phytoplankton, which is mediated by the hydrography of ECS. Indeed, the phytoplankton biomass also showed a significant relationship with nutrients (Table 1).

The change of the relationship between fish abundance and SST at about 25°C (Table 1, Figure 2) may be related to the physiology of fish, the change in fish assemblage, and food availability. When SST was below 25°C, the larval fish abundance was positively correlated with SST. This observed relationship could be underlain by the fact that fish assemblage changes along the temperature gradient (Hsieh et al. 2011; Lefébure et al. 2013) and different fish species may have different physiological responses. Note, however, that zooplankton biomass also play some roles when SST is below 25°C

Table 2 Generalized linear mixed-effects models for the abundance/biomass of each trophic level and the Z:P ratio

Model	AIC of final model	Variables in final model	Estimate	SE	t	P
Larval fish (SST ≤ 25°C)	231.4					
~ Zooplankton, SST, SSS		Zooplankton biomass	0.485	0.201	2.40	<0.05
		SST	0.279	0.052	5.32	<0.01
Larval fish (SST > 25°C)	198.1					
~ Zooplankton, SST, SSS		Zooplankton biomass	0.657	0.170	3.87	<0.05
Zooplankton (full range)	354.2					
~ Fish, phytoplankton SST, SSS		Phytoplankton biomass	0.565	0.143	3.94	<0.01
Zooplankton (SST ≤ 25°C)	195.8					
~ Fish, phytoplankton SST, SSS		Phytoplankton biomass	0.593	0.261	2.36	0.05
		SST	−0.065	0.047	−1.39	0.22
Zooplankton (SST > 25°C)	157.6					
~ Fish, phytoplankton SST, SSS		Phytoplankton biomass	0.643	0.190	3.38	<0.05
Phytoplankton (SST ≤ 25°C)	100.9					
~ Zooplankton, SST, SSS, nutrients		SST	−0.057	0.023	−2.52	0.05
		SSS	−0.149	0.048	−3.09	<0.05
		PO_4	0.673	0.488	1.38	0.22
Phytoplankton (SST > 25°C)	71.5					
~ Zooplankton, SST, SSS, nutrients		SST	−0.193	0.047	−4.07	<0.05
		SSS	−0.096	0.026	−3.76	<0.05
		PO_4	0.585	0.480	1.22	0.30
Zooplankton/phytoplankton ratio (full range)	342.9					
~ Fish, SST, SSS, nutrients		PO_4	−1.729	0.439	−3.87	<0.01

(Table 2, Figure 3a). When SST was above 25°C, the larval fish abundance decreased with increasing temperature. The change of fish assemblage along the temperature gradient could not provide a satisfactory explanation for this unexpected finding, as SST was not a significant predictor in the multivariate GLMMs (Table 2). The observed relationship between fish abundance and SST in warm waters could be best explained by food availability. Specifically, in the ECS, the biomass of zooplankton decreases with increasing water temperature, as a result of decreasing phytoplankton biomass through the bottom-up control. At SST above 25°C, the larval fish abundance is limited by zooplankton abundance and the increase of temperature exhibits no beneficial effects to the larval fish abundance. Our analyses are consistent with the results of Chen et al. (2014), which suggested that the abundance of larval fish in the ECS was mainly affected by SST in the cool season and by food availability in the warm season.

Our finding that the bottom-up control regulates the trophic structure in the East China Sea (Table 2, Figure 3) is consistent with the patterns found along the continental margin of western North America (Ware and Thomson 2005) and with the study using the cross-system comparative method (Iverson 1990). However, our finding is in contrast with some studies in the over-exploited ecosystems of northwest Atlantic (Frank et al. 2005), the Black Sea (Daskalov et al. 2007), and the Baltic Sea (Casini et al. 2008; Mollmann et al. 2008; Casini et al. 2009), where the top-down process and trophic cascade were found to control the trophic structure after the removal of top predators. Despite being heavily exploited (Chen et al. 1997; Liu and Zhan 1999), the top-down control did not dominate the ECS ecosystem in the trophic levels we examined. The current extent of exploitation and the more diverse community of the subtropical ECS compared to the other studied over-exploited systems could be part of the reason, as simpler food webs are more likely to be governed by top-down regulation (Bascompte et al. 2005; Frank et al. 2007). The regulating mechanism of the trophic structure may nevertheless change over time (Casini et al. 2009; Davis et al. 2010). Our results provide the scenario in the ECS between 2007 and 2010. It is worth further monitoring that, under the continuing natural and artificial changes into its environment and biological communities, whether the regulating mechanism in the ECS may alter.

We found that the Z:P ratio decreased with increasing PO_4 concentration (Table 2). Similar results have been found in phosphate-limited environments and could indicate a cascade effect that regulates trophic transfer

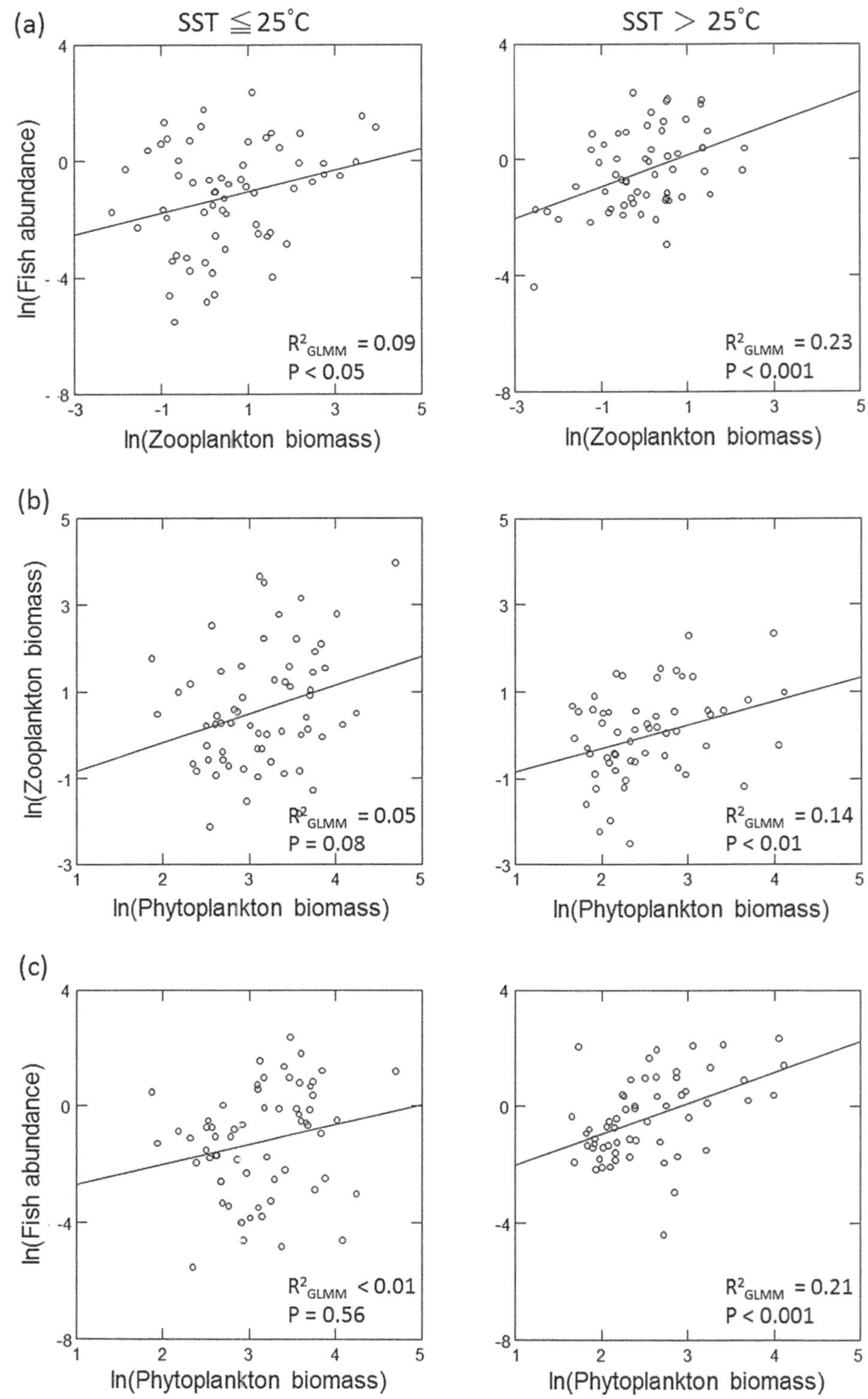

Figure 3 Relationships between the abundance/biomass of different trophic levels. (a) Larval fish abundance versus zooplankton biomass, **(b)** zooplankton biomass versus phytoplankton biomass, and **(c)** larval fish abundance versus phytoplankton biomass when SST ≤ 25°C and when SST > 25°C. The trend lines were based on generalized linear mixed models.

efficiency (Jeppesen et al. 1997, 2003; Dickman et al. 2008). Increasing concentration of PO_4 may boost the growth of fish via high food quality. This would result in a high predation rate of fish on zooplankton, which then further suppresses the grazing pressure of zooplankton on phytoplankton. Whether this is the actual mechanism operating in the ECS, though, remains to be further investigated.

We recognize that this study was based on a simplified tri-trophic food chain, whereas the real food web is much more complicated (Bascompte and Melian 2005). For example, the microbial loop may provide mesozooplankton with additional food sources (del Giorgio and Gasol 1995). The microzooplankton may have significant grazing impact on phytoplankton as well as contribute to the diet of mesozooplankton (Calbet 2008). However, we currently do not have the data to evaluate the roles of the microbial loop and microzooplankton. The current correlation analyses, nevertheless, reveal interesting information that may provide a first step for further investigation into the mechanisms.

Conclusions

In summary, our study suggested that the bottom-up control is the prevailing mechanism regulating the trophic structure of the East China Sea. We also demonstrated how temperature and nutrient status influence the abundance of each trophic level. Our study could offer a base in further understating how eutrophication and temperature change may influence the ecosystem of the East China Sea.

Competing interests
The authors declare that they have no competing interests.

Authors' contributions
CHH designed and conceptualized this study. GCC collected the nutrient and chlorophyll data. FSL and YCL quantified zooplankton and fish samples and performed the ZooScan analysis. MLB performed the statistical analysis. MLB and CHH drafted the manuscript. All authors read and approved the final manuscript.

Acknowledgements
This research is supported by the National Taiwan University and National Science Council, Taiwan. We thank the crews of the R/V 'Ocean Researcher I and II' for their assistance in the sampling.

Author details
[1]Institute of Oceanography, National Taiwan University, Taipei 10617, Taiwan. [2]Ocean Data Bank, Ministry of Science and Technology, Taipei 10617, Taiwan. [3]Center of Excellence for the Oceans, National Taiwan Ocean University, Keelung 20224, Taiwan. [4]Institute of Ecology and Evolutionary Biology, National Taiwan University, Taipei 10617, Taiwan.

References
Akaike H (1974) A new look at the statistical model identification. IEEE Trans Automat Contr 19(6):716–723, doi:10.1109/TAC.1974.1100705

Ballón M, Bertrand A, Lebourges-Dhaussy A, Gutiérrez M, Ayón P, Grados D, Gerlotto F (2011) Is there enough zooplankton to feed forage fish populations off Peru? An acoustic (positive) answer. Prog Oceanogr 91(4):360–381, doi:10.1016/j.pocean.2011.03.001

Bascompte J, Melian CJ (2005) Simple trophic modules for complex food webs. Ecology 86:2868–2873, doi:10.1890/05-0101

Bascompte J, Melian CJ, Sala E (2005) Interaction strength combinations and the overfishing of a marine food web. Proc Natl Acad Sci Unit States Am 102(15):5443–5447, doi: 10.1073/pnas.0501562102

Bolker BM, Brooks ME, Clark CJ, Geange SW, Poulsen JR, Stevens MHH, White J-SS (2009) Generalized linear mixed models: a practical guide for ecology and evolution. Trends Ecol Evol 24(3):127–135, doi:10.1016/j.tree.2008.10.008

Burnham K, Anderson D, Huyvaert K (2011) AIC model selection and multimodel inference in behavioral ecology: some background, observations, and comparisons. Behav Ecol Sociobiol 65(2):415–415, doi:10.1007/s00265-010-1084-z

Calbet A (2008) The trophic roles of microzooplankton in marine systems. ICES J Mar Sci 65(3):325–331, doi:10.1093/icesjms/fsn013

Casini M, Hjelm J, Molinero J-C, Lovgren J, Cardinale M, Bartolino V, Belgrano A, Kornilovs G (2009) Trophic cascades promote threshold-like shifts in pelagic marine ecosystems. Proc Natl Acad Sci Unit States Am 106(1):197–202, doi:10.1073/pnas.0806649105

Casini M, Lövgren J, Hjelm J, Cardinale M, Molinero J-C, Kornilovs G (2008) Multi-level trophic cascades in a heavily exploited open marine ecosystem. Proc Roy Soc Edinb B Biol 275(1644):1793–1801, doi:10.1098/rspb.2007.1752

Chang C-Y, Ho P-C, Sastri AR, Lee Y-C, Gong G-C, C-h H (2012) Methods of training set construction: towards improving performance for automated mesozooplankton image classification systems. Cont Shelf Res 36:19–28, doi:10.1016/j.csr.2012.01.005

Chassot E, Melin F, Le Pape O, Gascuel D (2007) Bottom-up control regulates fisheries production at the scale of eco-regions in European seas. Mar Ecol Prog Ser 343:45–55, doi: 10.3354/meps06919

Chen WY, Lee MA, Lan KW, Gong GC (2014) Distributions and assemblages of larval fish in the East China Sea during the northeasterly and southwesterly monsoon seasons of 2008. Biogeosciences 11(3):547–561, doi:10.5194/bg-11-547-2014

Chen WZ, Zheng YZ, Chen YQ, Mathews CP (1997) An assessment of fishery yields from the East China Sea ecosystem. Mar Fish Rev 59(4):1–7

Daskalov GM, Grishin AN, Rodionov S, Mihneva V (2007) Trophic cascades triggered by overfishing reveal possible mechanisms of ecosystem regime shifts. Proc Natl Acad Sci Unit States Am 104(25):10518–10523, doi:10.1073/pnas.0701100104

Davis JM, Rosemond AD, Eggert SL, Cross WF, Wallace JB (2010) Long-term nutrient enrichment decouples predator and prey production. Proc Natl Acad Sci Unit States Am 107(1):121–126, doi:10.1073/pnas.0908497107

del Giorgio PA, Gasol JM (1995) Biomass distribution in freshwater plankton communities. Am Nat 146(1):135–152

Dickman EM, Newell JM, González MJ, Vanni MJ (2008) Light, nutrients, and food-chain length constrain planktonic energy transfer efficiency across multiple trophic levels. Proc Natl Acad Sci Unit States Am 105(47):18408–18412, doi:10.1073/pnas.0805566105

Frank KT, Petrie B, Choi JS, Leggett WC (2005) Trophic cascades in a formerly cod-dominated ecosystem. Science 308:1621–1623, doi:10.1126/science.1113075

Frank KT, Petrie B, Shackell NL (2007) The ups and downs of trophic control in continental shelf ecosystems. Trends Ecol Evol 22(5):236–242, doi:10.1016/j.tree.2007.03.002

Gaedke U, Straile D (1994) Seasonal-changes of trophic transfer efficiencies in a plankton food-web derived from biomass size distributions and network analysis. Ecol Model 75:435–445, doi:10.1016/0304-3800(94)90038-8

García-Comas C, Chang C-Y, Ye L, Sastri AR, Lee Y-C, Gong G-C, Hsieh C-h (2013) Mesozooplankton size structure in response to environmental conditions in the East China Sea: how much does size spectra theory fit empirical data of a dynamic coastal area? Progr Oceanogr. doi:10.1016/j.pocean.2013.10.010

Geider RJ (1987) Light and temperature dependence of the carbon to chlorophyll a ratio in microalgae and cyanobacteria: implications for physiology and growth of phytoplankton. New Phytol 106(1):1–34, doi:10.2307/2434683

Gong G-C, Chang J, Chiang KP, Hsiung TM, Hung CC, Duan SW, Codispoti LA (2006) Reduction of primary production and changing of nutrient ratio in the East China Sea: effect of the Three Gorges Dam? Geophys Res Lett 33:L07610, doi:10.1029/2006GL025800

Gong G-C, Liu K-K, Chiang K-P, Hsiung T-M, Chang J, Chen C-C, Hung C-C, Chou W-C, Chung C-C, Chen H-Y, Shiah F-K, Tsai A-Y, C-h H, Shiao J-C, Tseng C-M,

Hsu S-C, Lee H-J, Lee M-A, Lin II, Tsai F (2011) Yangtze River floods enhance coastal ocean phytoplankton biomass and potential fish production. Geophys Res Lett 38(13), L13603, doi:10.1029/2011gl047519
Gong G-C, Wen Y-H, Wang B-W, Liu G-J (2003) Seasona variation of chlorophyll a concentration, primary production and environmental conditions in the subtropical East China Sea. Deep Sea Res Part II 50(5–7):1219–1236, doi:10.1016/S0967-0645(03)00019-5
Gripenberg S, Roslin T (2007) Up or down in space? Un ting the bottom-up versus top-down paradigm and spatial ecology. Oikos 116(2):181–188, doi:10.1111/j.0030-1299.2007.15266.x
Hairston NG Jr, Hairston NG Sr (1993) Cause-effect relationships in energy flow, trophic structure, and interspecific interactions. Am Nat 142:379–411, doi:10.1086/285546
Hoekman D (2010) Turning up the heat: temperature influences the relative importance of top-down and bottom-up effects. Ecology 91(10):2819–2825, doi:10.1890/10-0260.1
Hsieh CH, Kim HJ, Watson W, Emanuele DL, Sugihara G (2009) Climate-driven changes in abundance and distribution of larvae of oceanic fishes in the southern California region. Glob Chang Biol 15:2137–2152, doi:10.1111/j.1365-2486.2009.01875.x
Hsieh H-Y, Lo W-T, Wu L-J, Liu D-C, Su W-C (2011) Monsoon-driven succession of the larval fish assemblage in the East China Sea shelf waters off northern Taiwan. J Oceanogr 67(2):159–172, doi:10.1007/s10872-011-0015-6
Hunter MD, Price PW (1992) Playing chutes and ladders heterogeneity and the relative roles of bottom-up and top-down forces in natural communities. Ecology 73(3):724–732, doi: citeulike-article-id:3178334
Ichikawa H, Beardsley RC (2002) The current system in the Yellow and East China Seas. J Oceanogr 58(1):77–92, doi:10.1023/A:1015876701363
Iverson RL (1990) Control of marine fish production. Limnol Oceanogr 35:1593–1604
Jeppesen E, Jensen JP, Jensen C, Faafeng B, Hessen DO, Sondergaard M, Lauridsen T, Brettum P, Christoffersen K (2003) The impact of nutrient state and lake depth on top-down control in the pelagic zone of lakes: a study of 466 lakes from the temperate zone to the arctic. Ecosystems 6(4):313–325, doi:10.1007/s10021-002-0145-1
Jeppesen E, Peder Jensen J, Søndergaard M, Lauridsen T. Junge Pedersen L, Jensen L (1997) Top-down control in freshwater lakes: the role of nutrient state, submerged macrophytes and water depth. Hydrobiologia 342–343:151–164, doi: 10.1023/A:1017046130329
Lan Y-C, Lee M-A, Chen W-Y, Hsieh F-J, Pan J-Y, Liu D-C, Su W-C (2008) Seasonal relationships between the copepod community and hydrographic conditions in the southern East China Sea. ICES J Mar Sci 65(3):462–468, doi:10.1093/icesjms/fsm173
Lefébure R, Degerman R, Andersson A, Larsson S, Eriksson LO, Båmstedt U, Byström P (2013) Impacts of elevated terrestrial nutrient loads and temperature on pelagic food-web efficiency and fish production. Glob Chang Biol 19(5):1358–1372, doi:10.1111/gcb.12134
Li Y, Chen Y, Olson D, Yu N, Chen L (2009) Evaluating ecosystem structure and functioning of the East China Sea Shelf ecosystem, China. Hydrobiologia 636(1):331–351, doi:10.1007/s10750-009-9964-9
Liao CH, Chang WJ, Lee MA, Lee KT (2006) Summer distribution and diversity of copepods in upwelling waters of the southeastern East China Sea. Zool Stud 45:378–394
Liu K-K, Chao S-Y, Lee H-J, Gong G-C, Teng Y-C (2010) Seasonal variation of primary productivity in the East China Sea: a numerical study based on coupled physical-biogeochemical model. Deep Sea Res Part II: Top Stud Oceanogr 57(19–20):1762–1782, doi:10.1016/j.dsr2.2010.04.003
Liu KK, Gong GC, Lin SW, Yang CY, Wei CL, Pai SC, Wu CK (1992) The year-round upwelling at the shelf break near the northern tip of Taiwan as evidenced by chemical hydrography. Terr Atmos Ocean Sci 3(3):243–276
Liu WH, Zhan BY (1999) The dynamic analysis on the fishery stocks in the East China Sea. J Shanghai Fish Univ 8:19–24
Llope M, Licandro P, Chan K-S, Stenseth NC (2012) Spatial variability of the plankton trophic interaction in the North Sea: a new feature after the early 1970s. Glob Chang Biol 18(1):106–117, doi:10.1111/j.1365-2486.2011.02492.x
Mallin MA, Paerl HW (1994) Planktonic trophic transfer in an estuary: seasonal, diel, and community structure effects. Ecology 75(8):2168–2184, doi:10.2307/1940875
Mollmann C, Muller-Karulis B, Kornilovs G, St John MA (2008) Effects of climate and overfishing on zooplankton dynamics and ecosystem structure: regime shifts, trophic cascade, and feedback loops in a simple ecosystem. ICES J Mar Sci:fsm197. doi:10.1093/icesjms/fsm197
Nakagawa S, Schielzeth H (2013) A general and simple method for obtaining R2 from generalized linear mixed-effects models. Methods Ecol Evol 4(2):133–142, doi:10.1111/j.2041-210x.2012.00261.x
Okazaki Y, Nakata H (2007) Effect of the mesoscale hydrographic features on larval fish distribution across the shelf break of East China Sea. Cont Shelf Res 27:1616–1628, doi:10.1016/j.csr.2007.01.024
Omori M (1965) A 160 cm opening-closing plankton net. I. Description of the gear. J Oceanogr Soc Jpn 21:212–220
Pauly D, Christensen V, Dalsgaard J, Froese R, Torres F Jr (1998) Fishing down marine food webs. Science 279(5352):860–863, doi:10.1126/science.279.5352.860
Post DM (2002) The long and short of food-chain length. Trends Ecol Evol 17(6):269–277, doi:10.1016/S0169-5347(02)02455-2
Richardson TL, Jackson GA (2007) Small phytoplankton and carbon export from the surface ocean. Science 315:838–840, doi:10.1126/science.1133471
Sánchez F, Olaso I (2004) Effects of fisheries on the Cantabrian Sea shelf ecosystem. Ecol Model 172(2–4):151–174, doi:10.1016/j.ecolmodel.2003.09.005
Takahashi D, Morimoto A (2013) Mean field and annual variation of surface flow in the East China Sea as revealed by combining satellite altimeter and drifter data. Prog Oceanogr 111:125–139, doi:10.1016/j.pocean.2013.01.007
Toms JD, Lesperance ML (2003) Piecewise regression: a tool for identifying ecological thresholds. Ecology 84(8):2034–2041, doi:10.1890/02-0472
Tseng L-C, Souissi S, Dahms H-U, Chen Q-C, Hwang J-S (2008) Copepod communities related to water masses in the southwest East China Sea. Helgoland Mar Res 62(2):153–165, doi:10.1007/s10152-007-0101-8
Ware DM, Thomson RE (2005) Bottom-up ecosystem trophic dynamics determine fish production in the Northeast Pacific. Science 308(5726):1280–1284, doi:10.1126/science.1109049
Wiebe PH (1988) Functional regression equations for zooplankton displacement volume, wet weight, dry weight, and carbon: a correction. Fish Bull 86:833–835
Ye L, Chang CY, García-Comas C, Gong GC, Hsieh CH (2013) Increasing zooplankton size diversity enhances the strength of top-down control on phytoplankton through diet niche partitioning. J Anim Ecol 82(5):1052–1061, doi:10.1111/1365-2656.12067
Ye L, Chang CY, Hsieh CH (2011) Bayesian model for semi-automated zooplankton classification with predictive confidence and rapid category aggregation. Mar Ecol Prog Ser 441:185–196, doi:10.3354/meps09387

4

Cladistic reanalysis and historical biogeography of the genus *Lycinus* Thorell, 1894 (Araneae: Mygalomorphae: Nemesiidae) with description of two new species from western Argentina

Nelson Ferretti

Abstract

Background: The genus *Lycinus* includes ten species, one from Argentina, eight from Chile, and one from Brazil. Two new species from western Argentina are described, and some data on natural history are presented. A cladistic reanalysis with the newly described species is carried out, and a reconstruction of ancestral areas and primary Brooks Parsimony Analysis are performed. The origins of the main groups of *Lycinus* are also discussed in that light.

Results: *Lycinus lagigliai* sp. nov. and *Lycinus nevadoensis* sp. nov. are described. The distribution of these species comprised western Argentina at Mendoza province. The cladistic reanalysis showed *Lycinus* as monophyletic supported only by the synapomorphy of dorsal spines on male palpal tibia. The tree topology was as follows: (*Lycinus epipiptus*, *Lycinus domeyko* ((*Lycinus longipes*, *L. nevadoensis* sp. nov.) (*Lycinus portoseguro*, *L. lagigliai* sp. nov.)) (*Lycinus caldera* (*Lycinus gajardoi*, *Lycinus choros*)) (*Lycinus tofo*, *Lycinus quilicura*)). The biogeographical analysis showed a close relationship of the species from the Andean region and the South American Transition Zone. The reconstruction of ancestral areas recognized four vicariant nodes.

Conclusions: A hitherto two unknown *Lycinus* species are recognized and formally described, along with a cladistic reanalysis and historical biogeographical hypotheses.

Keywords: Andean; Mendoza; Spiders; Phylogeny; Taxonomy

Background

The Nemesiidae Simon, 1889 was raised to family status by Raven (1985) from the Nemesiae group proposed by Simon (1889). To date, the family comprises 366 species and 43 genera of medium-sized spiders that have a transverse fovea, eyes grouped on a tubercle, two to four short spinnerets, anterior tarsi without spines, tarsi III and IV with light or absent scopula, without claw tufts, and female's superior tarsal claws bipectinate with numerous teeth (Goloboff 1995; Montes de Oca and Pérez-Miles 2013). Goloboff (1995) revised the species of Nemesiidae from Peru, Chile, Argentina, and Uruguay since only a few contributions to the systematics of this family concluded (Indicatti and Lucas 2005; Indicatti et al. 2008; Lucas et al. 2008; Lucas and Indicatti 2010; Montes de Oca and Pérez-Miles 2013). Neotropical nemesiids usually live in silk tubes constructed under stones or logs or in burrows covered by a trapdoor (Capocasale and Pérez-Miles 1990; Goloboff 1995; Ferretti et al. 2011).

Recently, as the result of the examination of a lot of some mygalomorph spiders collected in Mendoza and deposited in the arachnological collection of the Instituto Argentino de Investigaciones de las Zonas Áridas (CAI), I found six mygalomorph species belonging to the Nemesiidae. The six species belong to the genera *Chaco* Tullgren, 1905, *Diplothelopsis* Tullgren, 1905, and *Lycinus* Thorell (1894), with the latter genus showing a remarkable predominance in abundance and diversity. Thus, the present study is focused on the genus *Lycinus*.

Correspondence: nferretti@conicet.gov.ar
Centro de Estudios Parasitológicos y de Vectores CEPAVE (CCT-CONICET-La Plata), Calle 2 N°584, La Plata B1902CHX, Argentina

Lycinus was established by Thorell (1894), with the type species, *Lycinus longipes* Thorell, 1984, based on a male from Córdoba, Argentina. Goloboff (1995) emended the diagnosis of the genus, mentioning that the 'only possible autapomorphy for the genus is the male bulb being coniform and gradually tapering to form the embolus. It can be quickly distinguished from *Diplothelopsis* Tullgren, 1905 by having the posterior median spinnerets present, and from *Flamencopsis* and *Chilelopsis* by the female patella IV having strong 1-1-1 prolateral spines'. The genus includes ten species, one from Argentina (*L. longipes*), eight from Chile, and one from Brazil (Goloboff 1995; Lucas and Indicatti 2010; Platnick 2014).

In this paper, I present the description of two new species of *Lycinus*: *Lycinus lagigliai* sp. nov. from Ñacuñan, Santa Rosa, central-eastern Mendoza, Argentina, and *Lycinus nevadoensis* sp. nov. from Cerro Nevado, Malargüe, southwestern Mendoza, Argentina. Also, new distributional data is presented for *L. longipes* and for *Lycinus epipiptus* (Zapfe, 1963), confirming the presence of this species in Argentina. Moreover, I present a cladistic reanalysis of *Lycinus* with newly described species and show data on natural history for some of the mentioned species. Finally, a reconstruction of ancestral areas and primary Brooks Parsimony Analysis are performed together with discussion concerning the origin of the main groups of *Lycinus*.

Methods

The examined material is deposited in the arachnological collection of the Instituto Argentino de Investigaciones de las Zonas Áridas (CAI, Susana Lagos Silnik). Specimens were examined using an Olympus SZ stereomicroscope (Shinjuku, Tokyo, Japan). Spine notation follows Petrunkevitch (1925). Terminology of genitalia follows Goloboff (1995). All measurements are in millimeters and were taken with a millimetric ocular lens. The leg segment length was measured between joints in dorsal view. Total body length excluded chelicerae and spinnerets. Left male palpal bulb was removed from the cymbium and photographed in prolateral, retrolateral, and ventral views. Specimens were photographed using a SONY Hx200v camera (Minato, Tokyo, Japan) attached to a stereomicroscope. Female genitalia was dissected and cleared in concentrated lactic acid for 60 to 120 min.

Cladistic analysis

I scored the newly described species (*L. lagigliai* sp. nov. and *L. nevadoensis* sp. nov.) and two recently described *Lycinus portoseguro* Lucas and Indicatti, 2010 and *Lycinus choros* Lucas and Indicatti, 2010 for 27 characters (Table 1) from Goloboff (1995), removing invariant characters and adding a new one (character 26). The ingroup comprised 12 taxa: *L. longipes* Thorell, 1894; *Lycinus gajardoi* (Mello-Leitão, 1940); *Lycinus caldera* Goloboff, 1995; *Lycinus tofo* Goloboff, 1995; *L. frayjorge* Goloboff, 1995; *L. epipiptus* (Zapfe, 1963); *Lycinus domeyko* Goloboff, 1995; *Lycinus quilicura* Goloboff, 1995; *L. portoseguro* Lucas and Indicatti, 2010; *L. choros* Lucas and Indicatti, 2010; *L. lagigliai* sp. nov., and *L. nevadoensis* sp. nov. The outgroup included the following: *Chilelopsis calderoni* Goloboff, 1995; *Diplothelopsis bonariensis* Mello-Leitão, 1938, and *Chaco obscura* Thorell, 1894. The outgroup taxa are considered close relatives of *Lycinus* (Goloboff, 1995). The tree was rooted using *C. obscura*. The data matrix was constructed using Nexus Data Editor ver 0.5.0 software (Page 2001). The cladistic analysis was carried out with the program TNT version 1.1 (Goloboff et al. 2003), using maximum parsimony as the optimality criterion. Tree searches were conducted using implicit enumeration and implied weighting (Goloboff 1993) with concavity indices (*K*) ranging from 1 to 6.

Characters (Table 1): (0) Sternum: 0, wide; 1, normal; 2, long. (1) Maxillary cuspules in females: 0, few (0 to 10); 1, medium (11 to 30); 2, many (over 30). (2) Maxillary cuspules in males: 0, few (0 to 10); 1, medium (11 to 30); 2, many (more than 30). (3) Female tarsi: 0, rigid; 1, flexible. (4) Scopula IV: 0, absent/very light; 1, light; 2, dense. (5) Basal bothrial plate: 0, convex; 1, flattened and with deeper ridges. (6) Trichobothria on male cymbium: 0, 1:3 M; 1, 1:2 B. (7) Preening combs: 0, absent; 1, present; 2, pseudocombs. (8) Dorsal spines on male palpal tibia: 0, absent; 1, present. (9) Spines on male cymbium: 0, present; 1, absent. (10) Proventral setae on male palpal tibia: 0, weak; 1, strong. (11) Dorsal spines on male tibia I: 0, absent; 1, present. (12) Dorsal spines on male posterior tibia: 0, few/absent; 1, numerous. (13) Spines on male patella I: 0, 0 R; 1, 1–1 R. (14) Shield of setae on male tibia I: 0, absent; 1, present. (15) Spines on male patellae I-II: 0, 0/1 P; 1, 1-1-1 P. (16) Male palpal tibia: 0, short; 1, long; 2, very long. (17) Male cymbium: 0, normal; 1, narrow, asymmetric, with posterior lobe longer. (18) Male cymbium: 0, normal; 1, large. (19) Setae on male cymbium: 0, thin, hairlike setae; 1, thickened setae. (20) Setae male cymbium: 0, directed forward; 1, directed backward. (21) Male bulb keels: 0, absent; 1, parallel keels or ridges along embolus base; 2, lateral keels or flanges. (22) Male bulb: 0, normal size; 1, large. (23) Male bulb shape: 0, piriform; 1, conical. (24) Male embolus: 0, short; 1, long; 2, very long. (25) Burrow entrance: 0, open; 1, hinged door; 2, flap door; 3, double flap; 4, stopper door. (26) Retrolateral megaspine on male tibia I: 0, absent; 1, present.

Biogeographical analysis

During the past few years, event-based methods are increasingly being used in historical biogeographic studies,

Table 1 Data matrix for the genus *Lycinus* and the outgroup

										1										2						
	0	**1**	**2**	**3**	**4**	**5**	**6**	**7**	**8**	**9**	**0**	**1**	**2**	**3**	**4**	**5**	**6**	**7**	**8**	**9**	**0**	**1**	**2**	**3**	**4**	**5**	**6**
C. obscura	1	1	1	1	1	0	0	0	0	1	0	0	0	0	0	0	1	0	0	0	0	1	0	1	-	-	0
D. bonariensis	0	0	0	1	2	0	1	0	0	1	0	1	1	1	0	1	1	0	0	1	0	0	0	0	1	-	0
C. calderoni	1	2	2	0	1	1	0	1	0	1	0	1	0	1	1	1	0	0	0	0	0	2	0	0	1	0	0
L. longipes	0	0	0	0	2	0	1	0	1	1	0	1	-	1	0	1	1	0	0	1	0	1	0	1	0	2	0
L. gajardoi	1	0	0	1	1	-	1	2	1	1	0	1	1	1	0	1	0	1	0	1	1	1	0	1	1	1	0
L. caldera	1	2	1	1	2	-	1	0	1	1	0	1	1	1	0	1	0	0	1	1	1	1	0	1	0	1	0
L. tofo	1	0	0	1	2	-	1	0	1	1	1	1	1	1	0	1	1	0	0	1	0	1	0	1	2	1	0
L. frayjorge	1	1	-	1	2	-	-	0	-	1	-	-	-	-	-	-	-	0	0	-	-	-	0	-	-	1	-
L. epipiptus	1	0	0	1	2	-	1	0	1	1	0	1	1	1	0	1	1	0	0	1	0	1	0	1	0	1	0
L. domeyko	1	-	0	1	2	-	1	0	1	1	0	1	1	1	0	1	1	0	0	1	0	1	0	1	0	1	0
L. quilicura	1	-	0	1	2	-	1	0	1	1	1	1	1	1	0	1	1	0	0	1	0	1	0	1	2	1	0
L. portoseguro	1	2	2	0	0	-	-	2	1	1	0	0	1	1	0	1	1	0	0	-	0	1	0	1	0	-	1
L. choros	1	-	0	-	1	-	-	0	1	1	0	1	1	1	0	1	0	0	0	1	1	1	0	1	1	-	0
L. lagigliai sp. nov.	1	0	0	0	2	-	1	2	1	1	0	1	1	1	0	1	1	0	0	1	0	1	0	1	1	2	0
L. nevadoensis sp. nov.	0	-	0	-	2	-	1	0	1	1	0	1	1	1	0	1	1	0	0	1	0	1	0	1	0	2	0

due to the noticeable increase in phylogenetic studies (Donato 2006; Guo and Wang 2007; Antonelli et al. 2010; Ferretti et al. 2012). These methods propose explicit models for the processes that affect the geographic distribution of living organisms (Crisci et al. 2003). To reconstruct ancestral distributions of *Lycinus*, the RASP 2.0 (Reconstruct Ancestral State in Phylogenies) software was used. This program determines statistical support for ancestral range reconstructions using the RASP value, and frequencies at a node are averaged over all trees with each alternative ancestral range weighted (Yu et al. 2010). In order to obtain a better resolution for ancestral reconstruction, RASP only accepts free polytomies trees; thus, the solved part of the cladogram was used. Distributional data of the genus *Lycinus* were taken from Goloboff (1995) and the records obtained in the present study. Four main areas defined in previous biogeographic schemes (Morrone 2006, 2014) were considered in the analysis. The following areas were used: Coquimbo (A) and Santiago (B) from the Andean region, Brazilian Atlantic Forest province (D) from the Neotropical region, and Monte province (C) from the South American Transition Zone. *L. longipes* has a wide distribution in western Argentina, occupying five biogeographic provinces; thus, for the purposes of RASP (to avoid many areas with only one species), we only consider the occupancy of the Monte province, where this species shows more abundance.

Primary Brooks Parsimony Analysis (BPA) was carried out in order to reconstruct area relationships according to the resolved phylogeny of the genus *Lycinus* (Brooks et al. 2001). The biogeographical analysis used distributional data of *Lycinus* reconciled to the phylogenetic hypothesis obtained in this study. For that, an individual area cladogram for the analyzed taxa constructed by replacing species names in the phylogeny with the areas where they occur was obtained. Internal and terminal nodes were numbered for later incorporation in the data matrix and were analyzed using a parsimony algorithm with TNT 1.1 (Goloboff et al. 2003). A hypothetical area with the absence of all taxa was added to the data matrix to root the cladogram. From this procedure, the area cladogram obtained was established as a general pattern of area relationships (Brooks et al. 2001).

Results and discussion

Cladistics

The phylogenetic analysis using implied weighting and implicit enumeration resulted in a single tree (Figure 1) with K values from 1 to 6 (51 steps, consistency index (CI) = 0.62, retention index (RI) = 0.56, $K = 1$, fit = 5.50, $K = 6$, fit = 2.03). The tree did not show the genus *Lycinus* as monophyletic and the lack of resolution is caused only by *Lycinus frayjorge* (a species for which males are unknown), and for which the only two synapomorphic characters for *Lycinus* had to be scored as missing entries. In the tree, *Lycinus* is paraphyletic by virtue of placing *L. frayjorge* as the sister group of *D. bonariensis* plus the rest of *Lycinus*. Goloboff (1995) obtained a tree where *L. frayjorge* is placed as a sister group of *L. caldera*, and as the author stated, 'it seems much more likely that the species actually belongs to the same group as *L. gajardoi* and *L. caldera*, also found in northern Chile'. Therefore, a single tree with the same topology

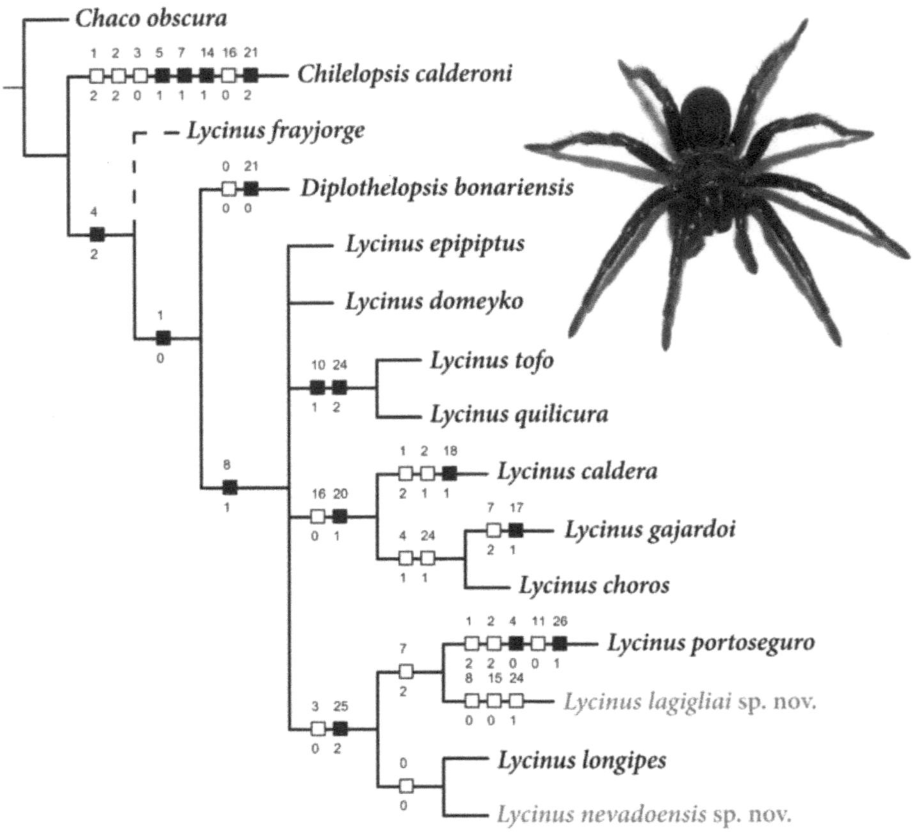

Figure 1 Results from cladistic analysis. Most parsimonious tree obtained using implied weighting (51 steps, CI = 0.62, RI = 0.56, *K* = 1, fit = 5.50, *K* = 6, fit = 2.03).

was obtained by physically excluding *L. frayjorge* from the matrix in which *Lycinus* was recovered as monophyletic. The monophyly of the genus *Lycinus* was supported only by the synapomorphy of dorsal spines on male palpal tibia. Although Goloboff (1995) proposed as an additional synapomorphy of *Lycinus* the coniform male palpal bulb, in this analysis, the character was shared by *C. obscura. L. gajardoi* and *L. choros* share the light scopula on metatarsus IV and male embolus long; the sister group of these two species is *L. caldera*, which shares the short male palpal tibia and the posteriorly directed cymbial setae. The group formed by *L. quilicura* and *L. tofo* share the strong proventral setae on male palpal tibia and the male embolus being long. The relationships found and the characters supporting are the same as those found by Goloboff (1995). Otherwise, the newly included *L. portoseguro* and *L. lagigliai* sp. nov. are sister groups supported by the presence of pseudocombs. *L. longipes* and *L. nevadoensis* sp. nov. (supported by the presence of a wide sternum) are the sister groups of *L. portoseguro* and *L. lagigliai* based on the rigid female tarsi and the construction of a burrow with a flap door. The previous analysis of Goloboff (1995) did not show a resolved position for *L. longipes*.

Taxonomy

Nemesiidae Simon, 1889

Lycinus Thorell, 1894

Diagnosis: See Goloboff (1995)

L. lagigliai sp. nov.

Type material Holotype ♂ (CAI 3375) from Argentina, Mendoza, Santa Rosa department, Ñacuñan, 20 October to 22 November 1997. Paratype ♀ (CAI 3495) from Argentina, Santa Rosa department, Ñacuñan, 34°2′47.2″ S, 67°54′26″W, 547 meters above sea level (m.a.s.l.), 12 February 2013, N. Ferretti leg.

Etymology Named in honor of Humberto Antonio Lagiglia (San Rafael, Mendoza, 1938 to 2009), an outstanding Argentinean anthropologist, archeologist, and naturalist born in San Rafael, Mendoza. This researcher contributed to the knowledge of the Ansilta culture and also was the founder and first director of the Museum of Natural History of San Rafael.

Diagnosis Males and females of *L. lagigliai* sp. nov. can be distinguished from those of their congeners by the color pattern, the shape of the palpal bulb, and the shape of the spermathecae. It resembles *L. longipes* by the general aspect of the bulb, but differs from it by the longer embolus with five low diagonal keels (instead of two strong keels); it also differs by the smaller body size, the sternum longer than wide, and metatarsi III and IV scopulated.

Description Male (CAI 3375)

Color in alcohol: entire spider brownish, carapace covered with many whitish hairs, abdomen lighter (Figure 2E,G). Total length 18.7. Carapace 9.3 long, 7.2 wide, fovea deep, procurve, 1.02 wide (Figure 2E). Clypeus narrow, 0.4. Anterior and posterior eye row procurve. Eyes sizes and interdistances: AME 0.37, ALE 0.31, PME 0.10, PLE 0.19, AME-AME 0.22, AME-ALE 0.09, PME-PME 0.82, PME-PLE 0.03, ALE-PLE 0.15. Eye quadrangle 0.9 long, 1.4 wide. Basal segment of chelicerae with five teeth, each in one row on promargin, with nine basal smaller teeth. Rastellum very weak. Intercheliceral tumescence large, rounded, pale yellow, with short thin setae. Labium 0.8 long, 1.1 wide. Labium and endites without cuspules. Sternum oval, slightly longer than wide, 4.5 long, 3.8 wide (Figure 2F). Six sternal sigilla, all away from margin by one length, posterior pair half of the size of anterior pair. Legs measurements are given in Table 2. Leg formula 4-3-1-2. Spination: tarsi I-IV and femur of palp: 0. Palp: patella 1 R; tibia 1-1-1-2 P. Legs: I: femur 1-1-1 D, 1–1 P; patella 1 P, 1 R; tibia 1-1D, 1 R, 1-1-1 P, 1 V; metatarsus 1–1 V, 1-1-1 P, 1-1-1 R, 2-1-1-1-2 D. II: femur 2-2-2 D, 1–1 P; patella 1 P; tibia 1–2 V, 1-1-1 R, 1-1-1-1 P, 1-1-1 D; metatarsus 1-1-2-1-1 D, 1–1 P, 1-1-1-1 R, 1-1-2-2 V. III: femur 1-2-1-1 D, 1–1 P; patella 1 R, 1-1-1 P; tibia 2-2-2-1 D, 1-1-1 P, 1-1-2 R, 2–1 V; metatarsus 1-1-3 V, 1-1-1 R, 1-1-2-1-1-2 P, 1-1-1-1 D. IV: femur 1-1-1-1-1 D, 1–1 P; patella 1-1-1 P, 1 R; tibia 1-1-1 D, 2-1-2-1 P, 1-1-1 R, 1-1-1 V; metatarsus 2-1-1 D, 1-1-1-3-2-1 P, 1-1-1-2 R, 1-1-2-2 V. Pseudopreening combs absent. Tarsi I-IV flexible. Scopulae on tarsi I-III full and entire, IV fully scopulated divided by a row of two setae. Scopulae entire on all metatarsi, I present on 1/2, II on 1/3, III on 1/4, IV on 1/2. Superior tarsal claws with two rows each of 9 to 11 teeth on all tarsi. Third claw absent on all tarsi. Four spinnerets, PLS three-segmented: basal, median, and apical (digitiform short) segments 1.7, 1.03, and 0.45 long, respectively. Cymbial setae long and apically directed (Figure 2D). Palp with coniform bulb gradually tapering to form embolus and with five very low diagonal keels (Figure 2A,B,C).

Female (CAI 3495)

Coloration pattern *in vivo*: carapace, coxae, and trochanters light brown with pinkish hairs, abdomen and legs dark with lighter color in metatarsus and tarsus

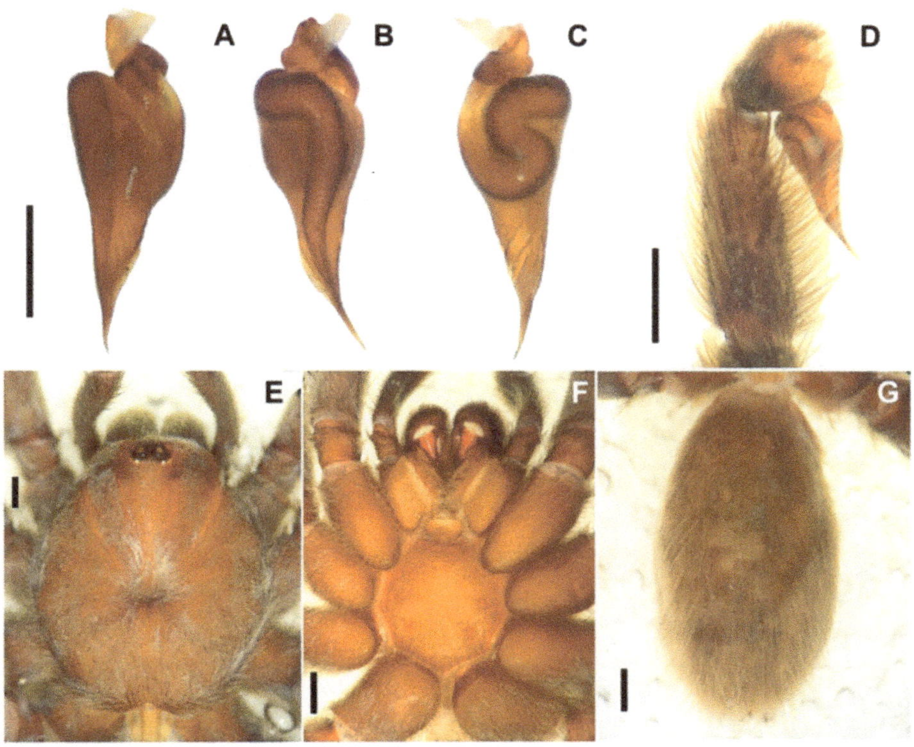

Figure 2 ***Lycinus lagigliai*** **sp. nov., holotype male (CAI 3375). (A-C)** Left palpal bulb: **(A)** prolateral, **(B)** ventral, **(C)** retrolateral. **(D)** Palpal tibia, retrolateral view. **(E)** Carapace, dorsal view. **(F)** Sternum, ventral view. **(G)** Abdomen, dorsal view. Scale bars = 1 mm.

Table 2 *Lycinus lagigliai* sp. nov., length of leg and palpal segments of male

	I	II	III	IV	Palp
Femur	8.7	10.3	9.6	11.0	4.3
Patella	3.9	4.0	5.3	5.0	2.2
Tibia	7.3	4.5	6.3	8.8	3.0
Metatarsus	10.1	9.3	11.8	14.5	-
Tarsus	6.6	6.5	7.5	8.3	1.0
Total	36.6	34.6	40.5	47.6	10.5

(Figure 3B). Color in alcohol: blackish with carapace lighter. Total length 16.3. Carapace 7.1 long, 6.1 wide, fovea deep, slightly procurve, 1.1 wide (Figure 3A). Clypeus narrow, 0.2. Anterior eye row procurve, posterior slightly procurve. Eyes sizes and interdistances: AME 0.29, ALE 0.16, PME 0.14, PLE 0.17, AME-AME 0.16, AME-ALE 0.07, PME-PME 0.55, PME-PLE 0.05, ALE-PLE 0.24. Eye quadrangle 0.8 long, 1.1 wide. Basal segment of chelicerae with five teeth in one row on promargin, with ten basal smaller teeth. Rastellum strong formed by numerous small, short cusps. Intercheliceral tumescence absent. Labium 0.4 long, 1.2 wide. Labium without cuspules, endites with three and one cuspules, respectively. Sternum oval, longer than wide, 3.8 long, 3.5 wide (Figure 3C). Six sternal sigilla, all away from margin by one length, posterior pair half of size of anterior pair. Legs measurements are given in Table 3. Leg formula 4-3-2-1. Spination: tarsi I-IV: 0. Palp: femur 1 P; patella 1-1-1 P, 1 R; tibia 1-1-1 P, 1–3 V, tarsus 1 R, 1–1 V. Legs: I: femur 1 D, 1–1 P; patella 1–1 P; tibia 1–2 V, 1–1 P; metatarsus 1-1-2 V. II: femur 1-1-1 P; patella 1 P; tibia 1 P, 1 V; metatarsus 1-1-1-2 V. III: femur 1-1-1 R, 1-1-1-2-1 P; patella 1-1-1 P, 1 R; tibia 1-1-1 R, 1-1-2-2-2 P, 1-1-2 V; metatarsus 1-1-1 D, 1-1-1-1-1-2 P, 1-1-2-2 V, 1-1-1-1 R. IV: femur 1–1 R, 1-1-1-2 P, patella 1-1-1 P; tibia 1-1-2 V, 1 P, 1-1-1 R; metatarsus 1-1-1-1-1 P, 1-1-1-1 R, 1-1-2-1-1-2 V. Pseudopreening comb on metatarsus IV. Scopulae on tarsi I-IV full. Scopulae on tarsus I divided by row of two setae, on II and III divided by row of four setae, on IV divided by row of four long and strong setae, basal and apical scopulae divided by many rows of setae. Scopulae entire on metatarsi I, II, and IV, I present on 3/4, II on 2/3, III on 1/2 divided by a line of setae, IV on 2/3. Superior tarsal claws with two rows each of 7 to 13 teeth on all tarsi. Third claw absent on all tarsi. Four spinnerets, PLS three-segmented: basal, median, and apical (digitiform short) segments 0.5, 0.6, and 0.4 long, respectively. Genitalia with two long thin and straight copulation ducts and oval-shaped spermathecae (Figure 3D).

Other materials examined Argentina, Mendoza, Santa Rosa department, Ñacuñán, 34°2′40.6″S, 67°54′26″W, 540 m.a.s.l., 10 February 2013, N. Ferretti leg., 1 juvenile (CAI 3496); Santa Rosa department, Ñacuñán, 20 October to 22 November 1997, 3♂ (CAI 3376, 3377, 3378).

Figure 3 *Lycinus lagigliai* sp. nov., paratype female (CAI 3495). (A) Carapace, dorsal view. **(B)** Female live *habitus*. **(C)** Sternum, ventral view. **(D)** Spermathecae, dorsal view. Scale bars **(A, C,** and **D)** = 1 mm. Scale bar (B) = 1 cm.

Table 3 ***Lycinus lagigliai*** **sp. nov., length of leg and palpal segments of female**

	I	II	III	IV	Palp
Femur	5.8	5.4	5.8	6.7	2.8
Patella	2.4	3.3	2.7	3.9	1.9
Tibia	3.7	4.0	3.7	5.7	3.1
Metatarsus	4.2	4.5	5.9	9.2	-
Tarsus	3.0	3.1	4.0	5.0	2.2
Total	19.1	20.3	22.1	30.5	10.0

Distribution and ecological notes Known only from the type locality, Biosphere Reserve of Ñacuñán, Santa Rosa department, Mendoza (Figure 4). The reserve is located in the lowland Monte Desert of northwestern Argentina that harbors a significant amount of the biotic diversity, landforms, and disturbance regimes of the arid lands of South America (Ojeda et al. 1998). The characteristic habitat (Figure 5A) where *L. lagigliai* sp. nov. was found comprised shrubby steppe and 'matorral' dominated by *Prosopis flexuosa*, *Geoffroea decorticans*, *Larrea divaricata*, and *Ligaria cuneifolia*; arboreal woodland with *P. flexuosa*; shrubby woodland dominated by *L. divaricata*, *Atriplex lampa*, *Verbena aspera*, *Lycium chilensis*, *Atamisquea enarginata*, and *Ephedra triandra*; and herbaceous woodland characterized by *Pappophorum caespitosum*, *Digitaria californica*, and *Setaria leucopila* (Ojeda et al. 1998). Spiders were active at night, even one was found preying on a juvenile scorpion about 50 cm away from the burrow entrance and when disturbed quickly returned to the burrow (Figure 5B). During observations at night, I saw that the burrows are closed by means of a door, without hinge, which lies over the burrow mouth during day (Figure 5C) but, when opened, spiders leave the door by the side of the burrow, facing downward (Figure 5D). When disturbed, they quickly hold the door with the two first pair of legs and covered the burrow entrance. The burrow resembles that of *L. longipes*, but in *L. lagigliai* sp. nov., it seems to be unstable with little silk cemented to the walls and when touched quickly collapsed. In this habitat, specimens of *Diplothelopsis ornata* Tullgren, 1905 (Nemesiidae), *Actinopus* sp. (Actinopodidae), and *Grammostola inermis* Mello-Leitão, 1941 and *Catumiri argentinense* (Mello-Leitão, 1941), both from the Theraphosidae, were also found.

L. nevadoensis sp. nov.

Type material Holotype ♂ from Argentina, Mendoza, Malargüe department, Cerro Nevado, 35°36′47.58″S, 68°31′39.6″W, 2,610 m, 8/17 February 2006, G. Debandi & E. Ruiz legs (CAI 3320). Paratypes: 2♂, same data as holotype (CAI 3321, 3322).

Etymology The name '*nevadoensis*' refers to the 'Cordón del Nevado' in Mendoza (Argentina), where the species was found.

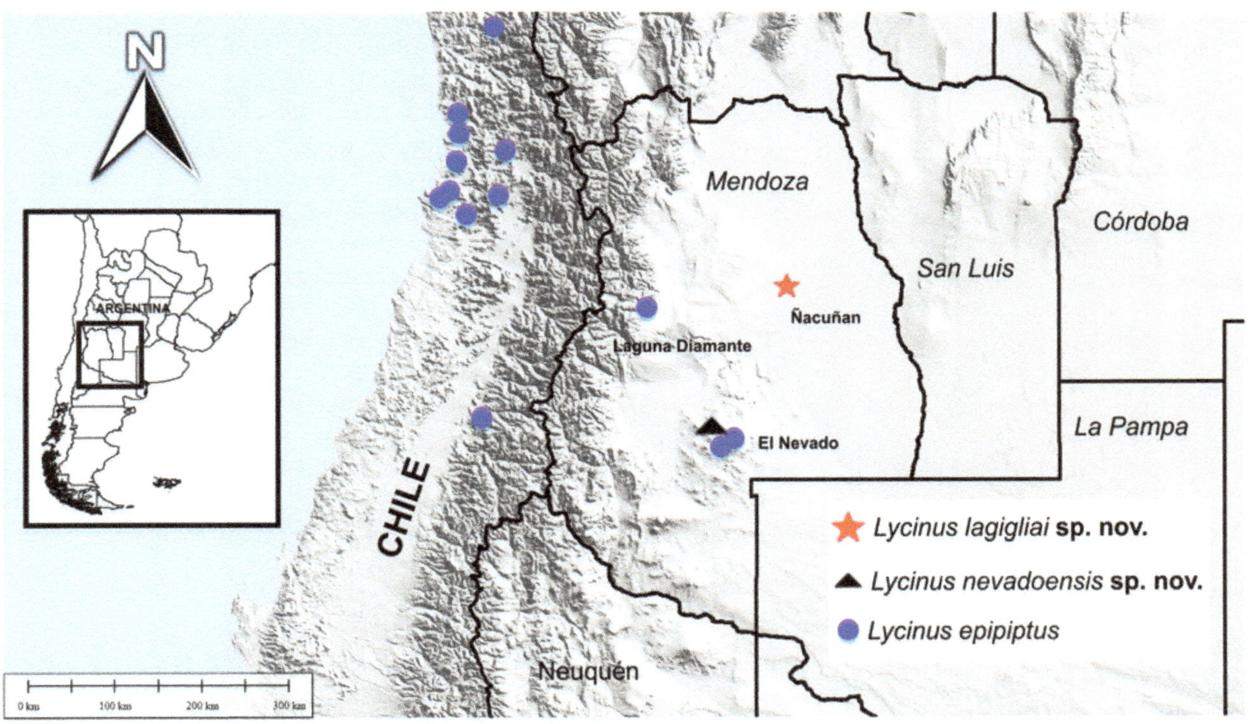

Figure 4 Map of central Argentina showing the distribution of *L. lagigliai* sp. nov., *L. nevadoensis* sp. nov., and *L. epipiptus*.

Figure 5 Natural history of *L. lagigliai* sp. nov. (A) Habitat characteristics where the species occur. **(B)** Juvenile returning to its burrow. **(C)** Female burrow with flap door closed. **(D)** Female burrow with opened flap door. Green arrows indicate the flap door.

Diagnosis Males of *L. nevadoensis* sp. nov. differ from those of the remaining species of the genus by the shape of the bulb and by the presence of spines on tarsi III and IV. They resemble *L. domeyko* in the body size and the palpal tarsus having thickened setae directed apically, but differ from it by the shape of the bulb, the absence of maxillary cuspules, and the scopula on tarsus IV being well developed. They resemble *L. epipiptus* by the light color and the patterned abdomen (Figures 6A,B,C and 7D), but in *L. nevadoensis* sp. nov., the abdomen is brownish with about six dark transverse lines (Figure 6B,C) (the abdomen of *L. epipiptus* is dark with lighter transverse lines).

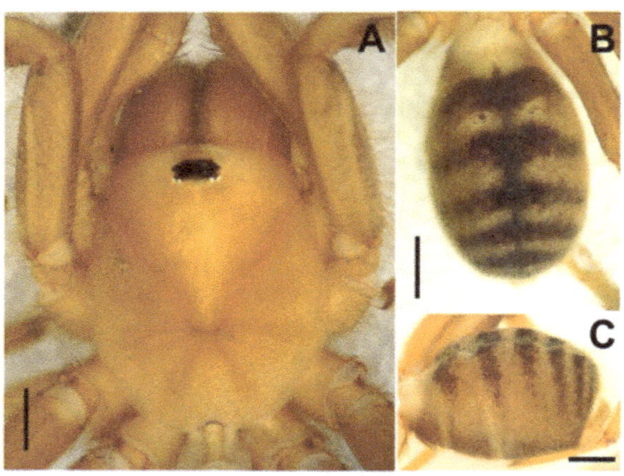

Figure 6 *Lycinus nevadoensis* sp. nov., juvenile (CAI 3353). (A) Carapace, dorsal view. **(B)** Abdomen, dorsal view. **(C)** Abdomen, lateral view.

Description Male (CAI 3320)

Color pattern in alcohol: entire spider brownish, carapace lighter, abdomen with clear-cut chevron occluded by hairs (Figure 8E,G). Total length 12.3. Carapace 6.7 long, 5.7 wide, fovea deep, slightly procurve, 0.9 wide. Clypeus narrow, 0.2. Anterior and posterior eye rows procurved. Eyes sizes and interdistances: AME 0.25, ALE 0.13, PME 0.07, PLE 0.09, AME-AME 0.20, AME-ALE 0.07, PME-PME 0.51, PME-PLE 0.02, ALE-PLE 0.07. Eye quadrangle 0.6 long, 0.9 wide. Basal segment of chelicerae with five teeth, each in one row on promargin, with 11 basal smaller teeth. Rastellum weak with thick setae. Intercheliceral tumescence large, rounded, pale yellow, asetose. Labium 0.4 long, 0.9 wide. Labium and endites without cuspules. Sternum oval, wider than long, 1.9 long, 2.7 wide (Figure 8F). Six small sternal sigilla, all away from margin by one length. Measurements are given in Table 4. Leg formula 4-3-1-2. Spination: tarsi I-II: 0. Palp: femur 1 P; patella 1–1 P; tibia 1 R, 2-1-1-2 D, 1-2-1-1 P, 1–1 V. Legs: I: femur 1-2-1-2 D, 1-1-1-1 P, 1 R; patella 1-1-1 P, 1 R; tibia 2-1-1 D, 2-1-1-2 R, 2-2-2 P, 2-2-1-2 V; metatarsus 1-1-1 V, 1-2-2-2-1 P, 1–1 R, 1-1-1-1-2 D. II: femur 2-1-1-2-1 D, 1-1-1 P, 1-1-1 R; patella 1-1-1 P, 1 R; tibia 1-2-2-2 V, 1-1-1 R, 1-2-1 P, 1-1-2 D; metatarsus 2–2 D, 2-1-2-2-1-1 P, 1-1-1-1 R, 2-1-1-1-2 V. III: femur 1-1-1-1 D, 1–1 P, 1-1-1-1 R; patella 1–1 R, 1-1-1 P; tibia 1-2-2-2 D, 2-2-2 P, 2-2-1 R, 1-1-1-1-1-2 V; metatarsus 2-2-1-1-2 V, 1-1-2-1-1 R, 2-1-2-1-2-2 P, 1-2-1-1-1 D; tarsus 1 D, 1 R, 1 P. IV: femur 1-1-1- P, 1-1-1-1 R, 1-1-1-1 D; patella 1-1-1 P, 1 R; tibia 2-2-2-2-1 D, 2-2-2-1 P, 2-1-1 R, 2-2-2 V; metatarsus 2-1-2-2 D, 1-1-1-1-2-

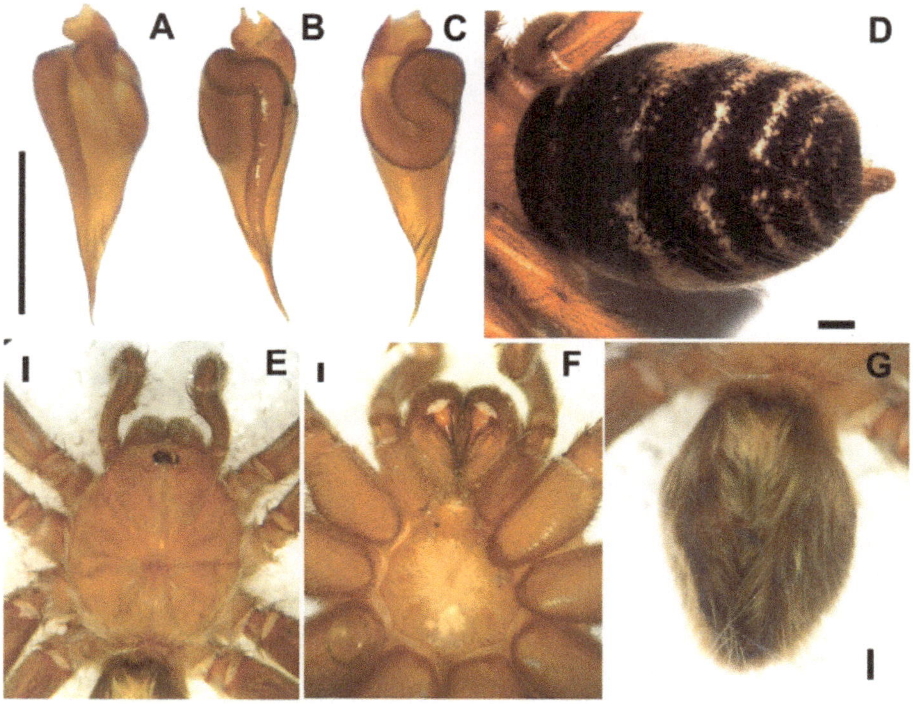

Figure 7 ***Lycinus epipiptus*****. (A-C, E-G)** Male (CAI 3277). **(D)** Female (MACN-Ar). **(A)** Left palpal bulb, prolateral. **(B)** Left palpal bulb, ventral. **(C)** Left palpal bulb, retrolateral. **(D)** Abdomen, dorsal view. **(E)** Carapace, dorsal view. **(F)** Sternum, ventral view. **(G)** Abdomen, dorsal view. Scale bars = 1 mm.

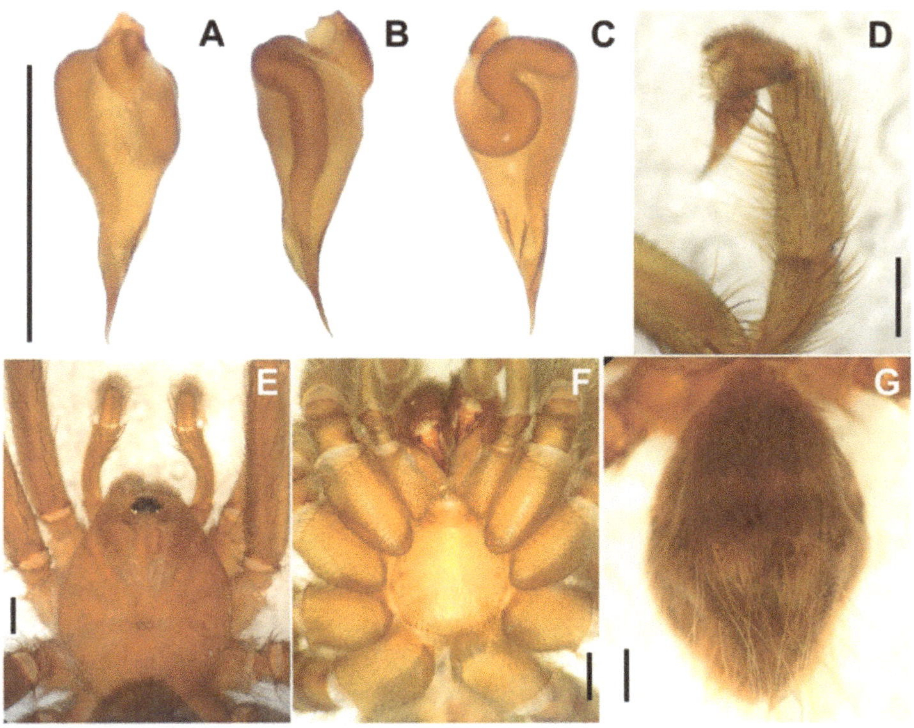

Figure 8 ***Lycinus nevadoensis*** **sp. nov., holotype male (CAI 3320). (A-C)** Left palpal bulb: **(A)** prolateral, **(B)** ventral, **(C)** retrolateral. **(D)** Palpal tibia, retrolateral view. **(E)** Carapace, dorsal view. **(F)** Sternum, ventral view. **(G)** Abdomen, dorsal view. Scale bars = 1 mm.

Table 4 *Lycinus nevadoensis* sp. nov., length of leg and palpal segments of male

	I	II	III	IV	Palp
Femur	6.8	7.1	6.3	7.4	3.1
Patella	3.5	3.4	3.1	3.0	1.6
Tibia	5.1	4.5	4.9	5.4	2.2
Metatarsus	6.7	7.0	9.2	10.5	-
Tarsus	5.0	4.8	5.8	6.3	1.3
Total	27.1	26.8	29.3	32.6	8.2

2 P, 1-1-1-1-1-2 R, 1-1-2-2-2 V, tarsus 1 D, 1-1-1 P. Pseudopreening combs absent. Tarsi I-IV flexible. Scopulae on tarsi I-II full and entire, III fully scopulated divided by one line of setae, IV fully scopulated divided by a row of two long setae. Scopulae entire on metatarsus I, divided by one line of setae on II-IV. Metatarsus I scopulated on 1/3, II on 1/4, III on 1/5, IV on 1/4. Superior tarsal claws with two rows of 10 to 11 teeth on each row on all tarsi. Third claw absent on all tarsi. Four spinnerets, PLS three-segmented: basal, median, and apical (digitiform short) segments 0.5, 0.3, and 0.2 long, respectively. Cymbium with thickened setae, apically directed (Figure 8D). Palp with coniform bulb gradually tapering to form a short embolus and with three high keels, almost with a vertical position (Figure 8A,B,C).
Female

Unknown.

Other materials examined Argentina, Mendoza, Malargüe department, Cerro Nevado, 35°36′47.58″S, 68°31′39.6″ W, 2,605 m.a.s.l., 15/24 February 2005, G. Debandi & R. Carrara legs., 2♂ (CAI 3331, 3332). Malargüe, Cerro Nevado, 35°37′43.56″S, 68°32′56.82″W, 2,333 m.a.s.l., 8/17 February 2005, G. Debandi & E. Ruiz Manzanos legs., 2♂ (CAI 3338, 3349). Malargüe, Cerro Nevado, cima, 35°35′45.6″S, 68°30′24.2″W, 3,130 m.a.s.l., 23/25 February 2006, M. Tognelli & R. Carrara legs., 3♂ (CAI 3292, 3293, 3296). Malargüe department, Cerro Nevado, 35°36′11.4″S, 68°30′48.96″W, 2,908 m.a.s.l., 4/14 December 2004, G. Debandi & G. Flores legs., 1 juvenile (CAI 3353) (Figure 6A,B,C).

Distribution and ecological notes Known only from the Cerro Nevado, Malargüe department, Mendoza (Figure 4). This species is known only from high-altitude habitats on the Cerro Nevado, an extra-Andean mountainous range located 200 km east of the Andes. The Nevado Range is separated from the Andean Range by a plateau of 1,800 m.a.s.l. It extends north–south between 34° and 36° S, parallel to the Andes, with a maximum altitude of 3,833 m.a.s.l (summit of Cerro Nevado). In the lower part of this range, the vegetation is a shrub steppe of *Neosparton aphyllum* and *Sporobolus rigens* on sandy and basaltic soils. All specimens were captured with pitfall traps at altitudes above 2,300 m.a.s.l. where the characteristic low vegetation has *Panthacantha ameghinoi* as a dominant species (Flores and Carrara 2006). Most of the individuals sampled corresponded to adult males and were caught in late summer (between 17 February 2005 and 24 February 2005; between 17 February 2006 and 25 February 2006).

L. epipiptus (Zapfe, 1963)

Diagnosis This species differs from *L. longipes* by its narrower sternum and lighter color (Figure 7E,F); the male palpal bulb has a long embolus with only three low keels (Figure 7A,B,C). Moreover, the male bulb embolus is shorter and distinguishes it from all Chilean species, except from *L. caldera* (which differs in the darker color and in having the cymbial setae directed backward; those setae are directed towards the apex in *L. epipiptus*). Cephalothorax, legs, and palpi yellowish brown, abdomen dark with clear-cut chevron (Figure 7D,G).

New records and distribution Known from Regions IV and V of Chile. New records for Argentina in Mendoza: Malargüe and San Carlos departments (Figure 4).

Material examined Argentina, Mendoza, Malargüe department, Cerro Nevado, 35°37′47″S, 68°32′15.42″ W, 2,353 m.a.s.l., 13/20 February 2006, M. Tognelli & R. Carrara legs., 1♂ (CAI 3277). Cerro Nevado, 35°37′ 43.56″S, 68°32′56.82″W, 2,333 m.a.s.l., 8/17 February 2005, G. Debandi & E. Ruiz Manzanos legs., 3♂ (CAI 3335, 3336, 3337). Cerro Nevado, 35°36′47.58″S, 68°31′ 39.6″W, 2,605 m.a.s.l., 15/24 February 2005, G. Debandi & R. Carrara legs., 1♂ (CAI 3333). Cerro Nevado, 35°36′ 47.58″S, 68°31′39.6″W, 2,610 m.a.s.l., 8/17 February 2006, G. Debandi & E. Ruiz Manzanos legs., 2♂ (CAI 3323, 3324). Cerro Nevado, cima, 35°35′45.06″S, 68°30′ 24.12″W, 3,130 m.a.s.l., 23/25 February 2006, M. Tognelli & R. Carrara legs., 2♂ (CAI 3297, 3276). San Carlos department, Reserva 'Laguna del Diamante', 34°14′6.17″S, 69°22′4.9″W, 8/17 February 2005, G. Debandi & S. Claver legs., 7♂ (CAI 3318, 3294, 3295, 3291, 3307, 3302, 3303). Reserva 'Laguna del Diamante', Alvarado, 34°14′6.17″S, 69°22′4.9″W, 2,347 m.a.s.l., 13/23 February 2006, S. Claver & R. Carrara legs., 2♂ (CAI 3328, 3329). Reserva 'Laguna del Diamante', Alvarado, 34°14′47.4″S, 69°22′ 58.2″W, 2,358 m.a.s.l., 23/25 February 2006, S. Roig & G. Debandi legs., 1♂ (CAI 3305). Reserva 'Laguna del Diamante', Alvarado, 34°14′46.9″S, 69°22′40.3″W, 2,310 m. a.s.l., 5/8 January 2006, S. Roig & G. Debandi legs., 1♂ (CAI 3301). Chile, Región V (Valparaíso), Quebrada El Tigre, Cachagua, November 1988, P. Goloboff, E. Maury & C. Szumik legs., 1♀ (MACN-Ar), pictures examined.

***L. longipes* Thorell, 1894**

Diagnosis This species differs from the Chilean species of *Lycinus* by the wider sternum (Figure 9F); the male palpal tibia long and cymbium with apically directed setae (Figure 9D), bulb tapering gradually, with short embolus and three very high diagonal keels (Figure 9A, B,C). Entire spiders uniformly dark black (Figures 9E,F,G and 10B).

Material examined Argentina, Mendoza, Las Heras department, Estancia Tambillos, 32°22′7.56″S, 69°26′3.26″W, 2,390 m.a.s.l., 4/14 December 2004, A. Scollo & S. Claver legs., 1♂ (CAI 3334). Same locality, 32°21′14.88″S, 69°25′52.38″W, 2,428 m.a.s.l., 20 December 2005/5 January 2006, S. Claver & A. Scollo legs., 2♂ (CAI 3325, 3326). Ruta prov. 52, 32°37′10.56″S, 68°54′49.02″W, 978 m.a.s.l., 4/14 December 2004, S. Claver & A. Scollo legs., 1♂ (CAI 3327). Capital department, Reserva Natural Divisadero Largo, 32°52′48.8″S, 68°56′02.3″W, 1,474 m.a.s.l., 17 February 2013, N. Ferretti leg., 2♀ (CAI 3497, 3498).

Distribution and ecological notes Previously known from Santiago del Estero, Salta, Córdoba, La Rioja, Neuquén, Catamarca, San Juan (Figure 11). In Mendoza, a single female (not examined) was recorded in Ñacuñan by Goloboff (1995). However, a recent campaign (involving day and night collections) on this Natural Reserve yielded only specimens of *L. lagigliai* sp. nov. There are new records of *L. longipes* for Mendoza in Las Heras and Capital departments (Figure 11). The only available information came from specimens collected in salt flats and in depressed, floodable areas, in which spiders live in deep burrows (up to 40 cm deep), about 20 mm wide, with smooth well-compacted and cemented walls, lined with a thin layer of silk. The burrows are closed by means of a door, without hinge, which lies over the burrow mouth (Goloboff 1995). I found specimens of *L. longipes* inhabiting non-flooding areas, comprising the Andean foothills at an elevation of about 1,500 m.a.s.l. (Figure 10A). Spiders captured at Divisadero Largo Natural Reserve were active at night, and open burrows (Figure 10C) are similar to those described by Goloboff (1995). One adult female (Figure 10B) was found walking; its burrow could not be located.

Biogeography

According to the RASP optimization, one tree of possible ancestral distributions was obtained (Figure 12A). The ancestral areas reconstructed at the root of the genera *Lycinus* and *Chilelopsis* are most likely to be either Coquimbo (Andean region) or both Coquimbo and Monte (South American Transition Zone). A vicariant

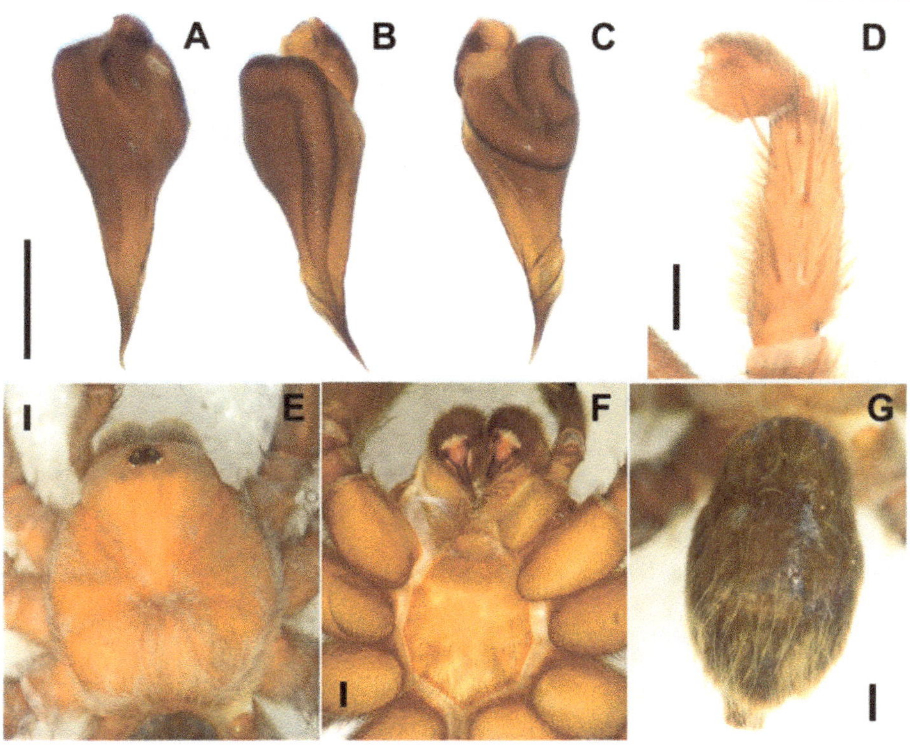

Figure 9 *Lycinus longipes*, male (CAI 3334). (A-C) Left palpal bulb: **(A)** prolateral, **(B)** ventral, **(C)** retrolateral. **(D)** Palpal tibia, retrolateral view. **(E)** Carapace, dorsal view. **(F)** Sternum, ventral view. **(G)** Abdomen, dorsal view. Scale bars = 1 mm.

Figure 10 Natural history of *L. longipes*. (A) Habitat characteristics where the species occur. **(B)** Female, live *habitus*. **(C)** Female burrow opened at night. Scale bar = 1 cm.

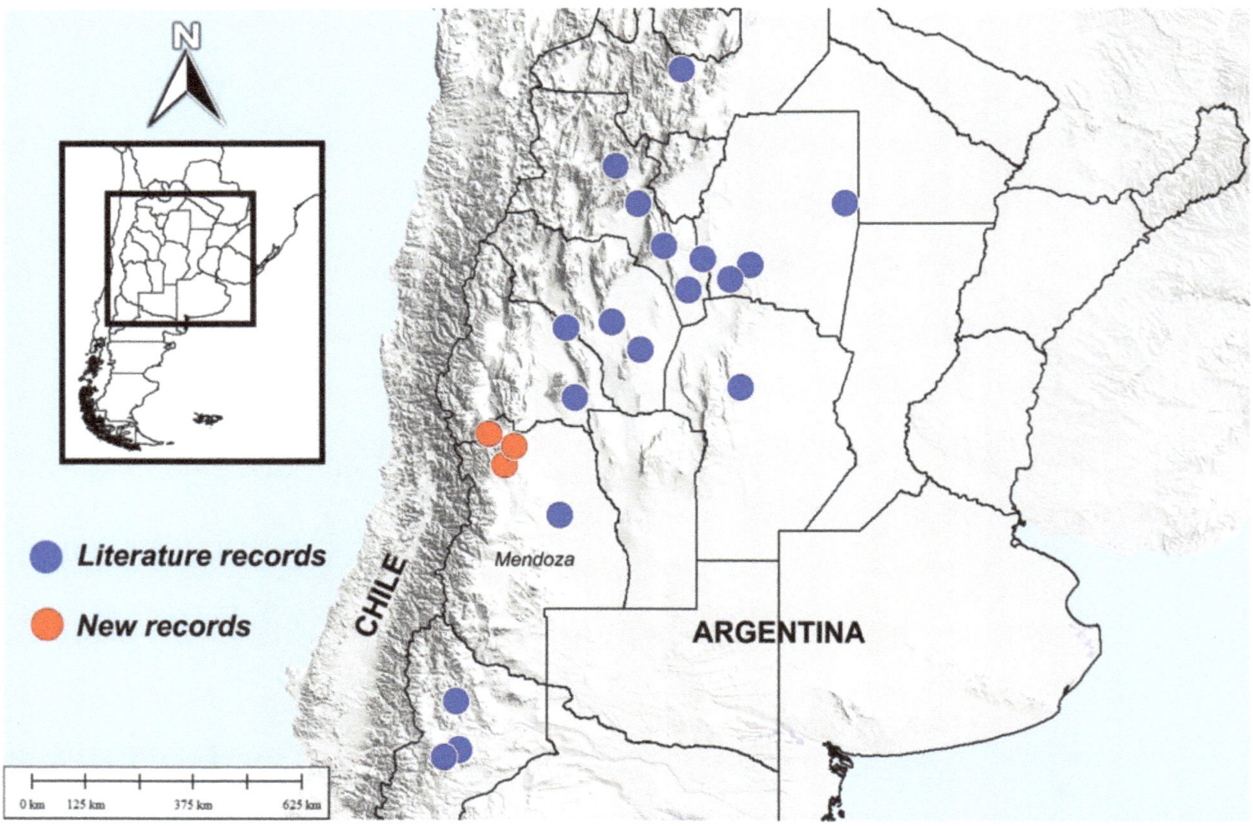

Figure 11 Map of western Argentina showing the known distribution of *L. longipes*.

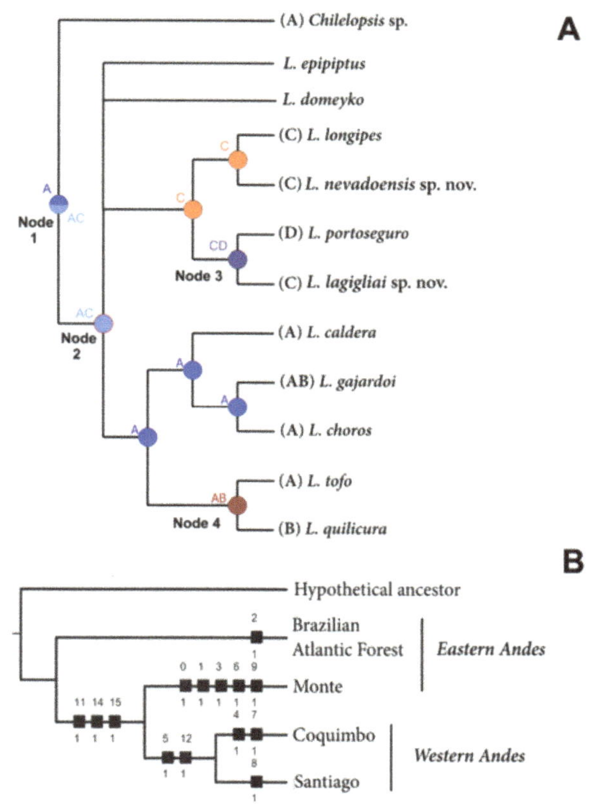

Figure 12 Biogeographic results. (A) Ancestral distributions reconstructed by RASP. Pie charts at the nodes give the relative frequencies of the ancestral area reconstruction. **(B)** Historical relationship between the areas found in the solved phylogeny of *Lycinus*.

event (node 1, Figure 12A) took place, resulting in two isolated groups: *Chilelopsis* (Coquimbo) and *Lycinus* (Coquimbo-Santiago-Monte-Brazilian Atlantic Forest). Subsequently, a second vicariant event (node 2, Figure 12A) separated the Argentinean-Brazilian species *L. longipes*, *L. nevadoensis* sp. nov., *L. portoseguro*, and *L. lagigliai* sp. nov. in Monte and Brazilian Atlantic Forest provinces from the clade containing the Chilean species *L. caldera*, *L. gajardoi*, *L. choros*, *L. tofo*, and *L. quilicura* in Coquimbo and Santiago provinces, with possible ancestral areas of the Coquimbo-Monte. Then, a new vicariant event (node 3, Figure 12A) separated *L. lagigliai* sp. nov. in the Monte from *L. portoseguro* in the Brazilian Atlantic Forest, with the possible ancestor located at the Monte. Finally, a new vicariant event (node 4, Figure 12A) separated *L. tofo* in the Coquimbo region from *L. quilicura* in the Santiago region. The possible ancestral areas for this node were the Coquimbo-Santiago. The primary BPA resulted in a single cladogram of length 19, with a consistency index of 0.89 and a retention index of 0.71 (Figure 12B).

According to the area cladogram obtained through BPA and based on the assumption of a vicariance model, the monophyly of areas at western and eastern of the Andes, such as Coquimbo, Santiago, and Monte, showed the close relationship between the fauna of the Andean region with that of the South American Transition Zone. From the results of RASP, the second ancestral distribution resulting from the analysis showed *Lycinus* inhabiting the areas today occupied by the Coquimbo (Andean region) and Monte (South American Transition Zone). The association of these two areas has been proposed by Morrone (2006) with examples of many species of Orthoptera, Coleoptera, Diptera, Hymenoptera, and Lepidoptera that are common to the Andean region and the South American Transition Zone. Moreover, most of the Andean biota originally evolved in Patagonia and then gradually spread northward into the South American Transition Zone during the Tertiary and Pleistocene (Morrone 1994). The events that influenced the separation of the Argentinean-Brazilian species from the Chilean could take place from the late Oligocene to the late Miocene. During this period, the present morphostructural configuration of the Andes of Central Chile began to develop (Donato et al. 2003). The southern South America landscapes began to be dominated by the processes that led to the present configuration of the Andean tectonic-magmatic belt (Uliana and Biddle 1988). Also, three successive Atlantic marine transgressions were recorded in southern South America. Therefore, during the middle and late Miocene, an open seaway separated the terrestrial environments of southern South America from those farther north, spreading over virtually all of eastern Argentina, western Uruguay, southern Paraguay, and southeastern Bolivia (Pascual et al. 1996; Donato et al. 2003). Finally, the events related to the split of *L. lagigliai* sp. nov. (Monte) from *L. portoseguro* (Brazilian Atlantic Forest) that could have been influenced during the late Miocene-late Pliocene span comprehend a new Andean diastrophic phase, named 'Diaguita phase' (Pascual et al. 1996; Donato et al. 2003) that resulted in the elevation of the Central Cordillera of Argentina and Chile, the eastern orographic systems of Argentina, and the uplift of the Mesopotamian region. This event led to the end of the Age of the Southern Plains that corresponded to widespread and varied plains, extending from northern Patagonia, reaching central and northern Argentina, Uruguay, the eastern slopes of the rising Andes in Bolivia, Peru, and Venezuela, and also in the Amazon basin (Marshall et al. 1983; Pascual et al. 1996) involving the ancestral areas of *L. lagigliai* sp. nov. and *L. portoseguro*.

The species *L. nevadoensis* sp. nov. described in this study could be another representative of the endemic fauna of El Nevado mountain chain. This orographic system comprises a relevant area of endemism for high-altitude arthropod fauna, with closely related species found at high-altitude locations in the Andean range.

This feature could be due to the high distance that separates it from the Andean mountain chain (200 km). Some of the endemic taxa described for the Nevado mainly include carabid beetles, such as *Cnemalobus nevado* Roig-Juñent et al., 2007, *Baripus nevado* Roig-Juñent et al., 2008, and *Trechisibus nevadoi* Roig-Juñent and Sallenave 2005, but recently, the scorpion *Orobothriurus grismadoi* Ojanguren-Affilastro et al., 2009 has been described as endemic for this mountain range (Ojanguren-Affilastro et al. 2009). As mentioned, it has been found that some of the sister species of these endemic taxa occurs in high-altitude locations in the Andean range (for example, Laguna del Diamante, Mendoza) (Roig-Juñent et al. 2007). This could be the case for *L. epipiptus* in Argentina, whose distribution corresponds to El Nevado range, and also at high Andean altitudes such as Laguna del Diamante, but unfortunately, the position of this species in the cladogram was not solved. The accumulation of diversity studies involving species that occur at the Andean mountains will shed light on the biogeographical hypotheses in relation to this relevant region.

Conclusions

Two new species are recognized in *Lycinus* and are diagnosed and described. According to the cladistic reanalysis, the main relationships found and the characters supporting are the same as those found by Goloboff (1995). Some events that could have influenced in the distributional patterns of the species of *Lycinus* comprised the following: i) the processes that led to the present formation of the Andean tectonic-magmatic belt, ii) three successive Atlantic marine transgressions that were recorded in southern South America, and iii) a new Andean diastrophic phase, named 'Diaguita phase', that resulted in the elevation of the Central Cordillera of Argentina and Chile, the eastern orographic systems of Argentina, and the uplift of the Mesopotamian region.

Abbreviations

AME: anterior median eyes; ALE: anterior lateral eyes; D: dorsal; P: prolateral; PLE: posterior lateral eyes; PLS: posterior lateral spinnerets; PME: posterior median eyes; R: retrolateral; V: ventral.

Competing interests

The author declares that he has no competing interests.

Author's information

NF is a researcher of the National Council of Scientific and Technical Research (CONICET) in Argentina. His research interest focuses on the systematics, biogeography, ecology, and ethology of mygalmorph spiders.

Acknowledgements

Thanks to Susana Lagos Silnik for loaning the collection material from IADIZA. The author thanks Cristian Grismado for supplying some images from MACN-Ar specimens. Thanks to Gabriel Pompozzi, Sofía Copperi, and Leonela Schwerdt for their help on field trips and photographs. Thanks to Alda González for her constant support on research experience. This work was partially funded by the American Arachnological Society and Vincent Roth Foundation to the project 'Diversity of mygalomorph spiders (Araneae) in Mendoza province, Argentina: How many species are there?' Special thanks to the park rangers from Ñacuñan and Divisadero Largo natural reserves for their help during field work. Thanks to the Instituto Argentino de Investigaciones de las Zonas Áridas (IADIZA), CCT-CONICET Mendoza, and the Dirección de Recursos Naturales Renovables, Mendoza for the collection permits. The author is a researcher of Consejo Nacional de Investigaciones Científicas y Técnicas (CONICET).

References

Antonelli A, Verola CF, Parisod D, Gustaffson ALF (2010) Climate cooling promoted the expansion and radiation of a threatened group of South American orchids (Epidendroideae: Laeliinae). Biol J Linn Soc 100:597–607

Brooks DR, Van Veller M, MacLennan DA (2001) How to do BPA, really. J Biogeog 28:343–358

Capocasale MR, Pérez-Miles F (1990) Behavioural ecology of *Acanthogonatus tacuariensis* (Pérez & Capocasale) (Araneae, Nemesiidae). Stud Neotrop Fauna Environ 25:41–47

Crisci JV, Katinas L, Posadas P (2003) Historical biogeography: an introduction. Harvard University Press, Cambridge

Donato M (2006) Historical biogeography of the family Tristiridae (Orthoptera: Acridomorpha) applying dispersal vicariance analysis. J Arid Environ 66:421–434

Donato M, Posadas P, Miranda-Esquivel DR, Ortiz-Jaureguizar E, Cladera G (2003) Historical biogeography of the Andean region: evidence from Listroderina (Coleoptera: Curculionidae: Rhytirrhinini) in the context of the South American geobiotic scenario. Biol J Linn Soc 80:339–352

Ferretti N, Pompozzi G, Pérez-Miles F (2011) Sexual behavior of *Acanthogonatus centralis* (Araneae: Mygalomorphae: Nemesiidae) from Argentina, with some notes on their burrows. J Arachnol 39:533–536

Ferretti N, González A, Pérez-Miles F (2012) Historical biogeography of the genus *Cyriocosmus* (Araneae: Theraphosidae) in the Neotropics according to an event-based method and spatial analysis of vicariance. Zool Stud 51:526–535

Flores G, Carrara R (2006) Two new species of *Nyctelia* Latrielle, from western Argentina with zoogeographical and ecological remarks on the high mountain habitat (Coleoptera: Tenebrionidae). Ann Zool 56:487–495

Goloboff PA (1993) Estimating characters weights during tree search. Cladistics 9:83–91

Goloboff PA (1995) A revision of the South American spiders of the family Nemesiidae (Araneae, Mygalomorphae). Part I: species from Peru, Chile, Argentina, and Uruguay. Bull Am Mus Nat Hist 224:1–189

Goloboff P, Farris J, Nixon K (2003) T.N.T.: tree analysis using new technology, program and documentation. http://www.lillo.org.ar/phylogeny/. Accessed 2 Apr 2012

Guo Y, Wang Y (2007) Partitioned Bayesian analyses, dispersal-vicariance analysis, and the biogeography of Chinese toad-headed lizards (Agamidae: Phrynocephalus): a re-evaluation. Mol Phylogen Evol 45:643–662

Indicatti RP, Lucas SM (2005) Description of a new genus of Nemesiidae (Araneae, Mygalomorphae) from Brazilian Cerrado. Zootaxa 1088:11–16

Indicatti RP, Lucas SM, Ott R, Brescovit AD (2008) Litter dwelling mygalomorph spiders (Araneae: Microstigmatidae, Nemesiidae) from Araucaria forests in southern Brazil, with the description of five new species. Rev Bras Zool 25:529–546

Lucas SM, Indicatti RP (2010) Description of two new species of *Lycinus* (Araneae: Nemesiidae). Zoologia 27:425–430

Lucas SM, Passanha V, Janini CRV, Indicatti RP (2008) On the genus *Neostothis* Vellard (Araneae, Nemesiidae). J Arachnol 36:472–475

Marshall LG, Hoffstetter R, Pascual R (1983) Mammals and stratigraphy: geochronology of the continental mammal-bearing Tertiary of South America. Paleovertebrata: Mémoire Extraordinaire 1983:1–93

Montes de Oca L, Pérez-Miles F (2013) Two new species of *Chaco* Tullgren from the Atlantic coast of Uruguay (Araneae, Mygalomorphae, Nemesiidae). Zookeys 377:73–87

Morrone JJ (1994) Distributional patterns of species of Rhytirrhinini (Coleoptera: Curculionidae) and the historical relationships of the Andean provinces. Global Ecol Biogeogr Lett 4:188–194

Morrone JJ (2006) Biogeographic areas and transition zones of Latin America and the Caribbean islands based on panbiogeographic and cladistic analyses of the entomofauna. Annu Rev Entomol 51:467–494

Morrone JJ (2014) Biogeographical regionalisation of the Neotropical región. Zootaxa 3782:1–110
Ojanguren-Affilastro AA, Fernández Campón F, Lagos Silnik S, Mattoni CI (2009) The genus *Orobothriurus* Maury in central Argentina with description of a new species from El Nevado mountain chain in Mendoza Province (Scorpiones: Bothriuridae). Zootaxa 2209:28–42
Ojeda RA, Campos CM, Gonnet JM, Borghi CE, Roig VG (1998) The MaB Reserve of Ñacuñán, Argentina: its role in understanding the Monte Desert biome. J Arid Environ 39:299–313
Page RDM (2001) NDE, Nexus Data Editor, ver 0.5.0. http://taxonomy.zoology.gla.ac.uk/rod/NDE/nde.html. Accessed 25 June 2010
Pascual R, Ortiz-Jaureguizar E, Prado JL (1996) Land mammals: paradigm of Cenozoic South American geobiotic evolution. Münchner Geowiss Abh 30:265–319
Petrunkevitch A (1925) Arachnida from Panama. Trans Conn Acad Arts Sci 27:51–248
Platnick, NI (2014) The world spider catalog, version 14.5. American Museum of Natural History. http://www.wsc.nmbe.ch/. Accessed 14 May 2014
Raven RJ (1985) The spider infraorder Mygalomorphae (Araneae): cladistics and systematics. Bull Am Mus Nat Hist 182:1–180
Roig-Juñent S, Carrara R, Ruiz-Manzanos E, Agrain F, Sackmann P, Tognelli FM (2007) Phylogenetic relationships and biogeographic considerations of four species of *Cnemabolus* (Coleoptera, Carabidae) from Patagonia. Insect Syst Evol 38:1–26
Simon E (1889) Voyage de M.E. Simon au Venezuela, Arachnides. Ann Soc Entomol Fr 6:169–220
Thorell T (1894) Decas aranearum in ins. Singapore a Cel. Th. Workman inventarum. Boll Soc Entomol Ital 26:321–355
Uliana MA, Biddle KT (1988) Mesozoic-Cenozoic paleogeographic and geodynamic evolution of southern South America. Braz J Geol 18:172–190
Yu Y, Harris AJ, He X (2010) S-DIVA (Statistical Dispersal-Vicariance Analysis): a tool for inferring biogeographic histories. Mol Phylogen Evol 56:848–850

5

Spatial and temporal changes in zooplankton abundance, biovolume, and size spectra in the neighboring waters of Japan: analyses using an optical plankton counter

Kaede Sato[1], Kohei Matsuno[2], Daichi Arima[1], Yoshiyuki Abe[1] and Atsushi Yamaguchi[1*]

Abstract

Background: An optical plankton counter (OPC) was used to examine spatial and temporal changes in the zooplankton size spectra in the neighboring waters of Japan from May to August 2011.

Results: Based on the zooplankton biovolume of equivalent spherical diameter (ESD) in 45 bins for every 0.1 mm between 0.5 and 5.0 mm, a Bray-Curtis cluster analysis classified the zooplankton communities into six groups. The geographical distribution of each group varied from each of the others. Groups with a dominance of 4 to 5 mm ESD were observed in northern marginal seas (northern Japan Sea and Okhotsk Sea), while the least biovolume with a dominance of a small-size class (0.5 to 1 mm) was observed for the Kuroshio extension. Temporal changes were observed along the 155° E line, i.e., a high biovolume group dominated by 2 to 3 mm ESD during May shifted to other size spectra groups during July to August. These temporal changes were caused by the seasonal vertical descent of dominant large *Neocalanus* copepods during July to August. As a specific characteristic of the normalized biomass size spectra (NBSS), the slope of NBSS was moderate (−0.90) for the *Neocalanus* dominant spring group but was at −1.11 to −1.24 for the other groups. Theoretically, the slope of the NBSS of the stable marine ecosystem is known to settle at approximately −1.

Conclusions: Based on the analysis by OPC, zooplankton size spectra in the neighboring waters of Japan were separated into six groups. Most groups had −1.11 to −1.24 NBSS slopes, which were slightly higher than the theoretical value (−1). However, one group had a moderate slope of NBSS (−0.90) caused by the dominance of large *Neocalanus* copepods.

Keywords: NBSS; *Neocalanus*; OPC; Zooplankton

Background

From the perspective of fisheries, mesozooplankton is an important food source for pelagic fish and larvae. The size of the mesozooplankton determines the bioenergetics of fish (Sheldon et al. 1977) and affects the growth and mortality rates of fish larvae (Van der Meeren and Næss 1993). From the perspective of oceanography, the size of the mesozooplankton is also important. In an oceanic region (>200 m depth, 92% of ocean area), mesozooplankton transport particulate organic matter vertically from the surface to deeper layers (Longhurst 1991; Boyd and Newton 1999). The activity of this process, termed "biological pump," is known to be correlated with the size of dominant mesozooplankton in the epipelagic layer (Michaels and Silver 1988; Ducklow et al. 2001). Thus, information on the size spectra of mesozooplankton community is important from the viewpoint of both fisheries and oceanography.

The size spectra of the mesozooplankton community were evaluated using NBSS (cf. Marcolin et al. 2013). The size of the mesozooplankton was accurately quantified

* Correspondence: a-yama@fish.hokudai.ac.jp
[1]Laboratory of Marine Biology, Graduate School of Fisheries Science, Hokkaido University, 3-1-1 Minatomachi, Hakodate, Hokkaido 041-8611, Japan
Full list of author information is available at the end of the article

using an optical plankton counter (OPC, Herman 1988). Recently, spatial and temporal changes in the size spectra of worldwide ocean mesozooplankton were evaluated by NBSS obtained using OPC measurements (cf. Huntley et al. 1995; Piontkovski et al. 1995; Zhou and Huntley 1997; Herman and Harvey 2006; Kimmel et al. 2006). The slope of NBSS is an index of bottom-up or top-down control of the marine ecosystem (Zhou 2006). In a nutrient-rich high productivity ecosystem, dominance of small-sized mesozooplankton induced a high intercept and slope of NBSS (bottom-up). However, high predation on the smaller size class may induce a low intercept and slope of NBSS (top-down) (Moore and Suthers 2006). Visual predators such as fish typically remove large particles, which act to steepen the slope and maintain the intercept (Suthers et al. 2006). Thus, information on spatial and temporal changes in mesozooplankton NBSS is highly valuable for evaluating structures of the marine ecosystem (cf. García-Comas et al. 2014).

The neighboring waters of Japan include the subarctic, transitional, and subtropical Western North Pacific and their adjacent seas: Okhotsk Sea, Japan Sea, and East China Sea. The oceanographic characteristics of these oceans vary greatly from each other. Various studies have been previously performed on mesozooplankton abundance, biomass, and community structure in these oceans. For example, studies were performed in the subarctic and transitional Western North Pacific (Odate 1994; Chiba et al. 2006, 2008, 2009), the subtropical Western North Pacific (Nakata et al. 2001; Nakata and Koyama 2003), and the Japan Sea (Hirota and Hasegawa 1999; Iguchi 2004). Based on these studies, regional, seasonal, and annual changes in mesozooplankton abundance, biomass, and community structure were evaluated. However, little information is available for their size spectra, and few attempts have been made regarding NBSS analysis. Because the NBSS of zooplankton connects phytoplankton with fisheries biomass, providing spatial and temporal change patterns of NBSS in this region will be valuable.

In the present study, spatial and temporal changes in the mesozooplankton size spectra in the neighboring waters of Japan (subarctic, transitional, and subtropical Western North Pacific and their adjacent seas: Okhotsk Sea, Japan Sea, and East China Sea) were evaluated using OPC measurements of net mesozooplankton samples collected by the same methods between May and August 2011. For all of the OPC data, NBSS analyses were performed and compared with the NBSS reported from various worldwide oceans. These comparisons revealed spatial and temporal changes in the size spectra of mesozooplankton in the neighboring waters of Japan; additionally, their characteristics were evaluated.

Methods

Field sampling

Mesozooplankton samplings were obtained on board the T/S *Oshoro-Maru* along the 155° E line (38° to 44° N) in the Western North Pacific from May 16 to 20, in the Okhotsk Sea from June 10 to 11, in the Japan Sea from June 8 to 24, in the East China Sea from July 1 to 9, in the subtropical Western North Pacific from July 11 to 12, and along the 155° E line (38° to 44° N) in the Western North Pacific from July 27 to August 2, 2011. The total number of stations was 78 (Figure 1). Samples were collected via the vertical hauls of a NORPAC net (mouth diameter 45 cm, mesh size 335 µm, Motoda 1957) from 150 m to the surface during the day and/or night. At stations where the depth was shallower than 150 m, vertical tows from 5 m above the bottom were performed. The volume of water filtered through the net was estimated from a reading of the flowmeter (Rigosha & Co., Ltd., Saitama, Japan) mounted on the net ring. The collected samples were immediately fixed with 5% borax-buffered formalin on board the ship. At each station, temperature and salinity were measured using a CTD system (Sea-Bird SBE 911 Plus, Sea-Bird Electronics, Bellevue, WA, USA). Because sampling depths varied from station to station (from 0 to 30 to 0 to 150 m), we applied temperature and salinity data at the euphotic zone (0 to 30 m) to evaluate their spatial and temporal changes (Figure 2).

OPC measurements

At the land laboratory, the mesozooplankton samples were divided into half-aliquots using a Motoda box splitter (Motoda 1959). For each half-aliquot, the zooplankton were filtered using a 100-µm mesh under low vacuum, and the wet mass was measured using an electronic microbalance with a precision of 10 mg. The remaining 1/2 sub-samples were used for OPC (Model OPC-1 L: Focal Technologies Corp., Dartmouth, NS, Canada) measurements using the flow-through system (CT&C Co. Ltd., Tokyo, Japan). OPC measurements were made at a low flow rate (*ca.* 10 L min^{-1}) and low particle density (<10 counts s^{-1}) without staining (Yokoi et al. 2008).

Abundance and biovolume

The abundance per cubic meter (N: ind. m^{-3}) for each of the 4,096 ESD size categories was calculated from the following equation:

$$N = \frac{n}{s \times F}$$

where n is the number of particles (=zooplankton ind.), s is the split factor of each sample, and F is the filtered volume of the net (m^3). The biovolume of the zooplankton community at 4,096 size categories was calculated from the ESD data, and the biovolume (mm^3 m^{-3}) was calculated by

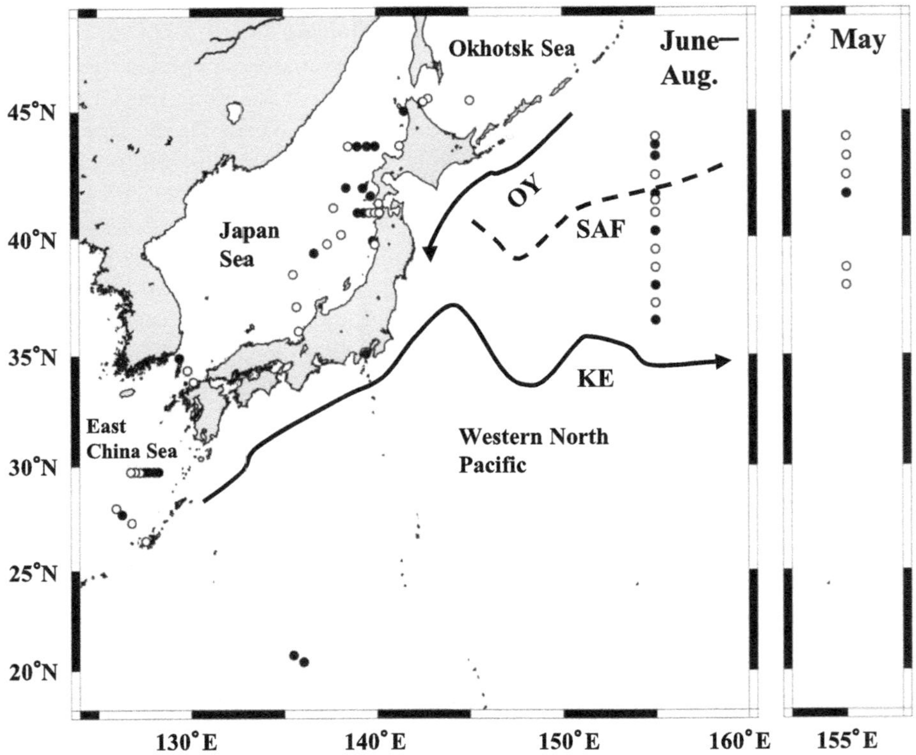

Figure 1 Location of sampling stations in the Western North Pacific and their adjacent seas. From May 16 to 20 (right panel) and June 8 to August 2 (left panel) 2011. Open and solid symbols denote stations where sampling was performed during the day and night, respectively. Approximate positions of OY: Oyashio, KE: Kuroshio Extension and SAF: Subarctic Front are superimposed (cf. Yasuda 2003). Samplings were conducted during the following periods: Western North Pacific along 155° E line (38° to 44°N) from May 16 to 20, Okhotsk Sea from June 10 to 11, Japan Sea from June 8 to 24, East China Sea from July 1 to 9, subtropical Western North Pacific from July 11 to 12 and along the 155° E line (38° to 44° N), in the Western North Pacific from July 27 to August 2, 2011.

multiplying N and volume (mm^3 $ind.^{-1}$) derived from ESD. Analyses on the mesozooplankton biovolume were performed with separation of six size classes (0 to 1, 1 to 2, 2 to 3, 3 to 4, 4 to 5, and >5 mm ESD). Day and night samplings accounted for 44 and 34 stations of all sampling stations, respectively (Figure 1). Day-night comparisons of the entire zooplankton abundance and biomass based on the whole sampling area showed no significant differences (U-test, abundance: $p = 0.567$, biomass: $p = 0.945$); thus, no day-night conversion for abundance or biomass was necessary.

Cluster analysis

To evaluate spatial and temporal changes in the size spectra of the zooplankton biovolume, cluster analysis was performed. Prior to the analysis, the biovolume data on 1,744 categories between 0.5 and 5.0 mm ESD were binned into 45 size classes at 0.1 mm ESD intervals (0.5 to 0.6, 0.6 to 0.7,..., 4.9 to 5.0 mm). Based on these biovolume data, similarities between the samples were evaluated using Bray-Curtis methods. To group the samples, similarity indices were coupled with hierarchical agglomerative clustering using a complete linkage method (Unweighted Pair Group Method using Arithmetic mean, UPGMA; Field et al. 1982). Non-metric multidimensional scaling (NMDS) ordination was performed to delineate the sample groups on a two-dimensional map (Field et al. 1982). To clarify which environmental parameters (latitude, longitude, integrated mean temperature and salinity at 0 to 30 m) exhibited significant relationships with the zooplankton sample groups, multiple regressions ($Y = aX_1 + bX_2 + c$, where Y is the environmental variable, X_1 and X_2 are axes 1 and 2 of NMDS, and a, b, and c are constants, respectively) were made using StatView (SAS Institute Inc., Cary, NC, USA).

Normalized biomass size spectra

From the OPC data, NBSS was calculated following Zhou (2006). First, zooplankton biovolume ($\bar{B}$: mm^3 m^{-3} [=μm^3 L^{-1}]) was averaged for every 100 μm ESD size class. To calculate the X-axis of NBSS (X: $\log_{10}$ zooplankton biovolume [mm^3 $ind.^{-1}$]), $\bar{B}$ was divided by the abundance of each size class (ind. m^{-3}) and converted to a common logarithm. To calculate the Y-axis of NBSS

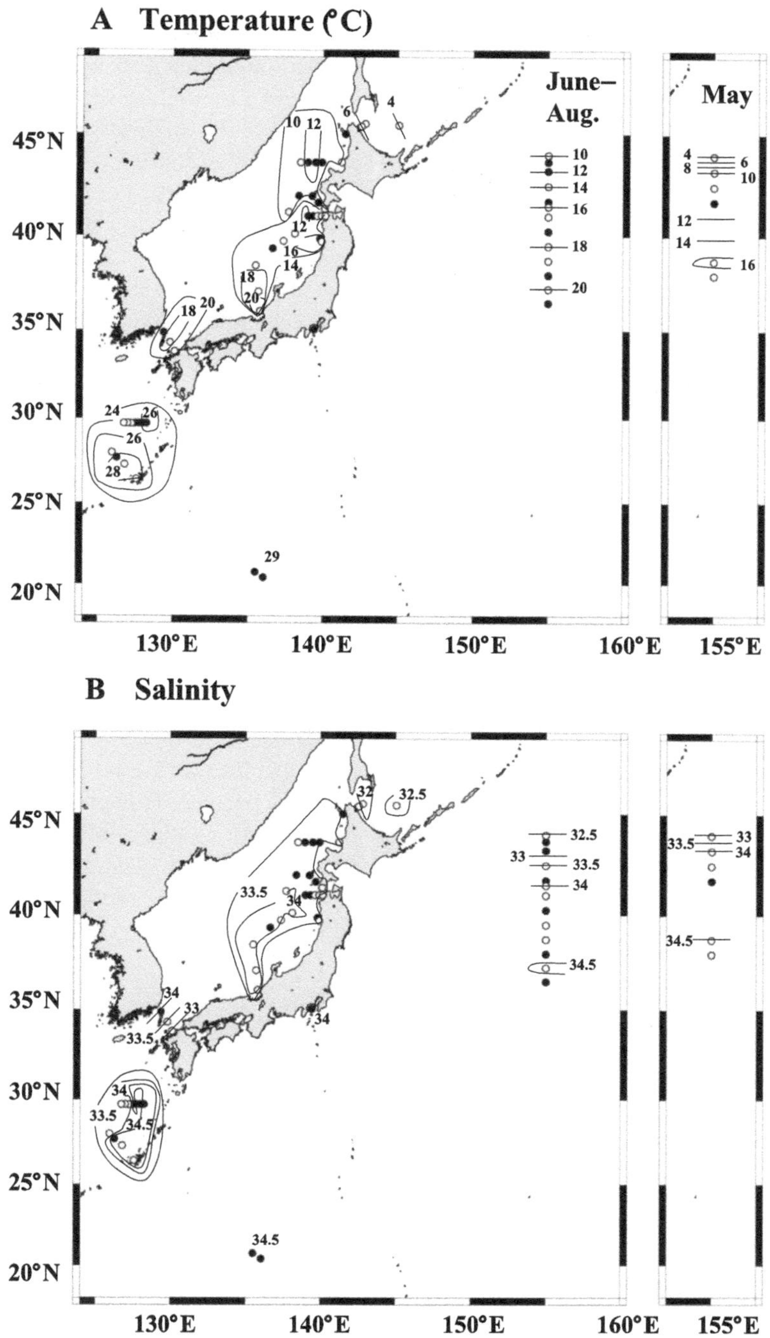

Figure 2 Horizontal distribution of the integrated mean temperature (A) and salinity (B). At 0 to 30 m in the Western North Pacific and their adjacent seas from May 16 to 20 (right panels) and June 8 to August 2 (left panels) 2011. Open and solid symbols denote stations where sampling was performed during the day and night, respectively.

(*Y*: $\log_{10}$ zooplankton biovolume [mm^3 m^{-3}]/Δbiovolume [mm^3]), $\bar{B}$ was divided by the interval of biovolume (Δbiovolume [mm^3]) and converted to a common logarithm. Based on these data, the NBSS liner model ($Y = aX + b$) was calculated, where a and b are the slope and intercept of NBSS, respectively.

Statistical analysis

Based on the mesozooplankton groups clustered based on their size spectra, inter-group differences in the zooplankton data (abundance, biovolume, and slope of NBSS) were tested using one-way analysis of variance (ANOVA) and Fisher's protected least-squares difference (PLSD) method. To determine the factors that govern the slope of NBSS, an analysis of covariance (ANCOVA) was performed using StatView, in which the intercept of NBSS and zooplankton group as independent variables.

Results

Hydrography

Throughout the entire sampling area and period, the integrated mean temperature at 0 to 30 m in the water column of each station ranged from 3.8 to 29.4°C (Figure 2A). Integrated mean temperatures were lower in the Okhotsk Sea and higher at the southern low-latitude station. With the temporal change between May and June to August, the temperature along the 155° E line in the Western North Pacific increased *ca.* 4°C within the same latitude from June to August. Integrated mean salinity ranged from 32.2 to 34.7 (Figure 2B) and showed a similar pattern to that of the integrated mean temperature and thus was lower in the Okhotsk Sea and higher in the southern low-latitude stations. Temporal changes in salinity along 155° E between May and June to August were not marked; this was comparable to the case of integrated mean temperature (Figure 2B).

OPC calibration

Comparison between OPC-derived wet mass (Y: mm^3 m^{-3}) and measured wet mass (X: mg m^{-3}) showed a highly significant correlation ($Y = 0.950X$, $r^2 = 0.691$, $p < 0.0001$, Figure 3). As an exception, one station (41° N, 155° E in 25 July) returned substantially higher values of direct measurement mass (marked with an open symbol in Figure 3). From microscopic observation, dominance of fragments of jellyfish and doliolids was the cause of the sample variation. We excluded the data of this station from the following analysis.

Zooplankton abundance, biovolume, and community

Zooplankton abundance ranged from 16.8 to 1,076 ind. m^{-3} and showed no clear spatial and temporal change pattern (Figure 4A). Zooplankton biovolume ranged from 2.24 to 1,007 mm^3 m^{-3} and was higher for the northern stations, particularly in the northern Japan Sea and north of the 155° E line (Figure 4B). Regarding temporal change, biovolume along the 155° E line was higher in May compared to June to August, with a factor of 1.4 to 9.8 times at the same latitude.

Based on the biovolume data of 45 size classes binned at every 0.1 mm, zooplankton communities were classified into six groups (A, B1, B2, B3, C, and D) using cluster analysis at 42% dissimilarities (Figure 5A). Each group contained 7 to 20 stations. Hydrographic variables showing significant relationships on the NMDS ordination were integrated mean temperature and integrated mean salinity at 0 to 30 m water column; these variables accounted for 18% and 15% of the changes, respectively.

Total abundance also significantly varied according to group and was the least for group D, followed by group A and was the highest for group B3 (Table 1). Total abundance was dominated by the 0.5 to 1 mm ESD size class for all groups. For the 1 to 2, 2 to 3, and 3 to 4 mm ESD size classes, the highest abundance was observed for group C. Moreover, the 4 to 5 mm ESD size class was the highest for group B1. Thus, the highest abundance group varied with size classes (Table 1).

For the total biovolume, the common group order was the least for group D followed by group A as observed for all size classes (Table 1, Figure 5C). The highest zooplankton biovolume was observed for group C; this was due to the dominance of biovolume at the 2 to 3 mm ESD size class (Figure 5C). Within the size class, group B3 was the highest in the 0.5 to 1 mm size class, while group B1 was the highest in the 4 to 5 mm size class. For the other size classes (1 to 2, 2 to 3, and 3 to 4 mm), group C was the highest biovolume group (Table 1).

The horizontal and temporal distribution of each group varied from those of the others (Figure 6). Groups A, B2, and B3 occurred in a broader region: Japan Sea, Western North Pacific, and East China Sea and had no geographical pattern. However, the horizontal distribution of groups B1, C, and D showed a clear geographical pattern. Group B1 was dominated by a large 4 to 5 mm ESD size class and was found in the Northern Japan Sea, Okhotsk Sea, and subarctic Western North Pacific. Group

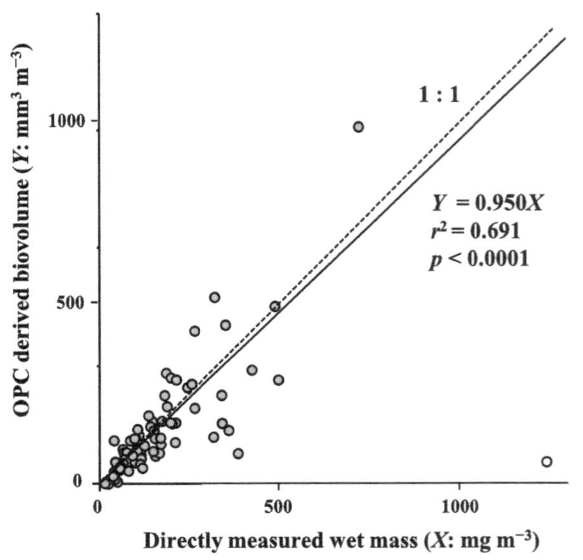

Figure 3 Comparison between OPC-derived biovolume and directly measured wet mass of whole samples. The long-dashed line indicates position 1:1. One datum at an anomalously high value of directly measured wet mass indicated by an open symbol was omitted from the following analysis.

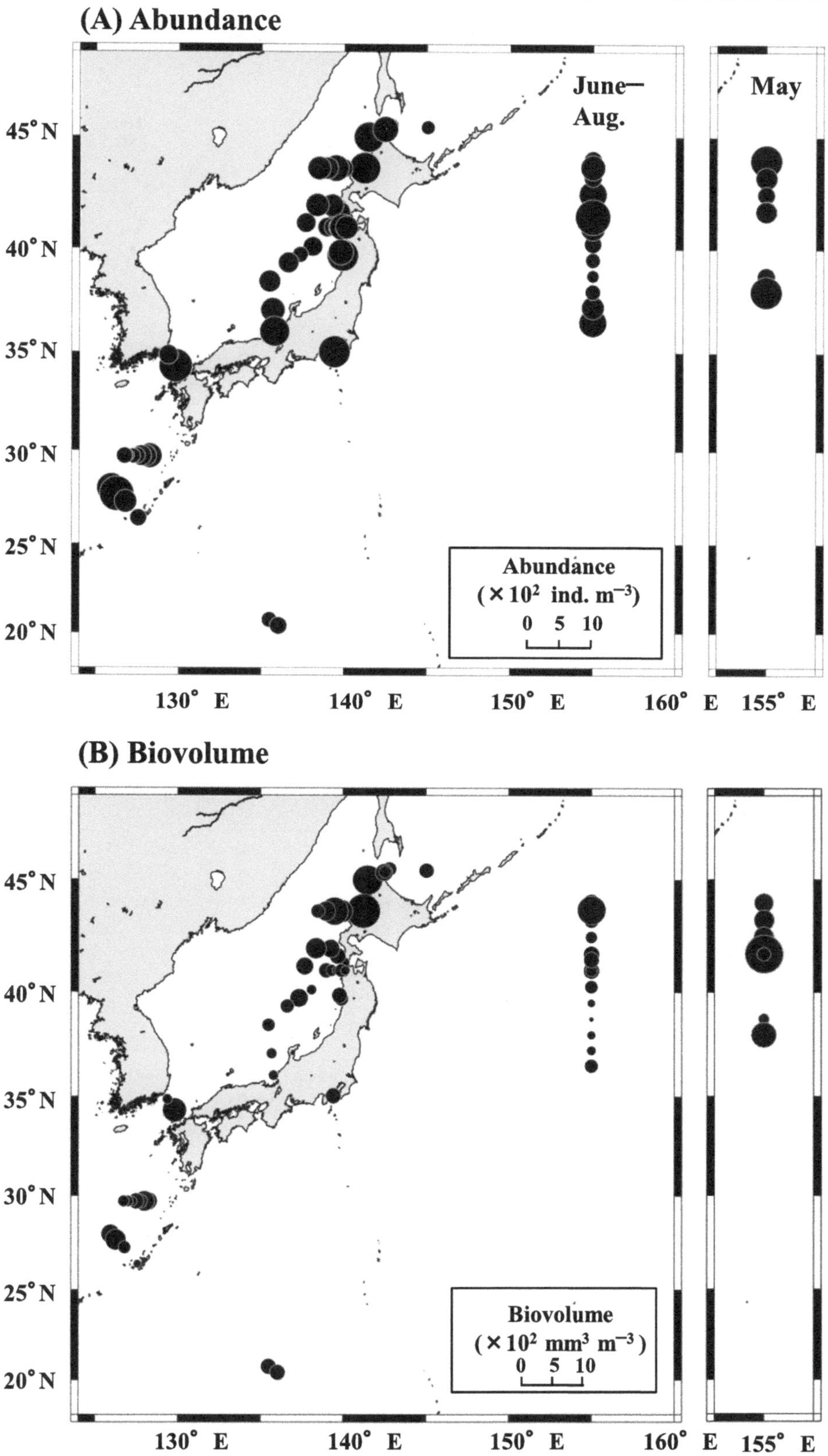

Figure 4 Horizontal distribution of abundance (A) and biovolume (B) of mesozooplankton in the Western North Pacific and adjacent seas. From May 16 to 20 (right panels) and June 8 to August 2, 2011 (left panels).

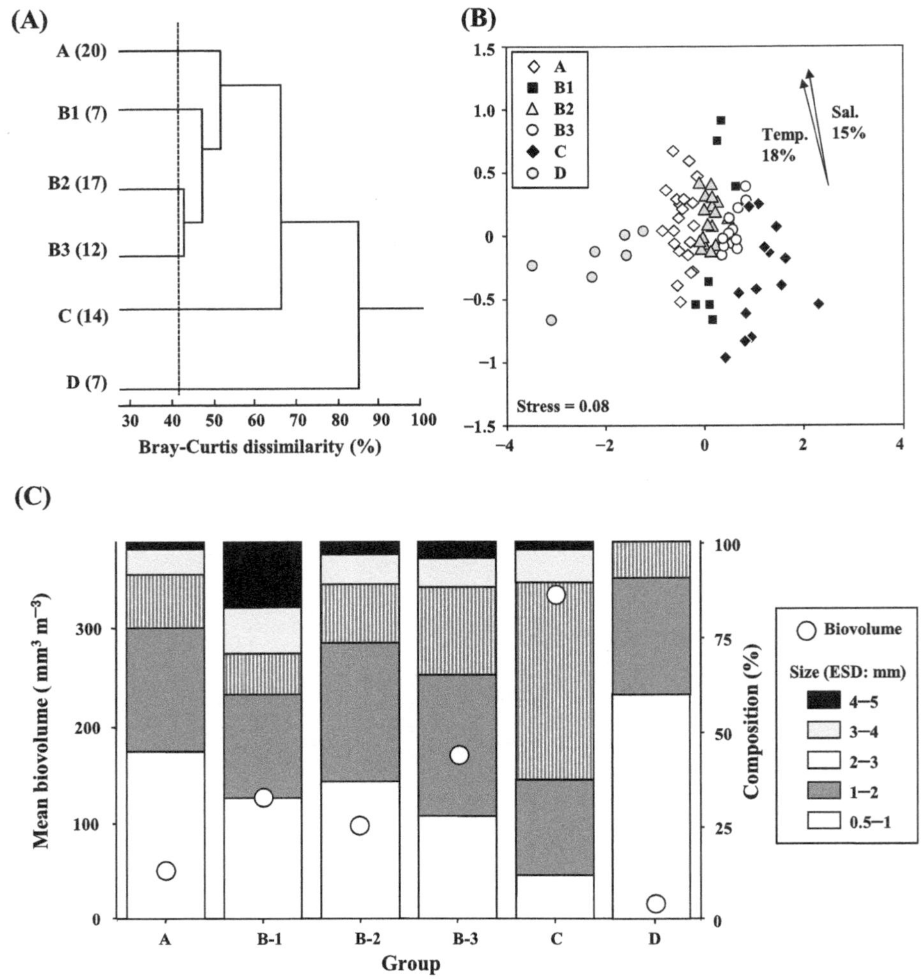

Figure 5 Results of cluster analysis based on mesozooplankton biovolume size spectra. In the Western North Pacific and adjacent seas from May and June to August 2011. **(A)** Six groups (A, B1, B2, B3, C, and D) were identified from Bray-Curtis dissimilarity connected with UPGMA. Numbers in parentheses indicate the number of stations contained in each group. **(B)** NMDS plots of each group. For correlation analyses with environmental parameters (temperature, salinity, latitude, and longitude), temperature and salinity showed a significant correlation (percentage indicates the coefficients of determination, r^2). **(C)** Mean biovolume and size composition (ESD, mm) of each group.

C, which was characterized by the highest biovolume and dominance of the 2 to 3 mm size class, occurred along the 155° E line during May and in the northern Japan Sea, Okhotsk Sea, and subarctic Western North Pacific from June to August. Group D, which was characterized by the least biovolume and was dominated by a small-sized 0.5 to 1 mm size class, was found at lower latitudes of the 155° E line (Kuroshio extension) (Figure 6).

NBSS

Results of the mean NBSS based on the complete data of each group are shown in Figure 7. For group C, the marked peak value on the *X*-axis ($\log_{10}$ zooplankton biovolume [mm^3 $ind.^{-1}$]) was observed at approximately 0.7; this corresponded with a 2 to 3 mm ESD size class and consisted of the copepodid stage 5 of the large copepod *Neocalanus* spp. (Figure 7). Significant inter-group differences were observed for the slope (*a*) and intercept (*b*) of NBSS (Table 1). The moderate slope (−0.90) of group C was significantly different from those of other groups (−1.11 ~ −1.24) (ANCOVA, $p < 0.001$, Table 2). For the intercept, the least was found for group D, while the highest was found for group C; the order of the intercept of each group corresponded to the order of total zooplankton biovolume (Table 1). From the ANCOVA analysis, there was no interaction between group and intercept, but significant relationships between slope and group were found ($p < 0.0001$, Table 2).

Discussion

OPC measurements

There have been several studies on OPC measurements for zooplankton in the western North Pacific. Table 3

Table 1 Comparison of zooplankton abundance, biovolume, and slopes (*a*) and (*b*) of NBSS ($Y = aX + b$) of each group

	Group						One-way ANOVA	Fisher's PLSD
Parameter	A(20)	B1(7)	B2(17)	B3(12)	C(14)	D(7)		
Abundance (inds. m^{-3})								
Total	331	503	476	528	469	101	**	D A C B2 B1 **B3**
0.25 to 1 mm	315	465	441	473	387	98	**	D **A** **C** **B2** **B1** **B3**
1 to 2 mm	15.1	34.0	32.3	48.5	52.3	2.9	***	D A B1 B2 **B3** **C**
2 to 3 mm	1.1	2.1	2.4	6.2	28.3	0.1	***	D A B1 B2 B3 **C**
3 to 4 mm	0.2	0.8	0.4	0.7	1.5	0.0	**	D A B2 B3 B1 **C**
4 to 5 mm	0.0	0.5	0.1	0.2	0.2	0.0	***	D A B2 C B3 **B1**
Biovolume (mm^{3}m^{-3})								
Total	57.0 ± 13.8	135.1 ± 34.1	106.7 ± 39.3	179.6 ± 44.7	348.7 ± 229.1	10.7 ± 7.5	***	D A B2 B1 B3 **C**
0.25 to 1 mm	25.7 ± 11.9	43.9 ± 30.9	39.3 ± 13.9	49.6 ± 21.5	41.3 ± 27.5	6.5 ± 4.7	***	D A **B2** **C** **B1** **B3**
1 to 2 mm	18.6 ± 7.1	37.0 ± 17.3	39.0 ± 9.6	66.9 ± 16.9	88.7 ± 46.2	3.3 ± 2.2	***	D A B1 B2 B3 **C**
2 to 3 mm	7.7 ± 4.1	14.5 ± 12.8	16.5 ± 6.1	41.5 ± 12.1	181.4 ± 160.9	1.0 ± 0.9	***	D A B1 B2 B3 **C**
3 to 4 mm	3.9 ± 4.1	16.5 ± 10.9	8.5 ± 7.5	13.9 ± 9.1	30.5 ± 42.7	0	**	D A B2 B3 B1 **C**
4 to 5 mm	1.1 ± 1.6	23.3 ± 24.1	3.5 ± 4.3	7.7 ± 9.9	6.7 ± 9.0	0	***	D A B2 C B3 **B1**
NBSS								
Slope (*a*)	−1.18 ± 0.14	−1.11 ± 0.16	−119 ± 0.94	−1.16 ± 0.94	−0.90 ± 0.16	−1.24 ± 0.11	**	D B2 A B3 B1 **C**
Intercept (*b*)	−0.23 ± 0.28	−0.06 ± 0.32	−0.04 ± 0.16	0.09 ± 0.27	0.29 ± 0.29	−0.65 ± 0.29	***	D A B1 B2 B3 **C**

Identified by Q-mode analysis (cf. Figure 5A) in the Western Pacific and their adjacent seas during May and June to August 2011. Differences between the group were tested by one-way ANOVA and *post hoc* test by Fisher's PLSD. Groups with bold letters indicate that the groups are significantly greater than the other groups ($p < 0.05$). Values are mean ± s.d. Numbers in the parentheses are the numbers of stations belonging to each group. **$p < 0.01$, ***$p < 0.0001$.

summarizes regressions between the OPC-derived mass and directly measured mass, ranges of zooplankton abundance, and biovolume from previous studies. The OPC-derived masses were 1.05 to 1.18 times the directly measured mass in previous studies (Yokoi et al. 2008; Matsuno and Yamaguchi 2010; Fukuda et al. 2012). In this study, the zooplankton biovolume estimated by OPC was 0.95 times the directly measured mass (Figure 3). This factor is slightly smaller than the previously reported values, but all these values (0.95 to 1.18) corresponded well at nearly 1:1. In the present study, a substantial underestimation of the OPC biovolume was caused by fragments of jellyfish and doliolids at one station (41° N, 155° E on 29 July). Regarding dominance of doliolids and gelatinous zooplankton in the Western North Pacific, Yokoi et al. (2008) reported that their dominance was observed at locations of thermocline-developed stations in the transitional domain. This condition may also have been present at 41° N, 155° E on July 29 of this study.

Based on this same method (OPC measurement of net samples), zooplankton abundance and biovolume in the Western North Pacific were reported ranging from 128 to 580 ind. m^{-3} and 96 to 880 mm^3 m^{-3}, respectively (Yokoi et al. 2008; Matsuno and Yamaguchi 2010; Fukuda et al. 2012). Both the zooplankton abundance (16.8 to 1,076 ind. m^{-3}) and biovolume (2.24 to 1,007 mm^3 m^{-3}) in this study fell in these ranges. This was shown by previous studies based on 0 to 150 m samplings obtained offshore in the North Pacific; this study included neritic to oceanic regions of the North Pacific and marginal seas, and sampling depths varied with each station (the shallowest station was 0 to 30 m sampling). These discrepancies in methodology (region and sampling depth) between previous studies and this study might have induced the extension of the range of abundance and biovolume data in this study.

Spatial changes

For spatial changes in zooplankton, the higher biovolume at high latitudes (Figure 4B) and horizontal distributions of groups B1, C, and D showed distinct geographical patterns (Figure 6).

Group B1 was observed only in areas north of the Japan Sea, in the Okhotsk Sea, and at the subarctic region along the 155° E line in the Western North Pacific (Figure 6). For zooplankton biomass in the Japan Sea, the southern part was reported to be less than that found in the north and was similar to that of the Kuroshio region of the Western North Pacific (Hirota and Hasegawa 1999; Iguchi 2004). Common to this study, the zooplankton biovolume in the Japan Sea was higher in the northern area (Figure 4B). In the northern Japan Sea, where the surface

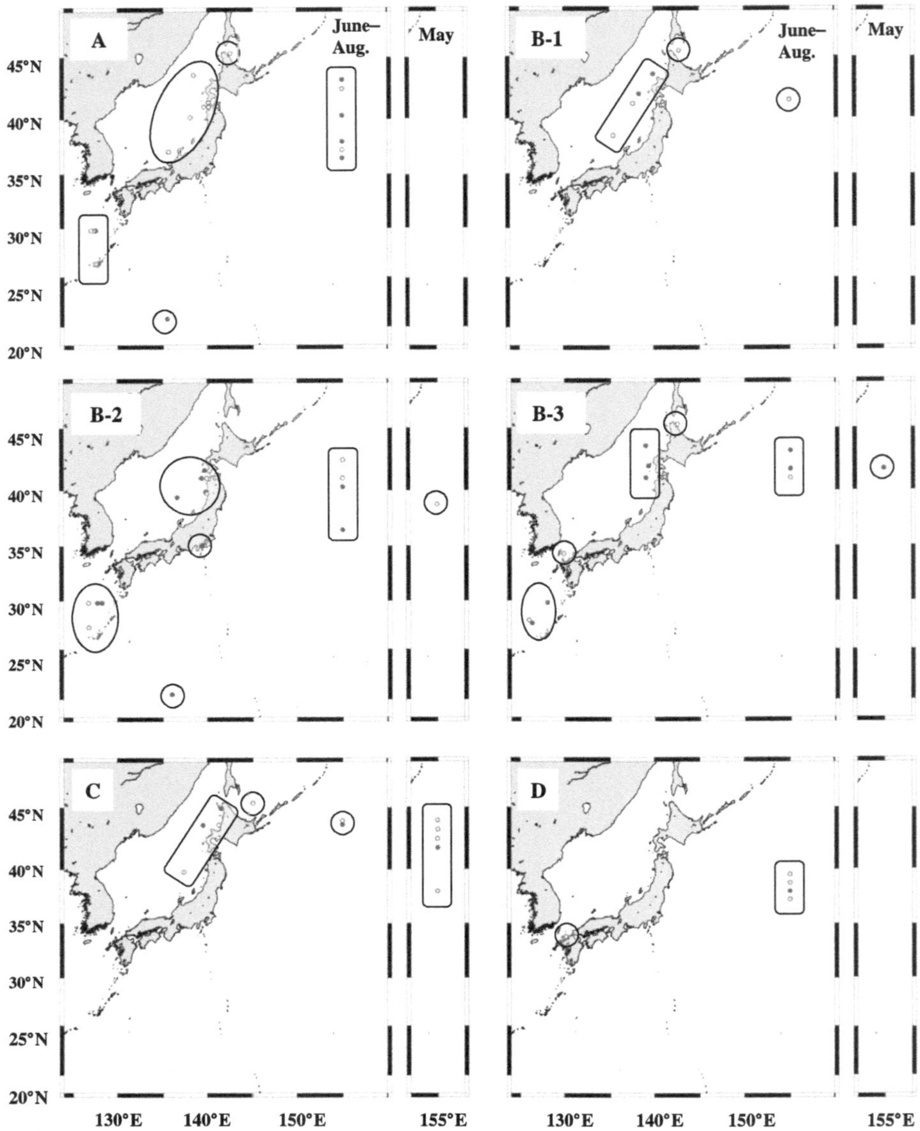

Figure 6 Horizontal distribution of six groups (A, B1, B2, B3, C, and D) identified from cluster analysis on mesozooplankton biovolume size spectra (cf. Figure 5A). In the Western North Pacific and adjacent seas from May and June to August 2011. Open and solid symbols denote day and night samples, respectively. The geographical ranges of each group are marked with boxes and circles.

temperature is low, the large-sized cold-water species, i.e., chaetognath *Parasagitta elegans*, amphipod *Themisto japonica* and euphausiid *Euphausia pacifica* are known to perform diel vertical migration and distribute on the surface at night (Ikeda et al. 1992; Iguchi et al. 1993; Terazaki 1993). The proportion of 4 to 5 mm ESD in group B1 was much higher than that in the other groups (Figure 5C). This size range (4 to 5 mm) exceeded the size of copepods (the size of the largest copepods in this region: *Neocalanus* spp. C5 is 2 to 3 mm ESD, Yokoi et al. 2008) and was considered caused by macrozooplankton such as amphipods, euphausiids, and chaetognaths. These taxa frequently occurred for samples belonging to group B1. The dominance of large-sized chaetognath *P. elegans* and amphipod *Themisto pacifica* was also reported for the Okhotsk Sea (Volkov 2008). Thus, group B1 was only observed for the northern area of the Japan Sea, Okhotsk Sea, and subarctic Western North Pacific, which were characterized by a higher proportion of the large 4 to 5 mm ESD size class consisting of macrozooplankton.

In contrast, the horizontal distribution of group D was only observed in the Kuroshio extension from June to August (Figure 6). Group D was characterized by the least biovolume and dominance of the small-size class (0.5 to 1 mm ESD) (Figure 5C). From microscopic observation, the samples belonging to group D were dominated

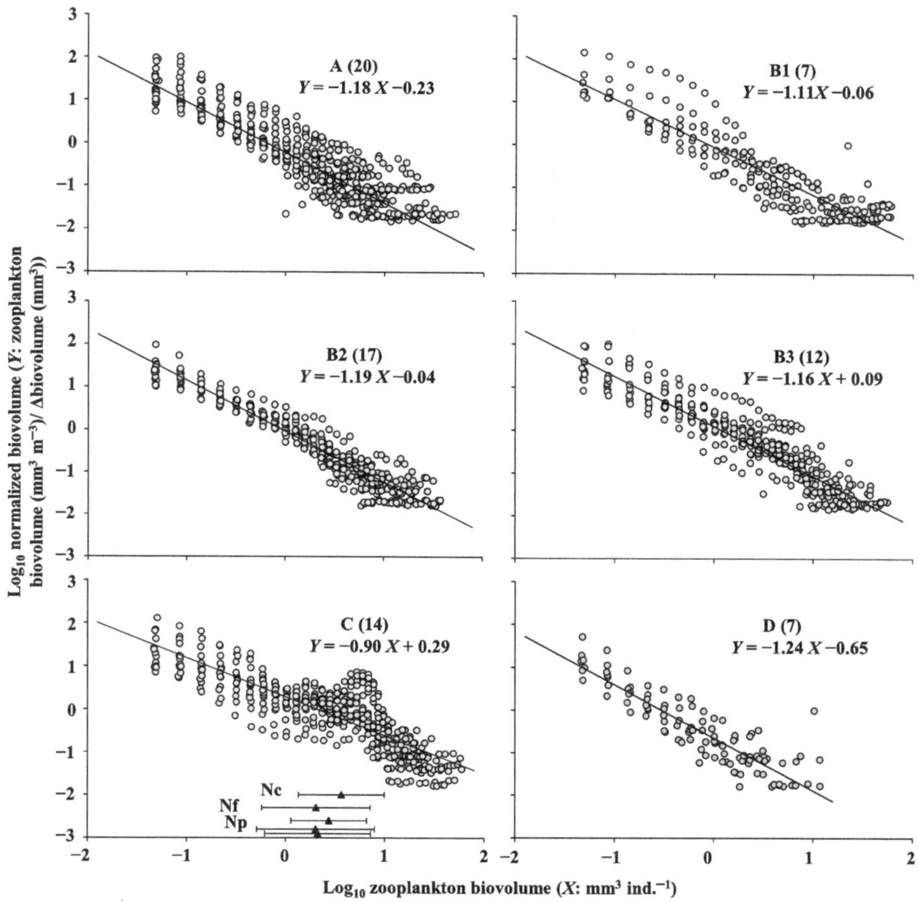

Figure 7 Mean NBSS of six groups (A, B-1, B-2, B-3, C and D) identified from cluster analysis on mesozooplankton biovolume size spectra (cf. Figure 5A). In the Western North Pacific and adjacent seas during May and June to August 2011. Numbers in parentheses indicate the number of stations belonging to each group. The mean and standard deviations of copepodid five stages of *Neocalanus* copepods (Nc, *Neocalanus cristatus*; Nf, *N. flemingeri*; and Np, *N. plumchrus*, Yamaguchi et al. 2014) are shown in panel for group **C**.

by small-sized copepods, e.g., *Paracalanus parvus* and poecilostomatoida. Hydrography of group D was also marked, and all of the stations showed high salinity (near 34.0, Figure 8), which is characteristic of the Kuroshio extension (Yasuda 2003). From the NMDS plot (Figure 5B), the direction of the two arrows indicates that groups A, B1, and C showed euryhaline and eurytherm, while the others (B2, B3, and D) were restricted, occurring in narrower ranges of temperature and salinity. These findings suggested that the horizontal distribution of zooplankton was regulated by water mass formation. In the Kuroshio region, it is well known that in an oligotrophic ocean, zooplankton fauna is dominated by small-sized copepods such as *Paracalanus* spp. and poecilostomatoida (Nakata et al. 2001; Nakata and Koyama 2003; Hsieh et al. 2004).

Table 2 Result of the ANCOVA for the slope (*a*) of NBSS ($Y = aX + b$)

Parameter	*df*	SS	*F*-value	*p*-value
Intercept	1	0.052	3.451	NS
Group	5	0.645	8.629	***
Group × Intercept	5	0.111	1.477	NS
Error	66	0.972	-	-

With the intercept (*b*) of NBSS and zooplankton group (cf. Figure 5A) applied as independent variables. *d.f.*, degrees of freedom; SS, sum of squares. N.S., not significant. ***$p < 0.0001$.

However, same ocean latitudes were adjusted with the Kuroshio extension; the East China Sea, southern coast of Japan, and southern Japan Sea were not as oligotrophic as the Kuroshio extension region, and middle-sized copepods, such as *Calanus sinicus*, are known to dominate in zooplankton fauna (Hirakawa et al. 1995; Shimode et al. 2006; Hsiao et al. 2011). Due to the dominance of middle-sized copepods (*C. sinicus*) for these marginal seas, zooplankton size spectra in these regions were dispersed into groups A, B2, and B3 (Figure 6). Thus, it is difficult to identify specific characteristics in the horizontal distribution of these groups.

Table 3 Comparison on regressions between OPC derived mass and directly measured mass, ranges of abudance, and biovolume of mesozooplankton

Location (period)	Regression (n, r^2, p)	Abundance (ind. m^{-3})	Biovolume (mm^3 m^{-3})	References
35° to 44° N, 155° E	$Y = 1.05X + 0.05$	351.7 to 579.5	177.1 to 678.9[a]	Yokoi et al. (2008)
(May to June 1993 to 2004)	(234, 0.63, $p < 0.0001$)			
22° to 53° 30′ N 165° E/165° W	$Y = 116X$	226.3 to 430.9	190.7 to 523.3[a]	Matsuno and Yamaguchi (2010)
(June to August 2003 to 2006)	(96, 0.79, $p < 0.0001$)			
35° to 51° N, 180°	$Y = 1.176X$	128.0 to 562.0	96.0 to 880.0[a]	Fukuda et al (2012)
(June 1981 to 2000)	(351, 0.521, $p < 0.0001$)			
20° to 45° N, 126° to 155° E	$Y = 0.950X$	16.8 to 1,075.9	2.2 to 1,006.7	This study
(May to Aug. 2011)	(78, 0.691, $p < 0.0001$)			

Evaluated by OPC analyses in the North Pacific. For regressions, Y, biovolume in mm^{-3} and X, mg m^{-3}.
[a]Biovolume data were calculated from dry mass unit in the references.

Temporal changes

In the present study, two time samplings in May and July to August were performed along the 155° E line in the Western North Pacific. In May, group C, which is characterized by a high biovolume and dominance of the 2 to 3 mm ESD size class, dominated; however, this changed to other groups during July to August (Figure 6). The 2 to 3 mm ESD zooplankton in this region corresponded with the C5 stages of large copepods *Neocalanus* spp., and their importance has been reported in previous OPC studies (Yokoi et al. 2008; Matsuno and Yamaguchi 2010; Fukuda et al. 2012). Along the 155° E line, group C was dominated by the 2 to 3 mm ESD size class and was characterized by a high biovolume during May and at the northern stations from July to August.

Neocalanus spp. is known to perform seasonal vertical migration. C1 of *Neocalanus* spp. grow to C5 near the surface from mid-March to June, descend to deep layers from June to August, and subsequently molt to an adult in the deep layers (Kobari et al. 2003). The large copepod *Eucalanus bungii* also perform seasonal vertical migration and reproduce near the surface from April to May during phytoplankton bloom (Shoden et al. 2005). Consequently, the total zooplankton biomass in this region peaked during May due to the dominance of *Neocalanus* and *Eucalanus* spp. near the surface (Odate 1994). The high biovolume dominated by the 2 to 3 mm ESD size class in May along the 155°E line in this study was caused by the dominance of these large copepods near the surface layer. After the descent of these large copepods into the deep layer, the biovolume decreased and size spectra characteristics changed to different groups from July to August.

The high biomass dominated by late copepodid stages of large copepods is a characteristic of a limited period (1 to 2 months in and near May) in the subarctic North Pacific (Odate 1994). This high zooplankton biomass season was reported to vary inter-annually owing to the decadal climate regime shift (Chiba et al. 2006, 2008). Chiba et al. (2006) observed that the zooplankton peak season in spring varied by one month depending on the decadal climate changes. Although there was a slight change in timing, the specific characteristics caused by seasonal changes in the zooplankton biomass of the neighboring waters of Japan were at high biomass caused by the dominance of large-sized copepods near the surface layer during spring.

NBSS

The slope of NBSS is known to be an index of productivity, transfer efficiency, and predation in each marine ecosystem (Zhou 2006; Zhou et al. 2009). From the theoretical mean, the slope of NBSS of the stable marine ecosystem settled at approximately −1 (Sprules and Munawar 1986). The steep slope of NBSS indicated high productivity but low transfer efficiency to a higher trophic level. However, the moderate NBSS slope may be caused by low productivity and high energy-transfer efficiency (Sprules and Munawar 1986). In the present study, the slopes of NBSS were slightly higher than −1 for most of the groups, except for group C (Table 1). These findings suggest that most of the zooplankton communities in neighboring Japanese waters were characterized by a bottom-up marine ecosystem. The intercept of NBSS is a reflection of the amount of primary production (Zhou 2006; Marcolin et al. 2013). The high correlation between the intercept of NBSS and the total zooplankton biovolume of each group in this study may confirm this theory (Table 1).

The slopes of NBSS of the zooplankton community reported from various oceans are summarized in Table 4. As previously mentioned, the slope of NBSS is an index of productivity, transfer efficiency, and predation. However, the treated size range varied with the study and ranged from microzooplankton (0.025 to 4.0 mm, Napp

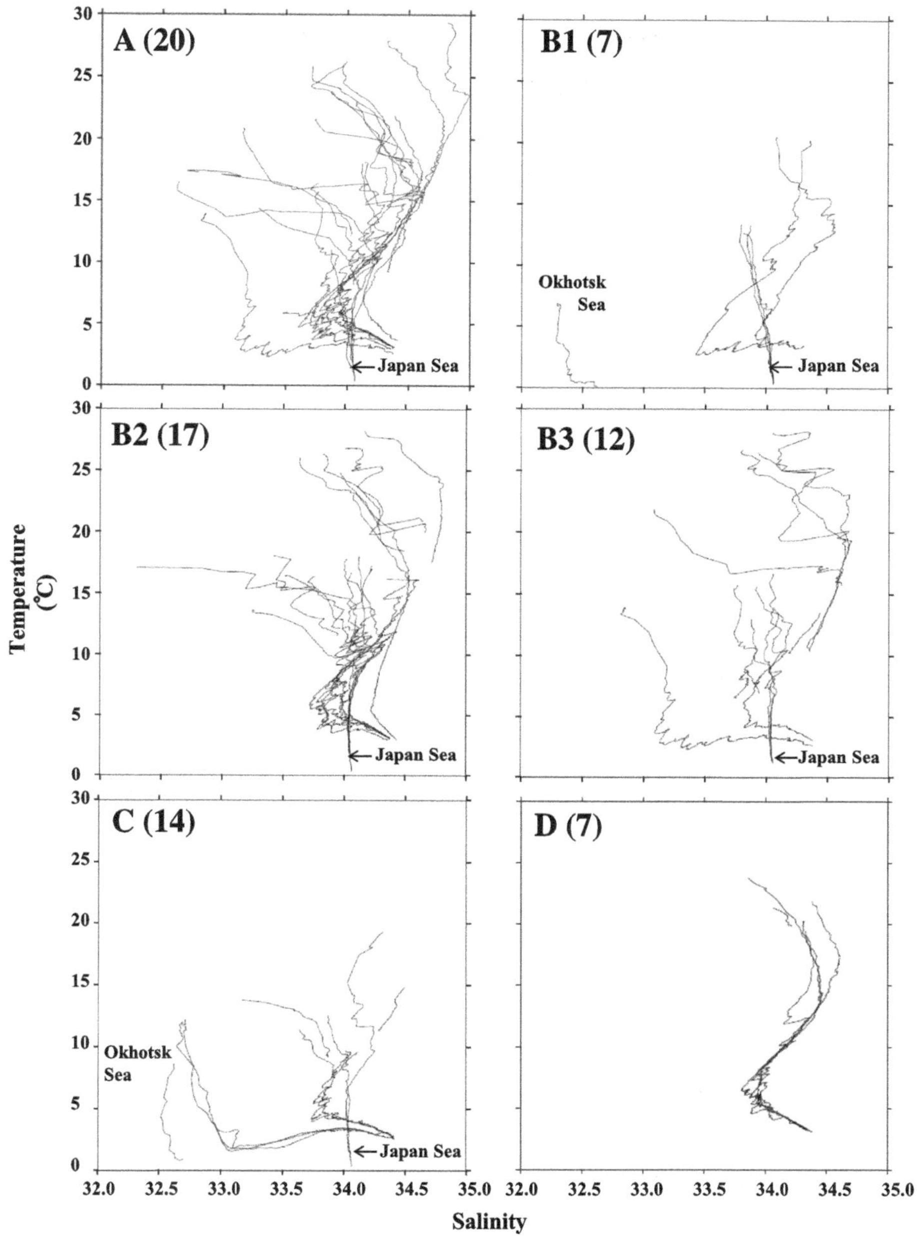

Figure 8 T-S diagrams of the six groups (A, B1, B2, B3, C, and D) identified from cluster analysis on mesozooplankton biovolume size spectra (cf. Figure 5A). In the Western North Pacific and adjacent seas from May and June to August 2011. To construct T-S diagrams, hydrographic data from 0 to 1,000 m were applied. The numbers in parentheses indicate the number of stations belonging to each group.

et al. 1993) to fish (20 to 1,200 mm, Macpherson et al. 2002). Thus, a careful, direct comparison of the NBSS slope is required. In the present study, specific characteristics of the NBSS slope include a moderate slope of group C (−0.90) caused by the dominance of C5 stages of large *Neocalanus* spp. (Figure 7). Along the 155° E line, group C during May varied with other groups from July to August; thus, the temporal changes were remarkable (Figure 6). A similar situation (alternation of NBSS slope by dominance of specific taxa) was reported for the dominance of barnacle larvae in the Chukchi Sea (Matsuno et al. 2012).

Conclusions

Through OPC analysis of zooplankton samples, zooplankton size spectra in the neighboring waters of Japan were separated into six groups. Most groups had −1.11 to −1.24 NBSS slopes, which were slightly higher than the theoretical value (−1). However, one group had moderate slope NBSS (−0.90) caused by the dominance of large

Table 4 Comparison of the slope (*a*) of NBSS ($Y = aX + b$) on the mesozooplankton community at various locations

Location/region	Unit	Size range (mm)	Slope	References
Chile, cold nutrient-rich conditions	Carbon	0.7 to 210	−0.44	Iriarte and González (2004)
Gulf of St. Lawrence (open water)	Biovolume	0.25 to 2	−0.47	Herman and Harvey (2006)
Barents Sea	Biovolume	0.25 to 14	−0.63	Basedow et al. (2010)
NW Mediterranean (protected area)	Wet weight	20 to 1,200	−0.65	Macpherson et al. (2002)
Bay of Biscay (oceanic stations)	Carbon	0.27 to 1.7	−0.71 to −0.64	Sourisseau and Carlotti (2006)
Tasman Sea	Biovolume	0.11 to 3.3	−0.69	Baird et al. (2008)
Unshant Tidal Front (stratified side/offshore)	Carbon	0.5 to 5	−0.73	Schultes et al. (2013)
Brazilian Continental Shelf (Oceanic Stations)	Biovolume	0.1 to 5	−0.86	Marcolin et al. (2013)
Chukchi Sea (less productive)	Biovolume	0.25 to 5	−0.86	Matsuno et al. (2012)
Gulf of St. Lawrence (estuary)	Biovolume	0.25 to 2	−0.90	Herman and Harvey (2006)
Western Antarctic Peninsula (fall)	Biovolume	0.25 to 14	−0.92	Zhou et al. (2009)
Chile-El Nino, oligotrophic conditions	Carbon	0.7 to 210	−0.93	(Iriarte and González 2004)
Coral Sea	Biovolume	0.11 to 3.3	−0.97	Baird et al. (2008)
Southwest Coral Sea	Wet weight	0.25 to 2.5	−1.00	Suthers et al. (2006)
Bay of Biscay (coastal stations)	Carbon	0.27 to 1.7	−1.05 to −0.91	Sourisseau and Carlotti (2006)
NW Mediterranean (unprotected area)	Wet weight	20 to 1,200	−1.04	Macpherson et al. (2002)
Scotia Sea (spring)	Carbon	Small phyto to large schyphomedusae	−1.09	Tarling et al. (2012)
North Iberian Shelf	Carbon	0.25 to 17	−1.11	Nogueira et al. (2004)
Chukchi Sea (more productive)	Biovolume	0.25 to 5	−1.11	Matsuno et al. (2012)
North Pacific Ocean	Carbon	0.18 to 4.0	−1.13	Rodriguez and Mullin (1986)
Northwest Atlantic Ocean	Carbon	0.07 to 8.0	−1.14	Quinones et al. (2003)
Ushant Tidal Front (mixed side/nearshore)	Carbon	0.5 to 5	−1.15	Schultes et al. (2013)
Brazilian Continental Shelf (Coastal Stations and Abrolhos Bank)	Biovolume	0.1 to 5	−1.25	Marcolin et al. (2013)
California Current	Carbon	0.2 to 3.3	−1.43	Huntley et al. (1995)
Ushant Tidal Front (mixed side/near)	Carbon	0.5 to 5	−1.76	Schultes et al. (2013)
Western Antarctic Peninsula (summer)	Biovolume	0.25 to 14	−1.8	Zhou et al. (2009)
Australian Estuary	Wet weight	0.25 to 1.6	−1.89	Moore and Suthers (2006)
California Bight	Biovolume	0.025 to 4.0	−2.30	Napp et al. (1993)
NW Pacific (group A)	Biovolume	0.50 to 5.00	−1.18	This study
NW Pacific (group B1)	Biovolume	0.50 to 5.00	−1.11	This study
NW Pacific (group B2)	Biovolume	0.50 to 5.00	−1.19	This study
NW Pacific (group B3)	Biovolume	0.50 to 5.00	−1.16	This study
NW Pacific (group C)	Biovolume	0.50 to 5.00	−0.90	This study
NW Pacific (group D)	Biovolume	0.50 to 5.00	−1.24	This study

Neocalanus copepods. Temporal changes in the slope of NBSS were observed from moderate slope (−0.90) in May back to high NBSS slope (−1.11 to −1.24) caused by a descent of *Neocalanus* copepods to deep layers from July to August. This temporal change in the NBSS slope was not caused by predator-prey interaction, but due to the seasonal vertical migration of dominant large-sized zooplankton species. This finding suggests that the ecology of dominant species (growth or seasonal vertical migration) also should be considered as a cause of temporal changes in NBSS slope.

Competing interests

The authors declare that they have no competing interests.

Authors' contributions

KS and AY wrote the manuscript. KM, DA, and YA participated in the design of the study and helped with OPC measurements. All authors read and approved the final manuscript.

Acknowledgements
We are grateful to the captain and crew of the T/S *Oshoro-Maru* for their help in our field sampling. This study was supported by a Grant-in-Aid for Scientific Research (A) 24248032 and a Grant-in-Aid for Scientific Research on Innovative Areas 24110005 from the Japan Society for the Promotion of Science (JSPS).

Author details
[1]Laboratory of Marine Biology, Graduate School of Fisheries Science, Hokkaido University, 3-1-1 Minatomachi, Hakodate, Hokkaido 041-8611, Japan. [2]Arctic Environmental Research Center, National Institute of Polar Research, 10-3 Midori-cho, Tachikawa, Tokyo 190-8518, Japan.

References

Baird ME, Timko PG, Middleton JH, Mullaney TJ, Cox DR, Suthers IM (2008) Biological properties across the Tasman Front off southeast Australia. Deep-Sea Res I 55:1438–1455
Basedow SL, Tande KS, Zhou M (2010) Biovolume spectrum theories applied: spatial patterns of trophic levels within a mesozooplankton community at the polar front. J Plank Res 32:1105–1119
Boyd PW, Newton PP (1999) Does planktonic community structure determine downward particulate organic carbon flux in different oceanic provinces? Deep-Sea Res I 46:63–91
Chiba S, Tadokoro K, Sugisaki H, Saino T (2006) Effects of decadal climate change on zooplankton over the last 50 years in the western subarctic North Pacific. Global Change Biol 12:907–920
Chiba S, Aita MN, Tadokoro K, Saino T, Sugisaki H, Nakata K (2008) From climate regime shifts to lower-trophic level phenology: synthesis of recent progress in retrospective studies of the western North Pacific. Prog Oceanogr 77:112–126
Chiba S, Sugisaki H, Nonaka M, Saino T (2009) Geographical shift of zooplankton communities and decadal dynamics of the Kuroshio-Oyashio currents in the western North Pacific. Global Change Biol 15:1846–1858
Ducklow HW, Steinberg DK, Buesseler KO (2001) Upper ocean carbon export and the biological pump. Oceanography 14:50–58
Field JG, Clarke KR, Warwick RM (1982) A practical strategy for analysing multispecies distribution patterns. Mar Ecol Prog Ser 8:37–52
Fukuda J, Yamaguchi A, Matsuno K, Imai I (2012) Interannual and latitudinal changes in zooplankton abundance, biomass and size composition along a central North Pacific transect during summer: analyses with an Optical Plankton Counter. Plank Benthos Res 7:64–74
García-Comas C, Chang C-Y, Ye L, Sastri AR, Lee Y-C, Gong G-C, Hsieh C-H (2014) Mesozooplankton size structure in response to environmental conditions in the East China Sea: how much does size spectra theory fit empirical data of a dynamic coastal area? Prog Oceanogr 121:141–157
Herman AW (1988) Simultaneous measurement of zooplankton and light attenuance with new optical plankton counter. Cont Shelf Res 8:205–221
Herman AW, Harvey M (2006) Application of normalized biomass size spectra to laser optical plankton counter net inter comparisons of zooplankton distributions. J Geophys Res 111:C05S05, doi:10.1029/2005JC002948
Hirakawa K, Kawano M, Nishihama S (1995) Seasonal variability in abundance and composition of zooplankton in the vicinity of the Tsushima Straits, southwestern Japan Sea. Bull Japan Sea Natl Fish Res Inst 45:25–38
Hirota Y, Hasegawa S (1999) The zooplankton biomass in the Sea of Japan. Fish Oceanogr 8:274–283
Hsiao S-H, Ka S, Fang T-H, Hwang J-S (2011) Zooplankton assemblages as indicators of seasonal changes in water masses in the boundary waters between the East China Sea and the Taiwan Strait. Hydrobiologia 666:317–330
Hsieh C-H, Chiu T-S, Shih C-T (2004) Copepod diversity and composition as indicators of intrusion of the Kuroshio Branch Current into the northern Taiwan Strait in Spring 2000. Zool Stud 43:393–403
Huntley ME, Zhou M, Nordhausen W (1995) Mesoscale distribution of zooplankton in the California Current in late spring, observed by optical plankton counter. J Mar Res 53:647–674
Iguchi N (2004) Spatial/temporal variations in zooplankton biomass and ecological characteristics of major species in the southern part of the Japan Sea: a review. Prog Oceanogr 61:213–225
Iguchi N, Ikeda T, Imamura A (1993) Growth and life cycle of a euphausiid crustacean (*Euphausia pacifica* HANSEN) in Toyama Bay, southern Japan Sea. Bull Japan Sea Natl Fish Res Inst 43:69–81
Ikeda T, Hirakawa K, Imamura A (1992) Abundance, population structure and life cycle of a hyperiid amphipod *Themisto japonica* (Bovallius) in Toyama Bay, southern Japan Sea. Bull Plankton Soc Japan 39:1–16
Iriarte JL, González HE (2004) Phytoplankton size structure during and after the 1997/98 El Niño in a coastal upwelling area of the northern Humboldt Current System. Mar Ecol Prog Ser 269:83–90
Kimmel DG, Roman MR, Zhang X (2006) Spatial and temporal variability in factors affecting mesozooplankton dynamics in Chesapeake Bay: evidence from biomass size spectra. Limnol Oceanogr 51:131–141
Kobari T, Shinada A, Tsuda A (2003) Functional roles of interzonal migrating mesozooplankton in the western subarctic Pacific. Prog Oceanogr 57:279–298
Longhurst AR (1991) Role of the marine biosphere in the global carbon cycle. Limnol Oceanogr 36:1507–1526
Macpherson E, Gordoa A, Garcia-Rubies A (2002) Biomass size spectra in littoral fishes in protected and unprotected areas in the NW Mediterranean. Estuar Coast Shelf Sci 55:777–788
Marcolin CR, Schultes S, Jackson GA, Lopes RM (2013) Plankton and seston size spectra estimated by the LOPC and ZooScan in the Abrolhos Bank ecosystem (SE Atlantic). Cont Shelf Res 70:74–87
Matsuno K, Yamaguchi A (2010) Abundance and biomass of mesozooplankton along north-south transects (165°E and 165°W) in summer in the North Pacific: an analysis with an optical plankton counter. Plank Benthos Res 5:123–130
Matsuno K, Yamaguchi A, Imai I (2012) Biomass size spectra of mesozooplankton in the Chukchi Sea during the summers of 1991/1992 and 2007/2008: an analysis using optical plankton counter data. ICES J Mar Sci 69:1205–1217
Michaels AF, Silver MW (1988) Primary production, sinking fluxes and microbial food web. Deep-Sea Res 35A:473–490
Moore SK, Suthers IM (2006) Evaluation and correction of subresolved particles by the optical plankton counter in three Australian estuaries with pristine to highly modified catchments. J Geophys Res 111:C05S04, doi:10.1029/2005JC002920
Motoda S (1957) North Pacific standard net. Inform Bull Plank Japan 4:13–15 (in Japanese with English abstract)
Motoda S (1959) Devices of simple plankton apparatus. Mem Fac Fish Hokkaido Univ 7:73–94
Nakata K, Koyama S (2003) Interannual changes of the winter to early spring biomass and composition of mesozooplankton in the Kuroshio Region in relation to climatic factors. J Oceanogr 59:225–234
Nakata K, Koyama S, Matsukawa Y (2001) Interannual variation in spring biomass and gut content composition of copepods in the Kuroshio current, 1971–89. Fish Oceanogr 10:329–341
Napp JM, Ortner PB, Pieper RE, Holliday DV (1993) Biovolume-size spectra of epipelagic zooplankton using a multi-frequency acoustic profiling system (MAPS). Deep-Sea Res I 40:445–459
Nogueira E, Gonzárez-Nuevo G, Bode A, Varela M, Morán XAG, Valdés L (2004) Comparison of biomass and size spectra derived from optical plankton counter data and net samples: application to the assessment of mesoplankton distribution along the Northwest and North Iberian Shelf. ICES J Mar Sci 61:508–517
Odate K (1994) Zooplankton biomass and its long-term variation in the western North Pacific Ocean, Tohoku Sea Area, Japan. Bull Tohoku Natl Fish Res Inst 56:115–173 (in Japanese with English abstract)
Piontkovski SA, Williams R, Melnik TA (1995) Spatial heterogeneity, biomass and size structure of plankton of the Indian Ocean: some general trends. Mar Ecol Prog Ser 117:219–227
Quinones RA, Platt T, Rodriguez J (2003) Patterns of biomass-size spectra from oligotrophic waters of the Northwest Atlantic. Prog Oceanogr 57:405–427
Rodriguez J, Mullin MM (1986) Relation between biomass and body weight of plankton in a steady state oceanic ecosystem. Limnol Oceanogr 31:361–370
Schultes S, Sourisseau M, LeMasson E, Lunven M, Marié L (2013) Influence of physical forcing on mesozooplankton communities at the Ushant tidal front. J Mar Syst 109–110:S191–S202
Sheldon RW, Sutcliffe WH Jr, Paranjape M (1977) Structure of pelagic food chain and relationship between plankton and fish production. J Fish Res Bd Can 34:2344–2353
Shimode S, Toda T, Kikuchi T (2006) Spatio-temporal changes in diversity and community structure of planktonic copepods in Sagami Bay, Japan. Mar Biol 148:581–197

Shoden S, Ikeda T, Yamaguchi A (2005) Vertical distribution, population structure and life cycle of *Eucalanus bungii* (Copepoda: Calanoida) in the Oyashio region, with notes on its regional variations. Mar Biol 146:497–511

Sourisseau M, Carlotti F (2006) Spatial distribution of zooplankton size spectra on the French continental shelf of the Bay of Biscay during spring 2000 and 2001. J Geophys Res 111, doi: 10.1029/2005JC003063

Sprules WG, Munawar M (1986) Plankton size spectra in relation to ecosystem productivity, size and perturbation. Can J Fish Aquat Sci 43:1789–1794

Suthers IM, Taggart CT, Rissik D, Baird ME (2006) Day and night ichthyoplankton assemblages and zooplankton biomass size spectrum in a deep ocean island wake. Mar Ecol Prog Ser 322:225–238

Tarling GA, Stowasser G, Ward P, Poulton AJ, Zhou M, Venables HJ, McGill RAR, Murphy EJ (2012) Seasonal trophic structure of the Scotia Sea pelagic ecosystem considered through biomass spectra and stable isotope analysis. Deep-Sea Res II 59–60:222–236

Terazaki M (1993) Deep-sea adaptation of the epipelagic chaetognath *Sagitta elegans* in the Japan Sea. Mar Ecol Prog Ser 98:79–88

Van der Meeren T, Næss T (1993) How does cod (*Gadus morhua*) cope with variability in feeding conditions during early larval stage? Mar Biol 116:637–647

Volkov AF (2008) Mean annual characteristics of zooplankton in the sea of Okhotsk, Bering Sea and Northwestern Pacific (annual and seasonal biomass values and predominance). Russ J Mar Biol 34:437–451

Yamaguchi A, Matsuno K, Abe Y, Arima D, Ohgi K (2014) Seasonal changes in zooplankton abundance, biomass, size structure and dominant copepods in the Oyashio region analysed by an optical plankton counter. Deep-Sea Res I 91:115–124

Yasuda I (2003) Hydrographic structure and variability of the Kuroshio-Oyashio transition area. J Oceanogr 59:389–402

Yokoi Y, Yamaguchi A, Ikeda T (2008) Regional and interannual changes in the abundance, biomass and community structure of mesozooplankton in the western North Pacific in early summer; as analysed with an optical plankton counter. Bull Plank Soc Japan 55:79–88 (in Japanese with English abstract)

Zhou M (2006) What determines the slope of a plankton biomass spectrum? J Plankton Res 28:437–448

Zhou M, Huntley ME (1997) Population dynamics theory of plankton based on biomass spectra. Mar Ecol Prog Ser 159:61–73

Zhou M, Tande KS, Zhu Y, Basedow S (2009) Productivity, trophic levels and size spectra of zooplankton in northern Norwegian shelf regions. Deep-Sea Res II 56:1934–1944

6

Biodiversity of Talitridae family (Crustacea, Amphipoda) in some Tunisian coastal lagoons

Raja Jelassi[1*], Hajer Khemaissia[1], Martin Zimmer[2], Dieter Garbe-Schönberg[3] and Karima Nasri-Ammar[1]

Abstract

Background: Although wetlands were remarkable habitats with their fauna and flora diversity, few studies have been devoted to the study of amphipod distribution in this type of environment.
To study both qualitatively and quantitatively amphipod community, surveys were conducted during the spring season in ten coastal lagoons ranging from subhumid to arid bioclimatic stage. At each station, eight quadrats of 50 × 50 cm were randomly placed. Amphipods were preserved in alcohol 70°C. In the laboratory, the specimens collected were identified and counted. Meanwhile, analyses of organic matter, particle size, and heavy metals from the soil taken from each station were made.

Results: A total of 1,340 specimens of amphipods were collected, and eight species belonging to Talitridae family were identified.
Species richness ranges from one species collected in the supralittoral zone of El Bcherliya (Ghar El Melh lagoon) and eight species in the supralittoral zone of Bizerte lagoon. In this last station, the relative abundance of amphipods was significantly higher (36.04%, $N = 483$). In addition, the diversity indices of Simpson, Shannon-Weaver, and equitability shows that the highest species diversity characterizes this same station while the community was more balanced in opposite El Boughaz (Ghar El Melh lagoon) ($J'' = 0.996$).

Conclusions: The spatial distribution of different amphipod species depends on edaphic (heavy metals, granulometry, organic matter) and climatic (temperature, humidity) factors.

Keywords: Tunisia; Biodiversity; Lagoon; Amphipoda; Talitridae; Heavy metals

Background

Coastal areas, the natural interface between water and land, were valuable ecosystems since they host high biodiversity levels (Defeo and McLachlan 2005; McLachlan and Brown 2006). Many human activities, such as fishing, land reclamation, engineering, shipping, and recreational activities, affect coastal ecosystems (McLachlan and Brown 2006; Schlacher et al. 2007). The impact of pollution in coastal areas was generally strong and the contaminants of main concern include persistent organic pollutants, oil, radionuclides, fertilizers, trace metals, and pathogens (Islam and Tanaka 2004). Trace metals were toxic and nondegradable elements that have adverse effects on living organisms (e.g., immunodeficiency, negative effect on metabolic processes and cell membrane permeability) (Ikem and Egiebor 2005). Many of these elements occur naturally, but their input has been enhanced through the millennia by various human activities (mining, metallurgical and tanning industries, chemical plants, paper mills) (Islam and Tanaka 2004).

In the Mediterranean, there was a high diversity of wetlands (lagoon, lake, sebkhas, oued, hill reservoir and dam) that were of great importance in conservation biology, and they were considered among the most biologically diverse and productive ecosystems (Medail and Quezel 1999). They offer a wide variety of natural habitats for plants and aquatic animals as well as semiterrestrial and terrestrial. The interactions of biological (plants, animals, microorganisms, etc.) and physicochemical components (granulometry, temperature, humidity, etc.) of wetlands enable them to perform many ecological functions such as shoreline stabilization and water purification.

In Tunisia, semi-closed shallow lagoons were among the most sensitive areas to environmental stresses

* Correspondence: djelassi.raja@gmail.com
[1]Université de Tunis El Manar II, Faculté des Sciences de Tunis, Unité de recherche Bio-Écologie et Systématique Évolutive, Tunis, Tunisie
Full list of author information is available at the end of the article

(Benrejeb-Jenhani and Romdhane 2002). In general, lagoon sediments were considered reservoirs of many chemical pollutants, especially heavy metals, which represent the most prominent marine pollution agents (Phillips and Rainbow 1994) affecting local communities and human health (Förstner and Wittmann 1981; Boucheseiche et al. 2002). Hence, quality management of marine coastal environments becomes a priority for many countries.

Peracarid crustaceans have received special attention, because many of them were important components of soft sediment faunas (Dauvin et al. 1994) and were considered good indicators of water and sediment quality (Corbera and Cardell 1995; Alfonso et al. 1998). Further, detritivorous peracarids play an important role in the degradation of organic matter, in both aquatic and terrestrial habitats. Talitrid amphipods were key species in the energy flow of sandy shore ecosystems (Griffiths et al. 1983). Feeding on terrestrial and marine material, these species integrate the different sources of contamination and constitute an important source of food for many species of invertebrates, fishes, and birds (Griffiths et al. 1983; Wildish 1988; Bergerard 1989; Koch 1989). This group constitutes one of the dominant macrofaunal groups in sandy beaches (Dahl 1946; McLachlan and Jaramillo 1995). Their ecological relevance has justified worldwide studies, for instance with respect to their behavioral plasticity (Scapini and Fasinella 1990; Scapini et al. 1993; Scapini et al. 1995), locomotor activity rhythms (Nasri-Ammar and Morgan 2005, 2006; Ayari and Nasri-Ammar 2012a, b; Jelassi and Nasri-Ammar 2013; Jelassi et al. 2013a, b), the factors influencing their spatial distribution and oriented movements in sandy beach (Scapini and Quochi 1992; Borgioli et al. 1999; Scapini et al. 1999; Ayari and Nasri-Ammar 2011; Jelassi et al. 2012; Jelassi et al. 2013c), their behavioral strategies (Fallaci et al. 1999), and genetic structure of different populations (De Matthaeis et al. 1995; Bulnheim and Schwenzer 1999). With regard to biodiversity, along the European coasts, talitrid populations have been compared genetically to assess inter- and intraspecific variations (De Matthaeis et al. 1995). In terms of applied research, a number of articles have been published on trace metal (Cu, Zn, Fe, Cd, Pb, Mn, and Ni) concentrations and bioaccumulation by talitrids and on their role in biomonitoring (Fialkowski et al. 2000; Rainbow et al. 1989; Weeks 1992). Studying the amphipod diversity in three different complexes, lagoons in Northern Tunisia, Jelassi et al. (2013c) showed that the abundance of different amphipod species at the three lagoon complexes in northern Tunisia could be best explained by the soil contents of several heavy metals.

In Tunisia, amphipod communities inhabiting the supralittoral zone of wetlands, other than sandy beaches (Jelassi and Nasri-Ammar 2013; Jelassi et al. 2012; Jelassi et al. 2013a), have not received much attention. Thus, aims of this study were to produce a list of amphipod species collected in the supralittoral zones of different lagoons, to estimate species richness, abundance, and density, and we focus on environmental factors that may control their spatial distribution.

Methods

Study stations

In total, ten stations belonging to the supralittoral zone of eight coastal lagoons distributed along the Tunisian coasts were studied (Figure 1). The lagoon is a stretch of salt water partially or completely separated from the open ocean by barriers of sand. These stations were S1: Bizerte lagoon (S1: 37°13′8″N009°55′1″E), three stations belonging to Ghar E Melh lagoon (S2: El Bcherliya (37°10′03″N 010°09′57″E), S3: the old harbor (37°10′04″N 010°11′40″E), S4: opposite El Boughaz (37°10′09″N 010°13′12″E), S5: Sidi Ali Mekki lagoon (37°09′50″N 010°14′45″E)), S6: North (36°48′01″N 010°12′27″E) and S7: South (36°47′59″N 010°12′26″E) lagoons of Tunis, S8: Korba lagoon (36°38′12″N 010°54′11″E), S9: Tazarka lagoon (36°32′20″N 010°50′38″E), and S10: El Bibane lagoon (33°15′57″N 011°08′28″E).

Sampling method and soil analysis

Quantitative samples of amphipods were manually collected in spring (April 2010) during ten consecutive days in the early morning hours. In the supralittoral zone of each site, eight quadrats of 50 × 50 cm were randomly placed and 20 min were devoted to collect amphipods inside the quadrat. The content of each quadrat (7 cm depth) was placed in individual boxes. Humidity and temperature of soil were measured *in situ* at each site. In the laboratory, specimens were preserved in 70% ethanol. Afterwards, they were identified and counted. The identification of these species was carried out under a Leica MS 5 binocular microscope (Leica Microsystems, Wetzlar, Germany), using the key of Ruffo (1993) and Ruffo et al. (2014) (for the new name of *Orchestia cavimana*).

At each station, a soil sample was taken from a depth of 0 to 10 cm. Grain size distribution of these composite samples was analyzed using different sieves in descending order (from 2 mm to 25 μm). A subsample was brought to the ICP-MS laboratory at University of Kiel and sieved to obtain the <250-μm grain size fraction which was then dried and milled (Bat and Raffaelli 1999). Heavy metals were extracted from a 250-mg sample of powder with 10 mL 7 N nitric acid on a hot plate at 80°C (2.5 h). The solution was made up to 20 mL, centrifuged at 3,500 rpm for 15 min, and the supernatant transferred to a 20-mL sample vial. The metals vanadium (V), chromium (Cr), manganese (Mn), cobalt (Co), nickel (Ni), copper (Cu), zinc (Zn), arsenic (As), cadmium (Cd), tin (Sn), thallium (Tl), lead (Pb), lithium

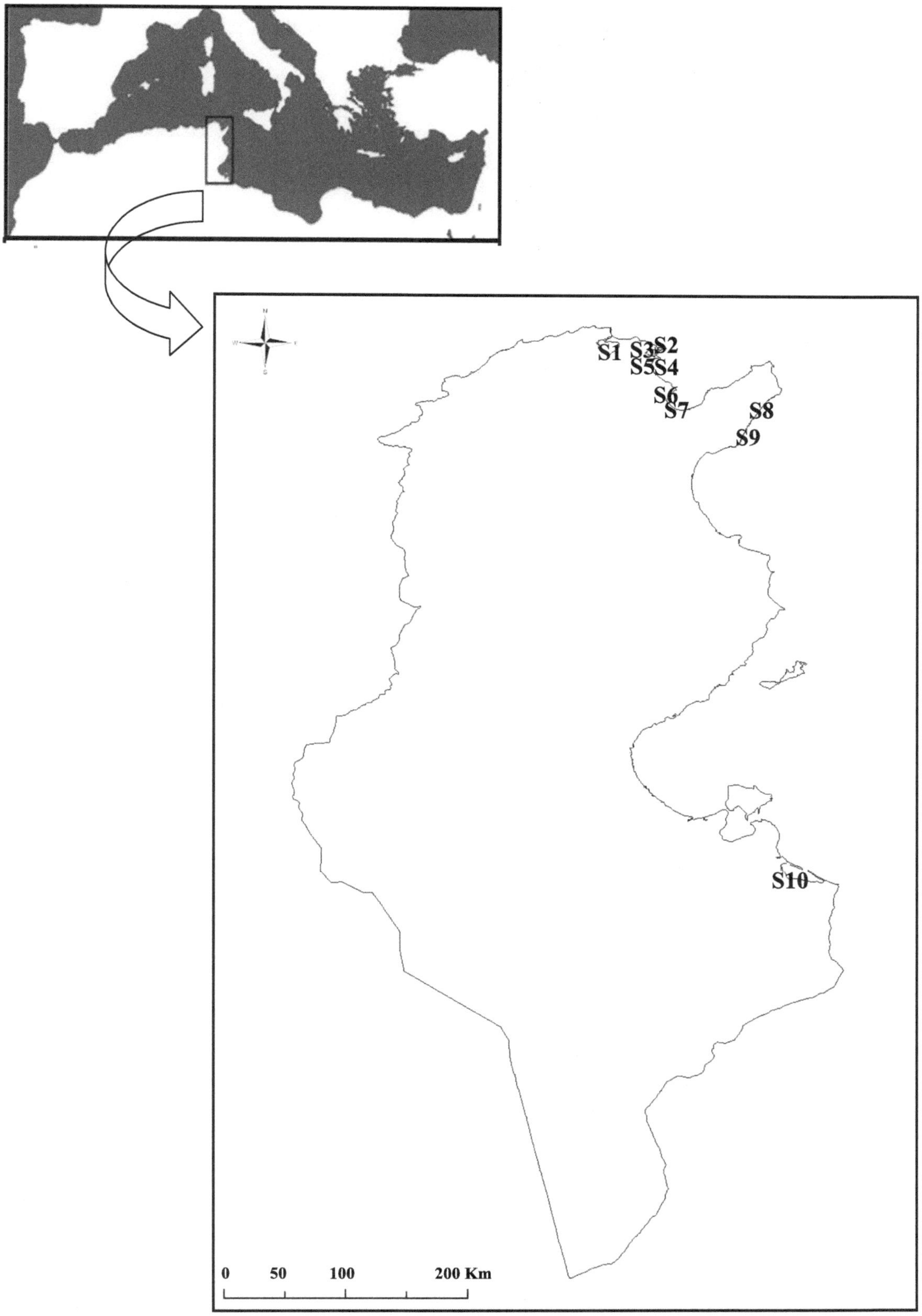

Figure 1 Localization of sampling lagoon stations. S1: Bizerte, S2: El Bcherliya, S3: Old harbor of Ghar El Melh, S4: Opposite El Boughaz, S5: Sidi Ali Mekki, S6: North lagoon of Tunis, S7: South lagoon of Tunis, S8: Korba, S9: Tazarka, S10: El Bibane.

(Li), rubidium (Rb), and strontium (Sr) were analyzed by inductively coupled plasma-mass spectrometry (ICP-MS). Average analytical reproducibility was estimated from replicate analyses of some samples and was found to be better than 2% RSD (1 sigma relative standard deviation) for all elements. The accuracy of analytical results was monitored by analyzing certified reference materials (CRM): GSMS-2 (marine sediment; Chinese Academy of Geological Sciences, PR China) and PACS-1 (coastal sediment; NRCC Canada) as unknowns along with the samples. Organic matter content was determined by weighing before and after ashing at 450°C for 3 h at the University of Salzburg.

Soil sodium content was determined by atomic absorption technique based on the methodology of the IC2MP (Institute of Chemistry of Poitiers: materials and natural resources). To obtain a usable sample, we put 5 g of soil in 50 ml of distilled water. After 1 h, the solution was filtered through a 50-lm mesh. To obtain reliable measurements, it was necessary to carry out successive dilutions.

Data analysis

In order to characterize the amphipod communities, the following ecological parameters were calculated: species richness (*S*) expressed by the number of species at each station, occurrence frequency ($F = (n/N) * 100$) where '*n*' was the number of times that species appears in the sample and '*N*' was the total number of samples, and relative species abundance ($A_r = (n_i/N) * 100$) where 'n_i' was the number of individuals of each species and '*N*' was the total number of amphipods at each station. Mean density of the amphipod community at each station and the mean density of each species at each station were expressed as number of individuals per m^2.

Species diversity and evenness were calculated by the Simpson index ($Is = 1/\Sigma\ p_i^2$) with $p_i = (n_i/N)$ (N = total number of amphipods; n_i = the number of individuals for each species) (Simpson 1949), the Shannon-Weaver index ($H' = -\Sigma\ p_i \log_2 p_i$) (Frontier 1983), and Pielou's evenness index ($J' = H' / \log_2 S$) (Pielou 1966).

The degree of similarity among sampling stations was evaluated using similarity cluster dendrograms. For this analysis, the data matrix consisting of total abundances of species at each site was converted into a symmetric matrix using the Bray-Curtis similarity index. The similarity matrix was agglomerately clustered using complete linkage based on presence/absence of species. The analysis above was performed with the PRIMER software package (Clarke and Warwick 1994).

Differences in abundance, species richness, and densities of amphipod species among lagoons were tested using ANOVA test. Principal correspondence analysis (PCA) of amphipod distribution was performed using the free version of XLSTAT.

Results

Heavy metals, sodium, organic matter, and granulometry of different stations

The highest concentration of the majority of heavy metals, namely vanadium, nickel, zinc, arsenic, cadmium, thallium, and lead, was observed in the supralittoral zone of Tunis North lagoon. For the chromium and manganese, the maximum of content characterizes the supralittoral zone of Bizerte lagoon with respectively 26.393 ppm and 281.748 ppm. The maximum content in copper (39.098 ppm) was observed in the supralittoral zone of El Bcherliya. The supralittoral zone of Korba lagoon revealed the highest concentration in cobalt (8.311 ppm) and rubidium (15.814 ppm). Whereas, the supralittoral zone of El Bibane lagoon was characterized by the highest content in lithium (29.087 ppm), in strontium (2,101.549 ppm), and in tin (7.340 ppm). Furthermore, the lowest concentration for all heavy metals studied was observed in the supralittoral zone of Sidi Ali Mekki.

The different heavy metals analyzed do not exceed the maximum value tolerated except for the lead that exceeded 100 ppm in the supralittoral zone of Tunis North lagoon (Henin 1983).

In these different stations, soil sodium content ranged between 1.26 mg/g and 7.2 mg/g of soil in the supralittoral zone of Bizerte lagoon and the old harbor of Ghar El Melh, respectively. For the organic matter, the percentage varied between 0.6% in the supralittoral zone of Sidi Ali Mekki and 9.46% in the supralittoral zone of Bizerte lagoon.

Concerning the granulometry, the supralittoral zones of Bizerte lagoon, El Bcherliya, old harbor, Tunis North, and Korba were characterized by loamy sand substrates. The supralittoral zones of Tunis South and El Bibane lagoons were characterized by sandy substrates. The sandy loam, sandy clay loam, and sandy silt loam substrates characterized respectively the supralittoral zones of opposite El Boughaz, Tazarka, and Sidi Ali Mekki lagoons.

Species richness and occurrence frequency

A total of eight species were collected in lagoons namely *Orchestia montagui* Audouin, 1826, *Orchestia mediterranea* Costa, 1853, *Orchestia gammarellus* (Pallas, 1766), *Orchestia stephenseni* Cecchini, 1928, *Cryptorchestia cavimana* Heller, 1865, *Platorchestia platensis* (Kroyer, 1845), *Deshayesorchestia deshayesii* (Audouin, 1826) and *Talitrus saltator* (Montagu, 1808). Species richness varied both between quadrats of the same station and between lagoons. It ranged from one species in the supralittoral zone of El Bcherliya and eight species in the supralittoral zone of Bizerte lagoon (Table 1). The differences observed between lagoons were highly significant ($F = 3.495$, DF = 7, $p = 0.004$).

Table 1 Species composition, mean density and diversity indices of amphipod community in different lagoons

	S1	S2	S3	S4	S5	S6	S7	S8	S9	S10
Orchestia montagui	41.5	0	0	0	0	15	30	0	0	28
Orchestia mediterranea	62	0	18	9	3.5	17	33.5	0	0	25
Orchestia gammarellus	45.5	0.5	19	10	3.5	34	34.5	0	0	25.5
Orchestia stephenseni	32	0	5	12	0	16	32.5	0	0	5.5
Cryptorchestia cavimana	17	0	0	0	0	0	0	0	0	0
Platorchestia platensis	19	0	4	10	0	15	22	0	0	0
Deshayesorchestia deshayesii	158	0	0	0	0.5	0	0	0	0	5.5
Talitrus saltator	9.5	0	0	0	0	0	0	0	0	9.5
Total number of amphipod (N)	483	1	92	82	15	194	275	0	0	198
S	8	1	4	4	3	5	5	0	0	6
Is	6.059	1	2.915	3.955	2.273	4.374	4.723	-	-	4.435
H'	2.771	-	1.711	1.992	1.287	2.232	2.278	-	-	2.309
J'	0.924	-	0.855	0.996	0.812	0.961	0.981	-	-	0.893

The study of the occurrence frequency of different species of amphipods on lagoon shores showed that *O. gammarellus* was qualified as common species ($F = 80\%$) and, *O. mediterranea* ($F = 70\%$) and *O. stephenseni* ($F = 60\%$) were classified as constant species. As for *P. platensis* ($F = 50\%$), *O. montagui* ($F = 40\%$), and *D. deshayesii* ($F = 30\%$), they were called accessory species. Finally, *T. saltator* ($F = 20\%$) and *C. cavimana* ($F = 10\%$) were qualified as rare species.

Relative abundance and density

A count of 1,340 amphipod individuals was collected in different lagoons. In this type of wetlands, the supralittoral zone of Bizerte lagoon revealed the highest abundance (36.04%) (Table 1). Moreover, the difference between lagoons was statistically significant (Anova test: $F = 4.371$; DF = 7; $p = 0.001$).

The study of relative abundance of the different species showed that *O. mediterranea* was the most abundant species in the supralittoral zone of Bizerte lagoon (25.7%). Whereas, in El Bcherliya, the old harbor of Ghar El Melh, the North and South lagoons of Tunis, *O. gammarellus* dominated the community. In the supralittoral zone of El Bibane lagoon, *O. montagui* was the most important species (28.3%). Anova test revealed that differences in species observed between lagoons were highly significant ($F = 4.371$; DF = 7; $p = 0.001$).

Concerning the mean density of amphipod community, it varied between lagoons but it was the supralittoral zone of Bizerte lagoon that presented the important density with 241.5 ind. m^{-2}. The lowest density was observed in the supralittoral zone of El Bcherliya (0.5 ind. m^{-2}) (Figure 2).

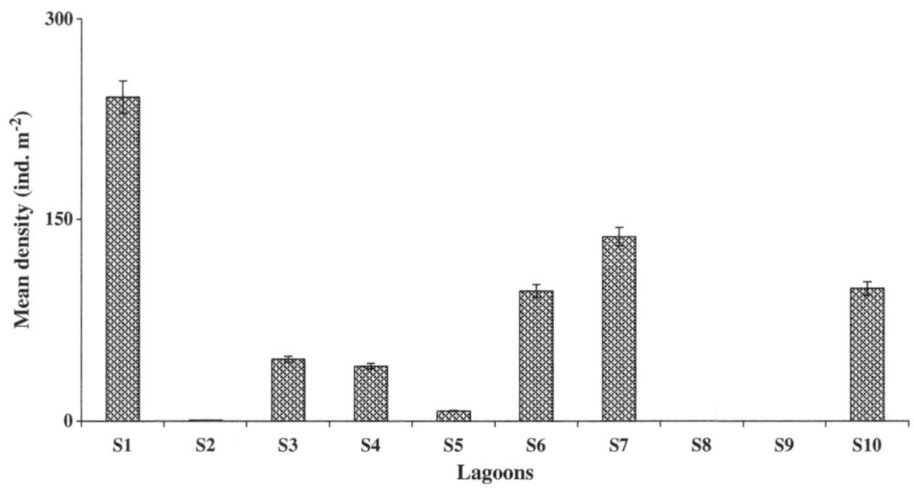

Figure 2 Variation of the mean density (ind. m – 2) of amphipod communities in the supralittoral zone of lagoons.

Furthermore, results showed that *O. mediterranea* presented the most important density in the supralittoral zone of Bizerte lagoon (62 ind. m^{-2}). Whereas, *O. gammarellus* showed the most important density in the supralittoral zones of El Bcherliya (0.5 ind. m^{-2}), the old harbor of Ghar El Melh (19 ind. m^{-2}), the North (34 ind. m^{-2}) and South (34.5 ind. m^{-2}) lagoons of Tunis.

Diversity and community similarity

According to Simpson index, the highest diversity was observed in the supralittoral zone of Bizerte lagoon (Is = 6.059) (Table 1). Unlike to the previous index, the Shannon-Weaver index takes into account much more the rare species, it confirmed this result and it ranged between 1.287 in the supralittoral zone of Sidi Ali Mekki lagoon and 2.771 in Bizerte lagoon. Concerning the equitability, it was important in opposite El Boughaz where we noted that the species were equitably distributed.

A dendrogram of similarity between lagoons, obtained by calculating the Bray-Curtis similarity index based on presence/absence of species, showed the presence of two clusters: the first associate Sidi Ali Mekki and El Bcherliya where we noted the lowest count of amphipods, the second associate the six others stations and it was divided into two sub-groups; the first combines the opposite El Boughaz and the old harbor of Ghar El Melh, and the other comprises North and South lagoons of Tunis, Bizerte lagoon, and El Bibane lagoon. The combination of these four stations could be explained on the one hand by the presence of the species *O. montagui* and on the other hand by a considerable concentration of lead exceeding 50 ppm (Figure 3).

Amphipod distribution according to environmental factors

Principal component analysis (PCA) was performed to better understand which of the environmental factors influenced the distribution of different species in lagoons (Figure 4).

The first three factorial axes (F1, F2, and F3) extract 54.39%, 19.57%, and 9.32% of the variance respectively; it was a cumulative percentage of 83.28% (Figure 4).

On the first axis (F1) extracting almost half of the total inertia, there were Bizerte lagoon, North and south lagoons of Tunis which were correlated positively with the majority of heavy metals namely vanadium, chromium, manganese, cobalt, nickel, copper, zinc, arsenic, rubidium, cadmium, thallium, and lead. These stations were characterized negatively by the soil temperature.

On the second axis which extracts only 19.57% of the total variance, there were El Bcherliya, El Bibane and Korba lagoons which were related to soil humidity, particle size, and the content of strontium, lithium, and tin.

On the third axis, were projected negatively the organic matter and positively the concentration on sodium. These two parameters do not characterize these different lagoon stations.

Discussion

The study of biodiversity of amphipods communities highlighted differences in species richness, relative abundance and diversity between and within lagoons.

Globally the species richness was higher in the supralittoral zone of Bizerte lagoon than in the other stations. In this station, eight amphipod species were identified. This would be related to the presence of important vegetation in spring as well as the *Cymodocea nodosa* leaf litter and a high percentage of organic matter. In this same station, the highest diversity, abundance, and density were observed. Moreover, it was *O. mediterranea* that presented the most important density in this supralittoral zone. In other type of wetlands such as oueds, species richness varied between one species in oueds Laakarit, Khniss, and El Fared and six species in Tinja (Korsi station) whereas in sebkhas, only two species were collected in sebkhas Gargour and Moknine (Jelassi 2014; Jelassi

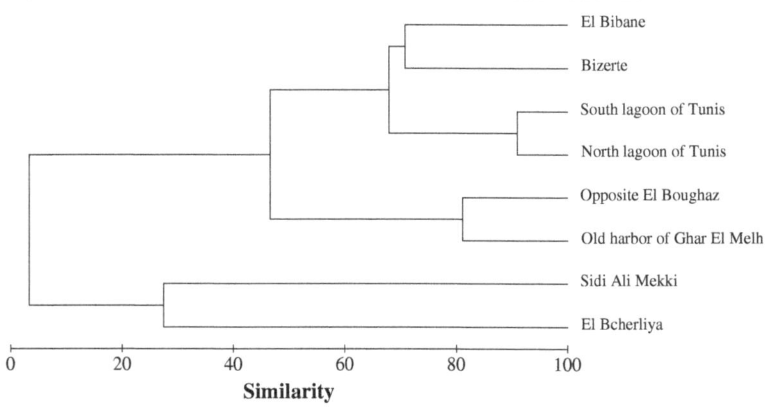

Figure 3 Clustering of sampling data, based on total number of amphipods collected (Bray-Curtis index, presence/absence, complete linkage).

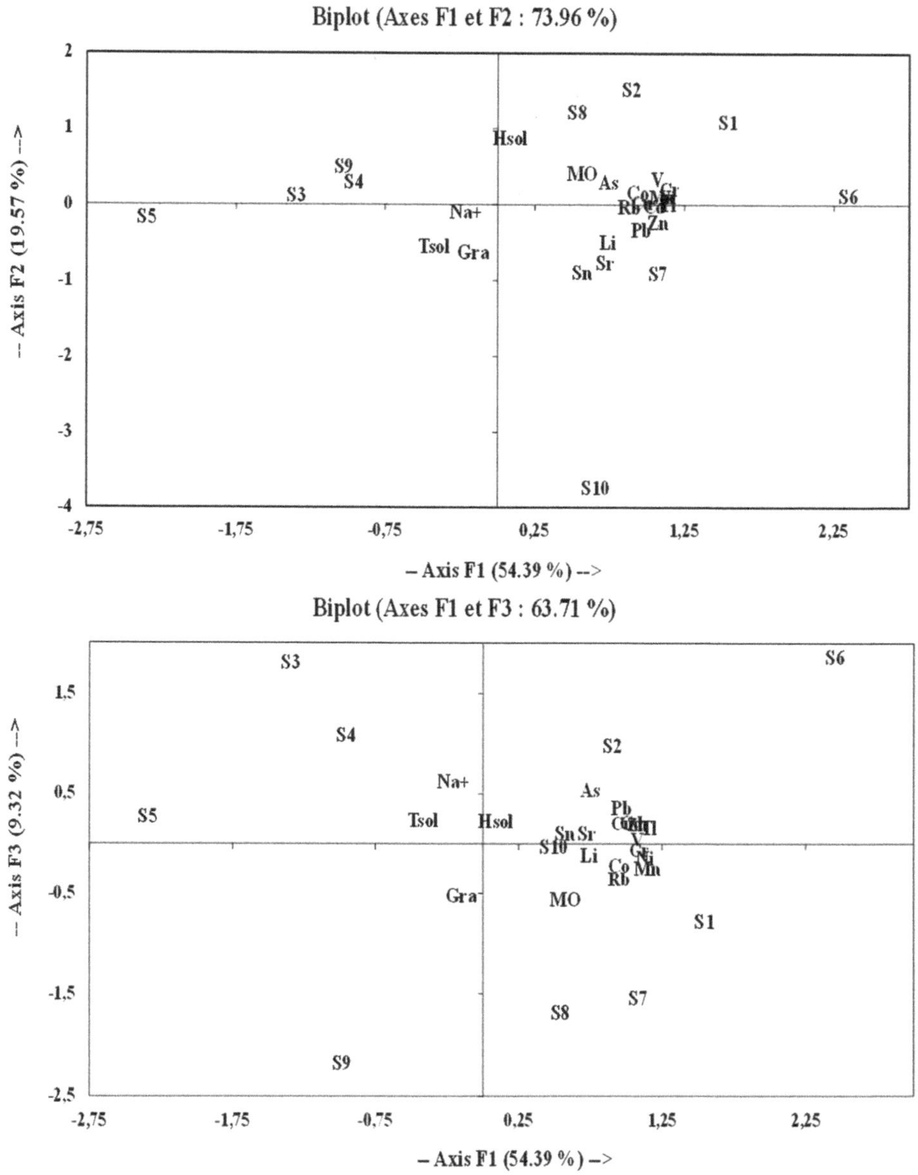

Figure 4 Principal Component Analysis (PCA) performed on the environmental factors. V: Vanadium, Cr: Chromium, Mn: Manganese, Co: Cobalt, Ni: Nickel, Cu: Copper, Zn: Zinc, As: Arsenic, Cd: Cadmium, Sn: Tin, Tl: Thallium, Pb: Lead, Li: Lithium, Rb: Rubidium, Sr: Strontium, Tsol: soil temperature, Hsol: soil humidity, Gra: granulometry, Na+: sodium, MO: organic matter.

et al. 2013c). In supralittoral zones of dams and hill lakes, no species has been found (Jelassi 2014). Using the same method of collection, Ayari (2012) showed the presence of five species namely *T. saltator, D. deshayesii, O. gammarellus, O. montagui,* and *O. mediterranea* in the supralittoral zone of Bizerte Corniche beach and only one species, *T. saltator,* in the supralittoral zone of Gabès beach. This author attributed the lowest species richness in the last station (Gabès) to the highest tidal amplitude reported in this beach and the highest species richness observed in Bizerte Corniche to the presence of *Posidonia oceanica* benches associated with other algae that provide shelter and food for different amphipod species. In Zouaraa beach, only two species were identified: *T. saltator* and *Talorchestia brito* living in sympatry with a total density generally low and never exceeded 860 ind.m^{-2} (Charfi-Cheikhrouha et al. 2000). At Ouderef, Gabes, and Zarrat beaches (Gulf of Gabes, Tunisia), the amphipod *T. saltator* was the most abundant species, present in the three beaches studied; with the highest densities in Ouderef beach (Pérez-Domingo et al. 2008). *O. montagui* and *D. deshayesii* in the Bay of Bou Ismail reached more than 45.000 ind. m^{-2} (Louis 1980), and *O. mediterranea* in the estuary of Bou Regreg, Morocco, attained 7,000 ind. m^{-2} (Elkaïm et al.

1985). On the Isle of Man, densities of *T. saltator* were estimated at 80 to 400 ind. m^{-2} (Williams, 1995). Marsden (1991) and Cardoso and Veloso (1996) showed that fluctuations in population density were frequent in talitrids and indicated periods of intense reproduction. They observed, in fact, similar patterns of variation for *Talorchestia quoyana* and *Pseudorchestoidea brasiliensis*, respectively, with highest densities in summer and late winter.

The distribution of amphipod species in the different lagoons was investigated according to environmental factors using PCA analysis. Our statistical analyses indicated that the variation in the spatial distribution of amphipod species depended on some climatic (temperature, humidity) as well as edaphic factors, which was influenced by soil quality. In the different lagoons, *T. saltator*, found only in the supralittoral zone of Bizerte and El Bibane lagoons, was correlated with temperature, humidity, particle size, and some heavy metals. In sebkhas and oueds, this species was correlated only with edaphic factors (unpublished data). Several studies were focused on the role of environmental factors and showed the influence of factors than others (Borgioli et al. 1999; Scapini et al. 1995; Scapini and Fasinella 1990; Scapini et al. 1999; Scapini and Quochi 1992). Jelassi et al. (2012) showed that Talitridae abundance was tightly controlled by air temperature. The highest abundance observed in this site could be also explained by the important recruitment in spring (Jelassi 2014). Studying the amphipod diversity at three Tunisian lagoon complexes, Jelassi et al. (2013c) showed that according to the constrained correspondence analysis, the abundance of *O. mediterranea* and *O. gammarellus* at the three lagoon complexes in northern Tunisia could be best explained by the soil contents of several heavy metals, namely zinc, thallium, and cadmium and the proportion of the coarse sand fraction. *T. saltator* abundance, by contrast, negatively corresponded to these station characteristics. Air and soil temperature were the best predictors for *O. stephenseni* abundance that negatively corresponded with the proportion of the fine sand fraction and the organic matter content of the soil. *O. montagui* and *C. cavimana* abundances corresponded positively with air humidity and the soil lithium and rubidium contents but negatively with the soil tin content and the proportion of the silt and clay fraction. *D. deshayesii* and *P. platensis* did not exhibit any clear correspondence with station characteristics. Bouslama et al. (2009) showed that the temperature was the most important factor influencing the zonation; indeed, the increase of temperature induced the migration of T. saltator population from the top to the bottom of beach. This result was similar to that found by Fallaci et al. (2003) that reported that the mean zonation of *T. saltator* was exclusively influenced by the temperature during the activity period. Marques et al. (2003) have highlighted a positive correlation between temperature and the density or biomass of *Talitrus saltator* population. These authors showed that the positive correlation may be interpreted as a cause-and-effect relation, with temperature favoring recruitment and, consequently, the increase in density and biomass. On the other hand, Colombini et al. (2002) noted the importance of the parameters of the sediment in the choice of specific distribution area especially for young individuals.

Conclusions

A total of 1,340 talitrids individuals was collected in different coastal lagoons. Abundance, density and diversity were more pronounced in the supralittoral zone of Bizerte lagoon characterized by the presence of eight amphipod species: *O. montagui, O. gammarellus, O. mediterranea, O. stephenseni, O. cavimana, P. platensis, D. deshayesii* and *T. saltator*. Moreover, according to principal component analysis, the spatial distribution of species in the different lagoons in Tunisia depends on edaphic (heavy metals, granulometry, organic matter) and climatic (temperature, humidity) factors.

Competing interests

The authors declare that they have no competing interests.

Authors' contributions

All the authors performed the field sampling, participated and coordinated in the design and analysis of the study, and drafted the manuscript. All authors read and approved the final manuscript.

Acknowledgements

The study was supported by the Research Unit of Bio-ecology and Evolutionary Systematics (UR11ES11), Faculty of Science of Tunis, University of Tunis El Manar, University of Salzburg, and the University of Kiel.

Author details

[1]Université de Tunis El Manar II, Faculté des Sciences de Tunis, Unité de recherche Bio-Écologie et Systématique Évolutive, Tunis, Tunisie. [2]FB Organismische Biologie, AG Ökologie, Biodiversität & Evolution der Tiere, Universität Salzburg, Salzburg, Austria. [3]Institut für Geowissenschaften, ICP-MS Labor, Universität zu Kiel, Kiel, Germany.

References

Alfonso MI, Bandera ME, Lopez-Gonzalez PJ, Garcia-Gomez JC (1998) The Cumacean community associated with a seaweed as a bioindicator of environmental conditions in the Algeciras Bay (Strait of Gibraltor). Cah Biol Mar 39:197–205

Ayari A (2012) Eco-éthologie de deux espèces sympatriques *Talitrus saltator* et *Deshayesorchestia deshayesi* (Crustacés, Amphipodes) au niveau de deux plages tunisiennes. Thèse de doctorat Faculté des Sciences de Tunis, Université de Tunis El Manar Tunisie, pp 210

Ayari A, Nasri-Ammar K (2012a) Seasonal variation of the endogenous rhythm in tow sympatrics amphipod: *Talitrus saltator* and *Deshayesorchestia deshayesii* from Bizerte beach (North of Tunisia). Biol Rhythm Res 43(5):515–526

Ayari A, Nasri-Ammar K (2012b) Locomotor rhythm phenology of *Talitrus saltator* from two geomorphologically different beaches of Tunisia: Bizerte (North of Tunisia) and Gabes gulf (South of Tunisia). Biol Rhythm Res 43(2):113–123

Ayari A, Nasri-Ammar K (2011) Distribution and biology of amphipods in two geomorphologically different sandy beaches of Tunisia. Crustaceana 84(5–6):591–599

Bat L, Raffaelli D (1999) Effects of gut sediment contents on heavy metal levels in the amphipod *Corophium volutator* (Pallas). Turk J Zool 23:67–71

Benrejeb-Jenhani A, Romdhane MS (2002) Impact des perturbations anthropiques sur l'évolution du phytoplancton de la lagune de Boughrara, (Tunisie). Bull Inst Natn Scien Tech Mer de Salammbô 29:65–75

Bergerard J (1989) Ecologie des laisses de marée. Ann Biol 28:39–54

Borgioli C, Martelli L, Porri F, D'Elia A, Marchetti GM, Scapini F (1999) Orientation in *Talitrus saltator* (Montagu): trends in intrapopulation variability related to environmental and intrinsic factors. J Exp Mar Biol Ecol 238:29–47

Boucheseiche C, Cremille E, Pelte T, Pojer K (2002) Bassin Rhône -Méditerranée-Corse. Guide technique n°7, Pollution toxique et écotoxicologie: notion de base. Agence de l'Eau Rhône – Méditerranée - Corse, Lyon

Bouslama MF, El Gtari M, Charfi-Cheikhrouha F (2009) Impact of environmental factors on zonation, abundance, and other biological parameters of two Tunisian populations of *Talitrus saltator* (Amphipoda, Talitridae). Crustaceana 82(2):141–157

Bulnheim HP, Schwenzer DE (1999) Allozyme variation and genetic divergence in populations of *Talitrus saltator* (Crustacea: Amphipoda) around the Atlantic coast, the Azores and the Canary Islands. Cah Biol Mar 40:185–194

Cardoso RS, Veloso VG (1996) Population biology and secondary production of the sandhopper *Pseudorchestoidea brasiliensis* (Amphipoda: Talitridae) at Prainha Beach, Brazil. Mar Ecol Progr Ser 142:111–119

Charfi-Cheikhrouha F, El Gtari M, Bouslama MF (2000) Distribution and reproduction of two sandhoppers, *Talitrus saltator* and *Talorchestia brito* from Zouaraa beach-dune system (Tunisia). Pol Arch Hydrobiol 43:621–629

Clarke KR, Warwick RM (1994) Change in marine communities: an approach to statistical analysis and interpretation. Nat Environ Res Council, Plymouth Marine Biological Laboratory, Plymouth, UK

Colombini I, Aloia A, Bouslama MF, El Gtari M, Fallaci M, Ronconi L, Scapini F, Chelazzi L (2002) Small-scale spatial and seasonal differences in the distribution of beach arthropods on the northern Tunisian coasts. Are species evenly distributed along the shore? Mar Biol 140:1001–1012

Corbera J, Cardell MJ (1995) Cumaceans as indicators of eutrophication on soft bottoms. Sci Mar 59:63–69

Dahl E (1946) The Amphipoda of the sound. I: terrestrial Amphipoda. Acta Univ Lundensis 42:1–53

Dauvin JC, Bellan G, Bellan-Santini D, Castric A, Comolet-Tirman J, Francour P, Gentil F, Girard A, Gofas S, Mahé C, Noël P, De Reviers B (1994) Typologie des ZNIEFF Mer. Liste des paramètres et des biocénoses des côtes françaises métropolitaines. Patrimoines Naturels 12:1–64

De Matthaeis E, Cobolli M, Mattoccia M, Scapini F (1995) Geographic variation in *Talitrus saltator* (Crustacea, Amphipoda): biochemical evidence. Boll Zool 62:77–84

Defeo O, McLachlan A (2005) Patterns, processes and regulatory mechanisms in sandy beach macrofauna: a multi-scale analysis. Mar Ecol Prog Ser 295:1–20

Elkaïm B, Irlinger JP, Pichard S (1985) Dynamique de la population d'Orchestia mediterranea L. (Crustacé, Amphipode) dans l'estuaire du Bou Regreg (Maroc). Can J Zool 63:2800–2809

Fallaci M, Aloia A, Audoglio M, Colombini I, Scapini F, Chelazzi L (1999) Differences in behavioural strategies between two sympatric talitrids (Amphipoda) inhabiting an exposed sandy beach of the French Atlantic coast. Estuar Coast Shelf Sci 48:469–482

Fallaci M, Colombini I, Lagar M, Scapini F, Chelazzi L (2003) Distribution patterns of different age classes and sexes in a Tyrrhenian population of *Talitrus saltator* (Montagu). Mar Biol 142:101–110

Fialkowski W, Rainbow PS, Fialkowska E, Smith BD (2000) Biomonitoring of trace metals along the Baltic Coast of Poland using the sandhopper *Talitrus saltator* (Montagu) (Crustacea: Amphipoda). Ophelia 52:183–192

Förstner U, Wittmann GTW (1981) Metal pollution in the aquatic environment, 2nd edn. Springer, Berlin

Frontier S (1983) Stratégies d'échantillonnage en écologie. Masson, Paris, p 494

Griffiths CL, Stenton-Dozey JME, Koop K (1983) Kelp wrack and the flow of energy through a sandy beach ecosystem. In: McLachlan A, Erasmus T (eds) Sandy Beaches as Ecosystems. Junk Publications, The Hague, pp 547–556

Henin S (1983) Les éléments traces dans les sols. Science Du Sol 2:67–71

Ikem A, Egiebor NO (2005) Assessment of trace elements in canned fishes (mackerel, tuna, salmon, sardines and herrings) marketed in Georgia and Alabama (United States of America). J Food Comp Anal 18:771–787

Islam MS, Tanaka M (2004) Impacts of pollution on coastal and marine ecosystems including coastal and marine fisheries and approach for management: a review and synthesis. Mar Poll Bull 48:624–649

Jelassi R (2014) Eco-éthologie des peuplements d'Amphipodes au niveau des zones humides de la Tunisie. Thèse de doctorat Université de Tunis, Tunisie, p 328

Jelassi R, Nasri-Ammar K (2013) Seasonal variation of locomotor activity rhythm of *Orchestia montagui* in the supralittoral zone of Bizerte lagoon (North of Tunisia). Biol Rhythm Res 44(5):718–729

Jelassi R, Khemaissia H, Nasri-Ammar K (2012) Intra-annual variation of the spatiotemporal distribution and abundance of Talitridae and Oniscidea (Crustacea, Peracarida) at Bizerte Lagoon (northern Tunisia). Afr J Ecol 50:381–392

Jelassi R, Ayari A, Nasri-Ammar K (2013a) Seasonal variation of locomotor activity rhythm of *Orchestia gammarellus* in the supralittoral zone of Ghar Melh lagoon (North-East of Tunisia). Biol Rhythm Res 44(6):956–967

Jelassi R, Akkari-Ayari A, Bohli-Abderrazak D, Nasri-Ammar K (2013b) Endogenous locomotor activity rhythm of two sympatric species of Talitrids (Crustacea, Amphipoda) from the supralittoral zone of Bizerte lagoon (Northern Tunisia). Biol Rhythm Res 44(2):265–275

Jelassi R, Zimmer M, Khemaissia H, Garbe-Schönberg D, Nasri-Ammar K (2013c) Amphipod diversity at three Tunisian lagoon complexes in relation to environmental conditions. J Nat His 47(45–46):2849–2868

Koch H (1989) The effect of tidal inundation on the activity and behavior of the supralittoral talitrid amphipod *Traskorchestia traskiana*. Crustaceana 57:295–303

Louis M (1980) Etude d'un peuplement mixte d'*Orchestia montagui* Audouin et d'*Orchestia deshayesii* Audouin dans la baie de Bou Ismail. Bull Ecol 11:97–111

Marques JC, Gonçalves SC, Pardal MA, Chelazzi L, Colombini I, Fallaci M, Bouslama MF, El Gtari M, Charfi-Cheikhrouha F, Scapini F (2003) Comparison of *T. Saltator* (Amphipoda, Talitridae) biology, dynamics and secondary production in Atlantic (Portugal) and Mediterranean (Italy and Tunisia) populations. Estuar Coast Shelf Sci 58:127–148

Marsden ID (1991) Kelp-sandhopper interactions on a sand beach in New Island. II. Population dynamics of *Talorchestia quoyana* (Milne-Edwards). J Exp Mar Biol Ecol 152:75–90

McLachlan A, Brown AC (2006) The ecology of sandy shores second ed. Elsevier, Amsterdam, p 392

McLachlan A, Jaramillo E (1995) Zonation on sandy beaches. Oceanogr Mar Biol Ann Review 33:305–335

Medail F, Quezel P (1999) Biodiversity hotspots in the Mediterranean basin: setting global conservation priorities. Conserv Biol 13:1510–1513

Nasri-Ammar K, Morgan E (2005) Variation saisonnière du rythme de l'activité locomotrice de *Talitrus saltator* issu de la plage de Korba (Cap Bon, Tunisie). Bull Soc Zool Fr 130(1):19–29

Nasri-Ammar K, Morgan E (2006) Seasonality of the endogenous activity rhythm in *Talitrus saltator* (Montagu) from a sandy beach in north-eastern Tunisia. Biol Rhythm Res 37:479–488

Pérez-Domingo S, Castellanos C, Junoy J (2008) The sandy beach macrofauna of Gulf of Gabès (Tunisia). Mar Ecol 29:51–59

Phillips DJH, Rainbow PS (1994) Biomonitoring of trace aquatic contaminants, Environmental Management Series. Chapman and Hall, London

Pielou EC (1966) The measurement of diversity in different types of biological collections. J Theor Biol 13:131–144

Rainbow PS, Moore PG, Watson D (1989) Talitrid amphipods (Crustacea) as biomonitors for copper and zinc. Estuar Coast Shelf Sci 28:567–582

Ruffo S (1993) The Amphipoda of Mediterranean. Part IV: Localities and map-Agenda to parts 1-3-Key to families-Ecology-Faunistics and zoogeography. Mem Inst Oceanogr de Monaco 13:959

Ruffo S, Tarocco M, Latella L (2014) *Cryptorchestia garbinii* n. sp. (Amphipoda: Talitridae) from Lake Garda (Northern Italy), previously referred to as *Orchestia cavimana* Heller, 1865, and notes on the distribution of the two species. Ital J Zool 81:91–98

Scapini F, Fasinella D (1990) Genetic determination and plasticity in the sun orientation of natural populations of *Talitrus saltator*. Mar Biol 107:141–145

Scapini F, Quochi G (1992) Orientation in sandhoppers from Italian populations: have they magnetic orientation ability? Bull Zool 59:437–442

Scapini F, Lagar MC, Mezzetti MC (1993) The use of slope and visual information in sandhoppers: innateness and plasticity. Mar Biol 115:545–553

Scapini F, Buiatti M, De Matthaeis E, Mattoccia M (1995) Orientation behaviour and heterozygosity of sandhopper populations in relation to stability of beach environments. J Evol Biol 8:43–52
Scapini F, Porri F, Borgioli C, Martelli L (1999) Solar orientation of adult and laboratory-born juvenile sandhoppers: inter- and intra-population variation. J Exp Mar Biol Ecol 238:107–126
Schlacher TA, Dugan J, Schoeman DS, Lastra M, Jones A, Scapini F, Mclachlan A, Defeo O (2007) Sandy beaches at the brink. Divers Distrib 13:556–560
Simpson EH (1949) Measurement of diversity. Nature 163:672–688
Weeks JM (1992) The use of the terrestrial amphipod *Arcitalitrus dorrieni* (Crustacea: Amphipoda: Talitridae) as a potential biomonitor of ambient zinc and copper availabilities in leaf-litter. Chemosphere 24:1505–1522
Wildish DJ (1988) Ecology and natural history of aquatic Talitroidea. Can J Zool 66:2340–2359
Williams JA (1995) Burrow-zone distribution of the supralittoral Amphipod *Talitrus saltator* on Derbyhaven beach, Isle of Man – a possible mechanism for regulating desiccation stress? J Crust Biol 15:466–475

7

Spatial changes in the vertical distribution of calanoid copepods down to great depths in the North Pacific

Atsushi Yamaguchi[1*], Kohei Matsuno[2] and Tomoe Homma[1]

Abstract

Background: Despite its ecological importance, little information is available regarding the spatial and vertical changes in the calanoid copepod community over large geographical regions. This study investigated the spatial and vertical patterns in calanoid copepod abundance and community structure using zooplankton samples collected between depths of 0 and 2,615 m across the North Pacific from 0° to 56°N.

Results: A total of 211 calanoid copepod species belonging to 66 genera and 24 families were identified. Calanoid copepod abundance decreased with increasing depth, and few latitudinal differences were detected. Across the entire region, species diversity peaked near 500 to 2,000 m in depth. The calanoid copepod community was separated into seven groups with distinct spatial and vertical distributions. For all groups, the number of species was low (28 to 37 species) in the subarctic region (north of 40°N) and high (116 to 121 species) in the subtropical-tropical region. The deepest group in the subtropical-tropical region was composed of cosmopolitan species, and this group was also observed in deep water in the subarctic region.

Conclusions: In deep water, most of the calanoid copepod community consisted of cosmopolitan species, while an endemic community was observed in the subarctic region. Because the food of deep-sea calanoid copepods originates from the surface layer, sufficient and excess flux in the eutrophic subarctic region may be responsible for maintaining the endemic species in the region.

Keywords: Zooplankton; Calanoid copepods; Community structure; Diversity; Deep-sea; North Pacific

Background

Calanoid copepods are the dominant component of the zooplankton biomass throughout the water column in the North Pacific. They consume much of the primary production and are an important link to higher trophic levels; thus, they have an important role in energy transport in marine ecosystems. Calanoid copepods are also an important component of the 'biological pump,' and they transport organic matter to the deep ocean in egested fecal pellets and through diel and seasonal vertical migration (Longhurst 1991; Hernández-León and Ikeda 2005; Kobari et al. 2008b). Because calanoid copepods are the major prey of meso- and bathypelagic fishes (Merrett and Roe 1974; Hopkins and Sutton 1998; Moku et al. 2000), it is important to clarify the calanoid copepod community structure at depth to better understand marine ecosystems.

Most studies on the community structure of calanoid copepods at great depths have collected samples using plankton nets (cf. Vinogradov 1968). Examples include studies in the Arctic Ocean (Kosobokova and Hirche 2000; Auel and Hagen 2002), Greenland Sea (Richter 1994), Bering Sea (Homma and Yamaguchi 2010), Atlantic Ocean (Roe 1972; Koppelmann and Weikert 1999), subarctic Pacific (Vinogradov 1962; Arashkevich 1972; Yamaguchi et al. 2002; Steinberg et al. 2008), Mediterranean Sea (Scotto di Calro et al. 1984; Weikert and Trinkaus 1990; Koppelmann and Weikert 2007), Arabian Sea (Madhupratap and Haridas 1990; Fabian et al. 2005; Koppelmann and Weikert 2005; Wishner et al. 2008), Red Sea (Weikert 1982; Weikert and Koppelmann 1993), and the Antarctic Ocean (Schnack-Schiel et al.

* Correspondence: a-yama@fish.hokudai.ac.jp
[1]Laboratory of Marine Biology, Graduate School of Fisheries Science, Hokkaido University, 3-1-1 Minatomachi, Hakodate, Hokkaido 041-8611, Japan
Full list of author information is available at the end of the article

2008). Most of these studies examined biomass, which is easy to measure, but few have identified and evaluated specimens down to species level. Furthermore, few studies have examined the abundance, species diversity, and community structure of calanoid copepods, and most that did were limited to a narrow geographical range. Throughout most of the North Pacific, little is known about the community structure throughout the water column.

During 1965 to 1967, detailed stratified zooplankton samplings from the surface to the deep sea (maximum depth 2,615 m) were made across the North Pacific (0° to 56°N), and the collected calanoid copepods were identified to the species level. In this study, we analyzed the calanoid copepod community down to great depths across the entire North Pacific. Based on this analysis, we discuss what determines the structure of calanoid copepod communities across the North Pacific.

Methods

Field sampling and data gathering

Field sampling was conducted at 12 stations between the equator and 56°N in the Bering Sea and Pacific Ocean during cruises of the T.S. *Oshoro Maru* and R.V. *Hakuho Maru* from 1965 to 1967 (Figure 1). Samples were collected simultaneously from six or seven discrete depths between 0 and 2,615 m by horizontal tows of 3 to 6 Motoda (MTD) horizontal closing nets (mesh size 350 μm, mouth diameter 56 cm; Motoda 1971) (Table 1). The volume of filtered water was calculated assuming a filtering rate of 0.6 (Motoda 1971). After collection, zooplankton samples were preserved in 10% borax-buffered formalin-seawater. In the terrestrial laboratory, 1/32 to 1/2 aliquot subsamples were examined under microscope to identify the calanoid copepods. Although the samplings were conducted at both day and night, the day-night differences are expected to be smaller than the differences due to the large geographical (0° to 56°N) and vertical (0 to 2,615 m) ranges in this study.

Water temperature and salinity were measured using a reversing thermometer and an inductive salinometer, respectively. Data collected by the T.S. *Oshoro Maru* were cited from 'Data Record of Oceanographic Observations and Exploratory Fishing' (Hokkaido University 1967), and those from the R.V. *Hakuho Maru* were provided by the Center for Cruise Coordination, Atmosphere and Ocean Research Institute, the University of Tokyo.

Data analysis

The species diversity index (H') (Shannon and Weaver 1949) for calanoid copepods was calculated as:

$$H' = -\sum n/Ni \ln n/Ni$$

where n is the abundance (individuals (inds.) m^{-3}) of each calanoid species at the ith layer, and Ni is the total abundance of calanoid copepods in the ith layer. This

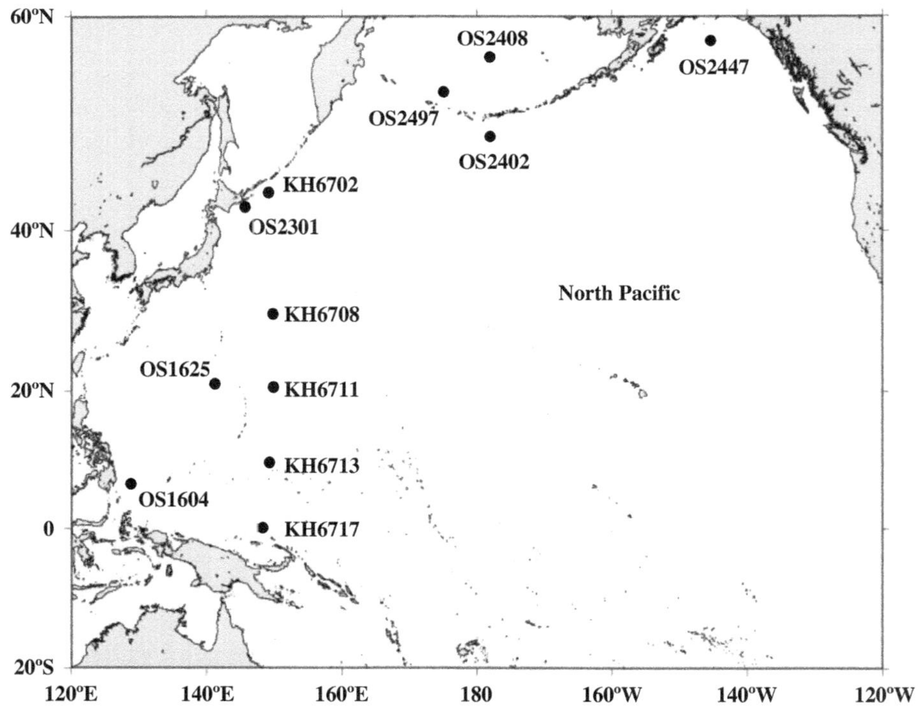

Figure 1 Locations of sampling stations in the Bering Sea and Pacific Ocean.

Table 1 Zooplankton sampling data used in the present analyses

Area	Station	Position	Sampling date	Local time	Sampled depth (m)
Bering Sea	OS2408	56°00′N 178°00′W	14 June 1967	08:54 to 09:53	0, 130, 270, 610, 983, 1,285
	OS2497	53°08′N 175°02′E	20 June 1967	15:42 to 18:24	0, 40, 105 167, 277, 555, 1,110
		53°08′N 175°02′E	21 June 1967	22:33 to 01:31	0, 85, 170, 260, 430, 860, 1,720
Eastern North Pacific	OS2447	57°33′N 145°56′W	20 August 1967	12:05 to 14:40	0, 65, 137, 200, 333, 665, 1,330
	OS2402	49°00′N 178°00′W	10 June 1967	22:06 to 23:39	0, 87, 177, 355, 432, 800
Western North Pacific	KH6702	43°47′N 149°55′E	6 December 1967	18:43 to 22:53	0, 435, 873, 1,360, 1,745, 2,180, 2,615
	OS2301	42°00′N 146°00′E	23 April 1967	12:06 to 13:61	7, 60, 120, 180, 240, 310, 360
		42°00′N 146°00′E	23 April 1967	22:12 to 23:36	7, 58, 117, 175, 233, 290, 345
	KH6708	29°57′N 150°06′E	13 December 1967	12:06 to 13:61	0, 83, 168, 392, 730, 1,670, 2,500
		29°57′N 150°06′E	13 December 1967	22:12 to 23:36	0, 83, 168, 392, 860, 1,170, 2,500
	OS1625	20°47′N 141°45′E	17 January 1966	09:50 to 13:49	7, 80, 153, 375, 740, 1,475, 2,200
	KH6711	20°04′N 150°07′E	17 December 1967	08:27 to 11:05	0, 75, 152, 380, 583, 835, 1,140
	KH6713	09°53′N 149°51′E	22 December 1967	09:26 to 13:35	0, 65, 122, 317, 637, 1,275, 1,910
		09°53′N 149°51′E	22 December 1967	23:45 to 04:05	0, 80, 140, 347, 685, 2,080
	OS1604	06°52′N 129°24′E	4 December 1965	10:38 to 19:65	9, 47, 85, 383, 755, 1,500
	KH6717	0°02′S 148°52′E	26 December 1967	18:40 to 20:40	0, 50, 105, 258, 518, 1,035, 1,555

All samples were collected with horizontal tows of Motoda horizontal closing nets (Motoda 1971).

diversity is an index of species richness, which includes qualitative evaluation. The value of the index increases when numerous species occur evenly and decreases when a few species dominate.

A *Q*-mode analysis, which evaluates the similarity between samples (Chiba et al. 2001), was also performed. For the *Q*-mode analysis, abundance data (X: inds. m^{-3}) were first log transformed ($\text{Log}_{10}[X + 1]$). A dissimilarity matrix between each sample was constructed based on the differences in species composition using the Bray-Curtis index (Bray and Curtis 1957). The matrix was analyzed by cluster analysis coupled with the unweighted pair-group method using arithmetic means to classify the samples into several groups with similar community compositions. To minimize the effect of minor species, data from only the 58 most dominant species (i.e., those that composed >10% in total abundance in any sampling layer) were utilized (Table 2). The computer software package BIOSTAT II was employed for this analysis. To verify whether the similarity matrix reflected differences in species composition between groups, each sample was plotted on a two-dimensional map by nonmetric multi-dimensional scaling (NMDS) methods. A close distribution in the NMDS plots indicates a similar community structure. We also conducted a multiple regression analysis to clarify which environmental variables (depth, latitude, and longitude) had a significant relationship with the NMDS plot of the calanoid copepod community. To determine the indicator species for each group, a one-way ANOVA and Fisher's protected least significant difference (PLSD) test were applied to the calanoid copepod abundance data.

Results

Hydrography

At the sampling stations, the surface temperature ranged from 6.0°C to 29.7°C, and salinity ranged from 32.4 to 35.0 (Figure 2). The study area was classified into three domains based on the locations of the subarctic front (where a 4°C vertical isotherm occurred below 100 m) and the subarctic boundary (where a 34.0 isohaline is stretched vertically). The subarctic domain occurred north of the subarctic front, the transition domain occurred between the subarctic front and the subarctic boundary, and the tropical/subtropical domain occurred south of the subarctic boundary (Favorite et al. 1976; Anma et al. 1990). According to the T-S diagrams, stations located at 42°00′N to 56°00′N occurred in the subarctic domain, and stations at 0°02′S to 29°57′N occurred in the tropical/subtropical domain (Figure 2).

Abundance and species diversity

Vertical changes in abundance are shown in Figure 3a. Calanoid copepod abundance ranged from 0.072 to 591.8 inds. m^{-3}; it was the greatest in the subarctic domain (40° to 50°N) and lowest in the tropical/subtropical domain (0° to 10°N). Calanoid copepod abundance decreased exponentially with increasing depth over the entire region (Figure 3a).

Table 2 List of families and species of calanoid copepods identified in this study

Acartiidae	Calanidae	Heterorhabdidae	Pontellidae
Acartia longiremis (Lilljeborg, 1853)	*_Calanus pacificus_ Brodsky, 1948	_Disseta palumbii_ Giesbrecht, 1889	*_Calanopia minor_ Scott A., 1902
Acartia negligens Dana, 1849	*_Canthocalanus pauper_ (Giesbrecht, 1888)	*_Heterorhabdus abyssalis_ (Giesbrecht, 1889)	_Labidocera acutifrons_ (Dana, 1849)
Acartia erythraea Giesbrecht, 1889	*_Cosmocalanus darwinii_ (Lubbock, 1860)	_Heterorhabdus clausi_ (Giesbrecht, 1889)	_Labidocera detruncata_ (Dana, 1849)
Aetideidae	_Megacalanus princeps_ Wolfenden, 1904	_Heterorhabdus norvegicus_ (Boeck, 1872)	_Labidocera japonica_ Mori, 1935
Aetideus armatus (Boeck, 1872)	_Mesocalanus tenuicornis_ (Dana, 1849)	_Heterorhabdus pacificus_ Brodsky, 1950	_Labidocera pavo_ Giesbrecht, 1889
*_Aetideus giesbrechti_ Cleve, 1904	*_Nannocalanus minor_ (Claus, 1863)	_Heterorhabdus papilliger_ (Claus, 1863)	_Pontella chierchiae_ Giesbrecht, 1889
Chiridius poppei Giesbrecht, 1893	*_Neocalanus cristatus_ (Krøyer, 1848)	_Heterorhabdus spinifrons_ (Claus, 1863)	_Pontella denticauda_ Scott A., 1909
*_Chiridius_ sp.	*_Neocalanus flemingeri_ Miller, 1988	*_Heterorhabdus tanneri_ (Giesbrecht, 1895)	_Pontella securifer_ Brady, 1883
Chirundina streetsii Giesbrecht, 1895	*_Neocalanus gracilis_ (Dana, 1852)	_Heterostylites longicornis_ (Giesbrecht, 1889)	_Pontella spinicauda_ Mori, 1937
Chirundina sp.	*_Neocalanus plumchrus_ (Marukawa, 1921)	_Heterostylites major_ (Dahl F., 1894)	*_Pontellina plumata_ (Dana, 1849)
Euchirella bitumida With, 1915	*_Neocalanus robustior_ (Giesbrecht, 1888)	_Paraheterorhabdus compactus_ (Sars G.O., 1900)	Rhincalanidae
Euchirella curticauda Giesbrecht, 1888	*_Undinula vulgaris_ (Dana, 1849)	_Paraheterorhabdus robustus_ (Farran, 1908)	*_Rhincalanus nasutus_ Giesbrecht, 1888
Euchirella messinensis (Claus, 1863)	Candaciidae	Lucicutiidae	*_Rhincalanus rostrifrons_ (Dana, 1849)
Euchirella pulchra (Lubbock, 1856)	_Candacia armata_ Boeck, 1872	_Lucicutia bicornuta_ Wolfenden, 1905	Scolecitrichidae
Euchirella rostrata (Claus, 1866)	_Candacia bipinnata_ (Giesbrecht, 1889)	_Lucicutia clausi_ (Giesbrecht, 1889)	_Amallothrix curticauda_ (Scott A., 1909)
Euchirella truncata Esterly, 1911	*_Candacia bispinosa_ (Claus, 1863)	*_Lucicutia curta_ Farran, 1905	_Amallothrix tenuiserrata_ (Giesbrecht, 1893)
Euchirella venusta Giesbrecht, 1888	*_Candacia catula_ (Giesbrecht, 1889)	*_Lucicutia flavicornis_ (Claus, 1863)	_Amallothrix valida_ (Farran, 1908)
Gaetanus armiger Giesbrecht, 1888	_Candacia columbiae_ Campbell, 1929	_Lucicutia grandis_ (Giesbrecht, 1895)	_Lophothrix frontalis_ Giesbrecht, 1895
Gaetanus brachyurus Sars G.O., 1907	_Candacia curta_ (Dana, 1849)	_Lucicutia longiserrata_ (Giesbrecht, 1889)	_Pseudoamallothrix inornata_ (Esterly, 1906)
Gaetanus brevicornis Esterly, 1906	*_Candacia ethiopica_ (Dana, 1849)	_Lucicutia magna_ Wolfenden, 1903	_Pseudoamallothrix ovata_ (Farran, 1905)
*_Gaetanus brevispinus_ (Sars G.O., 1900)	_Candacia longimana_ (Claus, 1863)	_Lucicutia maxima_ Steuer, 1904	_Pseudoamallothrix profunda_ (Brodsky, 1950)
Gaetanus miles Giesbrecht, 1888	_Candacia simplex_ (Giesbrecht, 1889)	_Lucicutia orientalis_ Brodsky, 1950	*_Racovitzanus antarcticus_ Giesbrecht, 1902
Gaetanus pileatus Farran, 1903	*_Candacia truncata_ (Dana, 1849)	*_Lucicutia ovalis_ (Giesbrecht, 1889)	_Racovitzanus levis_ Tanaka, 1961
Gaetanus tenuispinus (Sars G.O., 1900)	Centropagoidae	_Lucicutia pacifica_ Brodsky, 1950	_Scaphocalanus affinis_ (Sars G.O., 1905)
Gaetanus spp.	_Centropages abdominalis_ Sato, 1913	_Lucicutia polaris_ Brodsky, 1950	_Scaphocalanus brevicornis_ (Sars G.O., 1900)
Pseudochirella obesa Sars G.O., 1920	_Centropages bradyi_ Wheeler, 1900	_Lucicutia sewelli_ Tanaka, 1963	_Scaphocalanus echinatus_ (Farran, 1905)
Pseudochirella obtusa (Sars G.O., 1905)	*_Centropages calaninus_ (Dana, 1849)	Metridiidae	_Scaphocalanus elongatus_ Scott A., 1909
Undeuchaeta intermedia Scott A., 1909	*_Centropages elongatus_ Giesbrecht, 1896	*_Metridia asymmetrica_ Brodsky, 1950	_Scaphocalanus magnus_ (Scott T., 1894)
Undeuchaeta major Giesbrecht, 1888	_Centropages furcatus_ (Dana, 1849)	*_Metridia brevicauda_ Giesbrecht, 1889	_Scaphocalanus major_ (Scott T., 1894)
Undeuchaeta plumosa (Lubbock, 1856)	_Centropages gracilis_ (Dana, 1849)	*_Metridia curticauda_ Giesbrecht, 1889	_Scolecithricella bradyi_ (Giesbrecht, 1888)
Arietellidae	*_Centropages longicornis_ Mori, 1932	*_Metridia okhotensis_ Brodsky, 1950	_Scolecithricella dentata_ (Giesbrecht, 1893)
Arietellus setosus Giesbrecht, 1893	_Centropages violaceus_ (Claus, 1863)	*_Metridia ornata_ Brodsky, 1950	_Scolecithricella globulosa_ Brodsky, 1950
Arietellus simplex Sars G.O., 1905	Clausocalanoidae	*_Metridia pacifica_ Brodsky, 1950	_Scolecithricella longifurca_ (Giesbrecht, 1888)

Table 2 List of families and species of calanoid copepods identified in this study _(Continued)_

Augaptilidae	*_Clausocalanus arcuicornis_ (Dana, 1849)	_Metridia princeps_ Giesbrecht, 1889	*_Scolecithricella minor_ (Brady, 1883)
Augaptilus lamellifer Esterly, 1911	*_Clausocalanus pergens_ Farran, 1926	*_Metridia similis_ Brodsky, 1950	*_Scolecithricella orientalis_ Mori, 1937
Augaptilus longicaudatus (Claus, 1863)	_Microcalanus pusillus_ Sars G.O., 1903	*_Metridia venusta_ Giesbrecht, 1889	_Scolecithricella spinipedata_ Mori, 1937
Augaptilus spp.	_Microcalanus pygmaeus_ (Sars G.O., 1900)	*_Pleuromamma abdominalis_ (Lubbock, 1856)	_Scolecithricella vittata_ (Giesbrecht, 1893)
Centraugaptilus porcellus Johnson M.W., 1936	*_Pseudocalanus minutus_ (Krøyer, 1845)	*_Pleuromamma gracilis_ Claus, 1863	_Scolecithrix danae_ (Lubbock, 1856)
Centroaugaptilus sp.	_Pseudocalanus newmani_ Frost, 1989	_Pleuromamma quadrungulata_ (Dahl F., 1893)	_Scottocalanus longispinus_ Scott A., 1909
Euaugaptilus angustus (Sars G.O., 1905)	Eucalanidae	*_Pleuromamma scutullata_ Brodsky, 1950	_Scottocalanus persecans_ (Giesbrecht, 1895)
Euaugaptilus facilis (Farran, 1908)	*_Eucalanus bungii_ Giesbrecht, 1893	*_Pleuromamma xiphias_ (Giesbrecht, 1889)	_Scottocalanus securifrons_ (Scott T., 1894)
Euaugaptilus gracilis (Sars G.O., 1905)	*_Eucalanus californicus_ Johnson M.W., 1938	Nullosetigeridae	Spinocalanidae
Euaugaptilus gracilis (Sars G.O., 1905)	_Eucalanus elongatus_ (Dana, 1848)	_Nullosetigera bidentata_ (Brady, 1883)	_Monacilla typica_ Sars G.O., 1905
Euaugaptilus grandicornis Sars G.O., 1920	_Pareucalanus attenuatus_ (Dana, 1849)	*_Nullosetigera helgae_ (Farran, 1908)	_Spinocalanus longicornis_ Sars G.O., 1900
Euaugaptilus hecticus (Giesbrecht, 1893)	*_Subeucalanus crassus_ (Giesbrecht, 1888)	_Nullosetigera impar_ (Farran, 1908)	_Spinocalanus_ spp.
Euaugaptilus hyperboreus Brodsky, 1950	*_Subeucalanus mucronatus_ (Giesbrecht, 1888)	_Nullosetigera integer_ (Esterly, 1911)	Temoridae
Euaugaptilus nodifrons (Sars G.O., 1905)	_Subeucalanus subcrassus_ (Giesbrecht, 1888)	Paracalanidae	_Temora discaudata_ Giesbrecht, 1889
Euaugaptilus pacificus Matthews, 1972	_Subeucalanus subtenuis_ (Giesbrecht, 1888)	_Acrocalanus gibber_ Giesbrecht, 1888	_Temora stylifera_ (Dana, 1849)
Euaugaptilus palumboi (Giesbrecht, 1889)	Euchaetidae	*_Acrocalanus gracilis_ Giesbrecht, 1888	Tortanidae
Euaugaptilus propinquus Sars G.O., 1920	_Euchaeta concinna_ Dana, 1849	*_Acrocalanus monachus_ Giesbrecht, 1888	_Tortanus discaudatus_ (Thompson & Scott 1897)
Euaugaptilus squamatus (Giesbrecht, 1889)	_Euchaeta indica_ Wolfenden, 1905	*_Calocalanus pavo_ (Dana, 1852)	
Haloptilus acutifrons (Giesbrecht, 1893)	_Euchaeta longicornis_ Giesbrecht, 1888	_Mecynocera clausi_ Thompson I.C., 1888	
Haloptilus longicirrus Brodsky, 1950	*_Euchaeta marina_ (Prestandrea, 1833)	_Paracalanus parvus_ (Claus, 1863)	
*_Haloptilus longicornis_ (Claus, 1863)	_Euchaeta media_ Giesbrecht, 1888	Phaennidae	
Haloptilus mucronatus (Claus, 1863)	_Euchaeta plana_ Mori, 1937	_Cephalophanes frigidus_ Wolfenden, 1911	
Haloptilus pseudooxycephalus Brodsky, 1950	_Paraeuchaeta californica_ (Esterly, 1906)	_Cornucalanus chelifer_ (Thompson I.C., 1903)	
Haloptilus sp.	*_Paraeuchaeta elongata_ (Esterly, 1913)	_Onchocalanus affinis_ With, 1915	
Bathypontiidae	_Paraeuchaeta propinqua_ (Esterly, 1906)	_Onchocalanus magnus_ (Wolfenden, 1906)	
Temorites elongata (Sars G.O., 1905)	_Paraeuchaeta rubra_ Brodsky, 1950	_Talacalanus greenii_ (Farran, 1905)	
Temorites similis (Tanaka, 1965)	_Paraeuchaeta tonsa_ (Giesbrecht, 1895)	_Xanthocalanus pulcher_ Esterly, 1911	

*Species composing >10% of the total copepod abundance in at least one sample used for cluster and NMDS analyses (cf. Figure 4).

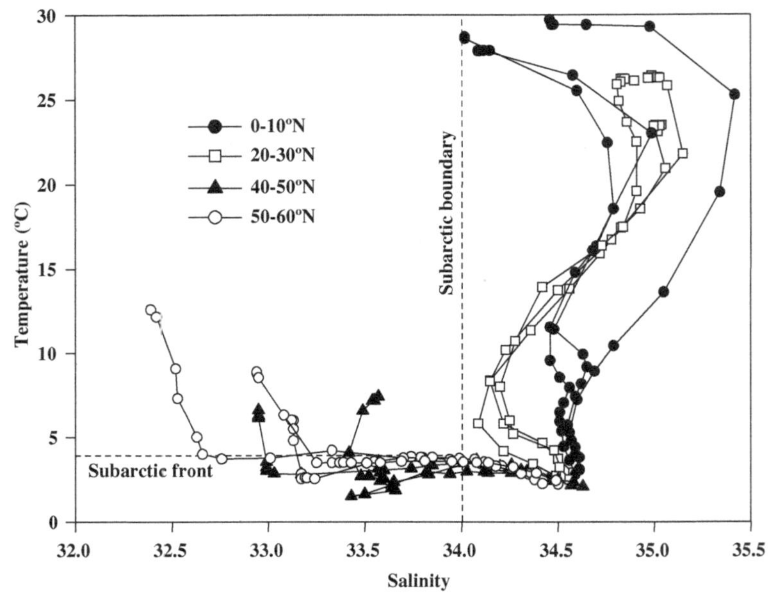

Figure 2 T-S diagram from all stations of this study in the Bering Sea and Pacific Ocean. Horizontal and vertical dashed lines indicate the locations of the subarctic front and boundary, respectively (Favorite et al. 1976; Anma et al. 1990).

Vertical changes in species diversity (*H′*) are shown in Figure 3b. *H′* varied from 0.04 to 3.4; it was low near the surface, high at depths of approximately 500 to 2,000 m, and decreased below 2,000 m. *H′* was high in the tropical/subtropical domain (0° to 10°N, 20° to 30°N) and low in the subarctic domain (40° to 60°N; Figure 3b).

A total of 211 calanoid copepod species belonging to 66 genera and 24 families were identified (Table 2). A total of 58 species composed >10% of the total abundance in at least one sampling layer; therefore, these species were applied in the cluster analysis. Thus, most species (153) were minor species that made up less than 10% of the total abundance in all sampling layers. The highest number of species (86 from 36 families) was observed at St. KH6708 in the subtropical domain, while the lowest number of species (20 from 14 families) was observed at St. OS2301 in the subarctic domain. This low number may have been caused by the limited sampling range (0 to 360 m) at this station (Table 1).

Community structure

Q-mode analysis classified the calanoid copepod community into seven groups (A to G) at 65% and 94% dissimilarity levels (Figure 4a). The number of samples in each group ranged from 4 to 23. Calanoid copepod abundance was greatest in group A with 229 ± 166 inds. m^{-3} (mean ± sd), followed by group E. Species diversity (*H′*) was lowest in group B (1.8 ± 0.3) and greatest in group F (2.4 ± 0.4). The number of species in each group varied between 28 and 121. Groups A to D had fewer species (28 to 37), and groups E to G had more (112 to 121; Figure 4a). In the NMDS plots, the distributions of groups A to D and E to G were clearly separated (Figure 4b). Depth, latitude, and longitude each showed a significant relationship with calanoid copepod group ordination. Group F occurred at low latitudes, groups A and B occurred at high latitudes, and group G occurred at great depths.

The vertical and horizontal distributions of each group are shown in Figure 5. In the subarctic domain, group A occurred at the surface, group B occurred at 100 to 500 m, group C occurred at 500 to 1,000 m, and group D occurred at 500 to 2,250 m. In the tropical/subtropical domain, group E occurred at 0 to 1,000 m, group F occurred at 500 to 1,500 m, and group G occurred at 750 to 2,615 m. The horizontal distributions of each group were clearly separated, and the greatest discrepancy was observed between 30° and 40°N. Group G was the only group collected from deep water in both the subarctic and tropical/subtropical domains.

An ANOVA and Fisher's PLSD test were used to determine indicator species for each group (Table 3). The indicator species were as follows: *Neocalanus plumchrus* and *Pseudocalanus minutus* for group A; *Eucalanus bungii*, *Racovitzanus antarcticus*, and *Scolecithricella minor* for group B; *Neocalanus cristatus* and *N. plumchrus* for group C; *Metridia pacifica*, *Metridia okhotensis*, *Paraeuchaeta elongata*, and *Pleuromamma scutullata* for group D; *Aetideus giesbrechti*, *Clausocalanus arcuicornis*, *Euchaeta marina*, *Cosmocalanus darwinii*, and *Undinula vulgaris* for group E; *Haloptilus longicornis*, *Metridia venusta*, and *Pleuromamma xiphias* for group F; and *Pleuromamma abdminalis* and *Pleuromamma gracilis* for group G.

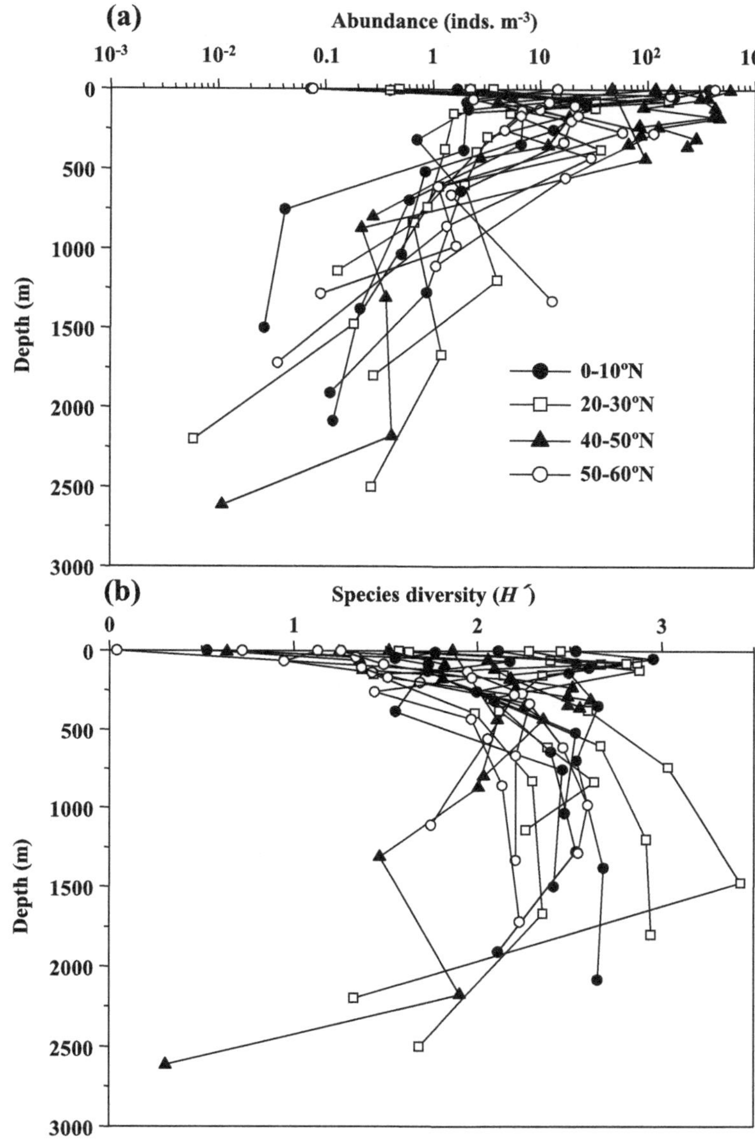

Figure 3 Vertical distribution of calanoid copepod abundance (a) and species diversity (b) in the Bering Sea and Pacific Ocean.

Discussion

Abundance and species diversity

Many studies have shown that zooplankton biomass decreases with depth, but few have examined abundance. This is partly because biomass is easier to measure and contains more useful information. Zooplankton biomass between the surface and 4,000 m in the North Pacific is greater in the subarctic than in the tropical/subtropical domain; in both regions, the biomass decreases with increasing depth (Vinogradov 1962). The rate of biomass decrease with increasing depth is similar in all domains, and the influence of the surface layer is known to extend over 4,000 m into the water column (Vinogradov 1962, 1968). Across the North Pacific, 65% of the zooplankton biomass between depths of 0 and 4,000 m occurs at 0 to 500 m, and this percentage is similar through all regions because zooplankton food in the deep sea depends on sinking particles provided by the upper layer. Therefore, biomass in the deep sea is proportionate to the biomass in the surface layer (Vinogradov and Tseitlin 1983; Vinogradov 1997). Although much is known about biomass, only limited information is available on abundance and species diversity. This is partly because of the difficulty in counting and identifying zooplankton under a microscope, which requires taxonomic knowledge and a significant time investment.

The most remarkable characteristic of the vertical changes in abundance described in this study was that while latitudinal differences are common in biomass (more biomass in the subarctic and less in the subtropical, cf.

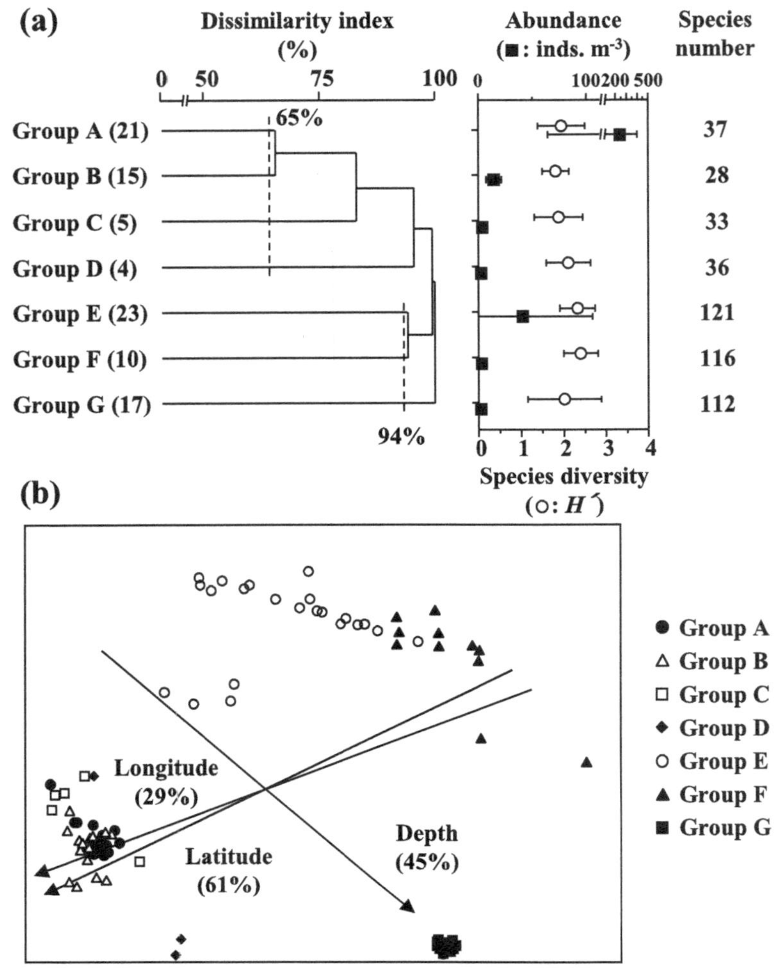

Figure 4 Results of cluster analysis (Bray-Curtis connected with UPGMA) on calanoid copepod community (a) and their NMDS plots (b). Seven groups (A to G) were identified from the cluster analysis, and the number of stations belonging to each group is shown in parentheses. The mean and standard deviation of their abundance and species diversity (cf. Figure 3) were calculated, and the number of species in each group is shown in the right column of (a). Arrows and values in (b) indicate the directions and r^2 values of the environmental parameters that showed a significant correlation with the NMDS ordination.

Yamaguchi et al. 2005), no differences were observed in abundance (Figure 3a). In all domains, calanoid copepods are the most abundant mesozooplankton taxa in terms of biomass (Yamaguchi et al. 2004). The latitudinal differences in zooplankton units may be caused by the latitudinal differences in individual mass. Thus, the individual size of calanoid copepods (=biomass) is expected to be larger in the subarctic domain than in the subtropical domain. In fact, Yamaguchi et al. (2004) reported that the individual size of calanoid copepods is smaller in the subtropics, while large calanoid copepods (i.e., *Neocalanus* spp. and *Eucalanus* spp.) are dominant in the subarctic, composing more than 50% of the mesozooplankton biomass in deep water.

Environmental factors that affect the vertical distribution of zooplankton abundance include the thermocline and the oxygen minimum zone (Sameoto 1986). In zooplankton biomass, a vertical minimum occurs below the thermocline throughout the year near the Kuril-Kamchatka Trench (100 to 200 m) (Bogorov and Vinogradov 1955) and in the Sea of Japan (Morioka and Komaki 1978). In the oxygen minimum zone of the Red Sea, the number of common calanoid copepod species is low, while the number of specialized species is high (Weikert 1982). In the present study, the abundance and species diversity of calanoid copepods did not show a clear decrease in either the thermocline or the oxygen minimum zone (Figure 3a,b). This may be an artifact of the sampling design in this study. Because MTD nets collect samples from irregular layers during simultaneous, multi-layer, horizontal tows, the small-scale vertical distribution changes in the thermocline or oxygen minimum zone would have been difficult to evaluate.

A unique finding of this study was the vertical change in species diversity, which was low at the surface and

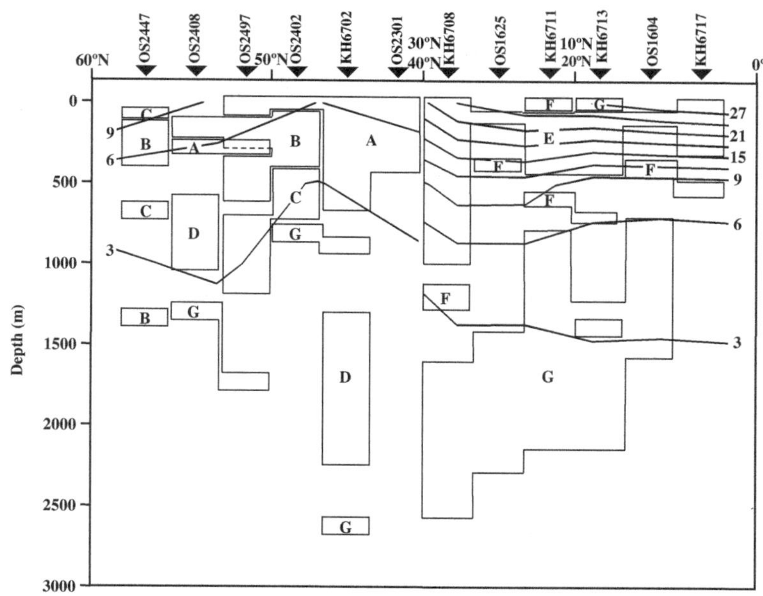

Figure 5 Horizontal and vertical distributions of seven calanoid copepod communities identified by cluster analysis (see Figure 4a). Stations are arranged from north (left) to south (right). Approximate positions of latitudes are shown in the upper margin. Temperature profiles along the latitudinal transect are superimposed.

Table 3 Mean abundance of copepods in each group identified through in-cluster analysis of the copepod community (see Figure 4a)

Species	Group and abundance (log10[inds. m^{-3} + 1])								Results of one-way ANOVA
	A	B	C	D	E	F	G	*F*	
Aetideus giesbrechti	0.00	0.00	0.00	0.00	*0.26*	0.10	0.04	3.37	**
Clausocalanus arcuicornis	0.00	0.00	0.l00	0.00	*0.67*	0.29	0.04	16.70	***
Eucalanus bungii	1.10	*1.68*	1.20	0.51	0.00	0.00	0.17	44.00	***
Euchaeta marina	0.00	0.00	0.00	0.00	*0.65*	0.42	0.11	10.10	***
Haloptilus longicornis	0.00	0.00	0.00	0.00	0.24	*0.40*	0.11	0.01	*
Metridia pacifica	1.03	0.84	0.52	*1.52*	0.00	0.00	0.43	12.10	***
Metridia okhotensis	0.33	0.00	0.00	*0.48*	0.00	0.00	0.06	4.93	***
Metridia venusta	0.00	0.00	0.00	0.00	0.00	*0.24*	0.21	3.90	**
Neocalanus cristatus	0.58	0.56	*0.94*	0.85	0.02	0.00	0.28	11.30	***
Neocalanus plumchrus	*1.13*	0.86	*1.13*	1.07	0.11	0.08	0.28	19.70	***
Paraeuchaeta elongata	0.01	0.11	0.16	*0.33*	0.00	0.00	0.16	2.63	*
Pleuromamma abdminalis	0.00	0.00	0.00	0.00	0.35	1.07	*0.36*	13.50	***
Pleuromamma gracilis	0.00	0.00	0.00	0.00	0.47	1.03	*0.48*	9.97	***
Pleuromamma scutullata	0.32	0.45	0.39	*0.61*	0.00	0.00	0.19	5.70	***
Pleuromamma xiphias	0.00	0.00	0.00	0.00	0.11	*0.35*	0.03	4.43	***
Pseudocalanus minutus	*0.67*	0.53	0.49	0.35	0.00	0.00	0.20	7.70	***
Racovitzanus antarcticus	0.19	*0.44*	0.00	0.02	0.00	0.00	0.20	3.90	**
Scolecithricella minor	0.37	*0.53*	0.14	0.00	0.00	0.00	0.00	11.20	***
Cosmocalanus darwinii	0.00	0.00	0.00	0.00	*0.64*	0.08	0.32	9.20	***
Undinula vulgaris	0.00	0.00	0.00	0.00	*0.36*	0.13	0.04	5.48	***
Total copepods	*2.24*	1.13	0.44	0.26	1.08	0.36	0.08	61.4	***

$*p < 0.05$, $**p < 0.01$, $***p < 0.001$. Groups where the significant greatest abundance was observed were italicized.

peaked at 500 to 2,000 m (Figure 3b). Increases in the number of species and species diversity in the mesopelagic zone (200 to 1,000 m) and bathypelagic zone (1,000 to 3,000 m) have been reported in the North Atlantic (Roe 1972), the Mediterranean Sea (Scotto di Calro et al. 1984), the western North Pacific (Yamaguchi et al. 2002), and the Bering Sea (Homma and Yamaguchi 2010). This suggests that an increase in the number of calanoid copepod species in the meso- and bathypelagic zones is a common phenomenon of the ocean worldwide. The differentiation of calanoid copepod species is reported to strongly influence their feeding habits, and an increased number of species in the meso- and bathypelagic zones are reported to reflect diverse feeding modes in these zones (Ohtsuka and Nishida 1997). In fact, detritivores attached to marine snow do not occur in the surface layer but constitute more than 50% of the total calanoid copepod abundance in the meso- and bathypelagic zones of the western subarctic Pacific and Bering Seas (Yamaguchi et al. 2002; Homma and Yamaguchi 2010). Large sinking particles (e.g., marine snow and giant larvacean house) play an important role as microcosms in the deep sea. The abundance of zooplankton attached to marine snow in the deep sea is reported to be 200 times higher than those in the surrounding waters (Steinberg et al. 1994, 1998).

Community structure

In the North Pacific, calanoid copepod communities in the surface layer generally correspond to the ocean current system and can be separated into subarctic and tropical/subtropical communities. The number of species is greater in the tropical/subtropical than in the subarctic (McGowan and Walker 1979; Ohtsuka and Ueda 1999). Much is known about calanoid copepod communities in the surface layer; however, little is known about those in the deep sea. In the present study, deep-sea calanoid copepods were divided into seven groups (A to G; Figure 4a), and the distribution of each group was clearly separated (Figure 5). Groups A to D occurred north of 40°N and consisted of a small number (28 to 37 species) of species. Groups E to G occurred mainly south of 30°N in the tropical/subtropical domain, and group G occurred in deep water in both the subarctic and tropical/subtropical domains where the temperature was below 3°C (Figure 5). Groups E to G contain a large number (116 to 121 species) of species, indicating that the number of species was high in both shallow and deep water in the tropical/subtropical domain.

The wide distribution of Group G indicates that this group consists of cosmopolitan species that occur widely throughout the uniform environmental condition of the deep sea (Figure 5). Concerning the regional distribution of the deep-sea calanoid copepod genus *Paraeuchaeta*, Park (1994) noted that the cosmopolitan species occurred widely in the oligotrophic region, while endemic species were found in the deep sea of the eutrophic region (e.g., subarctic Pacific and Antarctic). Park suggested that most food of deep-sea zooplankton sinks from the surface and that the low flux of sinking material in the oligotrophic region could maintain only cosmopolitan species; in contrast, the higher flux in the eutrophic region could sustain the endemic species (Park 1994). Group D in the deep sea of the subarctic domain consisted of endemic species in the subarctic Pacific (Table 3; *M. pacifica, M. okhotensis, P. elongata*, and *P. scutullata*) and corresponded to the endemic deep-sea species group observed in the eutrophic region (Park 1994). The uniform low temperature (≤3°C) in the deep sea indicates that there is no barrier to the distribution of deep-sea cosmopolitan species. This is presumably why group G (cosmopolitan species) was the only calanoid copepod community observed in both the subarctic and tropical/subtropical domains (Figure 5).

Some zooplankton have bipolar distributions; for instance, the chaetognath *Eukrohnia hamata* occurs at the surface layer at high latitudes and below 1,000 m near the equator (Pierrot-Bults and Nair 1991). For calanoid copepods, a bipolar distribution is considered to be possible for carnivorous or omnivorous species but may be difficult for herbivorous species that feed on phytoplankton in the epipelagic layer (Machida et al. 2006). In the present study, the species characterizing each group were identified (Table 3), but the limited study area (North Pacific) prevented an evaluation of bipolar distributions.

N. plumchrus was the only species that occurred in all groups (A to G) (Table 3). *N. plumchrus* is a subarctic species that occurs in the surface layer during the spring phytoplankton bloom and descends to deep water during the rest of the year (Miller and Clemons 1988; Kobari and Ikeda 2001). Various authors have reported that the diapause stages C5 and C6 of *Neocalanus* spp. are transported to the deep sea in the subtropical domain by the submerged Oyashio undercurrent (Omori and Tanaka 1967; Oh et al. 1991; Shimode et al. 2006; Kobari et al. 2008a). Within the tropical/subtropical community (groups E, F, and G), the order of depth distribution was E < F < G, and the shallowest group (E) was also occasionally observed at 1,000 m (St. KH6708, Figure 5). Thus, the transported *N. plumchrus* might have occurred in all tropical/subtropical community groups (groups E to G). Concerning the southern limit of transported *Neocalanus* spp., Yamaguchi et al. (2004) reported that they occur at 30°N, but not at 25°N; this agrees with the southern limit of their distribution (28°N) reported by Omori (1967). While *Neocalanus* spp. transported south of 40°N are important in terms of adding additional biomass to the deep-sea subtropical domain (Yamaguchi et al. 2004), this is considered to be an abortive transport, and they are not thought to reproduce in the subtropical domain (Oh et al. 1991).

Conclusions

In deep water, most of the calanoid copepod community consists of cosmopolitan species, while an endemic community was observed in the subarctic region. Because the food of deep-sea calanoid copepods originates on the surface layer, sufficient and excess flux in the eutrophic subarctic region may be responsible for maintaining endemic species in the region.

Competing interests

The authors declare that they have no competing interests.

Authors' contributions

AY and TH wrote the manuscript. KM participated in the design of the study and assisted with the statistical analyses. All authors read and approved the final manuscript.

Acknowledgements

We express our sincere thanks to Dr. Yasuhiro Morioka, who allowed us to use the calanoid copepod data in this study. This study was supported by a Grant-in-Aid for Scientific Research (A) 24248032 and a Grant-in-Aid for Scientific Research on Innovative Areas 24110005 from the Japan Society for the Promotion of Science (JSPS).

Author details

[1]Laboratory of Marine Biology, Graduate School of Fisheries Science, Hokkaido University, 3-1-1 Minatomachi, Hakodate, Hokkaido 041-8611, Japan. [2]Arctic Environmental Research Centre, National Institute of Polar Research, 10-3 Midori-cho, Tachikawa, Tokyo 190-8518, Japan.

References

Anma G, Masuda K, Kobayashi G, Yamaguchi H, Meguro T, Sasaki T, Ohtani K (1990) Oceanographic structures and changes around the transition domain along 180 degree longitude, during June 1979-1988. Bull Fac Fish Hokkaido Univ 41:73–88

Arashkevich EG (1972) Vertical distribution of different trophic groups of copepods in the boreal and tropical regions of the Pacific Ocean. Oceanology 12:315–325

Auel H, Hagen W (2002) Mesozooplankton community structure, abundance and biomass in the central Arctic Ocean. Mar Biol 140:1013–1021

Bogorov V, Vinogradov ME (1955) Some essential features of zooplankton distribution in the Northwestern Pacific Ocean. Trudy Inst Okeanol Acad Nauk USSR 18:113–23 (English translation in US Fish Wildl Serv Spec Sci Rep-Fish 192:6-15, Washington 1956)

Bray JR, Curtis JT (1957) An ordination of the upland forest communities of southern Wisconsin. Ecol Monogr 27:325–349

Chiba S, Ishimaru T, Hosie GW, Fukuchi M (2001) Spatio-temporal variability of zooplankton community structure off east Antarctica (90 to 160°E). Mar Ecol Prog Ser 216:95–108

Fabian H, Koppelmann R, Weikert H (2005) Full-depth zooplankton composition at two deep sites in the western and central Arabian Sea. Indian J Mar Sci 34:174–187

Favorite F, Dodimead AJ, Nasu K (1976) Oceanography of the subarctic Pacific region, 1960-71. Int North Pac Fish Comm Bull 33:1–187

Hernández-León S, Ikeda T (2005) A global assessment of mesozooplankton respiration in the ocean. J Plankton Res 27:153–158

Hokkaido University (1967) Data record of oceanographic observations and exploratory fishing No.11. Fac Fish Hokkaido Univ, Hakodate

Homma T, Yamaguchi A (2010) Vertical changes in abundance, biomass and community structure of copepods down to 3000 m in the southern Bering Sea. Deep-Sea Res I 57:965–977

Hopkins TL, Sutton TT (1998) Midwater fishes and shrimps as competitors and resource partitioning in low latitude oligotrophic ecosystems. Mar Ecol Prog Ser 164:37–45

Kobari T, Ikeda T (2001) Ontogenetic vertical migration and life cycle of *Neocalanus plumchrus* (Crustacea: Copepoda) in the Oyashio region, with notes on regional variations in body sizes. J Plankton Res 23:287–302

Kobari T, Moku M, Takahashi K (2008a) Seasonal appearance of expatriated boreal copepods in the Oyashio-Kuroshio mixed region. ICES J Mar Sci 65:469–476

Kobari T, Steinberg DK, Ueda A, Tsuda A, Silver MW, Kitamura M (2008b) Impacts of ontogenetically migrating copepods on downward carbon flux in the western subarctic Pacific Ocean. Deep-Sea Res II 55:1648–1660

Koppelmann R, Weikert H (1999) Temporal changes of deep-sea mesozooplankton abundance in the temperate NE Atlantic and estimates of the carbon budget. Mar Ecol Prog Ser 179:27–40

Koppelmann R, Weikert H (2005) Temporal and vertical distribution of two ecologically different calanoid copepods (*Calanoides carinatus* Krøyer 1849 and *Lucicutia grandis* Giesbrecht 1895) in the deep waters of the central Arabian Sea. Mar Biol 147:1173–1178

Koppelmann R, Weikert H (2007) Spatial and temporal distribution patterns of deep-sea mesozooplankton in the eastern Mediterranean- indications of a climatically induced shift? Mar Ecol 28:259–275

Kosobokova K, Hirche H-J (2000) Zooplankton distribution across the Lomonosov Ridge, Arctic Ocean: species inventory, biomass and vertical structure. Deep-Sea Res I 47:2029–2060

Longhurst AR (1991) Role of the marine biosphere in the global carbon cycle. Limnol Oceanogr 36:1507–1526

Machida RJ, Miya MU, Nishida M, Nishida S (2006) Molecular phylogeny and evolution of the pelagic copepod genus *Neocalanus* (Crustacea: Copepoda). Mar Biol 148:1071–1079

Madhupratap M, Haridas P (1990) Zooplankton, especially calanoid copepods, in the upper 1000 m of the south-east Arabian Sea. J Plankton Res 12:305–321

McGowan JA, Walker PW (1979) Structure in the copepod community of the North Pacific central gyre. Ecol Monogr 49:195–226

Merrett NR, Roe HSJ (1974) Patterns and selectivity in the feeding of certain mesopelagic fishes. Mar Biol 28:115–126

Miller CB, Clemons MJ (1988) Revised life history analysis for large grazing copepods in the subarctic Pacific Ocean. Prog Oceanogr 20:293–313

Moku M, Kawaguchi K, Watanabe H, Ohno A (2000) Feeding habits of three dominant myctophid fishes, *Diaphus theta*, *Stenobrachius leucopsarus* and *S. nannochir*, in the subarctic and transitional waters of the western North Pacific. Mar Ecol Prog Ser 207:129–140

Morioka Y, Komaki Y (1978) Seasonal and vertical distribution of zooplankton biomass in the Japan Sea. Bull Jpn Sea Reg Fish Res Lab 29:255–267

Motoda S (1971) Devices of simple plankton apparatus V. Bull Fac Fish Hokkaido Univ 22:101–106

Oh B-C, Terazaki M, Nemoto T (1991) Some aspects of the life history of the subarctic copepod *Neocalanus cristatus* (Calanoida) in Sagami Bay, central Japan. Mar Biol 111:207–212

Ohtsuka S, Nishida S (1997) Reconsideration on feeding habits of marine pelagic copepods (Crustacea) (in Japanese with English abstract). Oceanogr Jpn 6:299–320

Ohtsuka S, Ueda H (1999) Zoogeography of pelagic copepods in Japan and its adjacent waters (in Japanese with English abstract). Bull Plankton Soc Jpn 46:1–20

Omori M (1967) *Calanus cristatus* and submergence of the Oyashio water. Deep-Sea Res 14:525–532

Omori M, Tanaka O (1967) Distribution of some cold-water species of copepods in the Pacific water off east-central Honshu, Japan. J Oceanogr Soc Jpn 23:63–73

Park T (1994) Geographic distribution of the bathypelagic genus *Paraeuchaeta* (Copepoda, Calanoida). Hydrobiologia 292(293):317–332

Pierrot-Bults AG, Nair VR (1991) Distribution patterns in Chaetognatha. In: Bone Q, Kapp H, Pierrot-Bults AC (eds) The biology of chaetognaths. Oxford Univ. Press, Oxford

Richter C (1994) Regional and seasonal variability in the vertical distribution of mesozooplankton in the Greenland Sea. Ber Polarforsch 154:1–87

Roe HSJ (1972) The vertical distributions and diurnal migrations of calanoid copepods collected on the SOND cruise, 1965. I The total population and general discussion J Mar Biol Assoc UK 52:277–314

Sameoto DD (1986) Influence of the biological and physical environment on the vertical distribution of mesozooplankton and micronekton in the eastern tropical Pacific. Mar Biol 93:263–279

Schnack-Schiel SB, Michels J, Mizdalski E, Schodlok MP, Schröder M (2008) Composition and community structure of zooplankton in the sea ice-covered western Weddell Sea in spring 2004-with emphasis on calanoid copepods. Deep-Sea Res II 55:1040–1055
Scotto di Calro B, Ianora A, Fresi E, Hure J (1984) Vertical zonation patterns for Mediterranean copepods from the surface to 3000 m at a fixed station in the Tyrrhenian Sea. J Plankton Res 6:1031–1056
Shannon CE, Weaver W (1949) The mathematical theory of communication. The Univ, Illinois Press, Urbana
Shimode S, Toda T, Kikuchi T (2006) Spatio-temporal changes in diversity and community structure of planktonic copepods in Sagami Bay, Japan. Mar Biol 148:581–597
Steinberg DK, Silver MW, Pilskaln CH, Coale SL, Paduan JB (1994) Midwater zooplankton communities on pelagic detritus (giant larvacean houses) in Monterey Bay, California. Limnol Oceanogr 39:1606–1620
Steinberg DK, Pilskaln CH, Silver MW (1998) Contribution of zooplankton associated with detritus to sediment trap 'swimmer' carbon in Monterey Bay, California, USA. Mar Ecol Prog Ser 164:157–166
Steinberg DK, Cope JS, Wilson SE, Kobari T (2008) A comparison of mesopelagic mesozooplankton community structure in the subtropical and subarctic North Pacific Ocean. Deep-Sea Res II 55:1615–1635
Vinogradov ME (1962) Feeding of the deep-sea zooplankton. Rapp P-v Reun Cons Perm Int Exp Mer 153:114–120
Vinogradov ME (1968) Vertical distribution of the oceanic zooplankton. Academy of Science of the USSR, Institute of Oceanography (in Russian, English translation by Israel Program for Scientific Translations). Keter Press, Jerusalem
Vinogradov ME (1997) Some problems of vertical distribution of meso- and macroplankton in the ocean. Adv Mar Biol 32:1–92
Vinogradov ME, Tseitlin VB (1983) Deep-sea pelagic domain (aspects of bioenergetics). In: Rowe GT (ed) The sea vol. 8: deep-sea biology. John Wiley & Sons, New York
Weikert H (1982) The vertical distribution of zooplankton in relation to habitat zones in the area of the Atlantis II Deep, Central Red Sea. Mar Ecol Prog Ser 8:129–143
Weikert H, Koppelmann R (1993) Vertical structural patterns of deep-living zooplankton in the NE Atlantic, the Levantine Sea and the Red Sea: a comparison. Oceanol Acta 16:163–177
Weikert H, Trinkaus S (1990) Vertical mesozooplankton abundance and distribution in the deep Eastern Mediterranean Sea SE of Crete. J Plankton Res 12:601–628
Wishner KF, Gelfman C, Gowing MM, Outram DM, Rapien M, Williams RL (2008) Vertical zonation and distributions of calanoid copepods through the lower oxycline of the Arabian Sea oxygen minimum zone. Prog Oceanogr 78:163–191
Yamaguchi A, Watanabe Y, Ishida H, Harimoto T, Furusawa K, Suzuki S, Ishizaka J, Ikeda T, Takahashi MM (2002) Community and trophic structures of pelagic copepods down to greater depths in the western subarctic Pacific (WEST-COSMIC). Deep-Sea Res I 49:1007–1025
Yamaguchi A, Watanabe Y, Ishida H, Harimoto T, Furusawa K, Suzuki S, Ishizaka J, Ikeda T, Takahashi MM (2004) Latitudinal differences in the planktonic biomass and community structure down to the greater depths in the western North Pacific. J Oceanogr 60:773–787
Yamaguchi A, Watanabe Y, Ishida H, Harimoto T, Maeda M, Ishizaka J, Ikeda T, Takahashi MM (2005) Biomass and chemical composition of net-plankton down to greater depths (0-5800 m) in the western North Pacific Ocean. Deep-Sea Res I 52:341–353

8

Diversity and density of butterfly communities in urban green areas: an analytical approach using GIS

Cheol Min Lee[1*], Jin Wook Park[2], Tae-Sung Kwon[3], Sung-Soo Kim[4], Jae Won Ryu[5], Seung Jae Jung[6] and Sun Kyung Lee[7]

Abstract

Background: Urbanization is one of the leading causes of habitat loss, habitat degradation, and fragmentation. Urban development negatively affects biodiversity. We clarified changes in butterfly communities due to urbanization in urban green areas.

Results: In total, 59 species and 1,465 individuals of butterflies were observed in the four urban green areas – Namsan Park (NS), Ewha Womans University (EW), Bukseoul Dream Forest (BD), and Hongneung Forest (HF) – and a natural forest, Gwangneung Forest (GF). The categories of land use around the study sites were determined based on GIS data. Species richness and density of niche breadth and habitat type in the four urban green areas differed mostly from those in GF. Estimated species richness and species diversity (*H'*) in the four urban green areas were significantly lower than those in GF. Species richness and density of forest interior species and specialist species were positively correlated with paddy, field, and forest, whereas those of forest interior species and specialists were negatively correlated with urban area and road. Species composition and community structure of butterflies in the four urban were differed from those in GF.

Conclusions: These results suggest that decreases in paddy, field, and forest associated with the increased urban area and road negatively influenced species composition and changed butterfly communities.

Keywords: Butterfly; Urbanization; Urban green area; GIS

Background

Urbanization is usually considered to impact biodiversity (Blair and Launer 1997; McKinney 2002, 2006, 2008; Bergerot et al. 2011). Cities are generally grown in terms of human population size over time (Grimm et al. 2008). This growth is accompanied by an expansion of urban areas. Changes in habitat quality and quantity associated with urbanization have negative influences on biodiversity. Plants and animals have frequently experienced local extinction in urban area due to habitat loss, habitat degradation, and fragmentation (McKinney 2002; Clark et al. 2007). Studies on the effects of urbanization on biodiversity in urban areas compared with rural areas have shown different species compositions and less abundance of organisms: e.g., mammals (George and Crooks 2006), amphibians (Osawa and Katsuno 2005; Hamer and McDonnell 2008), birds (Motegi and Yanai 2005; Rubèn and Ian 2009), ground beetles (Niemelä et al. 2002; Lee and Ishii 2009; Lee and Kwon 2013), ants (Yamaguchi 2005; Lee and Kwon 2013), and butterflies (Imai 2004; Lee and Kwon 2012; Bergerot et al. 2011).

Butterflies are a useful bio-indicator because they can be readily surveyed, and they react rapidly to environmental changes due to their short generation time, high mobility, and specific habitat preferences (Clark et al. 2007; Lee et al. 2014). Butterflies are also well known both taxonomically and ecologically (Honda and Kato 2005; Kim et al. 2011). Butterfly communities are greatly affected by vegetation changes because most butterfly larvae have strong associations with host plants, and adults require a specific

* Correspondence: leecheolmin77@gmail.com
[1]Division of Forest Ecology, Korea Forest Research Institute, 57 Hoegi-ro, Dongdaemun-gu, Seoul 130-712, Republic of Korea
Full list of author information is available at the end of the article

range of nectar plants (Thomas 1995; Honda and Kato 2005). Butterflies are known to respond sensitively to the effects of urbanization (Clark et al. 2007; Lee and Kwon 2012; Lizée et al. 2012). An increase in urban features, such as buildings, roads, and population growth, correspond to a decreases in butterfly species richness, diversity, and abundance (Ruszczyk and De Araujo 1992; Blair and Launer 1997; Stefanescu et al. 2004; Dover and Settele 2009). Most butterfly studies on urbanization have investigated environmental variables using qualitative methods, aerial photographs, or Landsat TM data (Ishii et al. 1991; Kitahara and Fujii 1994, 1997; Blair and Launer 1997; Imai 2004; Mauro et al. 2007). However, a GIS method can more concretely analyze the land use conditions. Changes in environmental factors due to urbanization can be evaluated using a GIS method.

Since the 1960s, Korea has urbanized rapidly with economic development and the population growth concentrated in the cities (Kim 2008). Seoul, the capital, is the most urbanized area. Metro Seoul is divided into north and south by the Hangang River. Seoul is surrounded by Mt. Bukhan-san, Mt. Sulak-san, Mt. Kwanak-san, Mt. Cheongryang-san, and Mt. Cheoma-san. Recently, about 46.3% of the Korean population lives in Seoul and the greater metro capital area (Lee 2003). Most areas in Seoul are comprised of houses, apartments, and other buildings. However, there are also large urban green areas, such as ancient palace gardens, university campuses, urban parks, urban forests, and arboretums are scattered around Seoul. Urban green areas remained like green islands separated from other surrounding natural areas. Urban green areas play an important role in urban environments as habitats for organisms as well as recreational areas for humans (McKinney 2008). Urban green areas with significant ecological value should be protected and managed to maintain biodiversity and recreational value.

In this study, we explored the relationships between butterfly diversity and environmental variables based on GIS. Specifically, we sought to answer three questions. (1) How does the butterfly community in an urban green area differ from that in a natural forest? (2) Which species are more vulnerable to the effects of urbanization? (3) Which environmental factors influence (positively or negatively) the butterfly communities? Finally, we suggest an alternative to increase butterfly diversity in urban areas.

Methods

Study sites

We chose four urban green areas and one natural forest as a control (Figure 1). Namsan Park, Bukseoul Dream Forest, and Hongneung Forest are surrounded by apartment complexes, residential areas, and roads. Ewa Womans University is located near to natural forests. Gwangneung forest, a natural forest, is one of the most protected forests in Korea.

Namsan Park (NS: N 37° 33′, E 126° 59′) is a mountain that symbolizes Seoul. NS is one of the most famous

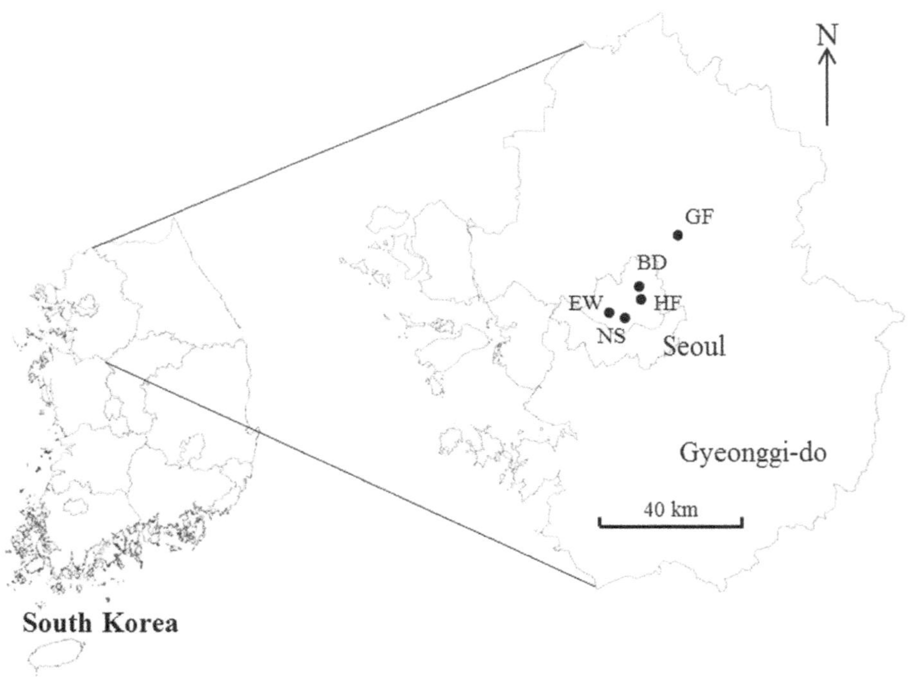

Figure 1 Map of the study sites. NS, Namsan Park; EW, Ewha Womans University; BD, Bukseoul Dream Forest; HF, Hongneung Forest; GF, Gwangneung Forest.

urban parks in Seoul, largely because of the N Seoul Tower (236.7 m). It is located in the heart of Seoul, hanging over Jung-gu and Yongsan-gu. About 10,000 people visit during the slow season and 30,000 during the peak season. In total, 552 species (191 woody plants, 361 herbaceous plants) belonging to 85 families grow on 103 ha (Parks of Seoul 2014). Broadleaf trees account for 76.6% and needle-leaf trees, like pine trees and pine nut trees, account for 23.3%. The major tree species are *Quercus mongolica*, *Robinia pseudoacacia*, and *Prunus serrulata* var. *spontanea* (Parks of Seoul 2014). The butterfly survey in NS was carried out on an asphalt road of about 2.5 km in length.

Ewha Womans University (EW: N 37° 33′, E 126° 56′) is located at Seodaemun-gu in Seoul. EW is located near to Mt. Bukak-san and Mt. An-san and its area is about 59 ha. EW is occupied mainly by buildings and paved roads. However, many gardens and natural forest remnants are managed. The major tree species are *Zelkova serrate*, *Q. mongolica*, and *P. serrulata* var. *spontanea*. The butterfly survey in EW was carried out on an asphalt road of about 2 km in length.

Bukseoul Dream Forest (BD: N 37° 37′, E 126° 2′) is located at Gangbuk-gu and Seongbuk-gu in Seoul. BD is composed of Mt. Opae-san (123 m) and Byeoko-san (135 m) (Dream Forest 2014). The area of BD is about 66 ha. The major tree species are *Q. acutissima*, *Pinus koraiensis*, *P. densiflora*, and *Robinia pseudoacacia*. Large natural forest remnants remain and many gardens, a zoo, and a fountain were created for visitors. The butterfly survey in BD was carried out on a forest road and a dirt road of about 2 km in length.

Hongneung Forest (HF: N 37° 35′, E 126° 2′) is located at Dongdamun-gu in Seoul. HF is an experimental forest established by the Korea Forest Research Institute. The area of HF is about 41 ha. As of 2011, 20,000 plants in 157 different families and 2,035 species (1224 woody plants and 811 herbaceous plants) grow in the 41 ha forest (Cho et al. 2011). The major tree species are *Alnus sibirica*, *Pinus rigida*, and *Robinia pseudoacacia*. In some areas, *Chamaecyparis pisifera* and native trees including *P. koraiensis* and *Abies holophylla* have been planted in a multi-layered structure (Cho et al. 2011). The butterfly survey in HF was carried out on a forest road of about 2.5 km in length.

Gwangneung Forest (GF: N 37° 44′, E 127° 9′) is located at Pocheon-si in Gyeonggi-do, mid-western Korea. The natural forest is older than 100 years and composed mainly of *Quercus serrata* and *Carpinus laxiflora* (Lee et al. 1990). In South Korea, old deciduous forests (>100 years) such as GF are now rare. The butterfly survey in GF was conducted on a forest road of about 2 km in length, and included the 'long-term ecological research' (LTER) site of Korea Forest Research Institute.

Butterfly surveys

The survey was carried out by steadily walking along the survey routes and recording butterflies observed within a 10 m width along the routes, using the line transect method (Yamamoto 1975; Pollard and Yates 1993). This method has been extensively used to survey and monitor butterfly populations and communities (Honda and Kato 2005; Clark et al. 2007; Lee et al. 2014). The survey was conducted during the adult flight season, one time in April and October and twice from May to September 2012. In total, 12 surveys were carried out on each survey route. The butterfly census was carried out from 10:00 a.m. to 14:00 p.m. in clear weather without clouds. When identifying species by sight was difficult, the butterflies were caught using a net, identified, and released.

Habitat type and niche breadth

Butterflies were classified according to their habitat type and niche breadth. Habitat type was determined based on their larva habitats, such as grassland species (GL), forest edge species (FE), and forest interior species (FI). If a species was found in more than one habitat, the habitat was considered the one in which the species was most often observed. Niche breadth modified the method of Kitahara and Fujii (2005) and used diet breadth and voltinism (number of generations per year) to determine the niche breadth of each species. Diet breadth was classified into two categories based on range of host plant species by larvae. Feeding specialists are species feeding on plants of one plant genus, whereas feeding generalists are species feeding on more than one plant family. Univoltine species are classified as seasonal specialists, whereas multivoltine species are classified as seasonal generalists. If a species was a feeding specialist and seasonal specialist, it was considered as specialist species. If a species was a feeding generalist and seasonal generalist, it was considered as generalist species. Intermediate species was species with specialist in one of the two ecological traits. Habitats, voltinism, and the range of host plant species of all butterfly species in South Korea are shown in Kim et al. (2012).

Land use analysis

The land use data were collected from a 1:25,000 digital map from the National Geographic Information Institute. The patterns of the sounding environment of survey routes at the five study sites were analyzed using GIS. Previous studies suggested that circle of radius 5 km was most appropriate for considering the influence of the landscape on butterfly communities (Hanski and Kuussaari 1995; Bergman et al. 2004). However, 5 km radius in the urban green areas in Seoul included some natural forest. Thus, land use within a width of 3 km along the survey route was analyzed. The category of land use was clarified as seven types: paddy, field, park and green space, forest, urban area, road, and river and pond. Urban area includes residential area, industrial area, commercial area, and public facilities. Seven types of land use were modified by the percentage for analysis.

Data analysis

Species richness (number of species) and species diversity were evaluated by 'Estimate S' (Colwell 2005). Species richness was estimated by 'Jack 1' (Heltshe and Forrester 1983; Palmer 1990) and species diversity was estimated by the Shannon diversity index (Shannon and Weaver 1949). Species richness (Jack 1) and species diversity (H') obtained from Estimate S were compared using a two-sample t-test (Zar 1999). The χ^2 test was used to compare differences in species richness and density (abundance per 1 km transect) of habitat type and niche breadth between study sites (Zar 1999). The relationships between categories of land use and the butterfly index were examined using a Pearson correlation analysis. Correspondence analysis (CA) was used for ordination of the butterfly communities. CA ordination used 23 butterfly species that appeared at two or more of the study sites. The number of butterfly individuals was log-transformed (ln N +1) to achieve a normal distribution of the data. The environmental factors used were percentage of paddy, field, park and green space, forest, urban area, road, and river and pond based on the GIS data. CA was conducted using PC-ORD (ver. 5.17) (McCune and Mefford 1999). Statistical analyses were performed using STATISTICA ver. 8.0 (StatSoft Inc. 2004).

Results

In total, 1,281 butterflies belonging to 57 species were observed (Table 1). There were 316 individuals of 38 species in GF, 342 individuals of 20 species in BD, 340 individuals of 19 species in HF, 161 individuals of 18 species in NS, and 122 individuals of 15 species in EW. Density was highest in BD (14.25), followed by GF (13.17), HF (11.33) and NS (5.37), and was lowest in EW (5.08). *Zizeeria maha* was the most abundant species, with 301 individuals (23.5% of all individuals). *Pieris canidia* (14.4%), *Pieris rapae* (11.6%), and *Libythea lepita* (10.4%) were also abundant. However, *Z. maha* and *P. rapae* were not observed in GF.

When habitat type was considered (Figure 2), species richness and density in the four urban green areas differed significantly from those in GF (species richness: χ^2 = 8.92-17.54, df = 2, P = 0.012-0.000; density: χ^2 = 8.70-18.52, df = 2, P = 0.013-0.0001). When niche breadth was compared between study sites (Figure 3), species richness and density in the four urban green areas differed significantly from those in GF (species richness: χ^2 = 7.30-8.16, df = 2, P = 0.026-0.017; density: χ^2 = 6.31-17.43, df = 2, P = 0.042-0.0002), except for species richness between EW and GF (χ^2 = 4.68, df = 2, P = 0.096).

The comparison of species richness and (Jack 1) and species diversity (H') using Estimate S are shown in Figure 4. Estimated species richness was highest in GF (58) and was lowest in EW (20). Estimated species richness in the four urban green areas was significantly lower than in GF (t = 3.87-5.04, df = 26, P < 0.001). Estimated species richness was significantly different between EW and BD (t = −2.37, df = 26, P = 0.026) and EW and HF (t = −2.10, df = 26, P = 0.045). Species diversity was highest in GF (2.41) and lowest in BD (1.81). Species diversity in the four urban green areas was significantly lower than that in GF (t = 2.47-5.35, df = 26, P = 0.02 - < 0.001), except for GF and EW (t = 1.128, df = 26, P = 0.27). Also, species diversity was significantly different between NS and BD (t = −2.20, df = 26, P = 0.037), EW and BD (t = 3.40, df = 26, P = 0.002), and BD and HF (t = −3.19, df = 26, P = 0.004).

The percentage of urban area in NS, EW, BD, and HF was over 70%, whereas that of forest in NS, EW, BD, and HF was below 17% (Table 2). The percentage of road in NS, EW, BD, and HF was over 5.1%, whereas that of GF was 0.1%. Butterfly species richness, and species richness and density of forest interior species and specialist species were positively correlated with paddy, field, and forest, whereas those of forest interior species and specialist species were negatively correlated with urban area and road (Table 3). Density of forest edge species was negatively correlated with park and green space. Species richness of intermediate species was positively correlated with paddy, field, and forest, whereas that of intermediate species was negatively correlated with urban area and road. Similarity relationships of the butterfly communities in the five study sites were visualized using CA ordination (Figure 5). Axis I (92%) and Axis II (5%) explained 97% of the total variation. Butterfly communities were separated along Axis I. The butterfly communities in the four urban green areas differed from that in GF and were more similar in BD, NS, and HF.

Discussion

Our data show that butterfly species richness was negatively correlated with increasing urbanization (urban area and roads), whereas it was positively correlated with paddy, field, and forest. Species richness and density of niche breadth and habitat type in the urban green areas differed mostly from those in GF, a natural forest. Species richness and density of forest interior species and specialist species were positively correlated with paddy, field, and forest.

In our study, 26 species, such as *Parnassius stubbendorfii*, *Luehdorfia puziloi*, *Wagimo signatus*, were observed only in GF. Among them, specialist species accounted for 57.7% and forest interior species accounted for 73.1%.

Table 1 Butterflies observed at five study sites

Species	NS	EW	BD	HF	GF	Total	%	Habitat type	Niche breadth
Parnassius stubbendorfii					0.04 (1)	0.01 (1)	0.1	GL	S
Luehdorfia puziloi					0.5 (12)	0.09 (12)	0.9	FI	S
Papilio xuthus	0.13 (4)	0.33 (8)	0.21 (5)	0.03 (1)		0.14 (18)	1.4	GL	G
Papilio macilentus				0.07 (2)		0.02 (2)	0.2	FE	G
Papilio bianor	0.13 (4)	0.04 (1)		0.07 (2)	0.13 (3)	0.08 (10)	0.8	FI	G
Papilio maackii					0.04 (1)	0.01 (1)	0.1	FI	G
Pieris melete	0.23 (7)	1.04 (25)	0.21 (5)	0.57 (17)	0.88 (21)	0.57 (75)	5.9	FE	G
Pieris canidia	1.90 (57)		1.17 (28)	3.10 (93)	0.25 (6)	1.39 (184)	14.4	FE	G
Pieris rapae	0.83 (25)	0.38 (9)	2.00 (48)	2.20 (66)		1.12 (148)	11.6	GL	G
Anthocharis scolymus	0.17 (5)		0.21 (5)	0.53 (16)	0.08 (2)	0.21 (28)	2.2	FE	I
Colias erate	0.10 (3)	0.21 (5)	0.38 (9)			0.13 (17)	1.3	GL	G
Spindasis takanonis		0.04 (1)				0.01 (1)	0.1	FE	S
Coreana raphaelis					0.04 (1)	0.01 (1)	0.1	FI	S
Japonica lutea					0.04 (1)	0.01 (1)	0.1	FI	S
Antigius butleri					0.04 (1)	0.01 (1)	0.1	FI	S
Wagimo signatus					0.04 (1)	0.01 (1)	0.1	FI	S
Favonius orientalis	0.13 (4)			0.10 (3)		0.05 (7)	0.5	FI	S
Favonius taxila					0.13 (3)	0.02 (3)	0.2	FI	S
Satyrium prunoides		0.04 (1)				0.01 (1)	0.1	FE	S
Callophrys ferreus	0.03 (1)			0.03 (1)		0.02 (2)	0.2	FE	I
Rapala caerulea	0.03 (1)		0.13 (3)	0.17 (5)		0.07 (9)	0.7	FE	G
Lycaena dispar		0.04 (1)				0.01 (1)	0.1	GL	I
Lycaena phlaeas		0.08 (2)				0.02 (2)	0.2	GL	I
Zizeeria maha	0.90 (27)	0.67 (16)	7.38 (177)	2.70 (81)		2.28 (301)	23.5	GL	I
Cupido argiades		0.46 (11)	0.04 (1)	0.07 (2)		0.11 (14)	1.1	GL	G
Tongeia fischeri	0.03 (1)		0.25 (6)	0.20 (6)	0.04 (1)	0.11 (14)	1.1	GL	G
Celastrina argiolus	0.23 (7)		1.08 (26)	0.83 (25)	0.08 (2)	0.45 (60)	4.7	FE	G
Libythea lepita	0.07 (2)		0.04 (1)	0.20 (6)	5.17 (124)	1.01 (133)	10.4	FI	S
Parantica sita					0.04 (1)	0.01 (1)	0.1	FE	G
Mycalesis francisca					0.25 (6)	0.05 (6)	0.5	FE	G
Mycalesis gotama					0.04 (1)	0.01 (1)	0.1	FE	G
Minois dryas			0.17 (4)			0.03 (4)	0.3	GL	I

Table 1 Butterflies observed at five study sites *(Continued)*

Argynnis paphia					0.33 (8)	0.06 (8)	0.6	FE	S
Argynnis laodice		0.04 (1)			0.04 (1)	0.02 (2)	0.2	GL	S
Argynnis ruslana	0.27 (8)		0.17 (4)	0.27 (8)	0.46 (11)	0.23 (31)	2.4	GL	S
Neptis sappho			0.04 (1)		0.08 (2)	0.02 (3)	0.2	FE	G
Neptis philyra					0.25 (6)	0.05 (6)	0.5	FI	I
Neptis philyroides					0.17 (4)	0.03 (4)	0.3	FI	I
Neptis speyeri					0.21 (5)	0.04 (5)	0.4	FI	S
Neptis alwina					0.08 (2)	0.02 (2)	0.2	FE	I
Neptis thisbe					0.04 (1)	0.01 (1)	0.1	FI	S
Neptis ilos					0.13 (3)	0.02 (3)	0.2	FI	S
Chitoria ulupi					0.88 (21)	0.16 (21)	1.6	FI	S
Dilipa fenestra					0.04 (1)	0.01 (1)	0.1	FI	S
Hestina assimilis					0.21 (5)	0.04 (5)	0.4	FI	I
Sasakia charonda					1.92 (46)	0.35 (46)	3.6	FI	S
Sephisa princeps					0.08 (2)	0.02 (2)	0.2	FI	I
Vanessa cardui		0.58 (14)	0.17 (4)			0.14 (18)	1.4	GL	G
Vanessa indica	0.10 (3)	0.13 (3)				0.05 (6)	0.5	GL	G
Polygonia c-aureum	0.03 (1)	1.00 (24)	0.17 (4)	0.03 (1)	0.04 (1)	0.23 (31)	2.4	GL	I
Burara aquilina					0.04 (1)	0.01 (1)	0.1	FI	I
Lobocla bifasciata				0.03 (1)		0.01 (1)	0.1	FE	I
Satarupa nymphalis					0.04 (1)	0.01 (1)	0.1	FI	S
Daimio tethys			0.13 (3)			0.02 (3)	0.2	FE	I
Erynnis montanus			0.04 (1)		0.25 (6)	0.05 (7)	0.5	FE	S
Thymelicus leoninus					0.04 (1)	0.01 (1)	0.1	GL	I
Parnara guttata	0.03 (1)		0.29 (7)	0.13 (4)		0.09 (12)	0.9	GL	G
Species richness	18	15	20	19	38	57			
Density	5.37 (161)	5.08 (122)	14.25 (342)	11.33 (340)	13.17 (316)	9.70 (1281)			

Density indicate abundance per 1 km transect. The numbers in parentheses indicate abundance. See text for classification methods of habitat type and niche breadth. Study sites, NS: Namsan Park, EW: Ewha Womans University, BD: Bukseoul Dream Forest, HF: Hongneung Forest, GF: Gwangneung Forest. Habitat type, GL: grassland species, FE: forest edge species, FI: forest interior species. Niche breadth, S: specialist species, I: intermediate species, G: generalist species.

Table 2 Area (ha) and percentage of land use categories within the range of 3 km from study route of five study sites

Land use category	Unit	NS	EW	BD	HF	GF
Paddy	ha	0	0	0	0	141.5
	%	0	0	0	0	4.5
Field	ha	0	0	0.8	0.5	231.7
	%	0	0	0	0	7.4
Park and	ha	161.4	151.7	152.8	151.2	130.1
green space	%	4.1	4.3	4.1	3.9	4.1
Forest	ha	392.5	485.9	625.9	403.9	2524.1
	%	9.9	13.8	16.9	10.3	80.5
Urban	ha	2855.7	2523	2681.7	3038.9	80.7
	%	71.8	71.7	72.3	77.4	2.6
Road	ha	282	254	188.2	284.5	2.3
	%	7.1	7.2	5.1	7.2	0.1
River and pond	ha	285.9	102.3	59.9	46.6	25.2
	%	7.2	2.9	1.6	1.2	0.8

Clark et al. (2007) showed that rare and specialized species were most affected by urbanization and the decrease in butterfly species richness was associated with increased urban areas. Butterfly species richness was lower as urban intensity became higher along an urban–rural gradient (Ruszczyk and De Araujo 1992; Fortunato and Ruszczyk 1997; Stefanescu et al. 2004; Posa and Sodhi 2006; Clark et al. 2007). Our results were consistent with those previous studies. Stefanescu et al. (2004) found loss of butterfly species richness was detectable in highly humanized landscapes although good butterfly habitats predominated along transects. The patterns of land use in urban areas contributed to declines in butterfly species richness (Casner et al. 2014). Posa and Sodhi (2006) reported that urbanization was detrimental to forest butterfly species. Butterflies specialized in both host plants and timing of reproduction disappeared with declining richness across sites 2.9-4.5 times faster than did generalists and less restricted specialists (Clark et al. 2007). Our study showed that species richness and density of specialist species in the four urban green areas were significantly lower versus those of specialist species in GF. Butterfly species with few larval host plants are more at risk than generalists, which can use a wider range of resources. Specialist species will be restricted to areas where they can lay their eggs because they are highly dependent on their host plants for survival. Six of seven *Neptis* species were found only in GF and they are specialist species and forest interior species except for *Nepis alwina*. *Nepits* species may be vulnerable in urban green area. One cause of the decline of *Nepits* species in urban green areas may be loss

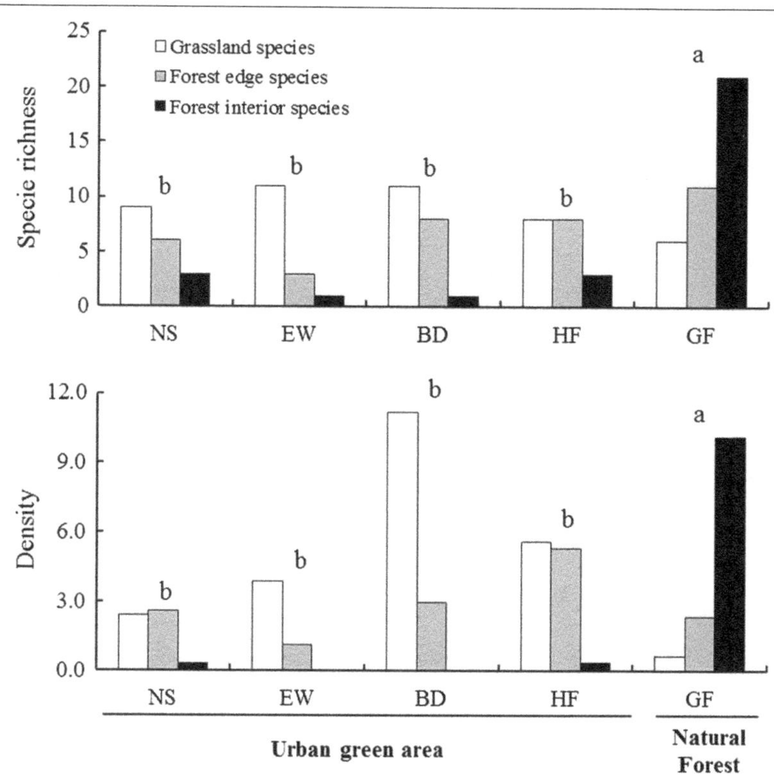

Figure 2 Species richness and density of habitat type of butterfly observed at five study sites. Species richness (number of species) and density (abundance per 1-km transect) of habitat type (forest interior species, forest edge species, and grassland species) of butterfly observed at five study sites. Different letters above the bars indicate significant difference.

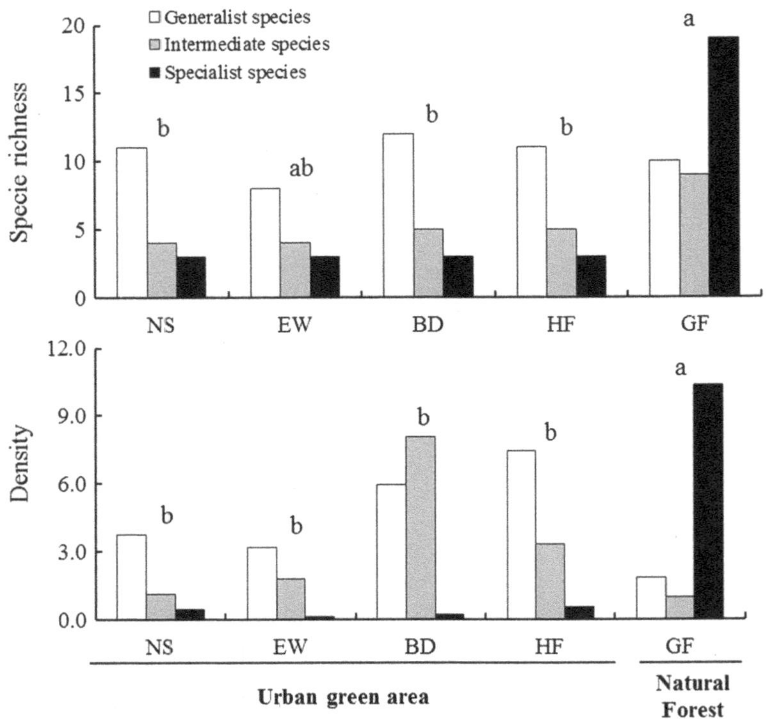

Figure 3 Species richness and density of niche breadth of butterfly observed at five study sites. Species richness (number of species) and density (abundance per 1-km transect) of niche breadth (specialist species, intermediate species, and generalist species) of butterfly observed at five study sites. Different letters above the bars indicate significant difference.

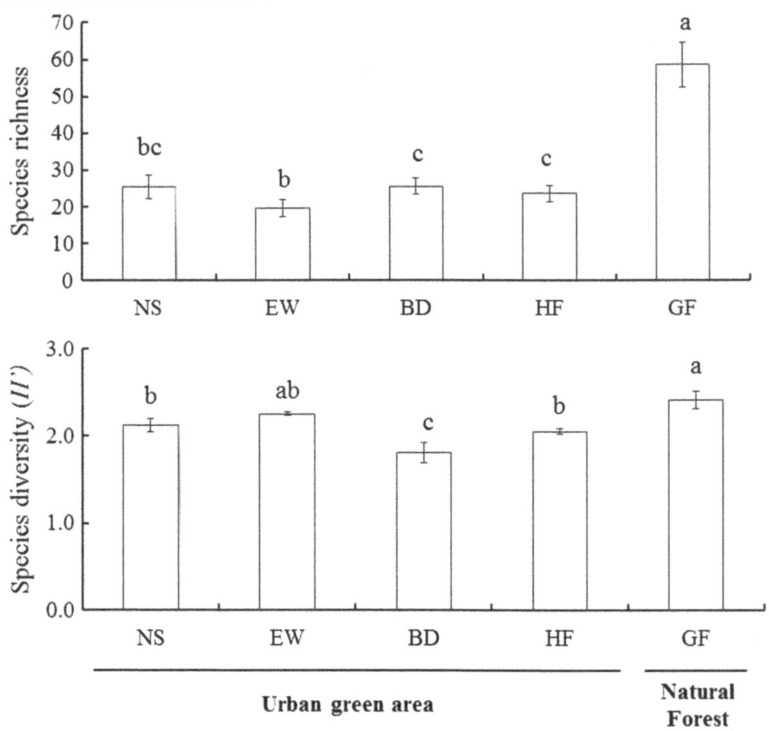

Figure 4 Estimation of species richness (Jack 1) and species diversity (*H'*) of butterfly at five study sites. Species richness and species diversity were obtained using Estimate S (Colwell et al. 2004). Different letters above the error bars indicate significant difference. The error bars indicate one standard deviation.

Table 3 Pearson correlation coefficient between butterfly index and land use category within the range of 3 km from study route of five study sites

Community index	Paddy	Field	Park and green space	Forest	Urban area	Road	River and pond
Species richness							
All species	0.98**	0.98**	0.02	0.98**	−0.97**	−0.97**	−0.45
Habitat type							
Forest interior species	0.99**	0.99**	0.07	0.99**	−0.99**	−0.93*	−0.37
Forest edge species	0.72	0.72	−0.43	0.72	−0.69	−0.79	−0.51
Grassland species	−0.79	−0.79	0.40	−0.74	0.75	0.65	0.25
Niche breadth							
Specialist species	1.00***	1.00***	0.17	1.00***	−1.00***	−0.96*	−0.42
Intermediate species	0.97**	0.97**	0.02	0.97**	−0.96*	−0.97**	−0.57
Generalist species	−0.15	−0.14	−0.66	−0.14	0.17	−0.03	0.02
Density							
All species	0.42	0.43	−0.32	0.47	−0.40	−0.61	−0.76
Habitat type							
Forest interior species	1.00***	1.00***	0.14	0.99**	−1.00***	−0.95*	−0.41
Forest edge species	−0.18	−0.17	−0.97**	−0.21	0.24	0.17	−0.28
Grassland species	−0.56	−0.56	−0.14	−0.50	0.56	0.31	−0.27
Niche breadth							
Specialist species	1.00***	1.00***	0.13	0.99**	−0.99**	−0.95*	−0.41
Intermediate species	−0.39	−0.39	−0.13	−0.32	0.39	0.12	−0.36
Generalist species	−0.66	−0.66	−0.71	−0.66	0.70	0.57	−0.18

$^{*}P < 0.05$; $^{**}P < 0.01$; $^{***}P < 0.001$.

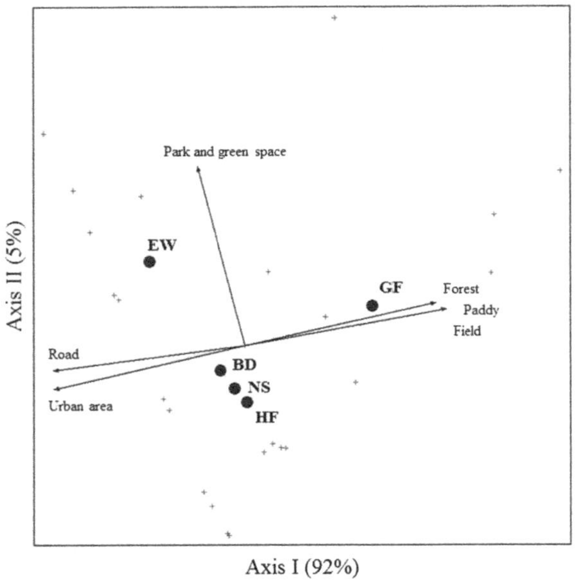

Figure 5 CA ordination of butterfly communities. Two axes explain 97% of total variation. Singleton species which occurred one site were excluded in CA ordination.

of their larval host plants. Host plants of *N. philyroides* and *N. speyeri* are *Carpinus cordata,Carpinus laxiflora*, and *Corylus heterophylla* var. *thunbergii* (Kim and Seo 2012). These trees were not located in urban green areas, Korea (S-SK, personal observation). In GF, *Carpinus cordata* and *Carpinus laxiflora* are one of dominant trees (Lee et al. 1990). Host plants of *N. thisbe* and *N. ilos* are *Quercus serrate* and *Quercus* sp., respectively. These trees were located in urban areas. However, Two *Neptis* species were not found in urban green areas because they may favour low temperature habitats as forests compared with urban green areas (S-SK, personal observation).

In Osaka City, Japan, the numbers of butterfly species richness decreased from 50 in the early 1930s to 30 in the late 1980s because of the effects of urbanization (Imai 1998). Imai (1998) found that the decrease in species richness was remarkable in univoltine and/or forest-preferring species groups. Similarly, the proportion of generalist species richness increased with the developing urban areas in Tsukuba City, Japan (Kitahara and Fujii 1994, 1997). Our study found that the proportion of generalist species richness was more than 53% in the urban green areas. Additionally, some studies have shown that butterfly species richness peaked in sites with intermediate disturbances (Blair and Launer 1997; Blair 2001; Hogsden and Hutchinson 2004). Our study was conducted relatively large urban green areas and a natural forest in a rural area. Further study will need to study butterflies in urban green areas of various sizes in order to clarify the effect of intermediate disturbances.

Responses to the effects of urbanization differ among butterflies. Hiura (1973) found that forest species, sedentary butterflies, univoltine species, and wild-tree feeders were greatly influenced by urbanization. In Osaka City, Japan, drastic decreases in species richness were first observed in Hesperiidae between the 1930s and 1950s, second in Satyridae between the 1950s and 1960s, and third in Pieridae between the 1970s and 1980s (Imai 1998). Species richness in Hesperiidae was most markedly decreased in urban green areas, suggesting that this family is most sensitive to habitat isolation through urbanization, except highly migratory species, such as *Parnara guttata* (Imai 2004). In our study, Hesperiid species richness in the urban green areas was lower than that in GF. *P. guttata* was observed in NS, BD, and HF. However, *Burara aquiline, Satarupa nymphalis*, and *Thymelicus leoninus* were only observed in GF. These three species have been not observed in urban green areas in Korea (Kim et al. 2012). The response of Satyridae to urbanization showed a similar trend with that of Hesperiidae (Imai 2004). In our study, *Mycalesis francisca* and *M. gotama* were absent from the four urban green areas. However, the Korean distribution of the two species has broadened recently because of vegetation changes due to the success of reforestation (Kwon et al. 2012). Ishii et al. (1991) noted that Satyrid species disappear in urban green areas because they are mainly forest species, sedentary butterfly, and wild-tree feeders. Furthermore, Ishii (2001) showed that the percentages of bamboo-grass feeders, univoltine species, and non-migratory species were lower in urban parks compared with those in Satoyama. However, urban landscapes often have high species diversity, even including rare and threatened species (Shepherd 1994). *Spindasis takanonis*, an endangered Korean species, has been observed in the Hongneung Forest (Lee and Kwon 2012). This species lives mainly at the forest edge of urban and suburban areas in Korea. Lee and Kwon (2012) suggested that there is a case that urban green areas play important roles in the conservation of rare species.

In contrast, Hiura (1973) suggested that grassland species, cultivated plant feeders, multivoltine species, and migratory species will be remained in urban areas. Papilionidae, Pieridae, and Lycaenidae did not differ greatly between urban and suburban areas, and species richness of the three families increased from small urban green areas to large urban green areas (Imai 1998). In the present study, *P. rapae, Papilio xuthus*, and *Z. maha* were observed in the four urban green areas, but not GF. These three species are included in the 'strong' species for urbanization suggested by Hiura (1973). *P. rapae* was one of the most common species in urban green areas in Korea (CML, personal observation). Kocher and Williams (2000) noted that *P. rapae* was found in higher abundance in disturbed than that in undisturbed habitats. Takami et al. (2004) suggested that *P. rapae* could tolerate high temperatures caused by urban warming, and could use ephemeral habitats within cities, such as gardens and vacant land. *P. xuthus* was observed in all urban parks of various sizes (2.0-126 ha) in and around Osaka City, Japan (Imai 1998). *P. xuthus* did not decrease during the grassland removal and deforestation period in Kanazawa Castle Park, Japan (Ohwaki et al. 2008). It is thought that *P. xuthus* may tolerate disturbances by humans because this species has a high dispersal ability and a broad range of food resources. In the present study, *Z. maha* was the most dominant species. Ishii et al. (1991) showed that *Z. maha* was a dominant species in urban parks, which had large areas of grassland. In Korea, this species was often observed on grassland in the urban green areas (JWR, personal observation). The host plant of *Z. maha* is *Oxalis corniculata*, which is very common weed in urban area and cities and is located nationally in Korea (Lee 2013). Recently, *Z. maha* seems to be abundant because this species can now survive the milder winters due to climate change in the central part of Korea (CML, unpublished data).

Butterfly species richness decreased as the distance from the nearest mountain increased (Imai 1998). In the present study, EW was located near to Mt. Ansan and Mt. Inwansan. However, butterfly species richness was lowest there, in this study. The butterfly community in EW was separated from that in BK, HF, and NS. Although there were gardens and trees, the butterfly community seemed to be negatively influenced by the many buildings and asphalt roads. The relationships between butterfly diversity and the influence of built elements have been shown to be significant (Lizée et al. 2012). In our study, butterfly species richness, species richness of forest interior species, specialist species, and intermediate species and density of forest interior species and specialist species were negatively correlated with urban area and road.

Conclusion

In conclusion, our results suggest that the decrease in paddy, field, and forest associated with the increased urban area and road negatively influenced species composition and changed butterfly communities. Biodiversity in urban area decreases due to the increase of areas such as residential area, industrial area, and commercial area associated with the decrease of natural environment. Urban areas homogenize the physical environment to satisfy the needs of humans. Therefore, improving biodiversity is important for comfortable human living in urban areas. We propose to plant trees and make biotope or butterfly garden in urban parks, factory sites or universities in order to increase biodiversity.

Competing interests
The authors declare that they have no competing interests.

Authors' contributions
CML designed the experiment for the present study, wrote the manuscript, and surveyed butterfly. JWP and T-SK analyzed the data. S-SK, JWR, SJJ, and SKL surveyed butterfly. All authors read and approved the final manuscript.

Author details
[1]Division of Forest Ecology, Korea Forest Research Institute, 57 Hoegi-ro, Dongdaemun-gu, Seoul 130-712, Republic of Korea. [2]Department of Landscape Architecture, Catholic University of Daegu, 13-13 Hayang-ro, Hayang-eup, Gyeongsan-si, Daegu 712-702, Republic of Korea. [3]Division of Forest Insect Pests and Diseases, Korea Forest Research Institute, 57 Hoegi-ro, Dongdaemun-gu, Seoul 130-712, Republic of Korea. [4]Research Institute for East Asian Environment and Biology, 293-27 Amsa 3 dong, Gangdong-gu, Seoul 143-203, Republic of Korea. [5]Ewha Womans University Natural History Museum, 52 Ewhayeodae-gil, Seodaemun-gu, Seoul 120-750, Republic of Korea. [6]Department of Biology, Kyung Hee University, 26 Kyungheedae-ro, Dongdaemun-gu, Seoul 130-701, Republic of Korea. [7]Entomology program, Department of Agricultural Biotechnology, Seoul National University, Seoul, Republic of Korea.

References

Bergerot B, Fontaine B, Julliard R, Baguette M (2011) Landscape variables impact the structure and composition of butterfly assemblages along an urbanization gradient. Landscape Ecol 26:83–94

Bergman KO, Askling J, Ekberg O, Ignell H, Wahlman H, Milberg P (2004) Landscape effects on butterfly assemblages in an agricultural region. Ecography 27:619–628

Blair RB (2001) Birds and butterflies along urban gradients in two ecoregions of the U.S. In: Lockwood JL, McKinney ML (eds) Biotic Homogenization. Kluwer, New York, pp 33–56

Blair RB, Launer AE (1997) Butterfly diversity and human land use: species assemblages along an urban gradient. Biol Conserv 80:113–125

Casner KI, Forister ML, O'brien JM, Thorne J, Waetjen D, Shapiro AM (2014) Contribution of urban expansion and a changing climate to decline of a butterfly fauna. Conserv Biol DOI:10.1111/cobi.12241

Cho JH, Park C-R, Kim S-K (2011) Hongneung Forest. Korea Forest Research Institute News 11–09

Clark PJ, Reed JM, Chew FS (2007) Effects of urbanization on butterfly species richness, guild structure, and rarity. Urban Ecosyst 10:321–337

Colwell RK (2005) Estimate S: Statistical estimation of species richness and shared species from sample. Version 7.5 User's Guide and application published at: [Cited 8 March 2005] Available from URL: http://viceroy.eeb.uconn.edu/estimates

Colwell RK, Mao CX, Chang J (2004) Interpolating, extrapolating, and comparing incidence-based species accumulation curves. Ecology 85:2717–2727

Dover J, Settele J (2009) The influences of landscape structure on butterfly distribution and movement: a review. J Insect Conserv 13:3–27

Dream Forest (2014) Dream Forest. Available at: http://dreamforest.seoul.go.kr

Fortunato L, Ruszczyk A (1997) Communities of fruit-feeding butterflies in urban and extraurban green areas of Uberlandia, Minas Gerais. Rev Bras Biol 57:79–87

George SL, Crooks KR (2006) Recreatin and large mammal activity in an urban nature reserve. Biol Conserv 133:107–117

Grimm NB, Faeth SH, Golubiewski NE, Redman CL, Wu J, Bai X, Briggs JM (2008) Global change and the ecology of Cities. Sciences 319:756–760

Hamer AJ, McDonnell MJ (2008) Amphibian ecology and conservation in the urbanizing world: a review. Biol Conserv 141:2432–2449

Hanski I, Kuussaari M (1995) Butterfly metapopulation dynamics. In: Cappuccino N, Price PW (eds) Population dynamics: new approaches and synthesis.. Academic Press, London, pp 149–171

Heltshe JF, Forrester NE (1983) Estimating species richness using the jackknife procedure. Biometrics 39:1–12

Hiura I (1973) Butterflies across the Sea. Soju-Shobo, Tokyo (in Japanese)

Hogsden KL, Hutchinson TC (2004) Butterfly assemblages along a human disturbance gradient in Ontario, Canada. Can J Zool 82:739–748

Honda K, Kato Y (2005) Biology of Butterflies. University of Tokyo Press, Tokyo, p 626pp
Imai C (1998) An ecological study for enrichment of biological diversity in urban areas. Jpn J Environ Entomol Zool 9:55–73
Imai C (2004) Ecological approaches for improvement of biological diversity in Osaka City, central Japan. Osaka and its Technol 44:13–22
Ishii M (2001) Studies on insects and their habitats in the Satoyama in a broader sense. Jpn J Environ Entomol Zool 12:187–193
Ishii M, Yamada M, Hirowatari T, Yasuda T (1991) Diversity of butterfly communities in urban parks in Osaka Prefecture. Jpn J Environ Entomolo Zool 3:183–195
Kim DS (2008) Geomorphological background and urbanized feature of Seoul. Ph.D. Thesis of Korea National University of Education. 180 pp
Kim S-S, Lee CM, Kwon T-S, Joo HZ, Sung JH (2012) Korean butterfly atlas 1996–2011. Research Note 461, Korea Forest Research Institute, Korea Disabled Human Good Life Pub. Co, Seoul, in Korean
Kim S-S, Seo YH (2012) Life histories of Korean butterflies. Publishing Sagyejel, Paju, 538 pp
Kitahara M, Fujii K (1994) Biodiversity and community structure of temperate butterfly species within a gradient of human disturbance: an analysis based on the concept of generalist vs. Specialist strategies. Res Popul Ecol 36:187–199
Kitahara M, Fujii K (1997) An island biogeographic approach to the analysis of butterfly community patterns in newly designed parks. Res Popul Ecol 39:23–35
Kitahara M, Fujii K (2005) Analysis and understanding of butterfly community composition based on multivariate approaches and the concept of generalist/specialist strategies. Entomol Sci 8:137–149
Kocher SD, Williams EH (2000) The diversity and abundance of North American butterflies vary with habitat disturbance and geography. J Biogeogr 27:785–794
Kwon T-S, Lee CM, Kim S-S, Sung JH (2012) Distribution change of Korean butterflies 1938–2011. Research Note 472. Korea Forest Research Institute. Samsung Adcom Pub. Co, Seoul, in Korean
Lee CM, Ishii M (2009) Species diversity of ground beetle assemblages at urban greeneries in southern Osaka, central Japan. Jpn J Environ Entomol Zool 20:47–58
Lee CM, Kwon T-S (2012) Characterization of the butterfly community of a fragmented urban forest, Hongneung Forest. Korean J Appl Entomol 51:317–323
Lee CM, Kwon T-S (2013) Community structure, species diversity of insects (ants, ground beetle), and forest health in the Hongneung Forest. J Korean For Soc 102:97–106
Lee CM, Kwon T-S, Kim S-S, Sohn J-D, Lee B-W (2014) Effects of forest degradation on butterfly communities in the Gwangneung forest. Entomol Sci 17:293–301
Lee DH (2013) Korean wild flowers. Ibirak Pub. Co, Seoul, p 581pp
Lee KJ, Jo JC, Lee BS, Lee DS (1990) The structure of plant community in Kwangnung Forest (I). J Korean For Soc 79:173–186
Lee S-K (2003) The change of migration in the Seoul metropolitan region, 1990–2003. MD. Department of Geography Education Graduate school of Education Sungshin Women's university
Lizée M-H, Manel S, Mauffrey JF, Tatoni T, Deschamps-Cottin M (2012) Matrix configuration and patch isolation influences override the species-area relationship for urban butterfly communities. Landscape Ecol 27:159–169
Mauro DD, Dietz T, Rockwood L (2007) Determining the effect of urbanization on generalist butterfly species diversity in butterfly gardens. Urban Ecosyst 10:427–439
McCune B, Mefford M (1999) PC-ORD. Multivariate analysis of ecological data. Version 4.0. MjM Software Design, Gleneden Beach, OR
McKinney ML (2002) Urbanization, biodiversity, and conservation. Bioscience 52:883–890
McKinney ML (2006) Urbanization as a major cause of biotic homogenization. Biol Conserv 127:247–260
McKinney ML (2008) Effects of urbanization on species richness: a review of plants and animals. Urban Ecosyst 11:161–176
Motegi N, Yanai S (2005) A study on the characteristics of bird distribution in rooftop vegetation in Tokyo Ward. Jpn Inst Landscape Architecture 68:597–600
Niemelä J, Kotze JD, Venn S, Penev L, Stoyanov I, Spence J, Hartley D, de Oca EM (2002) Carabid beetle assemblages (Coleoptera, Carabidae) across urban–rural gradients: an international comparison. Landscape Ecol 17:387–401
Ohwaki A, Tanabe S-I, Nakamura K (2008) Effects of anthropogenic disturbances on the butterfly assemblage in an urban green area: the changes from 1990 to 2005 in Kanazawa Castle Park, Japan. Ecol Res 23:697–708
Osawa S, Katsuno T (2005) Distribution pattern of amphibian on rural landscape in the middle reaches of the Oita river, northeast Kyusyu. Jpn Inst Landscape Architecture 68:563–566
Palmer MW (1990) The estimation of species richness by extrapolation. Ecology 71:1195–1198
Parks of Seoul (2014) Seoul metropolitan government. Available at: http://parks.seoul.go.kr/eng/common/park_info/introduce.jsp?park_id=namsan
Pollard E, Yates TJ (1993) Monitoring butterflies for ecology and conservation. Conservation biology series. Chapman & Hall, Institute of Terrestrial Ecology & Nature Conservation Committee, London, United Kingdom
Posa MRC, Sodhi NS (2006) Effects of anthropogenic land use on forest birds and butterflies in Subic Bay, Philippines. Biol Conserv 129:256–270
Rubèn OA, Ian MF (2009) Living in the big city: effects of urban land-use on bird community structure, diversity, and composition. Landsc Urban Plan 90:189–195
Ruszczyk A, De Araujo AM (1992) Gradients in butterfly species diversity in an urban area in Brazil. J Lepid Soc 46:255–264
Shannon CE, Weaver W (1949) A mathematical theory of communication. University of Illinois press, Urbana, IL, 117pp
Shepherd PA (1994) A review of plant communities of derelict land in the city of Nottingham, England and their value for nature conservation. Memorabilia Zoologica 49:129–137
StatSoft, Inc. (2004) Statistica forWindows. Version 6. Statsoft Inc., Tulsa, OK
Stefanescu C, Herrando S, Páramo F (2004) Butterfly species richness in the north-west Mediterranean Basin: the role of natural and human-induced factors. J Biogeogr 31:905–915
Takami Y, Koshio C, Ishii M, Fujii H, Hidaka T, Shimizu I (2004) Genetic diversity and structure of urban population of Pieris butterflies assessed using amplified fragment length polymorphism. Mol Ecol 13:245–258
Thomas CD (1995) Ecology and conservation of butterfly metapopulations in the fragmented British landscape. Chapman and Hall, London
Yamaguchi T (2005) Influence of urbanization on ant distribution in parks of Tokyo and Chiba City, Japan II. Analysis of species. Entomol Sci 8:17–25
Yamamoto Y (1975) Notes on the methods of belt transect census of butterflies. J Fac Sci Hokkaido Univ Ser VI Zool 20:93–116
Zar JH (1999) Biostatistical Analysis, 4th edn. Prentice Hall International, INC., Upper Saddle River, NJ

9

Geometric morphometric analysis of the head of *Microlophus atacamensis* (Tropiduridae) in a latitudinal gradient

Soledad Ibáñez[1*], Marcela A Vidal[1], Juan Carlos Ortiz[2] and Fernando Torres-Pérez[3]

Abstract

Background: Clinal variation is defined as gradual variation in a character associated with geographic distance among sites. *Microlophus atacamensis* is a medium large lizard species which inhabits the intertidal zone of northern Chile, distributed from Antofagasta (23° 39′ S) to Arrayán, La Serena (29° 41′ S) in a gradient which could show clinal variation. Geometric morphometrics analyzes differences in shape independent of size; information about shape allows a more complete biological interpretation than information on morphological variation. This study evaluated clinal variation in the head shape of *M. atacamensis* from five localities (Antofagasta, Paposo, Pan de Azúcar, Caldera, and Tres Playitas) using dorsal and lateral views, comparing form variation with latitudinal distribution. The heads of adults collected were photographed in lateral and dorsal views.

Results: The analysis did not find significant differences in form among the five localities, in contrast to the proposal of earlier studies, and no differences were recorded between the sexes. Possible reasons why these populations are not differentiated in the latitudinal gradient are discussed.

Conclusions: Our results show that there are no differences between the studied, among the explanations are that populations of this species are subjected to similar environments that promote convergence of the structures analyzed.

Keywords: Atacama Desert; Lizard; Shape morphometrics; Head

Background

The genus *Microlophus* (Squamata: Tropiduridae) has a disjunct distribution, rare in terrestrial vertebrates (Benavides et al. 2007). Twenty-one species are recognized; nine of these are found only in the Galapagos Islands, while the remainder are distributed along the coast and interior deserts of western South America, from southern Ecuador to northern Chile (Etheridge and De Queiroz 1988; Frost and Etheridge 1989; Frost 1992; Benavides et al., 2007). The species of the genus *Microlophus* form a monophyletic group of lizards, characterized by apical disks in the hemipenis (Frost 1992). According to Dixon and Wright (1975), two groups are recognized in the genus whose monophyly is justified by characters of body scales and skin folds, occipitalis and peruvianus; this was later corroborated by Frost (1992) with osteological characters. Two subgroups are recognized within the peruvianus group; the first group groups interior species which feed on insects and terrestrial plants with no direct ecological relation to the intertidal zone: *M. peruvianus*, *M. thoracicus*, *M. theresioides*, *M. tigris*, *M. yanezi*, *M. theresiae*, and *M. tarapacensis* (Ortiz and Serey 1979); the second group is composed of intertidal species which feed on crustaceans and algae: *M. heterolepis*, *M. atacamensis*, and *M. quadrivittatus* (Ortiz and Serey 1979). In Chile, this genus is currently represented by *M. atacamensis*, *M. quadrivittatus*, *M. yanezi*, *M. tarapacensis*, *M. theresioides*, and *M. heterolepis* (Ortiz 1980a), which are distributed from Arica to Arrayán (La Serena) (Donoso-Barros 1949; Sepúlveda et al. 2006).

According to Donoso-Barros (1960), the origin of the genus *Microlophus* in Chile began with *M. peruvianus*, which would have descended southward along the coast and then extended its range to the interior desert by way of the valleys. Ortiz (1980a) followed this logic but

* Correspondence: soledadibanez7@gmail.com
[1]Departamento de Ciencias Básicas, Facultad de Ciencias, Universidad del Bío-Bío, Casilla 447, 4051381 Chillán, Chile
Full list of author information is available at the end of the article

added that the distribution to the interior desert would have occurred through the Pampa del Tamarugal. Later, using isoenzyme data Victoriano et al. (2003) proposed that an ancestral lineage advanced from Peru towards the coast and later diverged into two groups; one was the ancestor of the *M. quadrivitattus-M. theresioides* group (one of whose subgroups still inhabits the interior desert) and the other was the ancestor of *M. atacamensis*. Based on a molecular phylogeny of data from many genes, Benavides et al. (2007) proposed that the origin of the Chilean clade was a migration from Peru to the interior desert; *M. theresioides* diverged and then dispersed along the Río Loa to the coast, giving rise to *M. quadrivittatus* in the north and *M. atacamensis* farther south. This last proposal is questionable, since the limit of distribution between *M. quadrivittatus* and *M. atacamensis* is near Antofagasta (Victoriano et al. 2003). Here, we describe some characteristics of *M. atacamensis*.

M. atacamensis is a lizard species which inhabits the intertidal zone of northern Chile from Antofagasta (23° 39′ S, 70° 22′ W) to Arrayán, La Serena (29° 41′ S, 71° 19′ W) (Ortiz 1980b; Heisig 1993, Sepúlveda et al. 2006). This is a medium large (mean: 102.7 mm length) (Donoso-Barros 1966; Ortiz 1980b), corpulent species with dark brown color and black blotches on the dorsum; crest tenuous in the vertebral zone and mouth narrow. According to Ortiz (1980a) and Vidal et al. (2002) there is a spatial segregation between juveniles and adults in the height of the perch used for thermoregulation, as well as a difference between sexes; although males are larger, there was no significant difference in their regulation, both were thermoconformers (Vidal et al. 2002). This species is omnivorous behavior, since they consumed mainly Diptera and algae obtained from the intertidal area (Ortiz 1980a; Vidal and Labra 2008; Fariña et al. 2008).

Clinal variation is defined as gradual variation of a character over geographic distance (Futuyma 1998). For example, Vidal et al. (2007) found latitudinal variation in coloration in *Liolaemus tenuis*; green increased and brown decreased from north to south. Sepúlveda et al. (2008) proposed a latitudinal gradient in the thermoregulatory ability of *M. atacamensis*, in which northern populations have low average to which lizards experience a Tb outside the selected temperature, while south this ability increases. Fariña et al. (2008) found a clinal pattern in the diet of *M. atacamensis*, in which algal consumption decreases and consumption of Diptera increases from north to south.

Many theoretical and empirical studies have focused on the adaptive significance of morphological clinal variation (Trussell 2000; Martínez-Freiría et al. 2009). While some analyses of linear morphometry have been often used to test hypotheses related to morphological polymorphism in a microevolutionary context (Endler 1977; Scolaro and Cei 1987; Quatrini et al. 2001), others are more holistic and quantitative analyses are then required to identify them and to appraise the selective forces responsible for their evolution (Adams and Rohlf 2000). In fact, due to scale problems with linear measures, these do not take into account the morphological complexity of biological structures (Humphries et al. 1981; Rohlf and Bookstein 1987; Mousseau 1991; Warheit 1992). In the case of traditional (non-geometric) measurements, statistical techniques for measuring distances, relations among distances, areas, volumes, or angles are applied (Bookstein et al. 1999). However, the geometric morphometry preserves the geometry configurations of landmarks, by allowing a statistical representation of real forms or forms only (Rohlf and Slice 1990; Rohlf et al. 1996; Rohlf 1999; Rohlf and Corti 2000), whereas denotes the geometric shape properties of a structure that is independent of size, position, and orientation of the same, while the form of an object includes both size and shape (Rohlf and Slice 1990; Adams and Rohlf 2000, Mitteroecker and Gunz 2009). This information on shape allows a more complete biological interpretation than morphological variation (Rohlf and Marcus 1993). In relation to clinal variation, results obtained in *Liolaemus* show that, although there is not a clear pattern of latitudinal variation of the shape in *L. tenuis* (Vidal et al. 2005), there are significant differences between two geographical areas (arid Mediterranean and oceanic with Mediterranean influence). These different geographic areas may act as partial barriers to gene flow in this species. Vidal et al. (2006) found divergence between *L. pictus* from the Isla de Chiloé and the Chilean continent in ocular extension and location of the labial commissure. Individuals from the island had more extended ocular orbits and a more posterior position of the labial commissure than individuals from the continent, which may be attributed to differences in diet between these two localities (Vidal et al. 2006).

In this study, we explored the morphological adjustment to clinal environment variation in the shape of the head of *M. atacamensis* in dorsal and lateral view in different localities, comparing over a latitudinal distribution from 23° to 28° S. Considering that environmental temperature decreases as latitude increases and previous data has shown variation in thermal and feeding in this species, we specifically attempted to test the hypotheses that the shape of the head varies latitudinally including more extended shape in the north for your diet and more compacted shape the south.

Methods

We examined 111 adult individuals of *M. atacamensis* from the following localities (Figure 1, Table 1): Antofagasta, Paposo, Pan de Azúcar, Caldera, and Tres

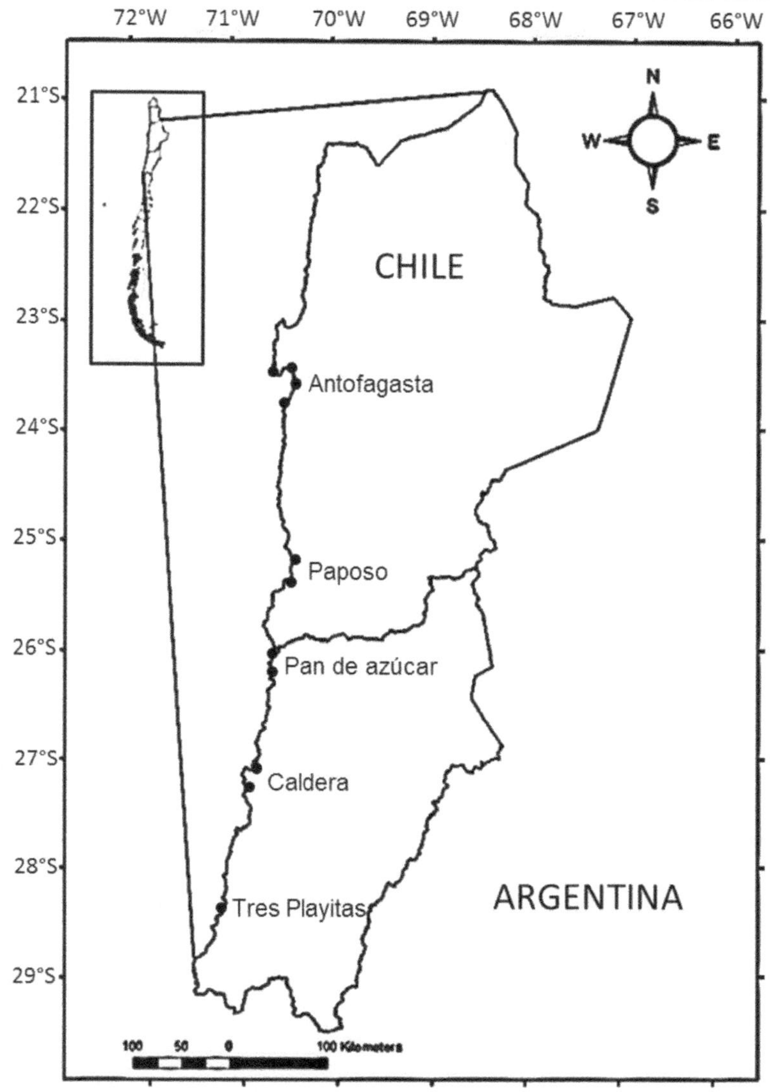

Figure 1 Map of the localities of *Microlophus atacamensis* analyzed in this study. These localities were grouped into five groups due to geographic proximity.

Playitas. The localities were grouped into five groups, due to their geographic proximity to the mentioned localities. All material is deposited in the Museo de Zoología of the Universidad de Concepción (MZUC). Digital photographs were taken of the head of each individual in dorsal and lateral views with a dSLR Canon T4 camera (Canon, Tokyo, Japan), using a 50-mm f/1.4 lens with a fixed focal length. In addition, specimens were placed 10 cm away from the camera using a fixed tripod. The quality of the photographs was optimized with the Microsoft Power Point 2010 program. Nine homologous landmarks were located in the dorsal view and eleven in the lateral view (Figure 2), coincident with intersections of cranial scales. The location of the homologous landmarks was performed using the proposal of Vidal et al. (2006), with the addition of new landmarks which are specific for the genus *Microlophus*. Coordinates were sampled from photographs in dorsal and lateral view by using tpsDig 1.20 (Rohlf 2003a). The analyses followed the procedures of Rohlf and Slice (1990) and Rohlf et al. (1996). The X and Y coordinates of the biologically homologous landmarks were aligned and

Table 1 Studied localities and their geographic coordinates

Locality	Geographic coordinates	*N* Dorsal view	*N* Lateral view
Antofagasta	23°37′ S, 70°24′ W	36	31
Paposo	25°15′ S, 70°23′ W	3	2
Pan de Azúcar	26°04′ S, 70°35′ W	15	14
Caldera	27°04′ S, 70°49′ W	26	25
Tres Playitas	28°27′ S, 71°13′ W	6	8

N = sample size.

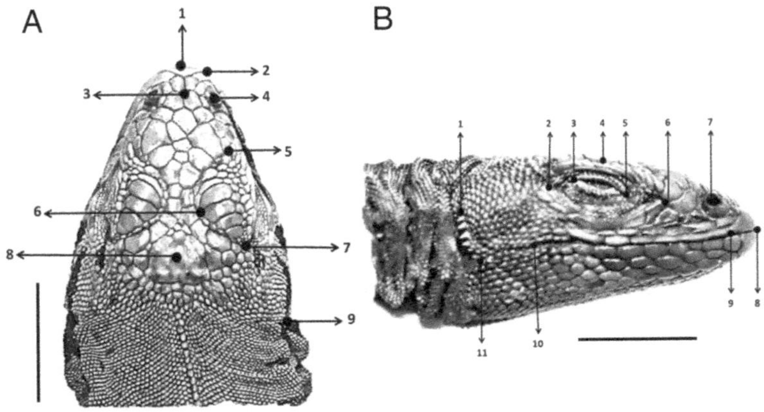

Figure 2 Landmarks selected for the dorsal (A) and lateral (B) views in *Microlophus atacamensis*. The lateral bar on each image represents 1 cm on the specimen.

superimposed using the minimum squares method based on the generalized Procrustes analysis (GPA), thus removing non-shape variation (Zelditch et al. 2004) in order to standardize the size and to translate and rotate the configurations of landmark coordinates using tpsRelw (Rohlf 2003b). We extracted relative warp scores, using the tpsRelw software according to Kaliontzopoulou et al. (2007). Using relative warps, we computed canonical scores for specimens of populations in order to visualize these variations. The deformation grids were produced by the regression of shape variables against canonical results (Rohlf 2003b). These variables are used in a multivariate analysis (Rohlf et al. 1996; Adams and Rohlf 2000). We used TpsRelw version 1.21 (Rohlf 2003b) to perform a principal component analysis.

To determine sexual dimorphism and variability among localities, we using an analysis of two-way MANOVA using sex and locality as factor on the matrix of relative warps. Centroid size was used as a size index, computed as the square root of the sum of the squared distances of a set of landmarks from their centroid (Marcus et al. 1996). The centroid size (log (CS)) of all individuals was compared with a two-way ANOVA using sex and locality as factor. The allometry calculations used in this work consisted of multivariate regression in which the explanatory variable is the CS, and the dependent variables are the variables of the shape (i.e., Procrustes residuals or PC scores, see Klingenberg 1996 and Depecker et al. 2006). According to Bookstein (1991), allometry is defined as changes in the shape related to the increase in size, where significant results in the multivariate regression of shape on CS indicate an allometric effect of size on shape (Zelditch et al. 2004).

Results

No significant differences were detected among localities, either in dorsal view (Wilk's lambda = 0.23; $p = 0.45$) or lateral view (Wilk's lambda = 1.77; $p = 0.24$), and no sexual dimorphism was found in dorsal (Wilk's lambda = 0.31; $p = 0.67$) and lateral (Wilk's lambda = 0.05; $p = 0.63$) views. The interaction between the two factors was not significant in both views, respectively (Wilk's lambda dorsal = 1.65; $p = 0.1$; Wilk's lambda lateral = 1.06; $p = 0.52$). The first three principal components of shape explained the majority of the variance for the dorsal and lateral views (96.9% and 86.6%, respectively, Figures 3 and 4). In the dorsal view (Figure 3), PC1 expresses a change in the shape of the head laterally (90.1% of the total Procrustes form variance), while in lateral view (Figure 4) it expresses the change in the posterior region (68.4% of the total Procrustes form variance). In both cases, PC1 shows changes in allometry. PC2 in dorsal view (5.6% of the total form variance) showing changes in the compression of the head on their sides; for the lateral view (12.6% of the total form variance), it shows the changes with a dorsal-ventral compression. The PC3 accumulates low variance, which is displayed on the projected surface by the three components (Figures 3 and 4). In centroid size case, no significant differences were detected among localities, either in dorsal view ($F_{(1,4)} = 6.57$; $p = 0.54$) or lateral view ($F_{(1,4)} = 0.21$; $p = 0.93$), and no sexual dimorphism was found in dorsal ($F_{(1,4)} = 0.38$; $p = 0.54$) and lateral views ($F_{(1,4)} = 0.06$; $p = 0.81$). The interaction between the two factors was not significant in both views, respectively ($F_{(1,4)} = 1.19$; $p = 0.15$; $F_{(1,4)} = 0.24$; $p = 0.91$). Landmark 9 of the dorsal view and landmarks 1 and 11 of the lateral view were those which showed the greatest variation, corresponding to the tympanic area. Although no significant differences among localities were detected in both views, there is a tendency to differentiate the locations of Antofagasta from Caldera - Three Playitas in dorsal view, and Caldera from Tres Playitas - and Pan de Azúcar in lateral view. The spatial ordering of the principal components analysis (PCA) showed superimposition

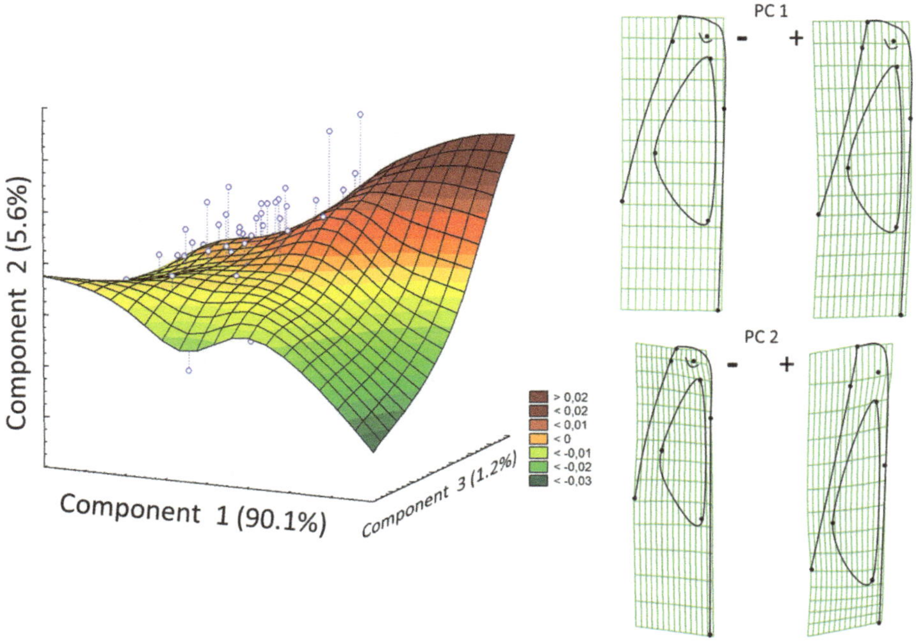

Figure 3 Form-space principal component analysis in *Microlophus atacamensis* in lateral view. The first three components account for 96.9% of the total form-space variance. Surfaces correspond to the shapes represented by negative and positive extremes of PC1 and PC2.

of all localities in the dorsal and lateral views (Figure 5). No allometric effect was detected in either dorsal or lateral view for the shape variables between the centroid size and the principal components ($r = 0.08$, $p = 0.65$ in dorsal view; $r = 0.02$, $p = 0.83$ in lateral view).

Discussion

A number of studies have been performed on chromosomes, coloration, diet, thermoregulation, and geometric morphometrics which have provided evidence for clinal variation of lizards in Chile (Lamborot 1991; Lamborot et al. 2012; Vásquez et al. 2007; Vidal et al. 2007; Sepúlveda et al. 2008; Fariña et al. 2008; Vidal et al. 2008). In spite of these evidences, our results do not show clinal variation in head shape among the five studied localities of *M. atacamensis*. The PCA showed a superposition of all the localities both in dorsal and lateral view, indicating that the different morphs may be found in all the studied localities. However, this analysis showed some tendency to group the Pan de Azúcar and Tres Playitas localities in dorsal view, and Caldera and Tres Playitas in the lateral view (Figure 3). Although the results of the *a posteriori* analysis show a tendency of segregation in two relative warps in dorsal view and three in lateral view, the majority of the

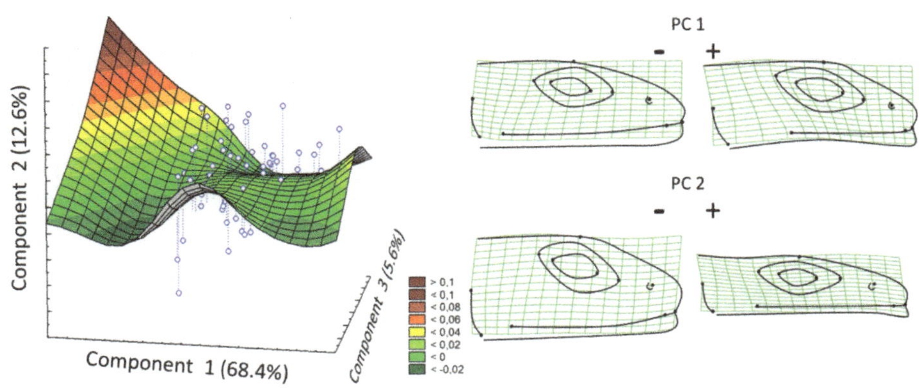

Figure 4 Form-space principal component analysis in *Microlophus atacamensis* in lateral view. The first three components account for 86.6% of the total form-space variance. Surfaces correspond to the shapes represented by negative and positive extremes of PC1 and PC2.

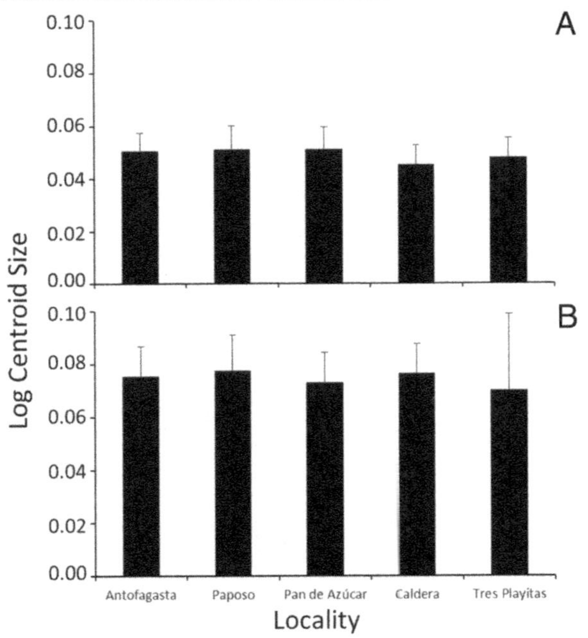

Figure 5 Centroid size (log) and standard deviation of the five localities in dorsal (A) and lateral (B) views.

results obtained did not show significant differences among localities. Although Fariña et al. (2008) showed a latitudinal gradient in the diet of *M. atacamensis*, we did not find a similar pattern in the shape of the head as shown in the study of Vidal et al. (2006) in *L. pictus*. Kaliontzopoulou et al. (2010) found significant different differences in the form of the head of *Podarcis bocagei*, related to the habitat utilized. In the present case, *M. atacamensis* uses the same habitat in its entire distribution range (Ortiz 1980b; Heisig 1993; Sepúlveda et al. 2006), which may explain in part the lack of clinal differences in head form. The same is true of *L. petrophilus* (Fontanella et al. 2012); although there are differences in the shape of the cranium from north to south, this is due to the fact that this species occupies different niches latitudinally. In *Strengeriana maniformis*, Pedraza and Campos (2007) did not find significant differences in the form of the gonopod among habits with and without mining contamination; they found, as in the present case, that the different morphs are distributed in both types of habitats, which may indicate that this species has some plasticity to mining contaminants. *M. atacamensis* may also have some plasticity in diet, depending upon the food available, explaining in part the lack of clinal variation in head form over a latitudinal range.

Also in this case, *L. tenuis* does not show a clinal variation pattern in its morphological characteristics (Vidal et al. 2005). By contrast, in *M. atacamensis*, Ortiz (1980a) showed that the number of mid-line scales and number of vertebral scales increased from Antofagasta to Huasco. Also, Sepúlveda et al. (2008) hypothesized that in this species, size is related to the thermoregulatory gradient, since juvenile individuals (smaller) were the first to emerge in the morning and adults (larger) were the last to take refuge in the evening, suggesting that populations in the north should be smaller than those in the south. The lizard plays an important role in linking the intertidal and terrestrial environment within desert systems, since it feeds in the intertidal area (eating algae, isopods, and Diptera) and thermoregulates in the terrestrial area, defecating in the latter zone and thus contributing to its biomass and maintaining the trophic networks (Sepúlveda et al. 2006; Fariña et al. 2008). In this case, at larger sizes, there is increased bite force, improved prey handling efficiency, and a consequent shift to larger and harder foods (Verwaijen et al. 2002), therefore it was expected to yield a variation in the shape of the head of *M. atacamensis.* On the other hand, much importance has been given to the three-dimensional representation (3D) to estimate the variations due to the shape in an evolutionary context. According Polly and MacLeod (2008), the success of our 3D eigensurface method, while qualified, is encouraging. Eigensurface analysis of the 3D topography of calcanea sorted them into the same functional spectrum that qualitative functional analysis would have. While 3D surface scans of complicated morphological structures (e.g., bones, teeth, and shells) are increasingly easy to generate, this technique is not attainable for everyone, but we hope to conduct new studies to incorporate 3D images for population comparisons (Polly and MacLeod 2008).

On the other hand, according to Bruner et al. (2005), it is assumed that during development the scales either grow or do not grow but that no absolute reduction takes place. Concerning the length of the frontal scales, Bruner et al. (2005) suggest that males show an absolute enlargement but a relative reduction when compared to females (i.e., negative allometry) in sexual dimorphism case. In relation to clinal variation, there are no studies indicating that the scales tend to differ with latitude. However, using the scales as a reflection of the bones suggests that they are good indicators of change at this level (Costantini et al. 2007). In the case of *M. atacamensis*, there are no differences in head shape connecting to clinal variation, which involves conducting new studies to understand in an evolutionary context.

Conclusions

Our results show that there are no differences between the studied, among the explanations are that populations of this species are subjected to similar environments that promote convergence of the structures analyzed.

Competing interests

The authors declare that they have no competing interests.

Authors' contributions
SI obtained photographs of museum specimens used in this study and contributed to the writing the manuscript. MV directed the photo shoots of the museum specimens used in this study. She directed the construction of the manuscript and the geometric and statistical analysis of the study. JCO provided the museum specimens that were photographed and provided input on improving the manuscript. FTP was involved in revising the manuscript and provided new ideas to improve the version sent. All authors read and approved the final manuscript.

Acknowledgements
We thank the Museo de Zoología of the Universidad de Concepción for supplying the samples used in this study. Finally, thanks to Lafayette Eaton, who made valuable suggestions which helped to improve this work. This study was financed by the Fondecyt project 1131009.

Author details
[1]Departamento de Ciencias Básicas, Facultad de Ciencias, Universidad del Bío-Bío, Casilla 447, 4051381 Chillán, Chile. [2]Departamento de Zoología, Facultad de Ciencias naturales y Oceanográficas, Universidad de Concepción, Casilla 160-C, Concepción, Chile. [3]Instituto de Biología, P. Universidad Católica de Valparaíso, Av. Universidad 330, Campus Curauma, Valparaíso, Chile.

References

Adams D, Rohlf FJ (2000) Ecological character displacement in *Pletodon*: biomechanical differences found from a geometric morphometric study. Proc Natl Acad Sci U S A 97:4106–4111

Benavides E, Baum R, McClellan D, Sites JW (2007) Molecular phylogenetics of the lizard genus *Microlophus* (Squamata, Tropiduridae): aligning and retrieving indel signal from nuclear introns. Syst Biol 56:776–797

Bookstein FL (1991) Morphometrics tools for landmark data: geometry and biology. Cambridge University Press, New York

Bookstein F, Schafer K, Prossinger H, Seidler H, Fieder M, Striger C, Weber G, Arsuaga JL, Slice D, Rohlf FJ, Recheis W, Mariam A, Marcus L (1999) Comparing frontal cranial profiles in archaic and modern *Homo* by morphometric analysis. Anat Rec 257:217–224

Bruner E, Costantini D, Fanfani A, Dell'Omo G (2005) Morphological variation and sexual dimorphism of the cephalic scales in *Lacerta bilineata*. Acta Zool 86:245–254

Costantini D, Bruner E, Fanfani A, Dell'Omo G (2007) Male-biased predation of western green lizards by Eurasian kestrels. Naturwissenschaften 94:1015–1020

Depecker M, Berge C, Penin X, Renous S (2006) Geometric morphometrics of the shoulder girdle in extant turtles (Chelonii). J Anat 208:35–45

Dixon J, Wright J (1975) A review of the lizards of the Iguanid genus *Tropidurus* in Perú. Contr. Sci. Nat. Hist. Mus. Contr Sci Nat Hist Mus Los Angeles 271:1–39

Donoso-Barros R (1949) Alimentación de *Tropidurus peruvianus* (Lesson). Biol Mus Nac Hist Nat Chile 24:213–216

Donoso-Barros R (1960) Ecología de los reptiles Chilenos. Invest Zool Chil 6:65–72

Donoso-Barros R (1966) Reptiles De Chile. Universidad de Chile, Santiago, Chile, 458

Endler J (1977) Geographic variation, speciation, and clines. Princeton University Press, New Jersey

Etheridge R, de Queiroz K (1988) A phylogeny of Iguanidae. In: Estes RR, Pregill G (eds) Phylogenetic relationships of the lizard families: essays commemorating Charles L. Camp. Stanford.Univ. Press, Stanford CA, pp 283–367

Fariña JM, Sepúlveda M, Reyna MV, Wallem KP, Ossa-Zazzali PG (2008) Geographical variation in the use of intertidal rocky shores by the lizard *Microlophus atacamensis* on the Atacama Desert coast. J Anim Ecol 77:458–468

Fontanella FM, Feltrin N, Avila LJ, Sites JW, Morando M (2012) Early stages of divergence: phylogeography, climate modeling, and morphological differentiation in the South American lizard, *Liolaemus petrophilus* (Squamata: Liolaemidae). Ecol Evol 2:792–808

Frost DR (1992) Phylogenetic analysis and taxonomy of the *Tropidurus* group of lizards (Iguania: Tropiduridae). Am Mus Novit 3033:1–68

Frost DR, Etheridge R (1989) A phylogenetic analysis and taxonomy of iguanian lizards (Reptilia: Squamata). Mus Nat Hist Univ Kansas Misc Publ 81:1–65

Futuyma D (1998) Evolutionary biology, 3rd edn. Sinauer Associates, Inc., Sunderland, Massachusetts

Heisig M (1993) An etho-ecological study of an island population of *Tropidurus atacamensis*. Salamandra 29:65–81

Humphries JM, Bookstein FL, Chernoff B, Smith GR, Elder RL, Poss SG (1981) Multivariate discrimination by shape in relation to size. Syst Zool 30:291–308

Kaliontzopoulou A, Carretero MA, Llorente GA (2007) Multivariate and geometric morphometrics in the analysis of sexual dimorphism variation in *Podarcis* lizards. J Morphol 268:152–165

Kaliontzopoulou A, Carretero MA, Sillero N (2010) Geographic patterns of morphological variation in the lizard *Podarcis carbonelli*, a species with fragmented distribution. Herpetol J 20:41–50

Klingenberg CP (1996) Multivariate allometry. In: Marcus LF, Corti M, Loy A, Naylor GJP, Slice D (eds) Advances in morphometrics. Plenum Press, New York, pp 23–49

Lamborot M (1991) Karyotypic variation among populations of *Liolaemus monticola* (Tropiduridae), separated by riverine barriers at the Andean Range. Copeia 1991:1044–1059

Lamborot M, Ossa CM, Vásquez M (2012) Population cytogenetics of the "Northern Mod 1" chromosomal race of *Liolaemus monticola* Müller & Helmich (Iguanidae) from Central Chile. Gayana 76:10–21

Marcus LF, Corti M, Loy A, Naylor GJP, Slice DE (1996) Advances in morphometrics. Plenum Press, New York

Martínez-Freiría F, Santos X, Pleguezuelos JM, Lizana M, Brito JC (2009) Geographical patterns of morphological variation and environmental correlates in contact zones: a multi-scale approach using two Mediterranean vipers. J Zool Syst Evol Res 47:357–367

Mitteroecker P, Gunz P (2009) Advances in geometric morphometrics. Evol Biol 36:235–247

Mousseau TA (1991) Landmarks in morphometrics, or the shape and size of morphometrics to come. Evolution 45:1991–1980

Ortiz JC (1980a) Estudios comparativos de algunas poblaciones de *Tropidurus* de la costa chilena. An Mus Hist Nat Valparaíso 13:267–280

Ortiz JC (1980b) Revisión taxonómica del género *Tropidurus* en Chile. Actas de la Primera Reunión Iberoamericana de Zoología de Vertebrados. Ediciones del Ministerio de Universidades e Investigación, La Rábida, España, pp 355–377

Ortiz JC, Serey I (1979) Análisis factorial de correspondencias de las especies del género *Tropidurus* en Chile. Rev Med Exp Chile 12:203–208

Pedraza M, Campos M (2007) Estudio de la variación morfológica del gonopodo de *Strengeriana maniformis* (Brachyura: Pseudothelphusidae) mediante aplicación de Morfometría Geométrica. Caldasia 29:143–152

Polly PD, MacLeod N (2008) Locomotion in fossil carnivora: an application of eigensurface analysis for morphometric comparison of 3D surfaces. Palaeontol Electron 11:1–13

Quatrini R, Albino A, Barg M (2001) Variación morfológica en dos poblaciones de *Liolaemus elongatus* Koslowsky 1896 (Iguania: Tropiduridae) del noroeste patagónico. Rev Chil Hist Nat 74:639–651

Rohlf FJ (1999) Shape statistics: Procrustes superimpositions and tangent space. J Classif 16:197–223

Rohlf FJ (2003a) TPSDIG. Version 1.22. Department of Ecology and Evolution, State University of New York, Stony Brook, NY, Website: http://life.bio.sunysb.edu/morph/

Rohlf FJ (2003b) TPSRELW. Version 1.21. Department of Ecology and Evolution, State University of New York, Stony Brook, NY, Website: http://life.bio.sunysb.edu/morph/

Rohlf FJ, Bookstein FL (1987) A comment on shearing as a method for size correction. Syst Zool 36:356–367

Rohlf FJ, Corti M (2000) Use of two-block partial least squares to study covariation in shape. Syst Zool 49:740–753

Rohlf JF, Marcus L (1993) A revolution in morphometrics. Trends Ecol Evol 8:129–132

Rohlf FJ, Slice D (1990) Extensions of the Procrustes method for the optimal superimposition of landmarks. Syst Zool 39:40–59

Rohlf FJ, Loy A, Corti M (1996) Morphometric analysis of old world Talpidae (Mammalia, Insectívora) using partial-warp scores. Syst Biol 45:344–362

Scolaro JA, Cei JM (1987) A multivariate analysis of morphometric and exosomatic characters of Iguanid lizards of the patagonian *Liolaemus kingi* complex. J Herpetol 21:343–348

Sepúlveda M, Vidal MA, Fariña JM (2006) *Microlophus atacamensis* (Atacama Desert Runner) predation. Herpetol Rev 37:224–225

Sepúlveda M, Vidal MA, Fariña JM (2008) Seasonal and geographic variation in thermal biology of the lizard *Microlophus atacamensis* (Squamata: Tropiduridae). J Thermal Biol 33:141–148

Trussell GC (2000) Phenotypic clines, plasticity, and morphological trade-offs in an intertidal snail. Evolution 54:151–166

Vásquez M, Torres-Pérez F, Lamborot M (2007) Genetic variation within and between four chromosomal races of *Liolaemus monticola* in Chile. Herpetol J 17:149–160

Verwaijen D, Van Damme R, Herrel A (2002) Relationships between head size, bite force, prey handling efficiency and diet in two sympatric lacertid lizards. Funct Ecol 16:842–850

Victoriano P, Torres F, Ortiz JC, Parra L, Northland I, Capetillo J (2003) Variación aloenzimática y parentesco evolutivo en especies de *Microlophus* del grupo "*peruvianus*" (Squamata: Tropiduridae). Rev Chil Hist Nat 76:65–78

Vidal MA, Labra A (2008) Dieta de anfibios Y reptiles. In: Vidal MA, Labra A (eds) Herpetología de Chile. Science Verlag, Santiago, pp 453–482

Vidal M, Ortiz JC, Labra A (2002) Sexual and age differences in ecological variables of the lizard *Microlophus atacamensis* (Tropiduridae) from Northern Chile. Rev Chil Hist Nat 75:283–292

Vidal MA, Ortiz JC, Ramírez CC, Lamborot M (2005) Intraspecific variation in morphology and sexual dimorphism in *Liolaemus tenuis* (Tropiduridae). Amphibia-Reptilia 26:343–351

Vidal MA, Veloso A, Méndez MA (2006) Insular morphological divergence in the lizard *Liolaemus pictus* (Liolaemidae). Amphibia-Reptilia 27:103–111

Vidal MA, Ortiz JC, Labra A (2007) Sexual and geographic variation of color patterns in *Liolaemus tenuis* (Squamata, Liolaeminae). Gayana 71:27–33

Vidal MA, Ortiz JC, Labra A (2008) Intraspecific variation in a physiological thermoregulatory mechanism: the case of the lizard Liolaemus tenuis (Liolaeminae). Rev Chil Hist Nat 81:171–178

Warheit KI (1992) Proceedings of the Michigan morphometrics workshop. Syst Zool 41:392–395

Zelditch ML, Swiderski D, Sheets D, Fink WL (2004) Geometric morphometrics for biologists: a primer. Elsevier, London

10

Guanaco abundance and monitoring in Southern Patagonia: distance sampling reveals substantially greater numbers than previously reported

Alejandro Travaini[1], Sonia C Zapata[2], Javier Bustamante[3], Julieta Pedrana[1,4], Juan I Zanón[2,5] and Alejandro Rodríguez[6*]

Abstract

Background: Guanacos (*Lama guanicoe*) are thought to have declined in Patagonia mainly as a result of hunting and sheep ranching. Currently accepted estimates of total population size are extrapolated from densities obtained through strip transects in local studies. We used road surveys (8,141 km) and distance sampling to estimate guanaco density and population size over major environmental gradients of Santa Cruz, a large region in southern Patagonia. We also calculated the survey effort required to detect population trends in Santa Cruz.

Results: We found considerable spatial variation in density (1.1 to 7.4 ind/km^2), with a mean value of 4.8 ind/km^2, which is more than twice the mean value guessed for central and northern Patagonia. Consequently, guanaco numbers in Santa Cruz were estimated at 1.1 million individuals (95% CI 0.7 to 1.6), which almost doubles current estimates of guanaco population size in South America. High guanaco abundance was found in arid lands, overgrazed and unable to support profitable sheep stocks. Detecting a 50% change in guanaco population size over a 10-year period requires substantial monitoring effort: the annual survey of between 40 and 80 30-km transects, which becomes up to 120 transects if trends are to be detected over 5 years.

Conclusions: Regional patterns in guanaco density can only be detected through large-scale surveys. Coupling these surveys with distance sampling techniques produce robust estimates of density and its variation. Figures so obtained improve currently available estimates of guanaco population size across its geographic range, which seem to be extrapolated from strip counts over small areas. In arid lands degraded by sheep overgrazing, sustainable use of guanaco populations would help harmonize guanaco conservation, socio-economic progress of rural areas, and eventually the restoration of shrub-steppes.

Keywords: Distance sampling; Guanaco; Livestock; Monitoring; Patagonian shrub-steppe; Sustainable uses

Background

Only three native ungulate species occur across Patagonian forests, arid grasslands, and shrub steppes: two cervids (Southern pudu, *Pudu pudu*, and South Andean huemul, *Hippocamelus bisulcus*) and one camelid (guanaco, *Lama guanicoe*; Mattioli 2011). Among them, the guanaco is the largest and most abundant species, as well as the one with the highest potential economic impact (Franklin et al. 1997). Guanacos range from Peru to Tierra del Fuego (Franklin 1983; Redford and Eisenberg 1992), although 90% of guanaco populations occur in Argentina (Franklin 1982), mainly throughout Patagonia (Puig 1995).

Despite the guanaco's flexible social behavior (Franklin 1983) and ecophysiological adaptations to harsh environments (Franklin and Fritz 1991), populations have declined in numbers and geographic range (Franklin 1982; Puig 1995; Baldi et al. 2010). This decline has been attributed primarily to intense and uncontrolled hunting (Franklin 1982; Cunazza et al. 1995; Donadio and Buskirk 2006) and competition with introduced wild and domestic herbivores, such as the European hare (*Lepus europaeus*;

* Correspondence: alrodri@ebd.csic.es
[6]Department of Conservation Biology, Estación Biológica de Doñana, CSIC, Américo Vespucio s/n, 41092 Sevilla, Spain
Full list of author information is available at the end of the article

Garrido 1985) and domestic sheep (*Ovis aries*; Puig et al. 1997; Baldi et al. 1997, 2001, 2004).

It is thought that Santa Cruz, the southernmost continental province of Argentine Patagonia, contains the largest guanaco population (Baldi et al. 2006). The suspected important contribution of this region to overall numbers contrasts with the paucity of available information about guanaco density and its spatial variation (Baldi et al. 2010). The only population estimation of guanacos in Santa Cruz is that of Amaya et al. (2001), who calculated a total of 223,847 guanacos based on strip counts along 2,800 km of aerial transects. Moreover, global population numbers are derived on the basis of densities reported in other parts of Patagonia (<600,000 individuals, about 90% in Argentina; IUCN 2014), where local guanaco densities were often below 2 ind/km^2 (Baldi et al. 2010). Excluding the Patagonian survey by Amaya et al. (2001), deemed unreliable by IUCN (2013), density estimates used to calculate global guanaco numbers come from local studies or surveys over relatively small regions (e.g., Baldi et al. 1997, 2001; Puig et al. 1997, 2003) that show considerable variation in guanaco density. Given this heterogeneity and the expected influential contribution of Santa Cruz populations (Baldi et al. 2006), estimates of population size in Patagonia could substantially depart from figures currently accepted after obtaining more reliable estimates of guanaco numbers in Santa Cruz. Moreover, large-scale surveys can help quantify spatial variation in guanaco density.

Most density estimates of guanaco across Patagonia have been derived from strip count methods over relatively small areas. A key limitation of this technique is that it does not allow the estimation of animals missed within the sampled area. In contrast, distance sampling yields an estimate of missed guanacos, an objective estimate of the detection band along line transects, more accurate density estimates and their variances, and the possibility of extrapolating densities outside the area effectively surveyed (Buckland et al. 2001; Thomas et al. 2010).

In Santa Cruz, sheep ranching has been the prevalent land use since the late 1800s when European settlers arrived to the southern Patagonia (Soriano and Paruelo 1990). From the beginning, guanacos were generally perceived by ranchers as a pest and competitors of sheep for forage. On the other hand, guanacos have the potential of becoming an important economic source of complementary income to sheep ranching on a sustainable basis because of its wool (Franklin 1983; Sacchero et al. 2006), meat (Soto et al. 1991; Franklin et al. 1997), leather (Ojeda and Mares 1982), appeal for ecotourism, and as a game species (Franklin et al. 1997; MACS 2002).

The sustainability of potential guanaco exploitation should be based not only on sound population estimates, but also on a well-designed monitoring protocol for measuring population size before starting any extractive activity and assessing its potential impact later on (Di Stefano 2001). Statistically powerful monitoring programs (Steidl et al. 1997) can help detect early trends in guanaco abundance, despite noise in the survey data associated with seasonal variations or unavoidable errors in field sampling (Gibbs and Melvin 1997; Elzinga et al. 2001). Statistical power is critical in the design phase of monitoring wildlife populations (Taylor and Gerrodette 1993).

In this paper, we report spatial variation in estimates of guanaco population size for Santa Cruz obtained with distance sampling techniques on data from an extensive, large-scale road survey. We also propose a population monitoring protocol to detect temporal changes in guanaco population size. Such a protocol has the potential to be an important management tool for detecting further declines in guanaco populations due to unsustainable sheep ranching or potentially unsustainable ill-planned extraction of guanacos in Patagonia (Montes et al. 2006; Rey et al. 2009).

Methods

Study area

Santa Cruz province (46° to 53° S; 65° to 73° W, Figure 1a) expands over 245,865 km^2 and accounts for 6.5% of the land area of Argentina (González and Rial 2004). With the exception of a western strip of Andean influence, this region is a cold semi-desert covered by shrub-steppes, where climate shows a marked gradient, with mean annual rainfall decreasing from west (800 mm) to east (down to 100 mm in some places) and mean annual temperatures decreasing from northeast to southwest in the range 5 to 10°C (Oliva et al. 2004). Aridity precludes shifting from sheep ranching to more profitable cattle ranching. Sheep overstocking (Golluscio et al. 1998) and overgrazing of mesic sites (Mazzoni and Vázquez 2004) has lead to widespread steppe degradation (León and Aguiar 1985). Unsustainable use of rangelands has resulted in land abandonment in 60% of the Santa Cruz province (Borrelli and Cibils 2005). A recent socioeconomic trend to resume sheep ranching has been reported but, unfortunately, practices responsible for previous damages to the steppe are being adopted again (Andrade 2005).

Guanaco social units

Guanacos are sexually monomorphic and aggregate in social groups during the breeding season (Sarno and Franklin 1999) and winter (Franklin 1983). The spatial distribution of individuals is strongly influenced by a mating system of resource defense polygyny, a territorial system wherein males compete for access to resources required by females (Franklin 1983; Young and Franklin 2004a). Guanacos can be found in three basic social units: family groups (a male with a group of females and young less than 1 year old called 'chulengos'), male groups composed mostly of

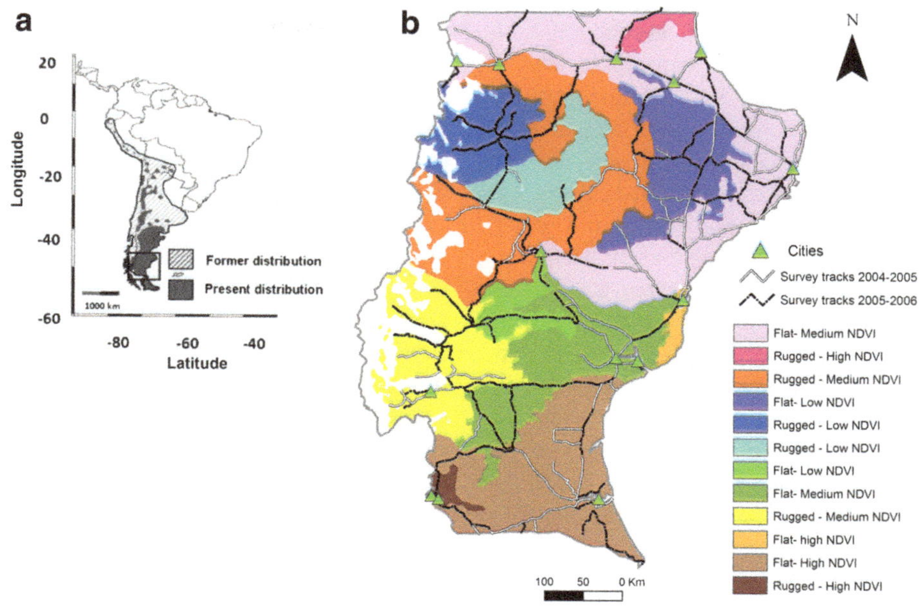

Figure 1 Guanaco geographic range, study area, and roads surveyed. **a** Former and present guanaco distribution throughout South America (adapted from Franklin et al. 1997) and study area, Santa Cruz province in southern Argentine Patagonia. **b** Initial stratification in 12 regions, based on productivity (three NDVI categories: low, medium, and high) and topography (two categories: flat and rugged terrain). The surveyed roads are shown.

non-breeding males, and solitary territorial males. There is a fourth potential social unit, female groups, observed during intensive studies in Torres del Paine National Park, Chile (Ortega and Franklin 1995). We did not consider this social unit because it may be less common (8% of all social units; Ortega and Franklin 1995) and could also include young (Bank et al. 2003) making them difficult to distinguish from family groups without closer examination. Pedrana et al. (2009) showed that only solitary males and family groups with chulengos could be unequivocally identified when conducting large-scale road surveys. This is because sex determination of every individual by their primary sexual characters is time consuming and hence impractical when, as it was our case, observers in a single vehicle had to survey thousands of kilometers during one breeding season. These difficulties increase for distant animals. Pedrana et al. (2009) also suggested that identification of male groups was challenging because an unknown proportion of family groups may have not contained chulengos.

When the goal is estimating population density, the relevant information is group size whereas recording group composition is of secondary importance. However, stratification helps to increase the precision of density estimates (Thomas et al. 2010) and, as social units differ in mean group size (Pedrana et al. 2009), social units represent a form of stratification that we considered during the survey design. Therefore, we opted to produce separate density estimates for three observable social units: breeding groups (defined as groups with detected chulengos), non-breeding groups (groups without chulengos or with undetected chulengos), and solitary animals. Guanaco contacts that could not be clearly assigned to any of these categories were placed in a fourth class called undetermined. A correct assignment of an observable group to a guanaco social unit could be possible (Franklin and Fritz 1991) but would require unavailable observation time, as in abundance surveys completing data collection in a period as short as possible (ideally a snapshot) is a priority. Finally, we also estimated density after pooling together all guanaco contacts.

Field surveys

The size, coloration, and behavior of guanacos make them easily detectable at distances of several hundred meters in the open Patagonian steppes where grasses and shrubs are generally <1 m high (Puig 1995). The prevalence of these conditions in the study area satisfies the key assumption of distance sampling methodology that all individuals that were on the road were detected (Bibby et al. 1992; Buckland et al. 2001; Thomas et al. 2010).

Road surveys were conducted from a vehicle during two consecutive birthing/breeding seasons (November 2004 to February 2005 and December 2005 to January 2006) in periods that lasted no more than 70 days per season. Given the extent of the area to be covered, we obtained data intensively, surveying during whole days with good visibility except periods within 1 h around twilight. Data were collected mostly at a speed of 20 km/h, by the driver and one observer, who exchanged their positions or were replaced every 2 h. Guanacos usually did not respond to our vehicle (with the exception of those that were on the road), so the distance sampling assumption that guanacos moved at much lower speeds than observers was also satisfied. A given road stretch was surveyed only during one breeding season and, in most cases, during a single day. All types of public roads were surveyed, although unpaved roads with scarce traffic (<10 vehicles/day) dominated (90%) the road network in Santa Cruz. When guanacos were detected, we stopped the vehicle, recorded group size (with binoculars when needed), and assigned the group to one of our three observable social units. We recorded the distance to the animal or the group center with a laser range finder (Leica LRF 1200 Rangemaster), thus fulfilling the distance sampling requirement of exact distance measurements. We also recorded our bearing relative to north obtained from the inertial compass reading of a GPS unit and the angle of the animal relative to our bearing. The perpendicular distance of contacts to the survey line was calculated from these data (Buckland et al. 2001).

Survey design

During the design of the study and in order to account for the expected regional variation in guanaco density associated with major environmental variability, stratified random sampling (Levy and Lemeshow 1999) was performed to establish which road segments would be surveyed each year (Figure 1b). We divided the study area into 12 geographical regions or strata, based on the combination of two environmental variables (Normalized Difference Vegetation Index, or NDVI, and mean slope) that we expected could influence the frequency of contacts with guanaco. Indeed, both variables accounted for guanaco distribution in Santa Cruz (Pedrana et al. 2010). NDVI was used because vegetation productivity could be an important driver of guanaco abundance, and mean slope because guanacos could prefer plains to rugged areas and their detectability could be affected by terrain irregularity. Further details about data sources and procedures are given by Travaini et al. (2007). We divided the values of mean NDVI into three classes and mean slope into two classes, which generated six possible environmental combinations. We selected the two largest polygons from each combination as seeds (12 seeds) and formed the final area by assigning the remaining polygons of a particular class to the closest seed.

Using vector coverage of roads, we randomly selected road segments, defined as complete or fractions of stretches between crossroads, up to a total length of 4,500 km during the first season (Figure 1b). To ensure that all strata would be properly sampled, 1,500 km were equally distributed among the 12 survey strata (125 km in each stratum in order to guarantee a minimum sampling effort in strata that could potentially cover small areas), and 3,000 km were subsequently allocated proportionally to the area of each stratum. During the second season, we randomly selected road segments not surveyed in the previous year up to completing a total of 8,141 km covered during the study (Figure 1b). Each road segment was surveyed just once, and segments surveyed in different years were typically hundreds of kilometers apart. The combination of randomization, exhaustive replication across strata and a balanced distribution among major environmental gradients led us to obtain abundance estimates representative for the whole study area. Again, environmental stratification helped to reduce the variance of density estimates (Thomas et al. 2010). Overall, our design allowed us to extrapolate density values beyond the sampled area and apply them to the whole extent of each stratum, i.e., to environmentally similar areas (Thomas et al. 2010), under the assumption that guanaco density tends to be homogeneous under similar ecological conditions.

Environments along and around roads seldom form a perfect random sample of ecosystems present in a region (Hawbaker et al. 2005; Bi et al. 2011). Using elevation as a proxy for environmental variation, we examined the extent to what mean altitude in 1 km cells overlapping roads (n = 23,230) differed from mean altitude in cells across Santa Cruz (n = 222,726), after excluding cells above 1,000 m which make up 8.6% of the study area. Along roads, frequencies expressed as proportions and computed for five 200 m classes of altitude between 0 and 1,000 m, ordered from low to high elevations, were as follows: 0.41, 0.31, 0.14, 0.11, and 0.03. Such values do not depart much from corresponding proportions of available elevations: 0.31, 0.30, 0.15, 0.15, and 0.09. This further suggests that major environmental variation across Santa Cruz was adequately captured by our road transects.

Abundance

Pedrana et al. (2010) modeled guanaco distribution in Santa Cruz using environmental predictors. Predicted probabilities of guanaco occurrence were divided in three classes and were plotted on a map (Figure 2). Assuming that predicted probabilities of detection are a positive function of guanaco abundance, this map can be read as a map with areas of expected low, medium, and high relative abundance of guanacos. Abundance estimates were calculated for each observable social unit, and for all guanaco contacts pooled together, within each area of expected relative abundance of guanacos and then for the whole study area. As predictions of guanaco occurrence were not made for water bodies, forests, areas above the treeline, and other habitats unsuitable for guanaco (Pedrana et al. 2010), the area for

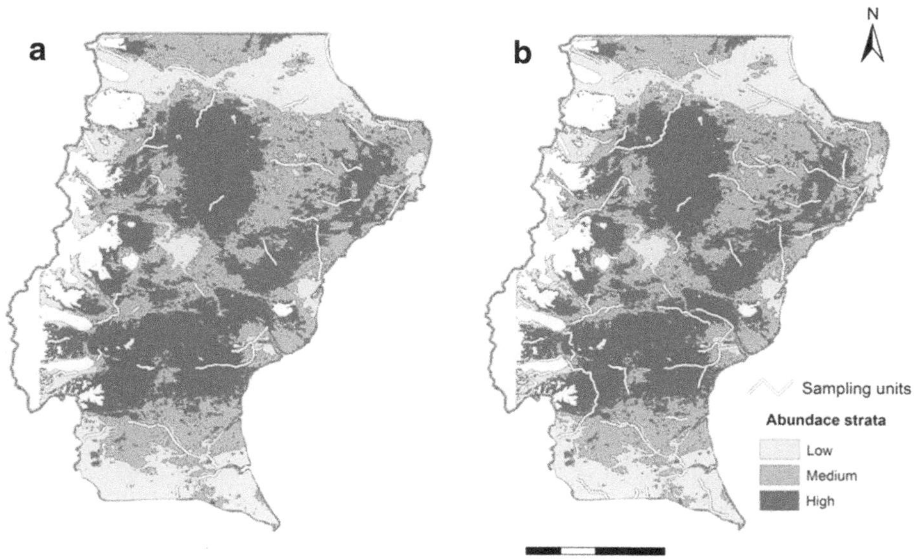

Figure 2 Distribution of road surveys for monitoring guanaco populations. Shading shows the expected relative abundance of guanacos assuming it is a positive function of the probability of guanaco occurrence. Probabilities of occurrence calculated by Pedrana et al. (2010) were divided in three classes: <0.4, 0.4 to 0.6, and >0.6. Lines denote one realization of the possible location of 30-km sampling units needed for monitoring guanaco numbers under a hypothetical objective of detecting a 50% population change in the next 10 years, with $a = 0.3$ and a minimum power of $1\text{-}\beta = 0.8$. **a** Example of 40 randomly placed sampling units required for detecting a population increase. **b** Example of 80 randomly placed sampling units needed to detect a population decrease.

which abundance was estimated (222,545 km^2; Table 1) was 91% of the surface of Santa Cruz province.

Line transect data were analyzed using program DISTANCE 5.0 (Buckland et al. 2001; Thomas et al. 2010). Estimates of guanaco densities were done by fitting a detection function to the perpendicular distances of guanaco groups to the survey line (Buckland et al. 2001). We pooled records across all transects and years within each area of guanaco relative abundance to estimate the detection function $g(x)$ for each social unit in order to attain the required minimum number of observations (Buckland et al. 2001). Although detectability varies among transects, the property of 'pooling robustness' ensures the reliability of abundance estimates when all data were analyzed together (Buckland et al. 2001; Thomas et al. 2010). Pooling helps even out fluctuations in guanaco density across surveyed transects as well as chance fluctuations in guanaco distribution (Fewster et al. 2005). Considering the extent of each area of guanaco relative abundance, density estimates were converted into estimates of population size. DISTANCE allows for identification of outlying observations at extreme distances, or sighting clumping, and suggests appropriate levels of truncation and grouping for fitting the detection function. We used the Akaike Information Criterion (AIC; Burnham and Anderson 1998) to choose the best competing model (key function + adjustment terms) for the detection function.

Monitoring program

We estimated the power of different monitoring designs to detect changes in guanaco population size with the aid of software MONITOR (Thomas and Krebs 1997; Gibbs and Ene 2010). MONITOR uses Monte Carlo simulations (Manly 1997) to model count surveys (encounter rate in our case) over time and generates detection rates derived from route-regression analyses (Geissler and Sauer 1990;

Table 1 Sampling effort and sightings across classes of expected relative abundance of guanacos

Guanaco relative abundance	Sampling effort (km)	Area (km^2)	Contacts with guanacos				
			Breeding groups	Non-breeding groups	Solitary individuals	Undetermined	Total
Low	2,314	59,306	13 (92)	21 (283)	31	8 (18)	73 (424)
Medium	4,149	86,789	119 (1,125)	152 (1,575)	167	100 (239)	538 (3,106)
High	1,678	76,450	71 (739)	123 (1,424)	126	60 (146)	380 (2,435)
Total	8,141	222,545	203 (1,956)	296 (3,282)	324	168 (403)	991 (5,965)

Distribution of sampling effort, in terms of road length and area surveyed, and number of contacts with guanacos (individuals within brackets) in three classes of expected guanaco relative abundance in Santa Cruz province, southern Patagonia, during the 2004 and 2005 breeding seasons.

Gibbs and Melvin 1997). This simulation procedure is useful for evaluating the trade-offs between sampling effort, logistical constraints, and power to detect trends (Gibbs et al. 1998; Field et al. 2005). To estimate power, we supplied the program with initial estimates of encounter rates and its variance for each scenario, that is, each combination of effect size (a pre-established percent increase or decrease in population size over a given period), time interval, class of guanaco relative abundance, and social unit. Monitoring output was defined in terms of sampling effort, survey frequency, and type I error. We arbitrarily defined sampling effort as the number of 30-km transects, but equally valid results could be obtained with shorter or longer transects. Based on our experience, a 30-km transect usually takes about 2 h to be completed, and 2 h was also the interval we used to switch observer and driver during field work. Additionally, random 30-km segments could be selected from the vector data of road coverage more easily than longer segments because distances between nodes of the road network were often >30 km. For each scenario, we estimated one value of statistical power after 1,000 replications. To select scenarios that we considered suitable for the design of a monitoring program, we set the minimum acceptable power at (1-β) = 0.80 (Di Stefano 2001), i.e., the highest probability of failing to detect an actual trend in guanaco abundance was 0.20.

As initial values, the program requires estimates of the mean and coefficient of variation of encounter rates which we derived empirically from our results.

Results

Guanaco abundance

We surveyed 8,141 km, approximately 93% of the public road network length in Santa Cruz. We recorded 991 contacts with guanaco social units that totaled 5,965 individuals (Table 1). Excluding contacts classified as undetermined, most guanacos were in groups, equally divided between breeding groups (47% of total contacts), and non-breeding groups (47%). The remaining 6% were solitary individuals (Table 1).

After considering competing detection functions, the Hazard-Rate key, with two-parameter cosine adjustment terms, was selected for all types of guanaco social units except solitary individuals, for which a function with four parameters was the most parsimonious. We truncated distance data at 500 m in all models except those for solitary individuals that were truncated at 600 m (Table 2), in

Table 2 Density and population size

Social unit	Relative abundance	Contacts	*g*(*w*)	Groups/km^2	*D* (ind/km^2)	Lower limit	Upper limit	*CV*	*N*
	Low	11	0.15	0.0431	0.4413	0.1289	1.5101	69.04	26,173
	Medium	103	0.13	0.2249	2.3047	0.8001	6.6393	57.77	200,020
	High	64	0.10	0.3455	3.5409	1.2066	10.3910	58.95	270,700
Breeding group	Pooled	178	0.12	0.2179	2.2328	0.7839	6.3600	57.04	496,900
	Low	18	0.14	0.0347	0.3712	0.1792	0.7681	37.81	22,013
	Medium	129	0.15	0.1413	1.5128	0.9988	2.2914	21.29	131,290
	High	100	0.19	0.2656	2.8437	1.6483	4.9059	27.77	217,400
Non-breeding group	Pooled	247	0.17	0.1556	1.6658	1.1253	2.4657	20.02	370,710
	Low	29	0.06		0.0814	0.0434	0.1643	36.72	4,828
	Medium	159	0.05		0.2538	0.1495	0.4309	27.42	22,029
	High	111	0.12		0.4297	0.2375	0.7774	30.76	32,849
Solitary individuals	Pooled	299	0.08		0.2683	0.1595	0.4512	26.90	59,706
	Low	7	0.13	0.0396	0.0910	0.0201	0.4130	89.33	5,402
	Medium	86	0.14	0.2766	0.6364	0.1479	2.7370	85.15	55,236
	High	47	0.22	0.3665	0.8434	0.1964	3.6225	85.04	64,481
Undetermined	Pooled	140	0.17	0.2443	0.5622	0.1328	2.3801	83.91	125,120
	Low	63	0.14	0.1920	1.1228	0.6663	1.8920	26.89	66,590
	Medium	464	0.14	4.7028	4.7028	3.1082	7.1156	21.32	408,160
	High	315	0.17	1.3236	7.7419	4.9303	12.1570	23.16	591,870
All contacts	Pooled	842	0.15	0.8194	4.7928	3.2704	7.0240	19.65	1,066,600

Estimates of guanaco density (*D*) and population size (*N*) for each observable social unit and in three classes of expected relative abundance of guanacos in Santa Cruz Province, southern Patagonia, during the 2004 and 2005 breeding seasons. The number of intervals used to fit the most parsimonious detection function was 10 in all cases. *g*(*w*) is the value of the detection function at the truncation distance (600 m for solitary guanacos, 500 m for other social units). Upper and lower limits of 95% confidence intervals for density estimates are reported. *CV* coefficient of variation of density estimates.

order to attain a robust line transect analysis after the elimination of outliers (Buckland et al. 2001). Our truncation distance closely satisfied $g(w) \approx 0.15$, where $g(w)$ is the detection function at the truncation distance w (Table 2). After truncation, the number of contacts used to estimate guanaco density in each observable social unit is shown in Table 2. Within truncation distances, sightings at distance zero accounted for 9.4% of contacts, whereas a further 31.1% of contacts were made within 50 m of the survey line. The band 0 to 50 m was the modal class in a uniform distribution of distance classes at 50-m intervals. Within the modal class, mean (±SD) perpendicular distance to contacts was 24 ± 12 m.

We estimated a mean density of 4.79 ind/km^2 with a coefficient of variation of 20% (Table 2). As expected, density estimates increased with the predicted probability of guanaco occurrence defined by Pedrana et al. (2010) as classes of expected relative abundance. Overall mean density greatly varied from areas with expected low relative abundance (1.12 ind/km^2) to areas with expected high relative abundance (7.74 ind/km^2; Table 2). Across social units, the highest value of mean density was found for breeding groups (2.23 ind/km^2), which was a 34% higher than that for non-breeding groups (1.67 ind/km^2; Table 2). High density values were also associated with high variances, which for breeding groups were about twice as high as for non-breeding groups or solitary guanacos (Table 2).

As classes of relative abundance were built with distribution models based on environmental predictors (Pedrana et al. 2010), assuming that guanaco densities fall within the confidence intervals shown in Table 2 in similar environments, we estimated a total guanaco population of 1,066,600 individuals in Santa Cruz (95% CI 727,800 to 1,563,200). Half of the estimated population size would appear in the form of breeding groups (Table 2). There was a remarkable match (a difference of only 1.33%) between estimates of guanaco population size obtained by summing up the estimated abundances for each guanaco social unit (N = 1,052,436 individuals) and those obtained after pooling all guanaco contacts together, although the confidence of the estimates using the former approach was lower (95% CI 489,900 to 2,594,200; Table 2).

Population monitoring

Detection of changes in guanaco population size over a 10-year period, either an increase or a decrease, would require considerable monitoring effort (Table 3). This effort grows significantly if a population trend in shorter periods, such as 5 years, is to be detected (Table 3). For instance, annual surveys of 40 30-km transects would be necessary for detecting a 50% increase in total guanaco

Table 3 Sampling effort needed to monitor guanaco population trends in Santa Cruz province, southern Patagonia

		Monitoring program scenario							
		50% in 5 years				50% in 10 years			
		Increase		Decrease		Increase		Decrease	
		α = 0.2	α = 0.3	α = 0.2	α = 0.3	α = 0.2	α = 0.3	α = 0.2	α = 0.3
	Relative abundance	80	50	>120	90	40	20	80	60
All guanaco groups	Low	110	80	>120	>120	80	40	120	80
	Medium	70	50	120	80	30	20	80	50
	High	40	30	70	50	20	20	40	30
Breeding groups	Low	>120	>120	>120	>120	>120	>120	>120	>120
	Medium	>120	>120	>120	>120	>120	>120	>120	>120
	High	60	40	110	80	30	20	70	40
Non-breeding groups	Low	>120	>120	>120	>120	120	>120	80	>120
	Medium	>120	70	>120	>120	60	40	120	90
	High	70	50	120	80	20	20	70	50
Solitary individuals	Low	>120	120	>120	>120	90	60	>120	120
	Medium	60	40	120	80	30	20	60	50
	High	50	30	90	60	30	20	50	40
Breeding groups		>120	>120	>120	>120	>120	>120	>120	>120
Non-breeding groups		>120	100	>120	>120	70	50	>120	90
Solitary individuals		90	60	>120	100	40	30	90	60

Sampling effort is expressed as the number of randomly located 30-km survey transects needed to attain the desired power of 1-β = 0.80 for each hypothetical scenario. Separate estimates of sampling effort are also given for areas with different combinations of classes of guanaco relative abundance and type of guanaco group. Estimates are also given after pooling all classes of relative abundance.

abundance during the next 10 years, keeping $\alpha = 0.2$ (Figure 2a). Detecting a 50% decrease in 10 years would require surveying twice as many transects every year (Figure 2b). The sampling effort required to detect a 50% increase during the next 5 years would be 80 transects. Finally, if we want to detect a 50% decrease within 5 years, we should relax type I error to $\alpha = 0.3$ and survey 90 transects every year. Otherwise, sampling effort would increase over 120 transects (Table 3). Monitoring efforts for specific combinations of strata of predicted probability of guanaco occurrence and types of guanaco groups range from 20 to >120 transects (Table 3).

Discussion

Our results confirm the expectation that a relatively high density of guanacos could be found in Santa Cruz (Baldi et al. 2010), where no comprehensive survey of abundance had been carried out before. Comparison with previously published figures for Santa Cruz is difficult because both ecological conditions and estimation methods used in available studies were quite different. For example, Manero et al. (1986), based on a single 119-km transect, estimated a summer density of 0.37 to 0.94 ind/km^2 for an unspecified area located in central Santa Cruz, which would correspond to a region with medium-high predicted probabilities of guanaco occurrence (Pedrana et al. 2010) and for which our density estimates were 5 to 8 times higher. The estimates of Manero et al. (1986) were based on an encounter rate of 0.28 ind/km, a value much larger than the detection rate we estimated in the medium and high abundance areas (0.11 and 0.19 ind/km, respectively). The low values reported by Manero et al. (1986) were contemporary with intensive sheep ranching and associated yearly culling of chulengos during the 1970s, but were also not derived from distance sampling which could also help explain this marked discrepancy.

Our density estimates (1.1 to 7.7 ind/km^2) fall within the range of values reported elsewhere in Patagonia. In northern areas, densities are also highly variable (1.0 to 13.9 ind/km^2; Puig et al. 1997), and extreme values over 40 ind/km^2 have been calculated (Puig et al. 2008). The same pattern can be found in central Patagonia where densities ranged between 0.6 and 8.8 ind/km^2 (e.g., Saba and Battro 1987; Baldi et al. 1997; Marino and Baldi 2008). Variability in higher latitudes outside Santa Cruz, including Tierra del Fuego, varied in a narrower range (0.3 to 4.7 ind/km^2; Raedeke 1982; Bonino and Fernández 1994; Ortega and Franklin 1995; Montes et al. 2000). In general, guanaco densities were lower in areas of high primary productivity if sheep occur in high density (Montes et al. 2000; Baldi et al. 2001, 2004), and our results agree with this pattern as we found high guanaco densities in areas with low sheep abundance identified by Pedrana et al. (2010).

In summary, guanaco density has been reported for about a dozen localities throughout Patagonia, generally very small in size as compared with Santa Cruz (e.g., Saba and Battro 1987; Ortega and Franklin 1995; Baldi et al. 1997; Puig et al. 1997). Baldi et al. (2010) suggested that guanaco density is rarely higher than 2 ind/km^2 in northern and central Patagonia, which would be consistent with the global population estimate of <600,000 individuals that is currently accepted (Franklin et al. 1997; Baldi et al. 2010, IUCN 2013). In agreement with the contrasting densities reported in the studies listed above, our results reveal large spatial variations in guanaco density, with the difference between extreme estimates being sevenfold within Santa Cruz. This pattern can be unveiled only through extensive, large-scale surveys covering major environmental gradients. As published average guanaco densities for Patagonia are extrapolations based on densities calculated over areas much smaller than Santa Cruz, the risk exists that large-scale variation in density is missing, introducing biases in population size estimates for the whole guanaco geographic range. Indeed, our estimates of population size in Santa Cruz, even when the lower endpoint of 95% confidence intervals was used, are as large or larger than estimates of global guanaco population size (Baldi et al. 2010; IUCN 2013), suggesting that these deserve revision. Large-scale surveys that use distance sampling techniques could produce more accurate estimates of guanaco density, population size, and its variation than estimated based on strip transects over relatively small areas.

A critical assumption of distance sampling in our road surveys was that the distribution of guanacos is independent of roads (Thomas et al. 2010). The assumption of independence would be violated if guanacos were closer or farther from roads than expected from a uniform or a random distribution. This possibility is entirely dependent on guanaco behavior and requires hypotheses about the underlying mechanism. Pedrana et al. (2010) found low probabilities of detecting guanacos from roads within 20 km of the few small cities that occur in Santa Cruz, and hypothesized that this pattern could represent a response to hunting from roads. This observation could be in agreement with the hypothesis by Donadio and Buskirk (2006) that recurrent exposure to hunting alters foraging and, consequently, guanacos could avoid these disturbed sites. However, >95% of our survey was conducted far away from cities, in depopulated areas and with scanty traffic during most of the year. In general, we did not notice any consistent avoidance behavior of guanacos in response to our presence, namely, a stopped car with people pointing instruments to them, even at close distances. Although the repulsion hypothesis seems unlikely in Santa Cruz, if road avoidance by guanacos were a general pattern, it would have contributed to underestimate currently accepted density and numbers of guanaco across Patagonia.

Another possibility is that guanacos were attracted to roads. No specific mechanistic hypothesis has been formally forwarded in this regard, but the potential presence of good quality forage at road verges, perhaps as a result of disturbance associated with road maintenance, could be one of them. This hypothesis predicts a concentration of guanaco observations at distance zero, i.e., the road verge, where the hypothesized valuable resource should occur. This prediction was rejected because <10% of observations were recorded on the road, whereas the modal distance class for all social units was 0–50 m, where guanacos were foraging on regular steppe vegetation. Moreover, in most sites, we were unable to detect major differences in the abundance or composition of plants at road verges as compared with surrounding vegetation, although examining this edge effect requires specific studies. Without surveys along truly random lines (randomly placed straight lines away from roads or smaller tracks) and without a plausible mechanistic explanation for the high frequency of guanaco sightings within 50 m, the hypothesis of guanaco attraction to roads and the hypothesis or a mere decrease in guanaco detectability with distance to roads cannot be distinguished with our data. Considering existing evidence, we assume that the distribution of guanacos was independent of roads over most of the road network in Santa Cruz, and preliminarily conclude that biases in our density estimates due to the violation of this assumption are implausible.

In Patagonia, two alternative survey methods could be used to test the assumption that guanaco distribution is independent of roads, but it is unclear that these methods would allow extensive distance sampling over large regions. Horseback distance sampling has a number of disadvantages: rugged terrain could hardly be sampled, surveying areas far away from human habitation would be unfeasible (with the consequent bias), and it would be difficult to cover large regions during a short session. Aircraft distance sampling would solve some of these problems and therefore is desirable but, on the other hand, logistic costs of aerial surveys make them much more expensive than road surveys (Travaini et al. 2007).

A second concern regarding randomization of transects was pointed out by Fewster et al. (2005): the full enumeration of sources of variation in detectability is generally impossible for animals because of factors such as habitat, sighting conditions, and observer experience and alertness. Survey design should ensure that all transects in the study area have the same probability of being included in the sample. We think that our distribution of road segments among 2 years of study was suitable considering the logistic and time constraints imposed by the large area to be covered. Moreover, as guanacos exhibit quite sedentary habits and long-distance movements are rare even for migratory individuals (Mueller et al. 2011), separation between segments surveyed in different years further ensure independence of observations.

Interest in sustainable use of the guanaco has increased in Argentina and Chile. National strategies are concerned with evaluating, conserving, and restoring wild populations in Argentina (Puig 1992; Baldi et al. 2006), Chile (Soto 1993; Skewes and Soto 2003), Bolivia (Velasco et al. 1992), Ecuador (Paucar 1992), Paraguay (Ríos 1992), and Perú (Hoces 1992). Large-scale assessments of variation in density, such as the one we present here, are a valuable tool for developing management plans that integrate conservation and sustainable exploitation of guanaco populations (Baldi et al. 1997). Our estimation of guanaco population size in Santa Cruz is only half the number of sheep currently raised in this province (2,607,000 heads; INDEC 2011), which is also the maximum number of sheep stocked in the province during the last decade (INDEC 2009). More than 60% of Santa Cruz area, including the central basaltic highlands, interspersed with marine and continental sediments of variable susceptibility to erosion in the lowlands, has a sheep carrying capacity of 5–20 ind/km^2 (Pablo Peri, personal communication). This low stocking rate corresponds to areas heavily overgrazed in the past where sustainable sheep ranging is no longer possible (Golluscio et al. 1998). Precisely, we estimated the highest guanaco abundances there (7.74 ind/km^2) suggesting that in such arid and degraded shrub-steppes, guanacos could be viewed as an opportunity for sustainable development based upon a profitable natural resource (Sacchero et al. 2006).

Guanaco densities vary across space and over time, depending on natural (Puig et al. 1997) and human factors (Baldi et al. 2001, 2004). This variability should be considered in any extractive activity and subsequent commercial use of populations. In our view, knowledge of sex and age composition as well as other demographic parameters is incompatible with the quick assessment of group composition imposed by surveying a very large area in a few months. This could be attempted by several teams working simultaneously in different parts of the area, as guanaco demographic details would be needed before planning any extractive activity in a local population. Survey techniques that can derive sex ratio and age structure of a population are advantageous, as these provide potential insight into recent history, current status, and future population trends (Dimmick and Pelton 1996). Sex ratios and age classes are necessary for modeling guanaco populations and for determining harvest rates (Franklin and Fritz 1991, Fritz and Franklin 1994; Franklin et al. 1997), but are not essential when only estimating population size and trends, which requires at most a quick assignment of contacts to social units. Failure to estimate reliable sex ratios of guanaco populations from road-survey data is not uncommon, as assignment of contacts to social units on the sole basis of guanaco behavioral traits is not always reliable (Pedrana et al. 2009). In applications where one needs to know sex and age structure, this problem can be minimized, if not overcome, if surveys are conducted towards the end of the birth season when chulengos are abundant. In addition, with training and experience, sexes can be identified even at moderate distance by morphological characteristics (testes/vulva, penis/mammary gland, and body shape) and behavioral cues, as demonstrated by field studies with a more acceptable 3 to 5% unclassified animals (Franklin and Fritz 1991, Fritz and Franklin 1994; Ortega and Franklin 1995; Sarno and Franklin 1999; Young and Franklin 2004b). Methodological improvements by guanaco researchers and managers need to be made to substantially increase the reliability of field identification of guanaco sexes and social groups.

Monitoring guanacos in Santa Cruz could be wrongly considered unaffordable by wildlife managers if not evaluated within a cost-benefit analysis. The economic cost of well interspersed survey transects throughout the territory, developed by a team of three experienced observers during 10 to 15 days, should be counterbalanced against the potential economic benefit of direct income derived from guanaco products and the large conservation benefits arising from land use diversification. Additionally, considering the relatively predictable guanaco local distribution due to male territorial fidelity over the years, the effect of management practices on guanaco populations could be easily surveyed and monitored to evaluate trends (Young and Franklin 2004a). The sustained yield harvesting of guanaco wool and/or meat, complementary or supplementary to sheep ranching, may be the most effective way to conserve guanacos (Fritz and Franklin 1994) and other wildlife, if land owners acknowledge the potential economic return of this species. Diversifying land uses and adopting sustainable forms of guanaco exploitation should be the shared objectives of Argentine wildlife agencies, as a potential mechanism pursuing both the reduction of the ecological damage of sheep overgrazing and the socioeconomic improvement of rural communities in Patagonia rangelands.

Conclusions

Our results show that Santa Cruz contains a large population of guanacos and confirms that guanaco density tends to be higher in areas with low sheep density. Estimates of population size of nearly one million individuals exceed currently accepted estimates for the guanaco geographic range which suggest that the latter need to be revised. Guanaco density shows considerable variation across large-scale environmental gradients. Therefore, detecting a 50% population change within 10 years would require annual surveys of up to 80 30-km road transects evenly distributed across areas with relatively homogeneous environmental conditions. Yet, large-scale road surveys, combined with distance sampling techniques, can be recommended as a cost-effective method to monitor guanaco population trends over large regions. Considering the growing interest in sustainable use of guanaco populations, monitoring is needed to quantify its effects, to inform management, and to plan conservation measures in less abundant or declining populations.

Competing interests

The authors declare that they have no competing interests.

Authors' contributions

AR, AT, and JB devised the study. All authors collected field data.
AT performed data analysis and wrote the first draft. JP prepared the figures. All authors commented on and contributed to the draft.
AR wrote the manuscript. All authors read and approved the final manuscript.

Acknowledgements

This work was primarily funded by the BBVA Foundation through a grant under the Conservation Biology Programme to A. Rodríguez and the Agencia Nacional de Promoción Científica y Tecnológica (PICTO Nr. 30723). Additional support was provided by Universidad Nacional de la Patagonia Austral, CONICET (PEI-6065), CONAE, and the Secretaría de Ambiente y Desarrollo Sustentable. Julieta Pedrana was supported by a CONICET (Argentina) predoctoral fellowship. Diego Procopio, Emilio Daher, Miguel Santillán, Martín Yaya, and Mara Brossman assisted during field work. Javier Bustamante acknowledges funding from the Visiting Academic Program of the Spanish Ministry of Education for a sabbatical stay at the Australian Commonwealth Scientific and Research Organization (CSIRO) Division of Marine and Atmospheric Research during the preparation of drafts. William Franklin, Pablo Carmanchahi, and four anonymous reviewers critically reviewed and improved the manuscript.

Author details

[1]Centro de Investigación de Puerto Deseado, Universidad Nacional de la Patagonia Austral, CONICET, Avenida Prefectura Naval s/n, 9050 Puerto Deseado, Santa Cruz, Argentina. [2]Centro de Investigación de Puerto Deseado, Universidad Nacional de la Patagonia Austral, Avenida Prefectura Naval s/n, 9050 Puerto Deseado, Santa Cruz, Argentina. [3]Department of Wetland Ecology, Remote Sensing and GIS Lab (LAST-EBD), Estación Biológica de Doñana, CSIC, Américo Vespucio s/n, 41092 Sevilla, Spain. [4]Recursos Naturales y Gestión Ambiental, Instituto Nacional de Tecnología Agropecuaria, EEA Balcarce, CC 276, 7620 Balcarce, Buenos Aires, Argentina. [5]Centro para el Estudio y Conservación de las Aves Rapaces en Argentina, UNLPam, INCITAP, CONICET, Avda. Uruguay 151, 6300 Santa Rosa, La Pampa, Argentina. [6]Department of Conservation Biology, Estación Biológica de Doñana, CSIC, Américo Vespucio s/n, 41092 Sevilla, Spain.

References

Amaya JN, von Thüngen J, de Lamo DA (2001) Relevamiento y distribución de guanacos en la Patagonia. Comunicación Técnica no. 109. Instituto Nacional de Tecnología Agropecuaria, Bariloche
Andrade L (2005) Sociología de la desertificación. Los productores ovinos extensivos de la Patagonia Austral. Minio & Dávila Editores, Buenos Aires
Baldi R, Campagna C, Saba S (1997) Abundancia y distribución del guanaco (*Lama guanicoe*), en el NE del Chubut, Patagonia Argentina. Mastozool Neotrop 4:5–15
Baldi R, Albon SD, Elston DA (2001) Guanacos and sheep: evidence for continuing competition in arid Patagonia. Oecologia 129:561–570

Baldi R, Pelliza Sbriller A, Lestón D, Albon SD (2004) High potential for competition between guanacos and sheep in Patagonia. J Wildl Manag 68:924–938

Baldi R, de Lamo D, Failla M, Ferrando P, Funes M, Nugent P, Puig S, Rivera S, von Thüngen J (2006) Plan nacional de manejo del guanaco (*Lama guanicoe*). Secretaría de Ambiente y Desarrollo Sustentable de la. Nación, Buenos Aires

Baldi R, Novaro A, Funes M, Walker S, Ferrando P, Failla M, Carmanchahi P (2010) Guanaco management in Patagonian rangelands: a conservation opportunity on the brink of collapse. In: du Toit J, Kock R, Deutsch J (eds) Wild rangelands. Wiley-Blackwell, Chichester, pp 266–290

Bank MS, Sarno RJ, Franklin LW (2003) Spatial distribution of guanaco mating sites in southern Chile: conservation implications. Biol Conserv 112:427–434

Bi X, Wang H, Zhou R (2011) The influence of landscape features on road development in a loess region, China. Environ Manage 48:774–780

Bibby CJ, Burgess ND, Hill DA (1992) Bird census techniques. Academic Press, London

Bonino N, Fernández E (1994) Distribución general y abundancia relativa de guanacos (*Lama guanicoe*) en diferentes ambientes de Tierra del Fuego, Argentina. Ecol Austral 4:79–85

Borrelli P, Cibils A (2005) Rural depopulation and grassland management in Patagonia. In: Reynolds SG, Frame J (eds) Grasslands: developments, opportunities, perspectives. Science Publishers, Enfield, NH, pp 461–487

Buckland ST, Anderson DR, Burnham KP, Laake JL, Borchers DL, Thomas L (2001) Introduction to distance sampling: estimating abundance of biological populations. Oxford University Press, Oxford

Burnham KP, Anderson DR (1998) Model selection and multimodel inference: a practical information-theoretic approach. Springer, New York

Cunazza C, Puig S, Villalba L (1995) Situación actual del guanaco y su ambiente. In: Puig S (ed) Técnicas para el manejo del guanaco. IUCN, Gland, Switzerland, pp 27–50

Di Stefano J (2001) Power analysis and sustainable forest management. Forest Ecol Manage 154:141–153

Dimmick RW, Pelton MR (1996) Criteria of sex and age. In: Bookhout TA (ed) Research and management techniques for wildlife and habitats. The Wildlife Society, Bethesda, MD, pp 169–214

Donadio E, Buskirk SW (2006) Flight behavior in guanacos and vicuñas in areas with and without poaching in western Argentina. Biol Conserv 127:139–145

Elzinga CL, Salzer DW, Willoughby JW, Gibas JP (2001) Monitoring plant and animal populations. Blackwell Science, Oxford

Fewster RM, Laake JL, Buckland ST (2005) Line transect sampling in small and large regions. Biometrics 61:856–861

Field SA, Tyre AJ, Possingham HP (2005) Optimizing allocation of monitoring effort under economic and observational constraints. J Wildl Manage 69:473–482

Franklin WL (1982) Mammalian biology in South America. In: Mares MA, Genoways HH (eds) Biology, ecology, and relationship to man of the South American camelids, vol 6, Pymatuning symposia in ecology, special publication series. University of Pittsburgh, Pittsburgh, PA, pp 457–489

Franklin WL (1983) Advances in the study of mammalian behavior. In: Eisenberg JF, Kleiman D (eds) Contrasting socioecologies of South America's camelids: the vicugna and guanaco. Special Publication No. 7. American Society of Mammalogists, Shippensburg, PA, pp 573–629

Franklin WL, Fritz MA (1991) Sustained harvesting of the Patagonia guanaco: is it possible or too late? In: Robinson JG, Redford KH (eds) Neotropical wildlife use and conservation. The University of Chicago Press, Chicago, IL, pp 317–336

Franklin WL, Bas F, Bonacic CF, Cunazza C, Soto N (1997) Striving to manage Patagonia guanacos for sustained use in the grazing agroecosystem of southern Chile. Wildl Soc Bull 25:65–73

Fritz MA, Franklin WL (1994) First estimates of guanaco male group harvestability in the Patagonia of South America. Vida Silvestre Neotrop 3:84–90

Garrido JL (1985) La utilización del guanaco (Lama guanicoe, Müller 1776), su situación en Argentina. Actas V Convención Internacional sobre Camélidos Sudamericanos. Cuzco, Peru

Geissler PH, Sauer JR (1990) Topics in route regression analysis. In: Sauer JR, Droedge S (eds) Survey designs and statistical methods for the estimation of avian population trends. Biological Report, vol 90 (1). US Fish and Wildlife Service, Washington DC, pp 54–57

Gibbs JP, Ene E (2010) Program monitor: estimating the statistical power of ecological monitoring programs., Version 11.0.0. http://www.esf.edu/efb/gibbs/monitor/ Accessed Sep 2013

Gibbs JP, Melvin SM (1997) Power to detect trends in waterbird abundance with call-response surveys. J Wildl Manage 61:1262–1267

Gibbs JP, Droedge S, Eagle P (1998) Monitoring populations of plant and animals. Bioscience 48:935–940

Golluscio RA, Deregibus VA, Paruelo JM (1998) Sustainability and range management in the Patagonian steppes. Ecol Austral 8:265–284

González L, Rial P (2004) Guía geográfica interactiva de Santa Cruz. INTA-CAP-UNPA, Río Gallegos, Argentina

Hawbaker TJ, Radeloff VC, Hammer RB, Clayton MK (2005) Road density and landscape pattern in relation to housing density, land ownership, land cover, and soils. Landscape Ecol 20:609–625

Hoces RD (1992) Situación del guanaco en Perú. In: Marchetti B, Oltremari Arregui J, Peters H (eds) Estrategias para el manejo y aprovechamiento racional del guanaco (*Lama guanicoe*). Oficina Regional de la FAO para América Latina y el Caribe, Santiago, Chile, pp 139–148

INDEC (2009) Censo Nacional Agropecuario. Instituto Nacional de Estadística y Censos, Government of Argentina, http://www.indec.mecon.ar Accessed

INDEC (2011) Censo Nacional Agropecuario. Instituto Nacional de Estadística y Censos, Government of Argentina, http://www.indec.mecon.ar Accessed

IUCN (2013) IUCN Red List of Threatened Species., Version 2013.2. http://www.iucnredlist.org Accessed Feb 2014

León RJC, Aguiar MR (1985) El deterioro por uso pastoral en estepas herbáceas patagónicas. Phytocoenologia 13:181–196

Levy PS, Lemeshow S (1999) Sampling of populations, methods and applications. Wiley series in probability and statistics. Wiley, New York

MACS (2002) Wild Camelid Management. Bulletin of Proyecto MACS-International Cooperation for improved management of the vicuña and guanaco, 1st edn., pp 1–12

Manero A, Amaya J, Fernández ME, Clarke R (1986) Recuento de guanaco (*Lama guanicoe*) en el centro de la provincial de Santa Cruz. Instituto Nacional de Tecnología Agropecuaria, Comunicación Técnica 29:58–66

Manly BFJ (1997) Randomization, boostrap and Monte Carlo methods in biology, 2nd edn. Chapman and Hall, London

Marino A, Baldi R (2008) Vigilance patterns of territorial guanacos (*Lama guanicoe*): the role of reproductive interests and predation risk. Ethology 114:413–423

Mattioli S (2011) Family Cervidae (Deer). In: Wilson DE, Mittermeier RA (eds) Handbook of the mammals of the world, vol 2, Hoofed mammals. Lynx Ediciones, Barcelona, Spain, pp 350–443

Mazzoni E, Vázquez M (2004) Ecosistemas de mallines y paisajes de la Patagonia Austral (Provincia de Santa Cruz). INTA-UNPA, Río Gallegos, Argentina

Montes C, De Lamo DA, Zavatti J (2000) Distribución de abundancias de guanacos (*Lama guanicoe*) en los distintos ambientes de Tierra del Fuego, Argentina. Mastozool Neotrop 7:23–31

Montes C, Carmanchahi PD, Rey A, Funes MC (2006) Live shearing free-ranging guanacos (*Lama guanicoe*) in Patagonia for sustainable use. J Arid Environ 64:616–625

Mueller T, Olson KA, Dressler G, Leimgruber P, Fuller TK, Nicolson C, Novaro AJ, Bolgeri MJ, Wattles D, DeStefano S, Calabrese JM, Fagan WF (2011) How landscape dynamics link individual- to population-level movement patterns: a multispecies comparison of ungulate relocation data. Global Ecol Biogeogr 20:683–694

Ojeda RA, Mares MA (1982) Conservation of South American mammals: Argentina as a paradigm. In: Mares MA, Genoways HH (eds) Mammalian biology in South America, vol 6, Pymatuning Symposia in Ecology. Special Publication Series. University of Pittsburgh, Pittsburgh, PA, pp 505–521

Oliva GL, González L, Rial P (2004) Áreas ecológicas. In: González L, Rial P (eds) Guía Interactiva de Santa Cruz. INTA-CAP-UNPA, Río Gallegos, Argentina, pp 14–15

Ortega IM, Franklin WI (1995) Social organization, distribution and movements of a migratory guanaco population in the Chilean Patagonia. Rev Chil Hist Nat 68:489–500

Paucar MA (1992) Situación del guanaco en Ecuador. In: Marchetti B, Oltremari Arregui J, Peters H (eds) Estrategias para el manejo y aprovechamiento racional del guanaco (*Lama guanicoe*). Oficina Regional de la FAO para América Latina y el Caribe, Santiago, Chile, pp 131–135

Pedrana J, Rodríguez A, Bustamante J, Travaini A, Zanón Martínez JI (2009) Failure to estimate reliable sex ratios of guanaco from road-survey data. Can J Zool 87:886–894

Pedrana J, Bustamante J, Travaini A, Rodríguez A (2010) Factors influencing guanaco distribution in southern Argentine Patagonia and implications for its sustainable use. Biodiv Conserv 19:3499–3512

Puig S (1992) Situación del guanaco en Argentina: Estado del conocimiento y perspectivas de manejo. In: Marchetti B, Oltremari Arregui J, Peters H

(eds) Estrategias para el manejo y aprovechamiento racional del guanaco (*Lama guanicoe*). Oficina Regional de la FAO para América Latina y el Caribe, Santiago, Chile, pp 79–95
Puig S (1995) Abundancia y distribución de las poblaciones de guanacos. Técnicas para el manejo del guanaco. IUCN, Gland, Switzerland
Puig S, Videla F, Cona MI (1997) Diet and abundance of the guanaco (*Lama guanicoe* Muller 1776) in four habitats of northern Patagonia, Argentina. J Arid Environ 36:343–357
Puig S, Ferraris G, Superina M, Videla F (2003) Distribution of densities of guanacos (*Lama guanicoe*) in the northern La Payunia protected area and its area of influence (Mendoza, Argentina). Multequina 12:37–48
Puig S, Videla F, Cona MI, Roig VG (2008) Habitat use by guanacos (*Lama guanicoe*, Camelidae) in northern Patagonia (Mendoza, Argentina). Stud Neotrop Fauna Environ 43:1–9
Raedeke KJ (1982) Habitat use by guanacos (*Lama guanicoe*) and sheep on common range, Tierra del Fuego, Chile. Turrialba 32:309–314
Redford KH, Eisenberg JF (1992) Mammals of the Neotropics: the Southern Cone. The University of Chicago Press, Chicago
Rey A, Carmanchahi PD, Puig S, Ghichón ML (2009) Densidad, estructura social, actividad y manejo de guanacos silvestres (*Lama guanicoe*) en el sur del Neuquén, Argentina. Mastozool Neotrop 16:389–401
Ríos AE (1992) Situación del guanaco en Paraguay. In: Marchetti B, Oltremari Arregui J, Peters H (eds) Estrategias para el manejo y aprovechamiento racional del guanaco (*Lama guanicoe*). Oficina Regional de la FAO para América Latina y el Caribe, Santiago, Chile, pp 137–138
Saba SL, Battro P (1987) Estimación de la densidad poblacional de Guanacos (*Lama guanicoe* Muller). Turrialba 37:113–118
Sacchero D, Maurino MJ, Lanari MR (2006) Diferencias de calidad y proporción de down en muestras individuales de vellones de guanaco (*Lama guanicoe*) en distintas ecorregiones de Argentina. Rev Argent Produc Anim 26:211–216
Sarno RJ, Franklin WL (1999) Maternal expenditure in the polygynous and monomorphic guanaco; suckling behaviour, reproductive effort, yearly variation, and influence on juvenile survival. Behav Ecol 10:41–47
Skewes O, Soto N (2003) The guanaco, *Lama guanicoe* M., of Tierra del Fuego, Chile. A program for conservation and use. Universidad de Concepción, Concepción, Chile
Soriano A, Paruelo JM (1990) El manejo de campos de pastoreo en Patagonia: aplicación de principios ecológicos. Ciencia Hoy 2:44–53
Soto VN (1993) Situación del guanaco en la XII región de Chile y rol de CONAF en la conservación del recurso. In: Bas F, Bonacic C (eds) Actas I Taller Binacional de Manejo Sustentable del Guanaco (*Lama guanicoe*) en la Patagonia Chileno-Argentina. Pontificia Universidad Católica de Chile, Santiago, Chile, pp 9–17
Soto VN, González F, Silva R (1991) Alternativas de elaboración de charqui de guanaco. In: Oltremari J (ed) Gestión en recursos naturales: un enfoque integrado para el desarrollo. Universidad Austral de Chile, Valdivia, Chile, pp 376–405
Steidl RJ, Hayes JP, Schauber E (1997) Statistical power analysis in wildlife research. J Wildl Manage 61:270–279
Taylor BL, Gerrodette T (1993) The uses of statistical power in conservation biology: the Vaquita and Northern Spotted Owl. Conserv Biol 3:489–500
Thomas L, Krebs CJ (1997) A review of statistical power analysis software. Bull Ecol Soc Amer 78:126–139
Thomas L, Buckland ST, Rexstad EA, Laake JL, Strindberg S, Hedley SL, Bishop JRB, Marques TA, Burnham KP (2010) Distance software: design and analysis of distance sampling surveys for estimating population size. J Appl Ecol 47:5–14
Travaini A, Bustamante J, Rodríguez A, Zapata S, Procopic D, Pedrana J, Martínez Peck R (2007) An integrated framework to map animal distributions in large and remote regions. Divers Distrib 13:289–298
Velasco A, Cardozo A, Alzerrega H (1992) Situación del guanaco en Bolivia. In: Marchetti B, Oltremari Arregui J, Peters H (eds) Estrategias para el manejo y aprovechamiento racional del guanaco (*Lama guanicoe*). Oficina Regional de la FAO para América Latina y el Caribe, Santiago, Chile, pp 105–111
Young JK, Franklin WL (2004a) Territorial fidelity of male guanacos in the Patagonia of Southern Chile. J Mamm 85:72–78
Young JK, Franklin JL (2004b) Activity budget patterns in family-group and solitary territorial male guanacos. Rev Chil Hist Nat 77:617–625

11

Heterochromatin distribution and localization of nucleolar organizing regions in the 2n = 52 cytotypes of *Nannospalax xanthodon* and *N. ehrenbergi* from Turkey

Atilla Arslan[1] and Jan Zima[2*]

Abstract

Background: The blind mole rats (Spalacinae) are fossorial rodents exhibiting exceptionally extensive variation in the karyotype. The taxonomy of this group is not definitively resolved, and the species discrimination is often not clear, even with the use of chromosomal analyses.

Results: We have studied the karyotype of three populations of blind mole rats (Spalacinae) from Anatolia classified tentatively as *Nannospalax xanthodon* (Bolu and Içel provinces) and *Nannospalax ehrenbergi* (Gaziantep province). C-banding and AgNOR staining were used in the cytogenetic analysis. In all populations, the karyotype comprised 52 chromosomes including 8 to 10 bi-armed and 17 to 15 acrocentric autosomal pairs. The centromeric position varied in the sex chromosomes between the populations. The C-positive centromeric bands occurred in most of the bi-armed autosomes and the sex chromosomes and in some of the acrocentric autosomes. The nucleolar organizer regions were localized in three (*N. xanthodon*) or five (*N. ehrenbergi*) autosomal pairs.

Conclusions: It is difficult to find particular chromosomal differences between the studied populations which could enable unequivocal species identification. This finding emphasizes the need of taxonomic revision of the species structure within the *Nannospalax* genus.

Keywords: Mole rats; Species; Systematics; Karyotypes

Background

The blind mole rats (Spalacinae; see Musser and Carleton 2005) are a group of fossorial rodents exhibiting exceptionally extensive variation in the karyotype (Kryštufek and Vohralík 2009, Arslan and Zima 2014 for review). This variation has been recorded within and between populations as well as within and between species. The species composition of Spalacinae, however, has not been yet definitively settled. Chişamera et al. (2014) demonstrated that the division of the extant species into two distinct genera, *Spalax* and *Nannospalax*, is justified and congruent with the pattern of phylogenetic divergence. The exceptional karyotype diversification is apparently confined to the genus *Nannospalax* which includes, in a traditional view, three nearly parapatric species, *Nannospalax leucodon*, *Nannospalax xanthodon*, and *Nannospalax ehrenbergi* (Kryštufek and Vohralík 2009). In respect of the remarkable chromosomal variation, various authors have proposed to recognize additional species, and the descriptions were mainly based on characteristic chromosomal features (Nevo et al. 2001, Coşkun 2004, Kankılıç and Gürpinar 2014). The taxonomy of the *Nannospalax* genus is thus not definitively resolved, and the species discrimination is often not clear, even with the use of chromosomal analyses.

In the present study, we aim to improve karyotypic characteristics revealed by C-banding and AgNOR staining in several cytotypes with 52 chromosomes included either in *N. xanthodon* or *N. ehrenbergi*. In Turkey, populations possessing 52 chromosomes were reported

* Correspondence: jzima@brno.cas.cz
[2]Institute of Vertebrate Biology, Academy of Sciences of the Czech Republic, Kvetna 8, 603 65 Brno, CZ, Czech Republic
Full list of author information is available at the end of the article

from northern Anatolia (e.g., Sözen 2004, Sözen et al. 2013) and from south-eastern Anatolia (Sözen and Kıvanç 1998, Nevo et al. 1995, Ivanitskaya et al. 1997, Coşkun et al. 2006, 2010). The former populations were recognized as *N. xanthodon* (formerly also as *leucodon* or *nehringi),* whereas those from south-eastern Anatolia, the populations with 52 chromosomes have commonly been classified within *N. ehrenbergi.* This cytotype was reported also from Iraq (Coşkun et al. 2012), and populations with 2n = 52 are known from Israel and were named here *Spalax galili* by Nevo et al. (2001). We attempt to investigate if the specific distinction among individual populations from Anatolia is reflected also in their karyotypic differentiation inferred from the C-banding and AgNOR staining pattern. To answer this question, we compare the results with other reports of chromosomal banding patterns found in Turkish populations of *Nannospalax* (Ivanitskaya et al. 1997, Ivanitskaya et al. 2008, Arslan and Bölükbaş 2010, Arslan et al. 2011a,b, Arslan and Zima 2013, Arslan et al. 2013, 2014).

Methods

Cytogenetic analyses were performed in six specimens of *N. xanthodon* and *N. ehrenbergi* from three Turkish populations. The specimens were caught with the use of a metal pipe-type trap (Arslan 2013), and this trapping method enabled to obtain living individuals without any injury. The study was done, and the specimens were obtained with the permission of Republic of Turkey, Ministry of Forest and Water Works (permit no. 89291). Standard voucher specimens (skins and skulls) are deposited at Selçuk University, Biology Department, Faculty of Science, Konya, Turkey. The number of specimens analyzed and location of the collection sites of mole rats are shown in Figure 1 and Table 1. The species discrimination was based on geographic distribution of the studied populations. The available maps of the distribution ranges of blind mole rats in Turkey (Bukhnikashvili et al. 2008, Schlitter et al. 2008, Kryštufek and Vohralík 2009) suggest that the populations from Bolu and Içel (Mersin) are situated within the range of *N. xanthodon,* whereas the population from Gaziantep within the range of *N. ehrenbergi.*

Karyotype preparations were obtained in the field from the bone marrow after colchicine treatment (Ford and Hamerton 1956). Air-dried preparations were stained conventionally by Giemsa. Constitutive heterochromatin and nucleolus organizer regions (NORs) were detected by the techniques of C-banding (Sumner 1972) and silver staining of nucleolar organizer regions (Howell and Black 1980), respectively. From each specimen, 10 to 20 slides were prepared, and at least 20 well-spread metaphase plates were analyzed. The system of classification of chromosomes according to the centromere position was adopted after Hsu and Benirschke (1967–1977), and the bi-armed (metacentric M, submetacentric SM, subtelocentric ST) and uni-armed (acrocentric A) chromosomes were distinguished. The fundamental number of autosomal arms (NFa) and the number of all chromosomal arms in the female complement (NF) were calculated.

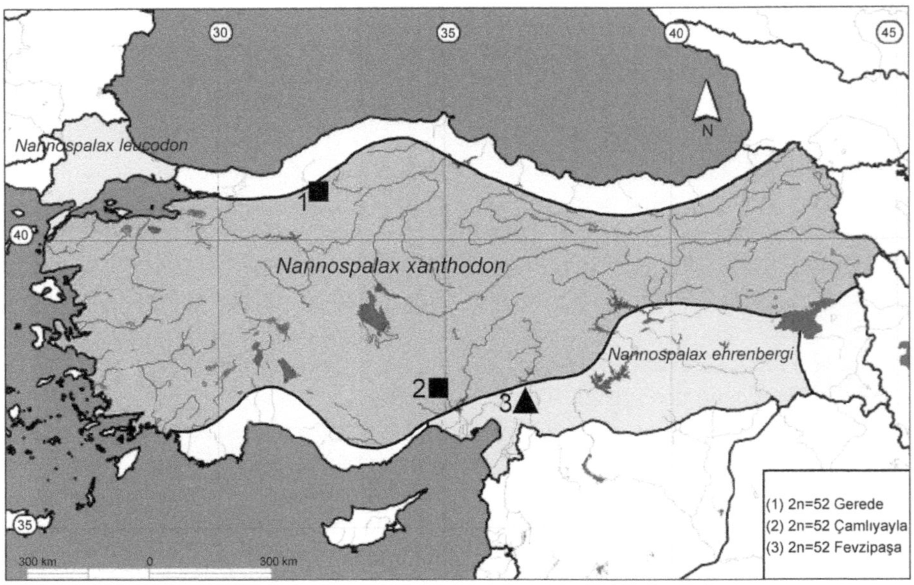

Figure 1 Collecting sites of *Nannospalax xanthodon* (*black square*) and *N. ehrenbergi* (*black triangle*) in Turkey. The numbering of sampling localities corresponds to data in Table 1. The approximate ranges of both species are indicated after Kryštufek and Vohralík (2009).

Table 1 Studied samples in Turkey and the basic characteristics of the examined specimens

No.	Species	Locality/province	Latitude/longitude	No. of specimens		2n	NF	NFa	X	Y
				Male	Female					
1	*N. xanthodon*	Gerede/Bolu	40° 46′ N, 32° 13′ E	2	-	52	70	66	ST	A/M
2	*N. xanthodon*	Çamlıyayla/İçel	37° 10′ N, 34° 36′ E	1	1	52	72	68	SM	A
3	*N. ehrenbergi*	Fevzipaşa/Gaziantep	37° 05′ N, 36° 38′ E	1	1	52	74	70	SM	A

The numbering of the sampling sites corresponds to data in Figure 1.

Results

Bolu province, Gerede population

The karyotype of the two males consisted of 52 chromosomes including a large subtelocentric (no. 1), two metacentric (nos. 2, 8), three submetacentric (nos. 3, 5, 7), two medium-sized subtelocentric (nos. 4, 6), and 17 acrocentric autosomal pairs of gradually diminishing size (nos. 9 to 25) (NFa = 66). The X chromosome was a medium-sized subtelocentric; the Y was a small metacentric (NF = 70) (Figure 2.1). The dark centromeric C-bands were observed in five biarmed (1 to 5) and some of acrocentric autosomes (nos. 10, 13, 14, 16). In the acrocentric autosomal pairs 13 and 14, the position of the dark C-bands appeared interstitial, not directly involving the centromeric region. C-heterochromatic short arms were not recorded in any autosomal pair in the complements. The sex chromosomes had distinct pericentromeric C-positive bands (Figure 3.1). The NORs were observed in the telomeric regions of the short arms of the autosomes 4, 6, and 8, with an occasionally heterozygous signal on the autosome 8 (Figure 4.1).

İçel province, Çamlıyayla population

The karyotype of a male and a female consisted of 52 chromosomes including a large submetacentric and a large subtelocentric pair (no. 1, 2), two metacentric pairs (nos. 3, 4), five submetacentric or subtelocentric pairs of medium size (nos. 5 to 9), and 16 acrocentric autosomal pairs of gradually diminishing size (nos. 10 to 25) (NFa = 68). The X chromosome was a large submetacentric

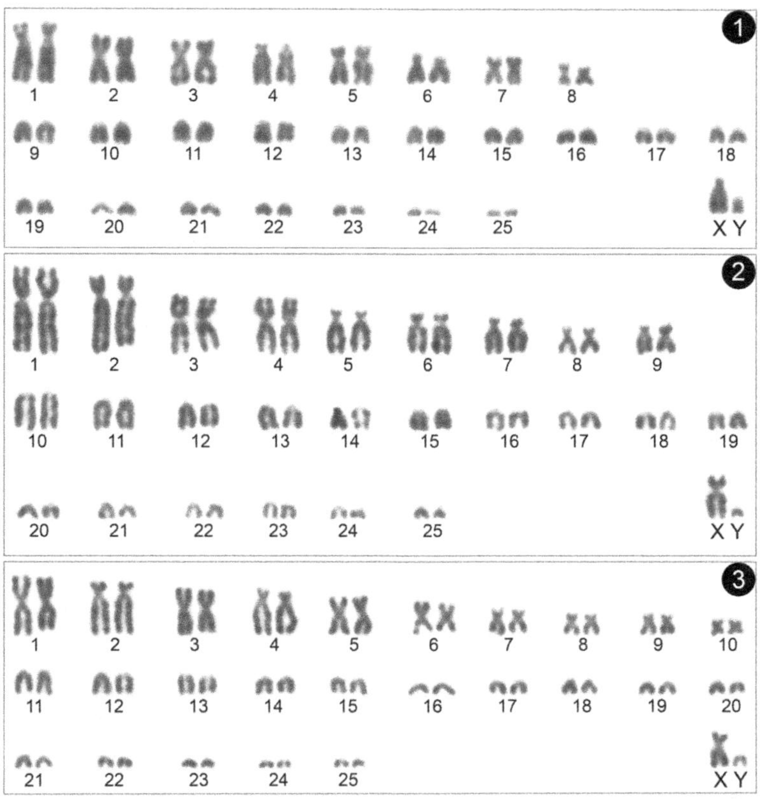

Figure 2 Standard karyotypes of specimens of *Nannospalax xanthodon* from Gerede (*1*) and Çamlıyayla (*2*) and *N. ehrenbergi* from Fevzipaşa (*3*).

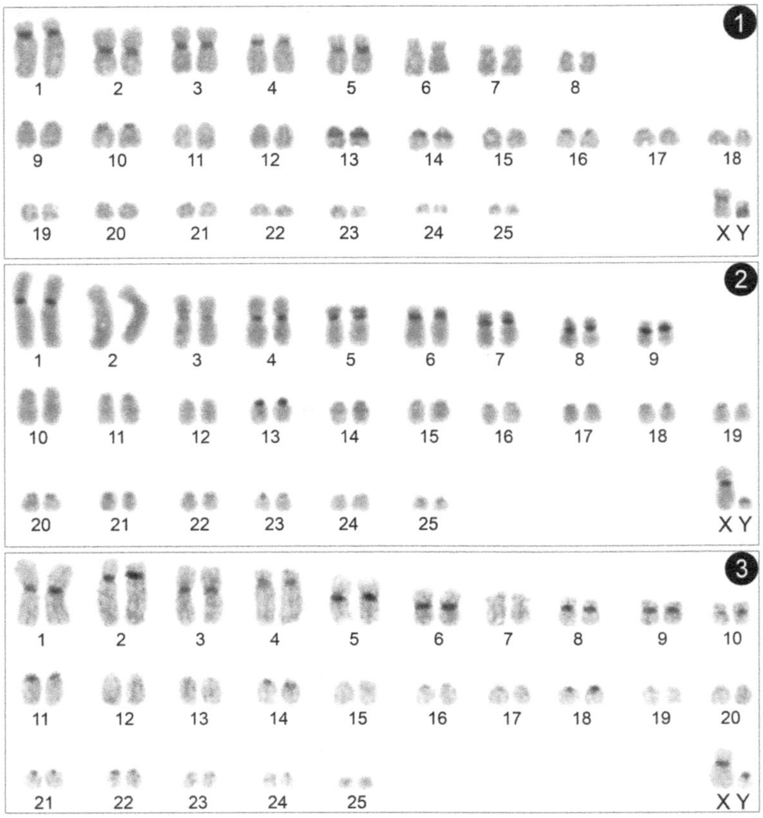

Figure 3 C-banded karyotypes of specimens of *Nannospalax xanthodon* from Gerede (*1*) and Çamlıyayla (*2*) and *N. ehrenbergi* from Fevzipaşa (*3*).

and the Y chromosome was a small acrocentric (NF = 72) (Figure 2.2). The dark C-bands were observed in the centromeric areas of all the bi-armed autosomes except of the pairs 2 and 3 and in the acrocentric autosome 13. Tiny positive centromeric bands were apparent also in some other acrocentric pairs. The sex chromosomes had a centromeric C-positive area (Figure 3.2). The NORs were observed in the telomeric regions of the short arms of the autosomes 5, 6, and 8 (Figure 4.2).

Gaziantep province, Fevzipaşa population

The karyotype of a male and a female consisted of 52 chromosomes including four metacentric pairs (nos. 1, 3, 5, 10), six submetacentric or subtelocentric pairs (nos. 2, 4, 6 to 9), and 15 acrocentric autosomal pairs of gradually diminishing size (nos. 11 to 25) (NFa = 70). The X chromosome was a medium-sized submetacentric, the Y a small acrocentric (NF = 74) (Figure 2.3). The dark C-bands were observed in the centromeric areas of all the bi-armed autosomes except of the pair 7 and in three acrocentric autosomes (nos. 11, 14, 18). Tiny positive centromeric bands were apparent also in some other acrocentric pairs (nos. 21, 22). The sex chromosomes had distinct centromeric C-positive bands (Figure 3.3). The NORs were observed in the telomeric regions of the short arms of the bi-armed autosomes 2, 4, 7, and 8. In the pairs 4 and 7, the positive signal was observed in only one homologue of the pair in some cells. The NOR was detected also in the acrocentric chromosome 15 (Figure 4.3).

Discussion

The comparison of chromosomal complements between the three studied populations shows great similarities between them. All the karyotypes contain 52 chromosomes with 8 to 10 bi-armed and 17 to 15 acrocentric autosomes. The sex chromosomes reveal variation in the centromeric position but their size seems similar in all individuals. The homology of the bi-armed autosomes between individual complements seems uncertain, and it is even improbable in some cases (e.g., in the two largest autosomal pairs). Unfortunately, we cannot evaluate the nature of homology and possible rearrangements involved in presumed diversification without the use of appropriate banding techniques. The pattern of the distribution of C-positive heterochromatin regions is also similar in the three karyotypes compared, with dark centromeric bands occurring in most of the bi-armed

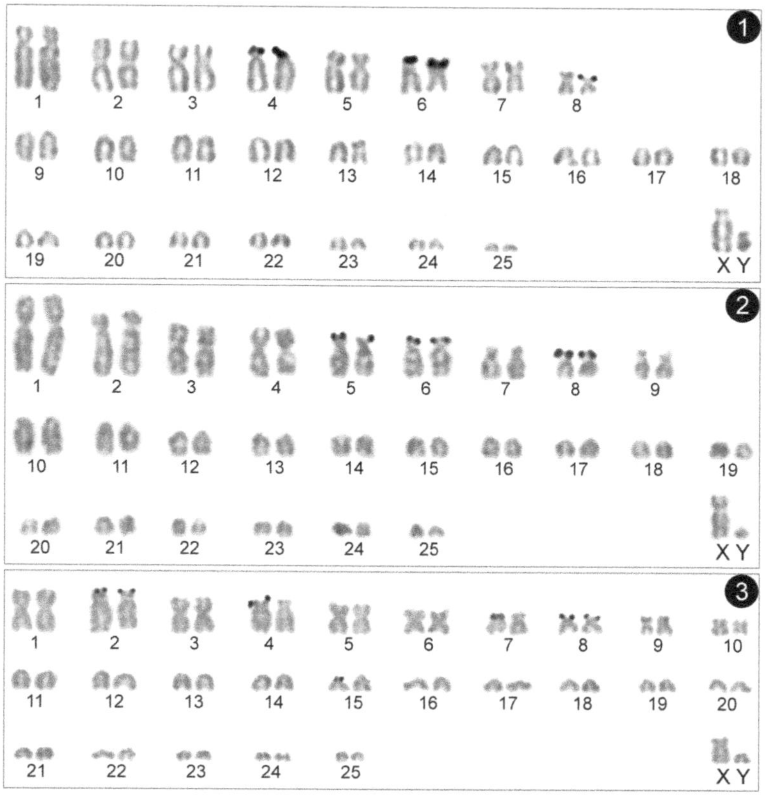

Figure 4 Silver-stained karyotypes of specimens of *Nannospalax xanthodon* from Gerede (*1*) and Çamlıyayla (*2*) and *N. ehrenbergi* from Fevzipaşa (*3*).

autosomes and the sex chromosomes and in some of the acrocentric autosomes.

In the karyotype with 52 chromosomes of blind mole rat populations from northern Anatolia, 8 to 9 bi-armed autosomal pairs were distinguished (Sözen 2004, Sözen et al. 2013), whereas in the karyotype of populations from south-eastern Anatolia, 11 or 13 bi-armed autosomal pairs were found (Yüksel 1984, Ivanitskaya et al. 1997, Coşkun et al. 2006, 2010). Ivanitskaya et al. (1997) found in this cytotype pericentromeric blocks of C-heterochromatin in one metacentric autosomal pair and one subtelocentric pair, and distinct dark C-bands were observed also in the pericentromeric area of all acrocentric autosomes. Additionally, these authors recorded dark C-bands in the telomeric regions of two pairs of subtelocentric or submetacentric autosomes. This C-banding pattern is thus different from that observed in our study in both *N. xanthodon* and *N. ehrenbergi* populations where the amount of pericentromeric C-heterochromatin was very low in the acrocentric autosomes and the telomeric positive C-bands were absent. Contrary to previous findings in other cytotypes (Ivanitskaya et al. 1997, Arslan et al. 2011a, Arslan et al. 2013), we have not recorded any C-heterochromatic short arms in the complements under study.

The pattern of the NORs distribution is rather uniform among populations of mole rats with varying chromosomal numbers. The number of the NOR-bearing autosomal pairs varies from two to five; however, their actual localization may involve various autosomal pairs (Ivanitskaya et al. 1997, 2008, Arslan et al. 2011a, 2013, 2014). Gülkaç and Küçükdumlu (1999) localized NORs in two pairs of subtelocentric autosomes in the 2n = 52 complement. Ivanitskaya et al. (1997) recorded NORs in the two largest subtelocentric pairs and, additionally, also in the small submetacentric pair possessing the telomeric dark C-band. In the Iraqi population with 52 chromosomes, Coşkun et al. (2014) localized NORs in three autosomal pairs, including a large and a small submetacentric pair, and the smallest acrocentric pair.

It is difficult to find particular chromosomal differences enabling the distinction between *N. xanthodon* and *N. ehrenbergi*. The studied populations differed in the proportion of bi-armed and uni-armed autosomes and in the number and distribution of NORs. In the population recognized as *N. ehrenbergi*, the number of detected NORs was five, and a site was localized in an acrocentric autosome. However, it is questionable if such variation could be used as a species-specific character for reliable species identification. The proportion of the bi-armed and acrocentric autosomes commonly varies between populations of mole rats with the same diploid number of chromosomes (Arslan et al. 2013, 2014) and such variation was reported also in the 52 chromosome race. The previously published karyological data on the 52 chromosome cytotype showed considerable variation also in the C-banding and AgNOR staining pattern, and this variability makes the search for species-specific cytogenetic characters difficult.

Conclusions

We can conclude that the populations of blind mole rats from Anatolia in Turkey classified as *N. xanthodon* and *N. ehrenbergi* reveal little difference in their karyotype structure, and hardly, any species distinguishing character can be defined based on the comparison of their chromosomal complements. This finding emphasizes the need of taxonomic revision of the species status and composition within the *Nannospalax* genus and cytogenetic studies using structural chromosome banding techniques will certainly be useful in this respect. The use of sensitive molecular markers such as microsatellites is obviously desirable to estimate the extent of gene flow and possible reproductive isolation between populations.

Competing interests

The authors declare that they have no competing interests.

Authors' contributions

AA collected the studied specimen, performed the cytogenetic analysis, designed the study, and contributed to interpretation of results. JZ contributed to the cytogenetic analysis and the arrangement of the karyotypes obtained, contributed to analysis and interpretation of results, and wrote the manuscript. Both authors read and approved the final manuscript.

Acknowledgements

This study was supported by grants from the Coordination Committee of Scientific Research Projects of Selçuk University (BAP, no. 14401098).

Author details

[1]Department of Biology, Faculty of Science, Selçuk University, 42031 Konya, TR, Turkey. [2]Institute of Vertebrate Biology, Academy of Sciences of the Czech Republic, Kvetna 8, 603 65 Brno, CZ, Czech Republic.

References

Arslan A (2013) A new live trap to catch blind mole rats (*Spalax* sp.). Folia Zool 62:130–132

Arslan A, Akan Ş, Zima J (2011a) Variation in C-heterochromatin and NORs distribution among chromosomal races of mole rats (Spalacidae) from Central Anatolia, Turkey. Mamm Biol 76:28–35

Arslan A, Arısoy A, Zima J (2013) The chromosome banding pattern in two cytotypes (2n = 36 and 38) of blind mole rats from Turkey (Mammalia: Spalacidae). Zool Middle East 59:95–100

Arslan A, Arısoy A, Zima J (2014) Comparison of the chromosome banding pattern in the 2n=56 cytotypes of *Nannospalax leucodon* and *N. xanthodon* from Turkey. Scientific World Journal 2014:121690, doi:10.1155/2014/121690

Arslan A, Bölükbaş F (2010) C-heterochromatin and NORs distribution of mole rat, *Nannospalax xanthodon* from Aksaray, Turkey. Caryologia 63:398–404

Arslan A, Toyran K, Gözütok S, Yorulmaz T (2011b) C- and NOR stained karyotypes of mole rat, *Nannospalax xanthodon* (2n = 54) from Kırıkkale, Turkey. Turk J Biol 35:655–661

Arslan A, Zima J (2013) The banded karyotype of the 2n = 58 chromosomal race of mole rat from Erzincan, Turkey. Folia Zool 62:19–23

Arslan A, Zima J (2014) Karyotypes of the mammals of Turkey and neighbouring regions: a review. Folia Zool 63:1–62

Bukhnikashvili A, Kryštufek B, Sozen M, Yigit N, Amori G (2008) *Spalax nehringi*. The IUCN Red List of Threatened Species. Version 2014.1. [www.iucnredlist.org] Downloaded on 17 July 2014

Chişamera G, Bužan EV, Sahlean T, Murariu D, Zupan S, Kryštufek B (2014) Bukovina blind mole rat *Spalax graecus* revisited: phylogenetics, morphology, taxonomy, habitat associations and conservation. Mamm Rev 44:19–29

Coşkun Y (2004) A new species of mole rat, *Nannospalax munzuri* sp. n., and karyotype of *Nannospalax tuncelicus* (Coşkun, 1996) (Rodentia: Spalacidae) in eastern Anatolia. Zool Middle East 33:153–162

Coşkun Y, Aşan Baydemir NA, Kaya A, Karöz AM (2014) Nucleolar organizer region distribution in *Nannospalax ehrenbergi* (Nehring, 1898) (Rodentia: Spalacidae) from Iraq. Turk J Zool 38:250–253

Coşkun Y, El Namee A, Kaya A (2012) Karyotype of *Nannospalax ehrenbergi* (Nehring, 1898) (Rodentia, Spalacidae) in the Mosul Province, Iraq. Hystrix It J Mamm (ns) 23:75–78

Coşkun Y, Ulutürk S, Kaya A (2010) Karyotypes of *Nannospalax* (Palmer 1903) populations (Rodentia: Spalacidae) from central-eastern Anatolia, Turkey. Hystrix It J Mamm (ns) 21:89–96

Coşkun Y, Ulutürk S, Yürümez G (2006) Chromosomal diversity in mole-rats of the species *Nannospalax ehrenbergi* (Mammalia: Rodentia) from South Anatolia, Turkey. Mamm Biol 71:244–250

Ford CE, Hamerton JL (1956) A colchicine, hypotonic citrate, squash sequence for mammalian chromosomes. Stain Technol 31:247–251

Gülkaç MD, Küçükdumlu İ (1999) Variation in the nucleolus organizer regions (NORs) in two mole rat species (*Spalax leucodon* and *Spalax ehrenbergi*). Turk J Biol 23:153–158

Howell WM, Black DA (1980) Controlled silver staining of nucleolar organizer regions with a protective colloidal developer: a 1-step method. Experientia 36:1014–1015

Hsu TC, Benirschke K (1967–1977) An atlas of mammalian chromosomes, vol 1–10. Springer Verlag, Berlin-Heidelberg-New York

Ivanitskaya E, Coskun Y, Nevo E (1997) Banded karyotypes of mole rats (*Spalax*, Spalacidae, Rodentia) from Turkey: a comparative analysis. J Zool Syst Evol Res 35:171–177

Ivanitskaya E, Sözen M, Rashkovetsky L, Matur F, Nevo E (2008) Discrimination of 2n = 60 *Spalax leucodon* cytotypes (Spalacidae, Rodentia) in Turkey by means of classical and molecular cytogenetic techniques. Cytogenet Genome Res 122:139–149

Kankılıç T, Gürpinar C (2014) Revised classification design of the Anatolian species of *Nannospalax* (Rodentia: Spalacidae) using RFLP analysis. Turk J Zool 38:68–78

Kryštufek B, Vohralík V (2009) Mammals of Turkey and Cyprus. Rodentia II: Cricetinae, Muridae, Spalacidae, Calomyscidae, Hystricidae, Castoridae. Univerzita za Primorskem, Koper

Musser G, Carleton M (2005) Superfamily Muroidea. In: Wilson DE, Reeder DM (eds) Mammal species of the world. A taxonomic and geographic reference, 2nd edn. The Johns Hopkins University Press, Baltimore, pp 894–1531

Nevo E, Filippucci MG, Redi C, Simson S, Heth G, Beiles A (1995) Karyotype and genetic evolution in speciation of subterranean mole rats of the genus *Spalax* in Turkey. Biol J Linn Soc 54:203–229

Nevo E, Ivanitskaya E, Beiles A (2001) Adaptive Radiation of blind subterranean mole rats: naming and revisiting the four sibling species of the *Spalax ehrenbergi* superspecies in Israel: *Spalax galili* (2n=52), *S. golani* (2n=54), *S. carmeli* (2n=58) and *S. judaei* (2n=60). Backhuys Publishers, Leiden

Schlitter D, Shenbrot G, Kryštufek B, Sozen M (2008) *Spalax ehrenbergi*. The IUCN Red List of Threatened Species. Version 2014.1. [www.iucnredlist.org] Downloaded on 17 July 2014

Sözen M (2004) A karyological study on subterranean mole rats of the *Spalax leucodon* Nordmann, 1840 superspecies in Turkey. Mamm Biol 64:420–429

Sözen M, Çolak F, Sevindik M, Matur F (2013) Cytotypes of *Nannospalax xanthodon* (Satunin, 1898) (Rodentia, Spalacidae) from western Anatolia. Turk J Zool 37:462–469

Sözen M, Kıvanç E (1998) Two new karyotypic forms of *Spalax leucodon* Nordmann, 1840 (Mammalia: Rodentia) from Turkey. Z Säugetierkunde 63:307–310

Sumner AT (1972) A simple technique for demonstrating centromeric heterochromatin. Exp Cell Res 75:304–306

Yüksel E (1984) Cytogenetic study in *Spalax* (Rodentia: Spalacidae) from Turkey. Comm Fac Sci Univ Ankara Serie 2:1–11

12

Adaptive flexibility in the feeding behaviour of brown trout: optimal prey size

Javier Sánchez-Hernández[1,2*] and Fernando Cobo[1,3]

Abstract

Background: Brown trout, *Salmo trutta* Linnaeus, 1758, is a species of significant conservation and socio-economic importance. A consequence of this importance is the enormous amount of literature that has been published on the species in the last few decades. In general terms, brown trout has been considered as a size-selective predator, even though it is able to feed on a wide range of prey sizes. Nevertheless, there are still some gaps in our knowledge, for example the theoretical relationship between prey numbers and prey sizes eaten by the fish need to be addressed. This research aimed to study optimal prey size in the environment (benthos and drift) as well as the potential relationship between prey size and two other feeding variables (prey numbers and stomach fullness). Additionally, ontogenetic shifts in these variables were addressed.

Results: Brown trout showed a clear preference for 4- to 6-mm-length prey, although the use of prey larger than 10-mm length may be feasible. The similarity of the prey size frequency distribution between the environment (benthos and drift) and the diet in some cases was considerable (from 57.7% to 95.9%). Moreover, the results revealed that the feeding strategy can be related to prey size and the numbers of prey eaten by the brown trout; as food size decreased, prey numbers increased. On the contrary, the correlation between the average prey size and fish length was positive but statistically nonsignificant. A significant ontogenetic shift, in terms of prey size sorted by age classes, was found in only two of eight studied populations. No clear relationship between prey size and stomach fullness was found.

Conclusions: The feeding strategy of this species is flexible and clearly influenced by the size frequency distribution of potential prey: trout fed on either small numbers of large prey or large numbers of small, and theoretically low energy, prey. Our approach covers a general subject in trophic ecology and animal behaviour that may be applicable to other fish species to improve our understanding of predator feeding behaviour.

Keywords: Trophic ecology; Foraging behaviour; Flexible behaviour; *Salmo trutta*; Prey size; Available prey; Benthos; Drift

Background

Ecologists have considered the prey size hypothesis, the relationships between prey size and handling efficiency by predators, as one of the main factors involved in feeding behaviour (e.g. Mock 1985; Török 1993; Denoël and Joly 2001). With regard to fish species, much research has focused on the relationship between prey size and handling efficiency (e.g. Mittelbach 1981; Reimchen 1991), and some fishes may be gape-limited predators, especially when they are young (e.g. Schmitt and Holbrook 1984; Schael et al. 1991; Sánchez-Hernández et al. 2011a). Indeed, prey size is a key variable in the feeding behaviour of fishes (see Keeley and Grant 2001 and references therein), which is usually considered to be size selective (e.g. O'Brien et al. 1976; Bannon and Ringler 1986). For example, studies under controlled laboratory conditions have demonstrated that fishes show a clear preference for large prey items (Ringler 1979; Wetterer 1989), which are normally the most profitable in energetic terms, even though handling costs increase with increasing prey size (e.g. Gill 2003). However, it is important to note that prey energy content may exhibit substantial variations depending on seasonal development, life history strategies or taxonomic group (e.g. Gupta and Pant 1983; Cobo et al. 1999; 2000). Factors other than prey size and handling efficiency, such as

* Correspondence: javier.sanchez@usc.es
[1]Department of Zoology and Physical Anthropology, Faculty of Biology, University of Santiago de Compostela, Campus Sur s/n, 15782 Santiago de Compostela, Spain
[2]Department of Arctic and Marine Biology, Faculty for Biosciences, Fisheries and Economics, UiT The Arctic University of Norway, N-9037 Tromsø, Norway
Full list of author information is available at the end of the article

some fish characteristics (e.g. prior experience, locomotor abilities, stomach fullness and sensory capabilities) and physical habitat characteristics (e.g. flow patterns and structural complexity of habitat) may also play an important role in the feeding behaviour of fishes (e.g. Gill and Hart 1994; Gerking 1994; Sánchez-Hernández et al. 2013). Although the feeding behaviour of fish species has received considerable attention from the scientific community (see above literature), to the best of our knowledge, the theoretical relationship between prey numbers and prey sizes eaten by fish has not been addressed so far.

Brown trout *Salmo trutta* Linnaeus, 1758 (henceforth simply trout), is a species of Eurasian origin but, at present, is naturalized in many other areas all over the world (Klemetsen et al. 2003). Fortunately, the feeding behaviour of trout has been well studied (e.g. Fochetti et al. 2003; Oscoz et al. 2008; Evangelista et al. 2014), and during their life history, trout undergo ontogenetic dietary shifts (Sánchez-Hernández et al. 2013 and references therein). With regard to changes in prey size during ontogeny, mean prey size usually increases with both trout size and age (e.g. Steingrímsson and Gíslason 2002; Montori et al. 2006; Jensen et al. 2008; Sánchez-Hernández and Cobo 2012; Sánchez-Hernández et al. 2013). In spite of this ontogenetic shift, several researchers have demonstrated that the influence of gape-limited prey ingestion in this species is insignificant (Newman 1987; Rincón and Lobón-Cerviá 1999). Notwithstanding, trout may be used as a model species in studies of feeding behaviour, regardless of the apparent lack of a relationship between mouth dimensions and prey size. Here, we studied optimal prey size in the environment (benthos and drift) as well as the potential relationship between prey size and other variables (prey numbers, stomach fullness, fish size and fish age). We hypothesized that prey numbers should be low when the predator feeds on large prey items, and *vice versa*. We further hypothesized that prey size may be highly dependent on fish size, fish age and stomach fullness.

Methods

For the purpose of the study, and in order to avoid possible differences in feeding behaviour among populations due to differences in physical habitat characteristics, samples were collected in wadeable riffle sections with similar environmental characteristics. In total, eight neighbouring rivers of Galicia (NW Spain) were sampled (Figure 1) during June 2003 (Rois, Santa Lucía, Sar and Traba rivers) and September 2007 (Anllóns, Furelos, Lengüelle and Tambre rivers). Prior to electrofishing, samples of potential prey items (benthic and drifting invertebrates) were collected to study prey availability in the environment. Benthic invertebrates were collected from riffles using a 0.1-m^2 Surber sampler ($n = 3$), and a Brundin net (250-μm mesh size, 1 m long, 30-cm mouth diameter) was used to collect drifting invertebrates. Drift nets were set at sunrise (8:00 a.m.) and retrieved after at least 2.5 h (ranging between 179 and 200 min). After collection, we fixed samples using 4% formalin and stored them for later processing. Information on prey availability is only provided for rivers surveyed in 2007.

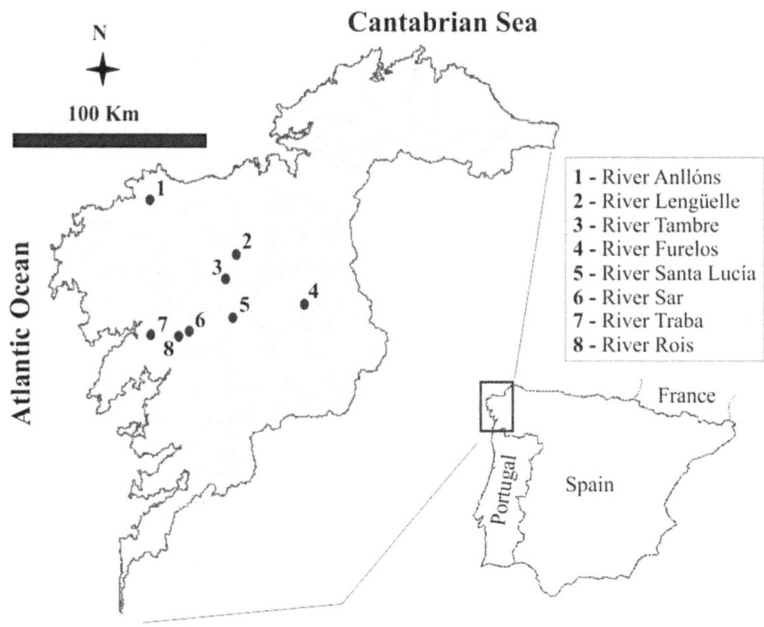

Figure 1 Maps of the Iberian Peninsula and north-western Spain showing the sampling sites.

Table 1 Size frequency (%) of the prey consumed by trout

	Size class (mm)										
	0 to 2	**2 to 4**	**4 to 6**	**6 to 8**	**8 to 10**	**10 to 12**	**12 to 14**	**14 to 16**	**16 to 18**	**18 to 20**	**>20**
Anllóns											
0+ ($n = 2$)	0	10.8	27.0	2.7	13.5	45.9	0	0	0	0	0
1+ ($n = 18$)	1.3	18.4	23.7	3.0	27.3	24.7	0	0.3	0	0.7	0.7
2+ ($n = 9$)	0.4	18.9	56.9	1.3	4.7	17.0	0	0.2	0	0	0.6
3+ ($n = 6$)	0	15.1	52.8	15.1	1.9	11.3	1.9	1.9	0	0	0
Pooled data	0.7	18.2	43.7	2.8	12.8	20.6	0.1	0.3	0	0.2	0.6
Furelos											
0+ ($n = 37$)	2.1	3.2	33.5	0	1.9	51.0	0	6.5	0	0	1.8
1+ ($n = 19$)	1.4	4.8	81.0	0.1	0.2	6.0	0	3.9	0	0	2.6
2+ ($n = 5$)	0.5	2.8	54.1	0	8.7	4.1	0	8.7	0	0	21.1
3+ ($n = 3$)	3.2	18.5	15.3	0.8	0	33.1	0	0	0	0	29.0
Pooled data	1.6	4.8	60.7	0.1	1.5	20.9	0	4.9	0	0	5.4
Lengüelle											
0+ ($n = 6$)	0	2.4	62.9	0	13.7	19.4	0	0.8	0	0	0.8
1+ ($n = 13$)	0	18.3	37.7	0.4	16.5	20.4	0	0	0	0	6.7
2+ ($n = 16$)	0.8	7.9	53.1	0.8	4.3	24.1	0	0.2	0	0	8.7
3+ ($n = 4$)	5.8	1.9	32.7	0	9.6	38.5	0	0	0	0	11.5
Pooled data	0.7	10.0	48.7	0.5	9.4	23.2	0	0.2	0	0	7.2
Tambre											
1+ ($n = 24$)	1.6	12.6	59.6	0.7	6.7	17.0	0	0.9	0	0	0.8
2+ ($n = 6$)	0.3	5.2	56.0	0.9	4.0	28.4	0	2.4	0	0	2.8
Pooled data	1.4	11.2	58.9	0.8	6.2	19.2	0	1.2	0	0	1.2
Rois											
1+ ($n = 31$)	2.7	7.7	33.8	3.2	14.3	33.9	0	2.7	0	0.2	1.4
2+ ($n = 2$)	2.0	6.0	28.0	6.0	4.0	46.0	0	2.0	0	0	6.0
Pooled data	2.7	7.6	33.6	3.4	13.8	34.5	0	2.7	0	0.2	1.6
Santa Lucía											
1+ ($n = 24$)	2.2	29.4	34.7	7.0	5.0	7.9	0	8.0	0.3	2.3	3.4
2+ ($n = 5$)	3.2	20.6	40.2	10.6	4.2	15.3	0	4.8	0	0.5	0.5
Pooled data	2.4	27.7	35.8	7.7	4.8	9.3	0	7.4	0.2	1.9	2.9
Sar											
1+ ($n = 25$)	0.6	33.8	4.0	0.9	1.1	57.8	0	0.2	0	0.1	1.6
2+ ($n = 4$)	1.0	17.7	3.0	0.5	3.4	70.4	0	1.5	0	0	2.5
Pooled data	0.6	32.2	3.9	0.9	1.3	59.0	0	0.3	0	0.05	1.7
Traba											
1+ ($n = 15$)	0.3	4.8	45.5	0.4	14.8	33.8	0	0.2	0	0	0.1
2+ ($n = 14$)	0.5	7.6	51.5	1.0	10.7	28.1	0	0.2	0	0	0.2
Pooled data	0.4	5.7	47.6	0.6	13.4	31.8	0	0.2	0	0	0.2
Total ($n = 288$)	3.44	5.00	44.44	3.42	8.27	31.97	0.01	1.30	0.03	0.17	1.97

Data are displayed for each sampling site, each age class and in total using pooled data. Prey were grouped into 2-mm-length classes. The sample size (n) of each age class is shown in brackets.

Trout were collected using pulsed D.C. backpack electrofishing equipment (ELT60II, Hans Grassl GmbH, Schönau am Königssee, Germany). Fishes were killed immediately with an overdose of anaesthetic (benzocaine) and transported in cool boxes (approximately 4°C) to the laboratory, where they were frozen at −30°C until processing. In the laboratory, fishes were measured for fork length (FL; nearest 1 mm) and weighed (nearest 0.01 g), and the stomachs were removed. Estimates of fish age were made by scale examination and by using Petersen's length-frequency method (Bagenal and Tesch 1978). Age-4+ individuals were not included in the diet analysis because only one specimen was captured in the River Furelos. No empty stomachs were found, and the stomach fullness index (f) was calculated as $f = (Ws/W) \times 100$, where Ws is the total stomach content mass (g) and W is the fish mass (g).

Potential (benthic and drifting invertebrates) and actual prey items were counted and measured (total length) with a digital micrometer (0.01-mm resolution, Mitutoyo Absolute, Mitutoyo Corporation, Takatsu-ku, Japan). The number of fragmented or partially digested invertebrates was estimated by counting body parts resistant to digestion. In those cases, prey length was estimated from the width of the cephalic capsule (see Rincón and Lobón-Cerviá 1999), which was normally the best-preserved part.

The similarity between the size distributions of potential prey in the environment (benthos and drift) and those consumed by trout was assessed using the Bray-Curtis similarity index. The data were first transformed by $Y = \log(x + 1)$, and the similarity index was calculated using the PRIMER statistical package version 5.0 (Clarke and Gorley 2001). In the present study, in order to assess whether prey size selection is dependent upon the size frequency distribution of available prey, we clustered all trout regardless of age at each sampling site. Additionally, to explore ontogenetic shifts, the similarity matrix was calculated separately for each age class.

Finally, with the aim of exploring the possibility of a nonlinear relationship between prey size and the other analysed variables (prey numbers, stomach fullness and fish size), the curve estimation procedure was used using pooled data, which compared eight different models (linear, logarithmic, inverse, quadratic, exponential, power, compound and S-curve). The model with the highest adjusted Pearson's rank correlations coefficient (R) was chosen. The data were not normally distributed, so in order to analyse differences among age classes in the studied variables (prey size, prey numbers and stomach fullness), nonparametric analyses (Mann-Whitney and Kruskal-Wallis tests) were used. The Mann-Whitney U-test was used to compare differences between two independent groups because in some rivers (Rois, Santa Lucía, Sar, Traba and Tambre rivers) only two age classes were analysed (1+ and 2+). Kruskal-Wallis test was used to detect differences among four groups (0+, 1+, 2+ and 3+) in the other rivers (Anllóns, Furelos and Lengüelle). Statistical analyses were conducted using the programme IBM SPSS Statistics 20 software (IBM Corporation, Armonk, NY, USA). All of these tests were considered statistically significant at P level < 0.05.

Results

A total of 288 trout (range = 48 to 300 mm) was examined in the present study, with 15,131 prey items (range = 1.1

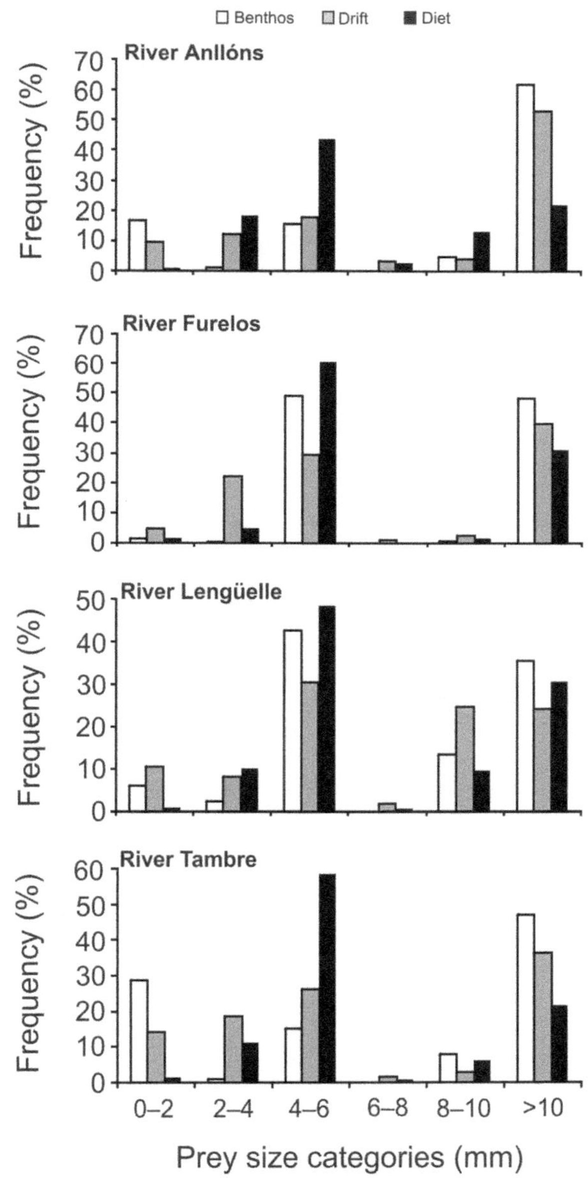

Figure 2 Size frequency of the benthos, drift and trout diet. Size frequency of the benthos, drift and trout diet from Anllóns, Furelos, Lengüelle and Tambre rivers surveyed in September 2007. Diet data were pooled regardless fish age for each sampling site.

Table 2 Values of the Bray-Curtis index of size frequency distribution similarity between the environment (benthos and drift) and diet

	Diet *versus* benthos				Diet *versus* drift			
	Anllóns	Furelos	Lengüelle	Tambre	Anllóns	Furelos	Lengüelle	Tambre
0+	70.6	89.1	88.0	-	85.3	84.8	78.2	-
1+	69.2	81.4	82.4	75.5	84.5	74.1	84.0	83.7
2+	68.3	82.7	83.5	71.8	80.7	77.8	83.3	81.6
3+	57.7	74.6	95.9	-	76.2	88.4	84.8	-
Pooled data	67.2	88.4	86.2	72.6	82.6	83.4	85.2	83.7

Similarities are shown as percentages. Data are only displayed for rivers surveyed in 2007.

to 60 mm), 3,855 benthic macroinvertebrates (range = 2 to 60 mm) and 980 drifting invertebrates (range = 0.5 to 30 mm) measured. Trout fed mainly on prey within the 2- to 6-mm size range, with prey of 4 to 6 mm being the most commonly consumed, except at two sampling sites (Rois and Sar) where it was 10 to 12 mm (Table 1). The observed prey size frequency distribution in the stomachs was not identical to the potential prey in the environment (benthos and drift) and varied among sampling sites (Figure 2). However, the Bray-Curtis similarity index (Table 2) showed that the similarity of the size frequency distribution between the environment (benthos and drift) and the diet in some cases was considerable and accounted for >55% in all cases, ranging from 57.7% to 95.9% (Table 2). In general, 4- to 6- and 6- to 8-mm size categories were more frequently encountered in the diet than in the environment, whereas invertebrates larger than 10 mm were more frequently found in the environment than in the diet (Figure 2).

With regard to ontogenetic shifts, the 4- to 6-mm size category was generally dominant in all age groups (Table 1), as previously observed from pooled data. There were only differences in the average prey size among age classes in two rivers (Table 3), where prey size increased with increasing fish age (Figure 3). Prey numbers were only statistically different among age classes in two rivers (Table 3), achieving the highest value in 2+ fish in both populations (Table 4). In most cases, stomach fullness varied among age classes (statistical analysis shown in Table 3), with young-of-the-year (YOY) fish having the highest values and fullness decreasing with age (Table 5).

The relationship between average prey size and stomach fullness was positive but only statistically significant for the exponential model ($R = 0.138$, $P = 0.019$). The correlation between mean prey size and fish length was positive but statistically nonsignificant ($P > 0.05$ in all cases). A noteworthy result of this study is the negative relationship between prey numbers and mean size ($P < 0.01$ in all cases); as prey numbers increased, prey size decreased (Figure 4; logarithmic regression model $R = -0.293$, $P = 0.001$).

Discussion

This study demonstrated that trout have a clear preference for certain prey size categories according to prey availability in the environment, corroborating the theoretical considerations predicted by Bannon and Ringler (1986) and field observations reported by several researchers (Newman and Waters 1984; Newman 1987; Rincón and Lobón-Cerviá 1999). In addition, the study exemplifies the feeding behaviour flexibility of this fish species with respect to the relationship between prey sizes and numbers eaten.

The size frequency distribution of potential prey can have a strong influence on prey size selection. Although

Table 3 Statistical comparisons of the mean prey size, prey numbers and stomach fullness index (*f*) among age classes

	Mean prey size		Prey numbers		Stomach fullness index	
	Test	*P* value	Test	*P* value	Test	*P* value
Anllóns	$H = 3.1$	0.373	$H = 12.3$	*0.006*	$H = 19.7$	*<0.001*
Furelos	$H = 8.7$	*0.033*	$H = 3.6$	0.303	$H = 46.2$	*<0.001*
Lengüelle	$H = 4.3$	0.226	$H = 1.5$	0.672	$H = 19.5$	*0.001*
Tambre	$U = 41.0$	0.108	$U = 69.5$	0.897	$U = 24.0$	*0.013*
Rois	$U = 18.0$	0.327	$U = 25.0$	0.650	$U = 4.0$	*0.042*
Santa Lucía	$U = 43.0$	0.325	$U = 24.5$	*0.040*	$U = 52$	0.644
Sar	$U = 12.0$	*0.016*	$U = 38.0$	0.448	$U = 46$	0.800
Traba	$U = 71.0$	0.138	$U = 63.0$	0.067	$U = 33.0$	*0.002*

Kruskal-Wallis (*H*) and Mann-Whitney (*U*) tests. Data are displayed for each sampling site. Statistically significant results are marked in italics.

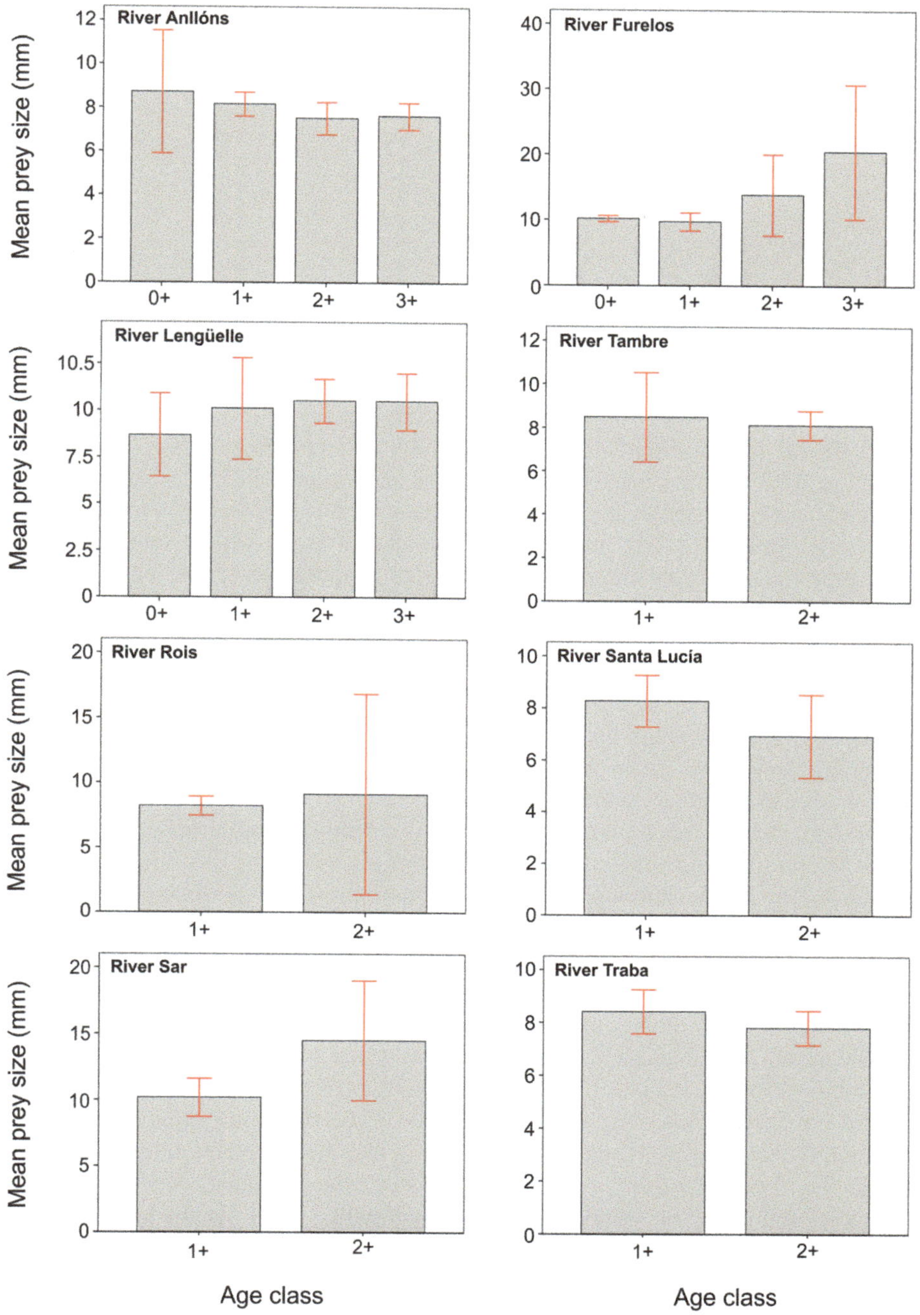

Figure 3 Mean prey size (mm) consumed by trout. Error bars represent the 95% confidence intervals.

the size frequency distribution of prey in the diet was not identical to that of potential prey in the environment, as anticipated, prey size selection was highly dependent upon the size frequency distribution of available prey (Rincón and Lobón-Cerviá 1999). Predictive models have estimated an optimal prey size of between 2.8 and 97 mm for trout (Bannon and Ringler 1986); however, a great variety of results, with respect to prey size, have been observed in the wild (e.g. McLennan and MacMillan 1984; Rincón and Lobón-Cerviá 1999; Montori et al. 2006; Sánchez-Hernández et al. 2011a; 2011b). In fact, studies have demonstrated that newly emerged trout fry mainly consume prey of 3 to 4 mm (Sánchez-Hernández et al. 2011a), whereas older age-0 individuals feed on prey of 5.5 mm mean size (Sánchez-Hernández et al. 2011b). McLennan and MacMillan (1984) found that trout preyed upon prey items varying in length between 6 and 10 mm. Rincón and Lobón-Cerviá (1999) demonstrated that organisms of 1 to 2 mm long were generally the most numerous in trout diets, while Montori et al. (2006) stated that 2 to 3 mm prey

Table 4 Mean prey numbers among age classes

	0+	1+	2+	3+	Total (pooled data)
Anllóns	18.5 (18 to 19)	16.9 (2 to 38)	52.3 (13 to 224)	8.3 (2 to 19)	24.7 (2 to 224)
Furelos	19.6 (4 to 82)	59.6 (3 to 363)	43.6 (7 to 139)	41.3 (3 to 100)	37.3 (3 to 363)
Lengüelle	20.7 (8 to 55)	22.4 (7 to 38)	30.9 (4 to 173)	13.0 (8 to 17)	24.6 (4 to 173)
Tambre	-	55.7 (1 to 214)	54.5 (6 to 117)	-	55.5 (1 to 214)
Rois	-	32.8 (7 to 85)	25.0 (24 to 26)	-	32.3 (7 to 85)
Santa Lucía	-	32.8 (7 to 62)	58.4 (32 to 103)	-	37.2 (7 to 103)
Sar	-	76.6 (12 to 309)	50.7 (33 to 88)	-	73.0 (12 to 309)
Traba	-	189.2 (8 to 388)	119.9 (11 to 235)	-	155.7 (8 to 388)

Data are displayed for each sampling site. Minimum and maximum are shown in brackets.

are commonly consumed. Steingrímsson and Gíslason (2002) found that the size of some prey items eaten by trout varied between 2.5 and 6.5 mm and between 1.5 and 12.5 mm for *Simulium vittatum* Zetterstedt, 1838, and *Radix* (=*Lymnaea*) *peregra* (Müller, 1774), respectively. However, the optimal prey size may vary ontogenetically, with mean sizes between 4.2 and 8.4 mm in 0+ and 2+ fish, respectively (Sánchez-Hernández and Cobo 2012). In addition, studies have demonstrated that optimal prey size is higher in lakes than rivers due to the piscivorous behaviour of trout in still waters (Keeley and Grant 2001 and references therein); generally, prey items range from 25 to 87 mm in lakes (e.g. L'Abée-Lund et al. 1992; Sánchez-Hernández and Amundsen 2015). Hence, based on the reviewed literature and the present study, the optimal prey size for trout appears to be 2 to 10 mm, although prey longer than 10 mm can be consumed.

Trout often undergo ontogenetic dietary shifts, and it has been demonstrated that mean prey size usually increases throughout ontogeny (e.g. Steingrímsson and Gíslason 2002; Montori et al. 2006; Sánchez-Hernández and Cobo 2012; Sánchez-Hernández et al. 2013). On the contrary, our study did not reveal consistent, significant increases in prey size with increasing fish length (no significant relationships) or age (statistically significant positive relationships were found for two of eight populations). The lack of allometric scaling in this study could partly be an artefact of our fish samples, which lacked large (>300 mm) individuals. Alternatively, the results may corroborate the suggestions of previous studies (Newman 1987; Rincón and Lobón-Cerviá 1999) that, in the absence of very small specimens, such as young larvae in which prey size may affect prey ingestion (Sánchez-Hernández et al. 2011a), gape-limited prey ingestion may not occur.

The sizes of prey consumed may affect the numbers of prey eaten by trout. A notable result of this study was that the relationship between mean prey size and prey numbers was significant and negative. A range of factors can influence whether large or small food items are consumed, but in the current study, it appeared that trout fed on either small numbers of large prey or large numbers of small, and theoretically low energy, prey. Stomach fullness, i.e. limitations of stomach volume, might therefore be a key variable in prey size selection. Within the limits imposed by gape size, fishes with big stomachs should be able to feed on a wider range of prey sizes than fishes with small stomachs (e.g. Gosch et al. 2009), but when the stomach is partially full, fish might choose small rather than large food items (Truemper and Lauer 2005). In this study, the relationship between average prey size and stomach fullness was positive, but was only statistically significant for the exponential model, suggesting that trout may be able to use relatively large food items regardless of their stomach fullness. This result could be because feeding intensity was low in the

Table 5 Mean feeding intensity (%), measured as stomach fullness index (*f*), among age classes

	0+	1+	2+	3+	Total (pooled data)
Anllóns	5.2 (2.2 to 8.2)	0.7 (0.3 to 1.6)	0.8 (0.1 to 3.5)	0.1 (0.1 to 0.3)	0.9 (0.1 to 8.2)
Furelos	9.0 (3.4 to 20.6)	1.0 (0.1 to 2.1)	1.0 (0.4 to 1.8)	1.2 (0.7 to 1.4)	5.6 (0.1 to 20.6)
Lengüelle	1.9 (0.5 to 6.6)	1.0 (0.3 to 1.9)	0.6 (0.1 to 2.8)	0.2 (0.1 to 0.3)	0.9 (0.1 to 6.6)
Tambre	-	0.3 (0.1 to 1.9)	0.3 (0.1 to 0.6)	-	0.7 (0.1 to 1.9)
Rois	-	4.8 (1.5 to 15.8)	1.9 (1.8 to 1.9)	-	4.6 (1.5 to 15.8)
Santa Lucía	-	5.1 (0.9 to 12.9)	4.3 (3.1 to 7.3)	-	5.0 (0.9 to 12.9)
Sar	-	10.5 (1.7 to 26.9)	7.7 (5.2 to 9.9)	-	10.1 (1.7 to 26.9)
Traba	-	16.6 (1.3 to 28.2)	7.3 (0.5 to 14.2)	-	12.1 (0.5 to 28.2)

Data are displayed for each sampling site. Minimum and maximum are shown in brackets.

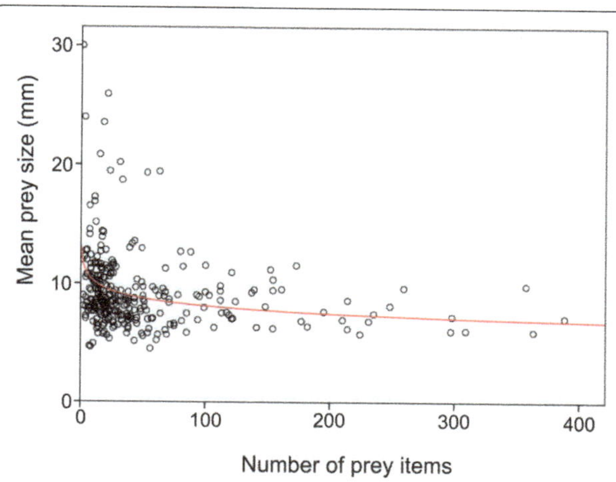

Figure 4 Relationship between mean prey size (mm) and numbers of prey eaten by trout using pooled data. Logarithmic regression equation is shown in red.

majority of the fishes; 85.1% of the fish had values of stomach fullness below 10%. Furthermore, the study confirmed a previous observation (Sánchez-Hernández and Cobo 2013) that summer feeding intensity of trout may decrease with fish age, being considerably higher in juveniles than in older age groups.

Conclusions

The feeding strategy of trout is flexible and clearly influenced by the size frequency distribution of potential prey: trout fed on either small numbers of large prey or large numbers of small, and theoretically low energy, prey. This study covers a general subject in trophic ecology and animal behaviour that may be applicable toward other fish species, especially other salmonids, to improve our understanding of feeding behaviour.

Competing interests

The authors declare that they have no competing interests.

Authors' contributions

FC participated in the design and coordination of the study, and JSH participated in field sampling and conducted the laboratory work. FC helped to draft the manuscript and JSH finalized the manuscript. Both authors read and approved the final manuscript.

Acknowledgements

The authors would like to thank the staff of the Station of Hydrobiology of the USC 'Encoro do Con' for their participation in the field work. We appreciate constructive comments from two anonymous referees, which considerably improved the quality of the manuscript. Thanks also to A.D. Nunn for English corrections. J. Sánchez-Hernández was supported by a postdoctoral grant from the Galician Plan for Research, Innovation, and Growth 2011–2015 (Plan I2C) and promoted by the Xunta de Galicia.

Author details

[1]Department of Zoology and Physical Anthropology, Faculty of Biology, University of Santiago de Compostela, Campus Sur s/n, 15782 Santiago de Compostela, Spain. [2]Department of Arctic and Marine Biology, Faculty for Biosciences, Fisheries and Economics, UiT The Arctic University of Norway, N-9037 Tromsø, Norway. [3]Station of Hydrobiology 'Encoro do Con', Castroagudín s/n, 36617 Vilagarcía de Arousa, Pontevedra, Spain.

References

Bagenal TB, Tesch FW (1978) Age and growth. In: Bagenal TB (ed) Methods for assessment of fish production in fresh waters, 3rd edn. Blackwell Science Publications, Oxford, pp 101–136

Bannon E, Ringler NH (1986) Optimal prey size for stream resident brown trout (*Salmo trutta*): tests of predictive models. Can J Zool 64:704–713

Clarke KR, Gorley RN (2001) PRIMER v5: user manual/tutorial. PRIMER-E, Plymouth

Cobo F, Mera A, González MA (1999) Proximate analysis and energy value of some families of freshwater heterometabolous insects. Boln Asoc Esp Ent 23:213–221 (In Spanish with abstract in English)

Cobo F, Mera A, González MA (2000) Proximate analysis and energy content of some families of freshwater holometabolous insects. NACC 10:1–12 (In Spanish with abstract in English)

Denoël M, Joly P (2001) Size-related predation reduces intramorph competition in paedomorphic alpine newts. Can J Zool 79:943–948

Evangelista C, Boiche A, Lecerf A, Cucherousset J (2014) Ecological opportunities and intraspecific competition alter trophic niche specialization in an opportunistic stream predator. J Anim Ecol 83:1025–1034

Fochetti R, Amici I, Agano R (2003) Seasonal changes and selectivity in the diet of brown trout in the River Nera (Central Italy). J Freshwat Ecol 18:437–444

Gerking SD (1994) Feeding ecology of fish. Academic, San Diego

Gill AB (2003) The dynamics of prey choice in fish: the importance of prey size and satiation. J Fish Biol 63:105–116

Gill AB, Hart PJB (1994) Feeding behaviour and prey choice of the threespine stickleback: the interacting effects of prey size, fish size and stomach fullness. Anim Behav 47:921–932

Gosch NJC, Pope KL, Michaletz PH (2009) Stomach capacities of six freshwater fishes. J Freshwat Ecol 24:645–649

Gupta PK, Pant MC (1983) Seasonal variation in the energy content of benthic macroinvertebrates of Lake Nainital U.P, India. Hydrobiologia 99:19–22

Jensen H, Kahilainen K, Amundsen P-A, Gjelland KØ, Toumaala A, Malinen T, Bøhn T (2008) Predation by brown trout (*Salmo trutta*) along a diversifying prey community gradient. Can J Fish Aquat Sci 65:1831–1841

Keeley ER, Grant JWA (2001) Prey size of salmonid fishes in streams, lakes, and oceans. Can J Fish Aquat Sci 58:1122–1132

Klemetsen A, Amundsen P-A, Dempson JB, Jonsson B, Jonsson N, O'Connell MF, Mortensen E (2003) Atlantic salmon, *Salmo salar* L., brown trout, *Salmo trutta* L., and Arctic charr, *Salvelinus alpinus* L.: a review of aspects of their life histories. Ecol Freshwat Fish 12:1–59

L'Abée-Lund JH, Langeland A, Sægrov H (1992) Piscivory by brown trout *Salmo trutta* L., and Arctic charr *Salvelinus alpinus* (L.) in Norwegian lakes. J Fish Biol 41:91–101

McLennan JA, MacMillan BWH (1984) The food of rainbow and brown trout in the Mohaka and other rivers of Hawke's Bay, New Zealand, New Zealand. J Mar Freshwat Res 18:143–158

Mittelbach GG (1981) Foraging efficiency and body size: a study of optimal diet and habitat use by bluegills. Ecology 62:1370–1386

Mock DW (1985) Siblicidal brood reduction: the prey size hypothesis. Am Nat 125:327–343

Montori A, Tierno de Figueroa JM, Santos X (2006) The diet of the brown trout *Salmo trutta* (L.) during the reproductive period: size-related and sexual effects. Int Rev Hydrobiol 91:438–450

Newman RM (1987) Comparison of encounter model predictions with observed size-selectivity by stream trout. J N Am Benthol Soc 6:56–64

Newman RM, Waters TF (1984) Size-selective predation on *Gammarus pseudolimnaeus* by trout and sculpins. Ecology 65:1535–1545

O'Brien WJ, Slade NA, Vinyard GL (1976) Apparent size as the determinant of prey selection by bluegill sunfish (*Lepomis macrochirus*). Ecology 57:1304–1310

Oscoz J, Leunda PM, Escala MC, Miranda R (2008) Summer feeding relationships of the co-occurring hatchling brown trout *Salmo trutta* and Ebro minnows *Phoxinus bigerri* in an Iberian river. Acta Zool Sinica 54:675–685

Reimchen TE (1991) Evolutionary attributes of headfirst prey manipulation and swallowing in piscivores. Can J Zool 69:2912–2916

Rincón PA, Lobón-Cerviá J (1999) Prey-size selection by brown trout (*Salmo trutta* L.) in a stream in northern Spain. Can J Zool 77:755–765

Ringler NH (1979) Selective predation by drift-feeding brown trout *Salmo trutta*. J Fish Res Board Can 36:392–403

Sánchez-Hernández J, Amundsen P-A (2015) Trophic ecology of brown trout (*Salmo trutta* L.) in subarctic lakes. Ecol Freshwat Fish 24:148–161

Sánchez-Hernández J, Cobo F (2012) Summer differences in behavioural feeding habits and use of feeding habitat among brown trout (Pisces) age classes in a temperate area. Ital J Zool 79:468–478

Sánchez-Hernández J, Cobo F (2013) Ontogenetic dietary shifts in the summer feeding intensity of brown trout in relation to fish condition. Folia Zool 62:110–114

Sánchez-Hernández J, Vieira-Lanero R, Servia MJ, Cobo F (2011a) First feeding diet of young brown trout fry in a temperate area: disentangling constrains and food selection. Hydrobiologia 663:109–119

Sánchez-Hernández J, Vieira-Lanero R, Servia MJ, Cobo F (2011b) Feeding habits of four sympatric fish species in the Iberian Peninsula: keys to understanding coexistence using prey traits. Hydrobiologia 667:119–132

Sánchez-Hernández J, Servia MJ, Vieira-Lanero R, Cobo F (2013) Ontogenetic dietary shifts in a predatory freshwater fish species: the brown trout as an example of a dynamic fish species. In: Türker H (ed) New advances and contributions to fish biology. InTech, Croatia, pp 271–298

Schael DM, Rudstam LG, Post JR (1991) Gape limitation and prey selection in larval yellow perch (*Perca flavescens*), freshwater drum (*Aplodinotus grunniens*), and black crappie (*Pomoxis nigromaculatus*). Can J Fish Aquat Sci 48:1919–1925

Schmitt RJ, Holbrook SJ (1984) Gape-limitation, foraging tactics and prey size selectivity of two microcarnivorous species of fish. Oecologia 63:6–12

Steingrímsson SÓ, Gíslason GM (2002) Body size, diet and growth of landlocked brown trout, *Salmo trutta*, in the subarctic River Laxá, North-East Iceland. Environ Biol Fish 63:417–426

Török J (1993) The predator-prey size hypothesis in three assemblages of forest birds. Oecologia 95:474–478

Truemper HA, Lauer TE (2005) Gape limitation and piscine prey size-selection by yellow perch in the extreme southern area of Lake Michigan, with emphasis on two exotic prey items. J Fish Biol 66:135–149

Wetterer JK (1989) Mechanisms of prey choice by planktivorous fish: perceptual constraints and rules of thumb. Anim Behav 37:955–967

13

Infection behavior, life history, and host parasitism rates of *Emblemasoma erro* (Diptera: Sarcophagidae), an acoustically hunting parasitoid of the cicada *Tibicen dorsatus* (Hemiptera: Cicadidae)

Brian J Stucky

Abstract

Background: 'Eavesdropping' parasitoids find their hosts by homing in on the communication signals of other insects. These parasitoids often exploit chemical communication, but at least some species of the sarcophagid genus *Emblemasoma* eavesdrop on the acoustic communications of cicadas. Despite considerable scientific interest in acoustic parasitoids, we know remarkably little about most species of *Emblemasoma*. To better understand the ecology and behavioral diversity of these flies, I used a combination of field and laboratory techniques to elucidate the infection behavior and life history of *E. erro*, which uses the cicada *Tibicen dorsatus* as a host, and I also investigated parasitoid loads and parasitism rates of *T. dorsatus* in multiple host populations in the central United States.

Results: Female *E. erro* used the acoustic signals of male *T. dorsatus* as the primary means of locating hosts, but they also required physical movement by the host, usually either walking or flight, to provide visual cues for the final larviposition attack. Larvae were deposited directly on the host's integument and burrowed through intersegmental membrane to enter the host's body. On average, *E. erro* larvae spent 88.0 h residing inside their host before leaving to pupariate, but residence time was strongly dependent on both ambient temperature and effective clutch size. Adult flies eclosed about 18 days after pupariation. Across all study sites, the mean parasitoid load of infected male *T. dorsatus* was 4.97 larvae/host, and the overall parasitism rate was 26.3%. Parasitism rates and parasitoid loads varied considerably among host population samples, and high parasitism rates were usually associated with high parasitoid loads.

Conclusions: Previously, detailed information about the infection behavior, life history, and host parasitism rates of sarcophagid acoustic parasitoids was only available for one species, *E. auditrix*. This study reveals that the infection behavior of *E. erro* is quite different from that of *E. auditrix* and, more broadly, unlike that known for any other species of acoustic parasitoid. The life histories of these two *Emblemasoma* are also divergent. These differences suggest that sarcophagid acoustic parasitoids are more behaviorally and ecologically diverse than previously recognized and in need of further study.

Keywords: Eavesdropping; *Emblemasoma*; Host defense; Host location; Infection behavior; Parasitoid; Parasitoid load; Phonotaxis; Superparasitism; *Tibicen*

Correspondence: stuckyb@colorado.edu
Department of Ecology and Evolutionary Biology, University of Colorado, Campus Box 334 UCB, Boulder, CO 80309, USA

Background

For female parasitoids, successful reproduction usually requires finding suitable hosts for their offspring. The problem, of course, is that potential hosts generally do their best *not* to be found. Sometimes, however, even well-hidden host insects must produce intraspecific communication signals, and these communication signals can be exploited by specialist parasitoids for use in efficient, long-range host location (Godfray 1994; Zuk and Kolluru 1998; Haynes and Yeargan 1999). Most often, such 'eavesdropping' parasitoids intercept chemical communications, but several species of flies (Diptera) from two families, Sarcophagidae and Tachinidae, use acoustic signals to find their hosts (Cade 1975; Soper et al. 1976; Lakes-Harlan and Lehmann 2015). Acoustically orienting tachinid parasitoids (tribe Ormiini) parasitize crickets and katydids (Orthoptera) (Lehmann 2003), while sarcophagid acoustic parasitoids, which are currently placed in the genus *Emblemasoma* (*sensu* Pape (1990)), parasitize cicadas (Hemiptera: Cicadidae) (Soper et al. 1976; Schniederkötter and Lakes-Harlan 2004). Because acoustic signals are often more amenable to experimental manipulation than pheromones, acoustic parasitoids have become valuable model organisms for investigating sexual signal exploitation and its consequences (e.g., Adamo et al. 1995; Allen 1998; Gray and Cade 1999; Müller and Robert 2002; Lehmann and Lehmann 2006; Beckers and Wagner 2011).

However, current knowledge of acoustic parasitoids is heavily biased toward the ormiine tachinids, which have received the majority of research (reviewed in Lehmann 2003). In comparison, our understanding of sarcophagid acoustic parasitoids is far more limited. *Emblemasoma* includes 16 described species (Pape 1996), but nearly everything known about the basic biology, behaviors, and ecology of these flies comes from study of a single species, *Emblemasoma auditrix* (Shewell), which is a specialist parasitoid of the cicada *Okanagana rimosa* (Say) (e.g., Lakes-Harlan et al. 2000; Köhler and Lakes-Harlan 2001; Schniederkötter and Lakes-Harlan 2004). No detailed information is available about the infection behaviors or life histories of any other *Emblemasoma*, and the only other record of phonotactic behavior comes from a study in which the species of *Emblemasoma* was not determined (Farris et al. 2008). Furthermore, no information about host parasitism rates or parasitoid loads is available for any species besides *E. auditrix* and its host *O. rimosa*.

Considering that *Emblemasoma* are frequently referenced in discussions of insect hearing and parasitoid biology (e.g., Godfray 1994; Feener and Brown 1997; Yager 1999; Yack 2004; Robert 2005; Hedwig and Robert 2014; Strauß and Lakes-Harlan 2014) and that their 'ears' have been the focus of multiple physiological investigations (Lakes-Harlan et al. 1999; Robert et al. 1999; Farris et al. 2008), it is perhaps surprising how little we actually know about the basic biology and ecology of any of these flies besides *E. auditrix*. As a consequence, it is nearly impossible to make meaningful generalizations about sarcophagid acoustic parasitoids, and drawing broader conclusions about acoustically orienting parasitoids in general is similarly difficult.

In 2008, I discovered that adults of the cicada *Tibicen dorsatus* (Say) (Figure 1), a large cicada that is common in the grasslands of central North America (Cole 2008), were sometimes infected with the larvae of a sarcophagid parasitoid. After a preliminary investigation revealed that this parasitoid was *Emblemasoma erro* Aldrich (Figure 2) and that these flies were most likely locating their hosts acoustically, I began a comprehensive study of the basic biology of this fly. The only information previously available about the biology of *E. erro* was a record of a single female fly that had been reared from a specimen of the cicada *Quesada gigas* (Olivier) in Brazil (Lopes 1981) and a report of rearings from *Tibicen* sp. in Texas, USA (Lakes-Harlan 2009).

In this paper, I 1) describe the host locating and larviposition behaviors of *E. erro*; 2) describe this parasitoid's life history; and 3) report the results of an investigation of parasitism rates and parasitoid loads in natural populations of the host cicada, *T. dorsatus*. I then discuss the following: the infection behaviors of *E. erro* in comparison with other acoustic parasitoids and other sarcophagid parasitoids, potential host defenses, and the causes of variation in host population parasitism rates and parasitoid loads, including empirical evidence that superparasitism might contribute to high parasitoid loads in some host populations. The results show not only that the behaviors and life histories of sarcophagid acoustic parasitoids are more diverse than previously recognized, but also that the infection behaviors of *E. erro* are unlike those known for any other acoustic parasitoid.

Methods

Study sites

Surveys of host populations, collections of adult hosts and parasitoids, and field behavioral observations were conducted at six primary study sites located in Ellsworth, Harvey, McPherson, and Reno counties in central Kansas, Hamilton County in western Kansas, and Prowers County in eastern Colorado (Figure 3). The central Kansas sites are located within the Central Great Plains level III ecoregion, while the western Kansas and eastern Colorado sites are located near the boundary between the High Plains and Southwestern Tablelands ecoregions (US Environmental Protection Agency 2013). All sites consisted of native midgrass or shortgrass prairie vegetation intermixed with riparian, woody vegetation, or

Figure 1 Male *Tibicen dorsatus*, Harvey Co., KS.

planted trees. Supplemental collections of host cicadas and adult flies for life history and behavioral study were made at five additional locations in Kansas with habitat that was similar to that at the primary study sites (Figure 3).

Host locating and larviposition behaviors of *E. erro*

The host locating and larviposition behaviors of *E. erro* were studied in three ways. First, natural cicada/fly interactions were observed directly in the field whenever possible during the summers of 2010 to 2014. Second, artificial broadcasts of acoustic stimuli were used to test for phonotactic behavior by female flies in the field; and third, cicada/fly interactions were observed in a controlled, outdoor laboratory environment. The methods for these latter two approaches are next described in more detail.

Field broadcasts of acoustic stimuli

Preliminary observations suggested that male cicadas' acoustic signals played a role in host location by *E. erro*, but such observations cannot assess whether acoustic cues by themselves are sufficient to attract female parasitoids. To separate acoustic stimuli from other possible sources of information about the location of potential hosts (e.g., visual or olfactory), a loudspeaker was used in the field to broadcast audio that mimicked the calling song of a typical male *T. dorsatus*.

Acoustic signals for attracting *E. erro* were generated by gathering audio recordings of calling male *T. dorsatus*, analyzing these recordings to estimate the mean values of several acoustic parameters, then constructing model acoustic signals that matched, as closely as possible, the mean calling song of the species. To ensure that the model acoustic signals were broadly representative of *T. dorsatus* from the general study area, I obtained recordings of 20 different individuals of *T. dorsatus* from six field sites in Kansas. All recordings were made as uncompressed, 16-bit PCM audio at a sampling rate of 44.1 kHz using a highly directional shotgun microphone (Sennheiser ME66 or ME67 with a matched windscreen; Sennheiser Electronic GmbH & Co. KG, Hannover, Germany) and a digital audio recorder (Sony MZ-M200 or PCM-M10; Sony Corporation, Tokyo, Japan). To minimize background noise while also avoiding near-field acoustic effects in the frequency range of the cicadas' calling songs (Michelsen and Nocke 1974; Peterson 1980), recordings were typically made with the microphone held at a distance of between 0.5 and 2 m from the calling cicada. Most recordings were made directly in the field, but in some cases, cicadas were captured and placed in mesh cages, then recorded once they resumed normal acoustic activity.

Each recording was analyzed to determine the values of three acoustic variables: peak frequency, pulse group (PG) length, and PG rate. A 'pulse group' is defined as a first-order assemblage of sound pulses (i.e., a train of sound pulses) that is separated by silence from the rest of the audio signal. It is the basic unit of temporal structure in the call of *T. dorsatus* (see Cole (2008) for a spectrogram and oscillogram of the *T. dorsatus* calling song; note that Cole refers to PGs as 'syllables'). Peak frequency was estimated by identifying the highest peak in a power spectral density plot generated by Audacity® (Audacity Team 2012) using a 512-sample fast Fourier transform with the Hann window function. If there were two or more peak frequencies that differed by less than 0.5 dB, their average was taken as the overall peak frequency. PG length and rate were determined using custom-written software to analyze 10 s of audio from the middle of each calling song recording. Following these analyses, a single model acoustic signal was constructed using the *T. dorsatus* recording that was as close as possible to the mean calling song observed for the species (mean peak frequency = 4,308 Hz [s = 444]; mean PG length = 20.3 ms [s = 1.45]; mean PG rate = 37.04 PG/s [s = 2.25]). I also generated synthetic acoustic signals constructed from amplitude-modulated sine waves that exactly matched the observed mean acoustic variable values.

Figure 2 Female *Emblemasoma erro*, Ellsworth Co., KS.

Acoustic signals were broadcast in the field with a custom-built, portable broadcasting system consisting of a 12-V audio amplifier and a high-output, horn-loaded tweeter speaker (PylePro PH44; Pyle Audio Inc., Brooklyn, NY, USA) mounted in the top of a wooden box. Acoustic signals were fed to the amplifier from either a portable CD player or a flash memory-based digital audio player. Broadcasts in the field were conducted either in the late morning or afternoon when cicadas were naturally active, usually for a duration of 4 to 8 min at one time. Flies that were attracted to the broadcast speaker were captured by hand.

Laboratory observations

Although natural cicada/fly interactions were occasionally observed in the field, opportunities for such observation were unpredictable and infrequent. Furthermore, close-range observation was often impossible, and cicadas could rarely be captured after an encounter with a fly to determine whether larviposition occurred. Consequently, observation of cicada/fly interactions in a more controlled setting was also necessary.

Initial attempts to observe the infection behavior of *E. erro* in 2010 used restrained cicadas and audio broadcasts in an approach similar to that used by Schniederkötter

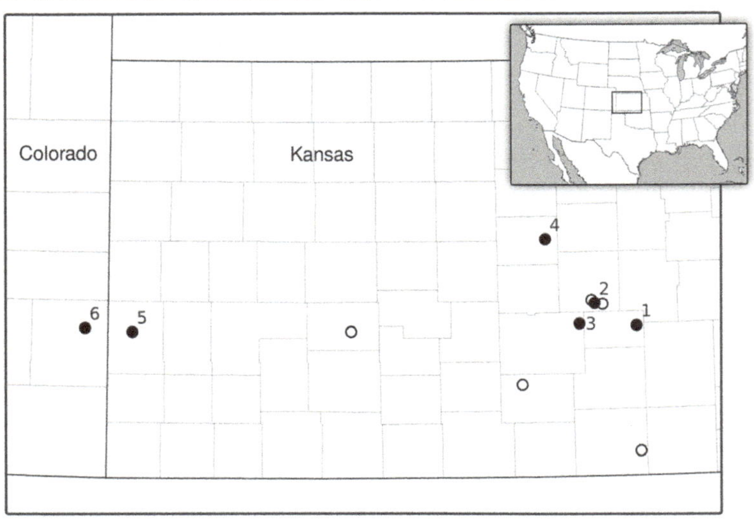

Figure 3 Locations of study sites. Filled circles indicate the primary sites used for estimating host parasitism rates, and open circles indicate secondary sites used for additional collections of cicadas and flies. Primary sites are referenced in the text by the counties in which they were located: 1) Harvey Co., 2) McPherson Co., 3) Reno Co., 4) Ellsworth Co., 5) Hamilton Co., and 6) Prowers Co. Both *T. dorsatus* and *E. erro* were found at all 11 sites. The inset map indicates the location of the main map in the United States.

and Lakes-Harlan (2004) for their study of *E. auditrix*. This technique was unsuccessful for *E. erro*, however, so further experiments with immobilized cicadas were abandoned.

Instead, unrestrained cicadas and flies were allowed to freely interact in outdoor cages during behavioral experiments in 2012 and 2013. For each infection behavior trial, one female fly was released into a mesh cage containing one or two uninfected male cicadas. Three types of cage were used: a cylindrical cloth mesh cage approximately 27 cm in diameter and 39 cm high; a larger cylindrical cloth mesh cage approximately 46 cm in diameter and 66 cm high; and a much larger, rectangular screen 'flight cage' with a square base and walls approximately 1.8 m on each side and just over 2.1 m high at the center of the top. The behaviors of the fly and cicada(s) were then observed carefully throughout the duration of the trial. If the fly appeared to directly contact the cicada with the tip of her abdomen or otherwise attack the cicada, the cicada was immediately removed and inspected for the presence of fly larvae. In most trials, the cicada and fly were not physically disturbed inside of the cage, but in some cases the cicada was induced to flight by the experimenter to observe the fly's response. Trials ended when a fly larviposited upon a cicada or the fly no longer showed interest in the cicada(s) in the cage. To avoid overly stressing the animals, trials were also usually terminated after a fly made several attempts to attack a cicada even if larviposition was not observed. If the infection status of a cicada could not be determined by visual inspection immediately after a trial, the cicada was not used in further trials for at least 48 h in order to verify whether it had become infected. No single female fly was used to infect more than two cicadas. Whenever a fly larviposited on a cicada, I attempted to immediately count the number of larvae deposited. This was not always possible, though, and in these cases, the total number of larvae was determined by rearing the parasitoids or dissecting the host.

All cicadas used for this part of the study were mature adult male *T. dorsatus* that were captured directly in the field. Captured cicadas were maintained outdoors in large cloth mesh cages placed over live branches of green ash (*Fraxinus pennsylvanica*), which provided the cicadas with a suitable food source. After capture, and before exposing them to parasitoids, all cicadas were closely monitored for up to 9 days to determine whether they had already been parasitized in the field by *E. erro*. Only unparasitized cicadas were used for studying cicada/fly interactions. Adult female *E. erro* were obtained by broadcasting the model call of *T. dorsatus* in the field, as described above, and collecting attracted female flies by hand. Flies were kept in small mesh cages in the laboratory and provided with sucrose and water *ad libitum*.

Life history of *E. erro*

The timings of key life history events for *E. erro* were estimated by rearing parasitoids from hosts that were naturally infected in the field, infected during the behavior studies described above, or artificially infected in the lab. To artificially infect cicadas in the laboratory, a female *E. erro* was first anesthetized by chilling the insect at approximately 4°C for several minutes. The fly was then decapitated, placed on a piece of moistened filter paper on a watch glass, and live first-instar larvae were carefully dissected from the fly's abdomen. Individual larvae were transferred to uninfected adult *T. dorsatus* cicadas using the moistened tip of a fine artist's brush. Most larvae were placed on the intersegmental membrane at the base of the cicada's wings, but some were placed at the lateral junction of the metathorax and mesothorax or the junction of the metathorax and the first abdominal tergum. A fine insect pin was sometimes used to make a small puncture in the membrane at the wing base in order to facilitate the larvae's entry into the host's body. Larvae from a single female fly were never used to infect more than two cicadas.

All infected cicadas were kept in outdoor, mesh cages as described above. Cicadas were checked several times daily, and any individuals that died or appeared moribund were moved indoors into small plastic emergence containers to capture emerging fly larvae. Once a host was moved to a larval emergence container, video recording was used to capture the precise time and location of larval egress from the host. If no larvae were observed in an emergence container approximately 48 h after host death, or if only undersized larvae emerged, the dead cicada was dissected to check for additional fly larvae.

Parasitoid larvae that emerged from their host were provided with fine sand in which to burrow and pupariate. Once pupariation was complete, bits of moist paper towels were placed in the emergence containers to help maintain suitable humidity, and the puparia were kept at room temperature (generally 24°C to 28°C) and exposed to the approximate natural daily photoperiod. The emergence containers were fitted with screen tops to capture any eclosing adult flies.

To better understand how biotic and abiotic factors influence larval development, I evaluated the effects of two key variables - effective clutch size (the number of larvae from a clutch that successfully develop inside a host) and the ambient temperature experienced by the host - on the total time larval parasitoids spent inside their host (the 'larval residence time'). These variables were chosen because temperature affects the development and growth rates of insects in general (Harrison et al. 2012), and the number of larvae inside a host might influence how rapidly the host is consumed. For this analysis, larval residence time was calculated as the total elapsed time, in hours, from the moment a larva was deposited on a cicada until the larva emerged from its host. Effective clutch size was taken as the total number of larvae that emerged from a host (because the host cicadas for this analysis were infected in the behavior studies or in the lab, all larvae inside a host were known to be from the same clutch). Temperature was calculated as the mean ambient air temperature experienced by each parasitized host. Temperature data were taken from the Daymet 1-km daily surface weather dataset (Thornton et al. 1997, 2014). The overall mean ambient air temperature experienced by a parasitized cicada was estimated by averaging the daily minimum and maximum temperatures for each day that the cicada was infected up to the time of larval egress from the host. The relationship among these three variables was analyzed using multiple linear regression with effective clutch size and temperature as the explanatory variables. To avoid potential non-independence problems caused by related larvae sharing the same host cicada, the data were summarized at the level of the host cicada; that is, for each host with multiple parasitoid larvae, the mean residence time for all larvae from the host was used in the analysis instead of the residence time for each parasitoid larva. Diagnostic plots of the standardized residuals were used to verify the fit of the regression model. This, and all other statistical analyses, was conducted in R version 3.1.1 (R Core Team 2014). Note that larvae from parasitized cicadas captured in the field could not be included in this analysis because it was not known when these larvae were deposited on their hosts.

Host parasitism rates and parasitoid loads

To obtain population samples for estimating host parasitism rates, adult *T. dorsatus* were surveyed by walking through the habitat at a study site and attempting to capture all *T. dorsatus* that were observed perched in the vegetation or disturbed into flight. Teneral or recently emerged cicadas were excluded because male cicadas do not develop full calling capabilities or begin sexual acoustic behaviors until several days after eclosion (Maier 1982; B. Stucky, unpublished data). Captured cicadas were maintained in captivity to rear the parasitoids from all infected cicadas, determine the total number of infected cicadas in each sample, and determine the parasitoid load of each host, following the methods described above.

Cicada population surveys were conducted at the six primary study sites in July, August, or early September of 2011 to 2014, although not all sites were sampled all 3 years (Table 1). The survey dates were limited by when adult *T. dorsatus* were actually present in the field, which varied from year to year. In 2012, for example, adult *T. dorsatus* were abundant in central KS by July 1, but they did not reach similar abundance in 2013 until the latter half of July.

By the time the first population samples were collected in 2011, I had established that the male cicada's calling song was a critical cue used by female parasitoids to locate their hosts. Consequently, surveys from 2011 to 2013 focused on male cicadas only (females do not produce sound) in order to use the limited space available for housing these large insects as efficiently as possible. In 2014, both female and male cicadas were sampled at the field sites in Harvey, McPherson, and Reno counties in central Kansas.

Logistic regression (generalized linear models with binomial-distributed response and logit link function) was used to evaluate whether host parasitism rates varied among the field sites and whether sample year or sample date also influenced parasitism rates. The proportion of parasitized cicadas in population samples was modeled with field site, year, and ordinal sample date as possible predictor variables. Both field site and year were treated as categorical variables. To test the effects of individual predictors and decide which variables to retain in the model, nested models were compared using the difference of their deviance statistics (i.e., likelihood-ratio tests) (Dobson and Barnett 2008). Standardized residuals plots were examined to check for any problems with model specification. To further assess the final model fit, the likelihood-ratio (also known as McFadden) pseudo R^2 was calculated (McFadden 1974; Menard 2000).

Early in this study, it became clear that parasitoid loads varied among the population samples. One possible cause of such variation is superparasitism, which, for gregarious parasitoids such as *E. erro*, is expected to occur more often when unparasitized hosts are rare (Godfray 1994). Unparasitized hosts are rare when parasitism rates are high, so to test for this causal relationship, I used simple linear regression to evaluate whether high host parasitism rates corresponded with high parasitoid loads. For this analysis, each data point was the estimated mean host parasitoid load and parasitism rate for a single study site in a given year. Yearly population samples for which fewer than three parasitized cicadas were available to estimate the mean parasitoid load were excluded from the analysis. Parasitism rate was used as the explanatory (i.e., x-axis) variable, and because parasitism rates were estimated from population samples, some of which were small, there was the possibility of substantial measurement error. Consequently, the regression analysis was likely to suffer from slope attenuation bias, in which the slope estimator is biased to be less than the true slope (Bulmer 1979; Smith 2009). To compensate for this, I used the sizes of each population sample to estimate the mean measurement error variance across all population samples. I then used this estimate of

Table 1 Observed parasitism rates of male *T. dorsatus* in the field

Study site	Dates	Infected	Uninfected	Total	% infected	95% CI
McPherson Co.	Site summary (2012 to 2014)	3	45	48	6.3	2.1 to 16.8
	2012: July 2, 4	0	12	12	0.0	
	2013: Aug. 2, 20	0	10	10	0.0	
	2014: Aug. 3, 12	3	23	26	11.5	
Prowers Co.	Site summary (2013 to 2014)	27	9	36	75.0	58.9 to 86.2
	2013: Aug. 22, 28	11	4	15	73.3	
	2014: Aug. 21, Sept. 4	16	5	21	76.2	
Hamilton Co.	Site summary (2013 to 2014)	10	3	13	76.9	49.7 to 91.8
	2013: Aug. 22, 28	6	3	9	66.7	
	2014: Aug. 21	4	0	4	100.0	
Harvey Co.	Site summary (2011 to 2014)	11	51	62	17.7	10.2 to 29.0
	2011: Aug. 12	1	9	10	10.0	
	2012: July 7, 12	5	13	18	27.8	
	2013: Aug. 5, 12	3	14	17	17.6	
	2014: Aug. 9, 12	2	15	17	11.8	
Reno Co.	Site summary (2011 to 2014)	8	65	73	11.0	5.7 to 20.2
	2011: Aug. 14, 15	1	3	4	25.0	
	2012: July 16, Aug. 17	1	13	14	7.1	
	2013: July 23	0	15	15	0.0	
	2014: Aug. 4, 11	6	34	40	15.0	
Ellsworth Co.	Site summary (2011 to 2013)	11	23	34	32.4	19.1 to 49.2
	2011: Aug. 11	5	3	8	62.5	
	2012: Aug. 19	2	1	3	66.7	
	2013: Aug. 10, 17	4	19	23	17.4	
	Overall	70	196	266	26.3	21.4 to 31.9

Observed parasitism rates are given for each study site for all sample years combined, with the yearly observations for each site given below the site summary rows. The overall totals for all sites and years combined are given at the bottom of the table. 'Infected' is the number of parasitized cicadas that were captured, 'Uninfected' is the number of unparasitized cicadas, 'Total' is the total number of cicadas captured, and '95% CI' is the Wilson 95% confidence interval for the population estimate of the percentage of infected male cicadas. Refer to Figure 3 for study site locations.

the error variance with the method of moments estimator (MME) of the bias correction factor (Carroll and Ruppert 1996; Smith 2009) to calculate an attenuation-corrected slope estimate. Diagnostic plots of the standardized residuals were used to verify the fit of the simple linear regression model.

Additional statistical methods

All confidence intervals (CIs) for population proportion estimates were calculated using the Wilson method (also known as the score confidence interval) because of its good performance across a broad range of sample sizes (Wilson 1927; Agresti and Coull 1998). CIs for the estimates of population means were constructed using the standard *t*-distribution method when possible (Whitlock and Schluter 2009), but in cases where the population distribution appeared to be non-normal (as determined by examining plots of sample distributions), CIs were calculated using the bootstrap-*t* resampling method with 1,000,000 replicates (Efron and Tibshirani 1993; Carpenter and Bithell 2000). Bootstrap-*t* resampling with 1,000,000 replicates was also used to compare the means of non-normally distributed populations. Throughout this paper, '*s*' is used to indicate the sample standard deviation.

Results

Host locating and larviposition behaviors of *E. erro*

Host locating behavior

Field and laboratory observations of cicada/fly interactions and field broadcasts of the *T. dorsatus* call all confirmed that *E. erro* uses the calling songs of male cicadas as the primary cue for locating potential hosts. In the field, I was able to observe, at relatively close range, the interactions between 14 *T. dorsatus* cicadas and *E. erro* flies. All of the cicadas involved in these interactions were males, 13 of which were acoustically active while I observed them. Seven individual flies were observed in the process of locating a perched cicada (either by flight or walking), and in every case, the perched cicada was calling while the fly was moving toward it. In the other observed cicada/fly interactions, the flies were already perched near the cicada when I first saw them and I did not observe how, or when, the flies actually arrived near these cicadas. I never saw flies near perched female cicadas, which cannot produce sound.

During the infection behavior trials in outdoor cages in 2012 and 2013, female flies in the experiment cages typically showed an immediate, strong phonotactic response to calling cicadas. When a cicada in the cage began to call, a fly would either walk towards the calling cicada, fly to within several centimeters of the cicada and then walk towards it, or, in some cases, fly directly to, and land on, the calling cicada. Thus, both field and laboratory observations of cicada/fly interactions provided strong circumstantial evidence that *E. erro* use the acoustic calls of cicadas to locate their hosts, but such observations cannot definitively rule out the possibility that some other source of information was actually being used, such as visual or olfactory cues.

Field broadcasts of the model *T. dorsatus* calling song furnished unambiguous evidence that acoustic cues, by themselves, are sufficient to attract female flies. The model *T. dorsatus* calling song was broadcast at least once at all six primary field sites. *E. erro* were attracted to the broadcast speaker at every location. Flies often arrived within a few seconds of the start of a broadcast, and it was not uncommon to see multiple flies perched on the top of the speaker box at the same time. At least two or more flies were collected at each primary field site, and more than 60 *E. erro* in total were captured during this study. Often, many more flies arrived at the broadcast speaker than could be captured by hand. I did not attempt to quantify the number of flies that arrived and were not captured, however, because individual flies will sometimes arrive at and leave the speaker multiple times during a single broadcast (B. Stucky, personal observation), making any such counts unreliable. The broadcast apparatus never attracted any flies when the loudspeaker was not operating.

Larviposition behavior

Acoustic cues and phonotaxis are clearly critical for *E. erro* to locate its hosts, but the calling song by itself never induced *E. erro* to larviposit in the absence of a host. Despite the large numbers of flies that were attracted to the calling song broadcasts, no fly larvae were ever found on the speaker or surface of the speaker box following a broadcast of the model *T. dorsatus* call.

Furthermore, laboratory observations of infection behaviors revealed that, even when a potential host was present, the cicada's calling song was still not the stimulus that ultimately triggered larviposition. Although cicadas often called during the infection behavior trials in 2012 and 2013, in no case did this directly result in larviposition by a female fly. Instead, once a fly had moved to within a few centimeters of a calling cicada by orienting to the cicada's calling song, it would typically remain more or less stationary next to the cicada with its head facing toward the cicada's body. At this point, repeated calls from the cicada usually resulted in relatively little additional movement from the fly. However, if the cicada moved either by walking or flight, the fly usually attempted to maintain its proximity to the potential host. Thus, if the cicada began walking, the fly would typically follow it from a short distance (e.g., 2 to 3 cm away). If the cicada took flight, the fly almost always immediately took flight as well and attempted to follow the cicada in the air.

Sometimes, cicada locomotion resulted in a larviposition attack by the fly, and all evidence suggested that at least some movement by the cicada was essential for larviposition. Seventeen incidents of successful larviposition were obtained during the infection behavior trials in 2012 and 2013, and in every case, larviposition was only observed when cicadas were in motion. I was able to determine the moment of larviposition for 15 of these attacks. Of these, five (33.3%) occurred while the cicada was in flight; the remaining ten attacks (66.7%) occurred while the cicada was either walking, flapping its wings, or both. In a few cases, the cicadas that were attacked never called at all during the time the fly was in the cage. These cicadas were 'discovered' by the flies purely due to their physical movement in the cage, further indicating that movement by a potential host, not sound, provides the visual cues that ultimately trigger larviposition. As a further example, in 2012, I experimented with releasing a female *T. dorsatus* (which cannot produce sound at all) in the air in front of a female fly. The fly eventually followed the cicada and larviposited on it in flight.

Successful larviposition attempts resulted in the deposition of one or more tiny first-instar larvae directly upon the exterior of the cicada's body (Figure 4). A sticky secretion also usually accompanied the larvae, presumably to help them adhere to the host. Flies appeared to briefly contact the cicada with their abdomens during larviposition, but high-speed video recording would be needed to reveal the exact mechanics of this process. After larviposition, the larvae immediately began searching for an area of intersegmental membrane through which to burrow and enter the host's body. The larvae typically entered the host quite rapidly, disappearing after anywhere from a matter of seconds to a few minutes.

Although the location of larvae deposition could not be determined in all cases because the larvae sometimes disappeared into the host's body before they could be observed, the evidence suggests that flies prefer to attack

Figure 4 Larviposition by *E. erro*. A first-instar larva of *E. erro* on the right fore wing of a *T. dorsatus* moments after larviposition (larva indicated by blue arrow). The cicada's head and foreleg are at top center.

the base of a host's wings. Of 15 attacks for which the exact location of larviposition was determined (out of 17 total successful attacks), 1 (6.7%) was on the base of a fore leg, 2 (13.3%) were on the abdomen, and 12 (80%) were either directly on the wings (usually near the base) or on the pterothorax or the first two abdominal segments next to the base of the wings. Left/right orientation was recorded for 14 of the 15 attacks, and of these, 9 (64.3%) were on the left side of the cicada's body, 3 (21.4%) were on the right side, and 2 (14.3%) were approximately medial. In their studies of the infection behaviors of *E. auditrix*, Schniederkötter and Lakes-Harlan (2004) discovered that *E. auditrix* preferentially attacked the left side of potential hosts. Most laterally oriented attacks by *E. erro* also occurred on the left side of the cicada's body (9 of 12, or 75%), but this asymmetry was not statistically significant for this sample size (exact binomial test, $p = 0.146$).

The number of larvae deposited by a single female fly on a host cicada during the infection behavior trials (i.e., the clutch size) varied from a minimum of one to a maximum of six, but more than 80% of the time, flies (14 of 17) deposited three or fewer larvae. The mean clutch size was 2.53 (95% bootstrap-*t* CI: 1.85 to 3.45 larvae/host, $s = 1.50$, $n = 17$ hosts) and the median was 3.

Observations of fly behavior in the field appeared to corroborate the infection behaviors in the lab. I observed eight male *T. dorsatus* that each produced one or more complete calling songs (as many as five in one case) while an *E. erro* was perched next to the cicada. In no case did the calling song appear to trigger an attempt at larviposition. Just as in the laboratory, flies often waited, nearly motionless, next to calling cicadas, and if a cicada in the field crawled up or down the vegetation it was perched on, the fly usually followed it. If the cicada took flight, the fly usually also took flight and followed the cicada in the air.

As in the cage infection behavior trials, flies in the field only seemed to attack a cicada if the cicada was in motion or had just moved, and most apparent attacks occurred when the cicada was in flight. On two such occasions, cicadas that were evidently struck in the air by *E. erro* had their flight disrupted to such an extent that the cicadas crashed to the ground. Unfortunately, I could not determine with absolute certainty whether any of the flies I observed in the field actually larviposited on the attacked cicadas. I was able to capture eight cicadas shortly after their interactions with *E. erro*, but I was unable to locate first-instar fly larvae on any of them. Considering how rapidly larvae can burrow into their host's body, it is likely that they had already disappeared from view by the time I was able to look for them. Parasitoids were reared from all eight cicadas, though, so it is very likely that at least some became infected during the observed cicada/fly interactions.

Life history of *E. erro*

From the moment of larviposition until they completely exited the host's body, *E. erro* larvae spent, on average, 88.0 h residing inside their host (95% CI: 81.19 to

94.76 h, $s = 17.1$, $n = 27$ larvae from 13 host cicadas and 10 female flies, range = 61.3 to 116.0). Multiple regression analysis of these data revealed that both temperature and effective clutch size had significant effects on larval residence time. Together, these two variables explained more than 93% of the observed variation in residence times ($R^2 = 0.934$; p values for the coefficients of both explanatory variables were <0.002). Increases in either ambient temperature or the number of larvae in a host were associated with a decrease in residence time (the estimated relationship was *residence_time* = 211.2 – 4.14 *temperature* – 5.50 *effective_clutch_size*) (Figure 5). By the time all larvae left an infected host, it was common to find all soft tissues inside the cicada's body entirely consumed so that nothing but the exoskeleton remained.

To exit their host, larvae used their oral hooks to burrow through intersegmental membrane, and they usually emerged by squeezing between one of the cicada's opercula and its abdomen (Figure 6). The exact location of egress was observed for 83 larvae from 28 *T. dorsatus* hosts, and of these, 64 (77.1%) exited from behind one of the cicada's opercula. Of the remaining larvae, 16 (19.3%) exited next to the pygofer or terminal abdominal segments at the apex of the abdomen, and 3 (3.6%) burrowed through the membrane between the head and prothorax.

After leaving their host, larvae immediately burrowed into the soil (or sand, in the case of the emergence containers) to pupariate. Although more than 300 *E. erro* larvae were obtained from infected *T. dorsatus* specimens during the course of this study, relatively few of these were successfully reared to the adult stage. Fifty-three flies survived to adulthood, and the times of both larval egress from the host and adult eclosion were obtained for 31 of these flies. Adult flies eclosed 18.4 days, on average, after leaving their host (95% CI: 18.02 to 18.69 days, $s = 0.91$, $n = 31$ flies from 15 host cicadas, range = 16 to 20 days). The lifespan of adult flies in the field is unknown. Adult flies maintained in the laboratory survived as long as 92 days.

The lifetime reproductive potential of female *E. erro* was not determined, but I did dissect 14 gravid female flies that were collected at audio broadcasts of the *T. dorsatus* calling song in 2013 and 2014 and counted all larvae contained within their abdomens. These flies carried as few as 3 and as many as 174 larvae in their incubatory pouches, with a mean of 60.7 larvae per fly (s = 57.5). The observed distribution of larvae counts was strikingly bimodal: Three flies had more than 150 larvae, while all of the rest had fewer than 80. The larvae of the four flies with the largest larvae counts were noticeably smaller than those from the remaining flies and generally had less well-developed bristles. Remnants of eggshell were still visible in the incubatory pouches of three of these flies, suggesting that the larvae had recently hatched.

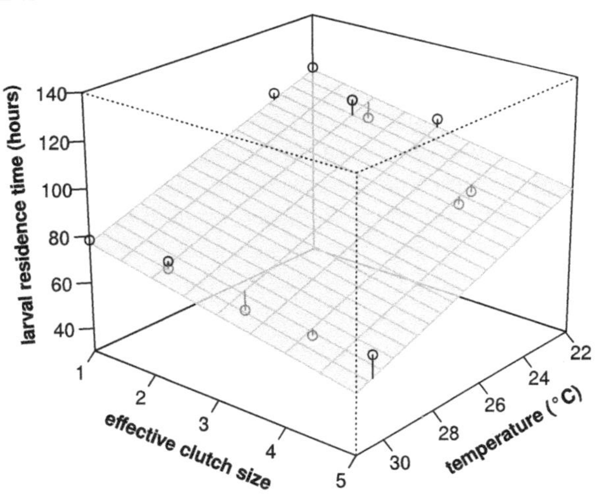

Figure 5 Relationship of effective clutch size and temperature to larval residence time. Each data point represents the mean residence time of the parasitoid larvae inside a single host cicada along with the effective clutch size (number of larvae emerging from the host) and the mean air temperature experienced by the host during parasitoid development. The planar surface represents the multiple linear regression model of the effects of temperature and effective clutch size on larval residence time. Lines connected to the data points indicate the vertical distance of each data point from the regression surface (i.e., the residuals).

Host parasitism rates and parasitoid loads

Parasitism rates

The results of the *T. dorsatus* population surveys for the prevalence of *E. erro* infection are presented in Table 1. Parasitized *T. dorsatus* were collected at all six of the primary study sites, although infected cicadas were not detected in all population surveys. All parasitoids that were reared to the adult stage were identified as *E. erro*, and the morphologies of all other larvae and puparia that were obtained were also consistent with *E. erro*. No hyperparasitoids of *E. erro* were observed.

Across all four sampling years (2011 to 2014) and all six primary study sites, the overall observed parasitism rate for *T. dorsatus* males was 26.3% (95% CI: 21.4 to 31.9%, $n = 266$ cicadas). The surveys in 2014 also included a sample of 28 female *T. dorsatus* from the central KS field sites (in Harvey, McPherson, and Reno counties), and of these, one female cicada was infected with *E. erro* larvae (3.7%; 95% CI: 0.7% to 18.3%).

There was substantial variation in observed parasitism rates among the population samples (summarized in Table 1). The results of the logistic regression analysis suggested that much of this variation was due to differences among field sites, with sampling year possibly also having a small effect (likelihood-ratio tests of *field site*

Figure 6 Emergence of *E. erro* from its host. A mature larva of *E. erro* emerges from between the left operculum and the abdomen of a deceased male *T. dorsatus* from Prowers Co., CO.

and *year* as predictors: $p < 0.00001$ and $p = 0.0663$, respectively). The model including these two predictor variables seemed to explain the data reasonably well, with pseudo $R^2 = 0.549$. Nevertheless, this result must be interpreted with caution. The westernmost field sites often had the highest sample parasitism rates, but because of logistical constraints, these sites were always surveyed later in the summer than the other field sites (Table 1). Thus, the variable *field site* was at least partially collinear with *sample date*. Consequently, the high parasitism rates observed at these sites could have been due, at least in part, to seasonal effects rather than inherent site differences.

Parasitoid loads

The mean parasitoid load of all field-collected infected cicadas was 4.97 larvae/host (95% bootstrap-*t* CI: 4.23 to 5.92 larvae/host, $s = 3.95$, $n = 91$ hosts, range = 1–19 larvae/host) and the median was 4, reflecting the strong right skew of the distribution (Figure 7).

The parasitoid loads of field-collected infected cicadas were often much higher than the clutch sizes of larvipositing females in the laboratory infection trials. A bootstrap-*t* comparison of means confirmed that the mean clutch size of female parasitoids (2.53 larvae/host) was significantly less than the mean parasitoid load of hosts in the field (4.97 larvae/host) (95% bootstrap-*t* CI: 1.32 to 3.52 fewer larvae/host, $p < 0.0001$).

Overall, there was a strong, positive relationship between host cicada parasitism rates and mean parasitoid loads per host (Figure 8), with parasitism rate explaining about 65% of the variation in mean parasitoid load (simple linear regression: $b = 8.29$, $R^2 = 0.650$, $p = 0.0048$). The estimated slope of the relationship was 8.29, but the imprecision of the parasitism rate estimates meant that this slope estimate likely suffered from attenuation bias. The estimated bias correction factor was approximately 1.117, giving an attenuation-corrected slope estimate of 9.25.

Discussion

The results of this study provide the first detailed information about the infection behaviors and life history of any species of *Emblemasoma* besides *E. auditrix*. Both laboratory and field observations reveal that *E. erro* find their hosts by eavesdropping on the sexual communication signals of male cicadas. When a female *E. erro* locates a calling cicada, she waits to attack until the host is in motion, and larviposition on flying cicadas is not uncommon. The results also show that male *T. dorsatus* are commonly parasitized by *E. erro* and that there can be substantial variation in population parasitism rates and parasitoid loads. I next discuss the behavior and life history of *E. erro*, especially in comparison to other acoustic parasitoids and other sarcophagid parasitoids; assess possible host defenses; and discuss possible causes of variation in host parasitoid loads and parasitism rates.

Host locating and infection behaviors of *E. erro*

E. erro's use of phonotaxis to locate potential hosts is similar to that reported for other acoustically hunting parasitoids (Soper et al. 1976; Lehmann 2003; Lakes-Harlan and Lehmann 2015), but *E. erro*'s preference for attacking moving targets is apparently unique among known acoustic parasitoids. For example, *E. auditrix*, the only other *Emblemasoma* for which larviposition behaviors are known, will aggressively attack stationary or restrained cicadas. Upon finding a male cicada, female *E.*

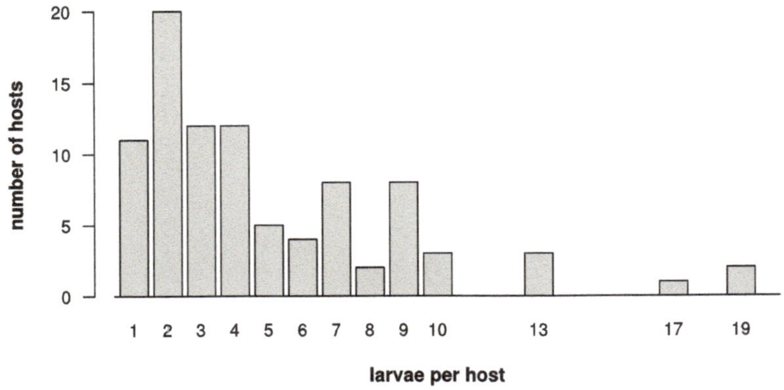

Figure 7 The distribution of parasitoid loads (larvae per host) of infected cicadas in the field.

auditrix exhibit a stereotyped behavioral sequence in which the female fly immediately attempts to squeeze underneath the perched cicada's wings to gain access to the cicada's timbal region. She then uses specialized terminal abdominal sternites to cut through the cicada's timbal membrane and injects larvae directly into the host's body (Schniederkötter and Lakes-Harlan 2004). Female *E. erro* lack any comparable abdominal modifications, but larvipositing through the host's timbal would likely be impossible for *E. erro* anyway, because male *Tibicen dorsatus* have timbals that are fully protected by well-developed timbal covers. In contrast, *E. auditrix*'s host cicada, *O. rimosa*, lacks timbal covers entirely.

Tachinid acoustic parasitoids of the tribe Ormiini will also attack stationary hosts, and they will even larviposit without visual or tactile confirmation of a host's location. For example, *Homotrixa alleni* Barraclough, *Ormia depleta* (Wiedemann), and *O. ochracea* (Bigot) will all deposit larvae at a sound source regardless of whether or not a potential host insect is actually present (Cade 1979; Fowler 1987; Allen et al. 1999). For *E. erro*, the host's calling song was never sufficient by itself to trigger larviposition, even when a potential host was present. In contrast to *E. erro*, ormiine tachinids are all nocturnal parasitoids of Orthoptera, and their willingness to larviposit in the absence of a host probably reflects an almost total reliance on acoustic cues at night. For acoustic parasitoids such as *E. erro* that are active during the day, requiring visual confirmation of a suitable host prior to larviposition allows for more precise placement of larvae and undoubtedly decreases the number of larvae that are wasted by the female fly.

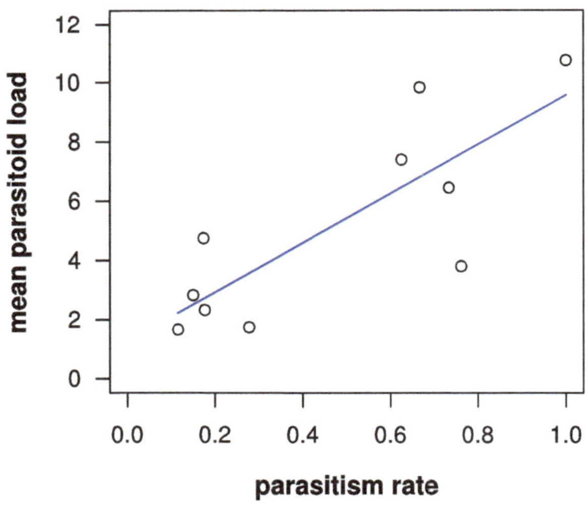

Figure 8 Relationship between host parasitism rate and mean parasitoid load per host. Each data point represents 1 year of host population sampling data for a single study site. The solid line (blue in the color figure) represents the linear regression model for the data.

In comparison to the larviposition behaviors of other acoustic parasitoids, *E. erro*'s tendency to attack flying cicadas is especially striking. One third of the successful attacks observed in the experiment cages took place while the cicada was in flight, but this is almost certainly an underestimate of the true frequency of flight-based attacks in nature. Due to the size of the cages used in the trials, most attempts by flies to follow cicadas in the air resulted in failure because the cicada crashed into a side of the cage before the fly could approach and orient itself to the flying cicada. It was hoped that the large'-flight cage' would alleviate this problem, but even it appeared to be too small for most aerial attacks to succeed. Nevertheless, flies seemed much more reluctant to attack potential hosts that were not in flight.

This conclusion is further supported by observations in the field, where nearly all apparent larviposition attacks occurred while cicadas were in flight. Flies sometimes even followed a single cicada from perch to perch, waiting patiently next to the cicada each time it landed, but never attempting to attack while the cicada was not flying. As an example, in 2013, I observed a male *T. dorsatus* calling from a grass flowering culm with a female *E. erro* perched on the opposite side of the stalk near the cicada's abdomen. The fly was nearly motionless until the cicada backed a short distance down the stalk, causing the fly to move with him nearly in unison, but the fly made no move to attack the cicada. When the cicada flew a short distance (approximately 1 to 2 m) to a new perch, the fly closely followed him in the air, landed next to the cicada, and again remained nearly motionless while the cicada began calling. The cicada flew twice more, with the fly following both times, and after the final flight of at least 30 m, I captured the cicada and later reared two *E. erro* larvae from it.

While *E. erro*'s behavior of larvipositing on hosts while they are in flight or otherwise in motion might be different from *E. auditrix* and tachinid acoustic parasitoids, it is remarkably similar to the larviposition behaviors reported for some sarcophagid parasitoids of the genus *Blaesoxipha* that parasitize acridid grasshoppers. *B. aculeata* (Aldrich), *B. caridei* (Brethes), *B. kellyi* (Aldrich), *B. redempta* (Pandellé), and *B. reversa* (Aldrich), among others, have all been reported to attack grasshoppers while in flight (Coquillett 1892; Kelly 1914; Aldrich 1916; Lloyd 1951; Rees 1973; Povolný and Verves 1997). Kelly (1914) provided a detailed description of the larviposition behaviors of *B. kellyi*, reporting that grasshoppers were attacked either on the wing or on the ground, and that grasshoppers were only attacked when they were in motion (but see Smith 1915). Furthermore, both *B. kellyi* and *B. reversa* typically place larvae near the base of a host's wings, much like *E. erro* (Kelly 1914; Rees 1973).

It is worth noting that early last century, Beamer (1928) and Kelly (1914), both working in Kansas, reported seeing cicadas pursued by flies while in flight. Beamer noted that 'the flies follow but a few inches away, and sometimes seem almost to alight on the body of the cicada.' Although their observations were largely adventitious and incidental, and neither author identified the flies involved, it seems plausible in retrospect that their papers might have been the first published records of *E. erro*'s host infection behavior.

Infection of female hosts

Given *E. erro*'s primary host-finding mechanism, male cicadas are clearly the primary targets of infection by this parasitoid. However, the observation of a fly larvipositing on a female cicada in the laboratory, along with the 2014 survey of female *T. dorsatus* in the field, confirms that female cicadas are also sometimes attacked.

Since female cicadas are silent, how are they discovered by *E. erro* in the field? One possibility is that, simply by chance, they happen to fly within the visual range of a perched female *E. erro*. Perhaps more likely, though, female *T. dorsatus* and female *E. erro* might sometimes encounter one another while seeking male cicadas. Like *E. erro*, female cicadas perform phonotaxis in response to males' calls, so female cicadas could become parasitized if they were attracted to the same calling male as a female *E. erro*. In any case, despite many hours spent observing cicadas in the field, I never witnessed any interactions between female *E. erro* and female *T. dorsatus*, so such encounters must be rare in comparison to encounters between male cicadas and female *E. erro*. However, *E. erro*'s occasional use of female hosts is not unique. Several other species of acoustic parasitoids that primarily attack male hosts are also known to sometimes parasitize females (Soper et al. 1976; Lehmann 2003).

Phenology and fecundity of *E. erro*

Little is known of the seasonal phenology of *E. erro*. In this study, adult flies were observed in the field as early as June 13 (in 2012) and as late as September 4 (in 2014), and these were also the earliest and latest dates that I attempted to find them. The rearing data strongly suggest that *E. erro* is multivoltine in the geographic area covered by this study. With a total development time from larviposition to adult eclosion of about 22 days, it seems possible that there could be at least three generations per year. *E. auditrix*, in contrast, is apparently univoltine (Soper et al. 1976; de Vries and Lakes-Harlan 2005).

Female *E. erro* were observed with as many as 174 first-instar larvae, nearly 3.5 times the maximum of 50 observed for *E. auditrix* (De Vries and Lakes-Harlan 2005). The apparently large difference in fecundity between these two species might be at least partially explained by their larviposition behaviors and life histories. *E. auditrix* deposits larvae directly inside a host's body, one larva per host, and all available evidence suggests that *E. auditrix* is a solitary parasitoid (Soper et al. 1976). By injecting larvae into its hosts, *E. auditrix* likely 'wastes' relatively few larvae during larviposition, and as a solitary parasitoid, it is plausible that multiple larvae inside a single host would physically attack one another (Godfray 1994). Under these conditions, females might benefit by producing fewer, larger larvae to increase their chances of survival. In contrast, because *E. erro* deposits its larvae on the exterior of a host, it is likely that some percentage of these larvae never manage to make it inside the host's body. Moreover, *E. erro* is a gregarious parasitoid, and as such, larvae probably face little direct physical aggression from conspecifics (Godfray 1994). For *E. erro*, then, investing fewer resources in more larvae might increase a female's lifetime reproductive success. Some tachinid acoustic parasitoid species, which deposit their larvae even more haphazardly, also have large larval complements (Wineriter and Walker 1990; Allen et al. 1999; Kolluru and Zuk 2001), and although behavioral data for other sarcophagid parasitoids is extremely limited, at least some parasitoid species in the genus *Blaesoxipha* also appear to follow this pattern (Middlekauff 1959).

Host defenses and mortality

Once discovered by a female *E. erro*, male *T. dorsatus* appeared to have relatively few viable options to defend themselves. When approached by a parasitoid fly, calling male *T. dorsatus* cicadas responded either by flying, immediately terminating their call and remaining motionless on their perch (hereafter referred to as 'hiding'), or simply continuing their calling behavior. The latter seemed to be the most common. Cicadas often called repeatedly and walked freely about the walls of the experiment cages despite being followed by a fly only a few centimeters away. Cicadas sometimes even called with a fly perched right on top of them. However, stationary cicadas that were directly contacted by a fly would often vigorously flick their wings to try to repel the parasitoid. Unfortunately, given the relatively small space inside the cages, evaluating the effectiveness of any of these behaviors was nearly impossible because a cicada could never truly escape from the fly.

Nevertheless, observations in the field suggested that both the flight and hiding strategies do sometimes work. In at least one case, a fly lost interest in a hiding cicada and left before the cicada resumed calling, and in another, a cicada that was contacted by an approaching fly managed to escape by flying away. Most of the time, though, flies simply waited until a hiding cicada became active again, and they usually had little difficulty in following a flying cicada from one perch to another. As a defensive strategy, flying seems especially risky given *E. erro*'s aptitude for aerial larviposition.

After being larviposited upon, cicadas had yet another option for defending themselves. I repeatedly observed cicadas perform 'wing flipping' behavior immediately after being attacked, characterized by rapidly flapping their wings several times while perched. In this way, one cicada managed to completely dislodge the single larva that had been deposited on the cicada's right fore wing, thus avoiding infection completely. This was the only case for which I confirmed that a cicada was able to remove all larvae from its body, but it is possible that some larviposition events were not detected during the behavioral experiments. Wing flipping by *T. dorsatus* appears to be functionally similar to the grooming behaviors used by the cricket *Gryllus texensis* Cade and Otte to prevent infection by the larvae of *Ormia ochracea* (Vincent and Bertram 2010).

Although the hosts of some other sarcophagid parasitoids have been reported to occasionally survive parasitism (Spencer and Buckell 1957; Danyk et al. 2000), infection by *E. erro* appears to be invariably fatal for *T. dorsatus*. In most cases, hosts died several hours before the parasitoid larvae emerged. Host death was usually preceded first by loss of wing function, then loss of leg function beginning with the hind legs and ending with the fore legs. Prior to death, a cicada's antennae were typically the last appendages to display a visible response to external touch. After a cicada died, small, rhythmic movements of the legs or head capsule were often visible as the parasitoid larvae used their oral hooks to scrape muscle and other soft tissue from the integument.

Sometimes, though, when a cicada was infected with only a single larva, the larva emerged before the cicada died, leaving the host in a severely weakened, moribund state. Cicadas in this condition usually succumbed after a few hours. In one exceptional case, a large male *T. dorsatus* from the Prowers Co., CO site that was infected with a single *E. erro* larva survived for more than 24 h following parasitoid emergence. Although sluggish, it was still able to cling to and crawl on a perch, weakly flutter its wings (but not fly), and was even observed attempting to feed before its movements became uncoordinated and it, too, died. Overall, *E. erro* must be a major cause of mortality for adult male *T. dorsatus*, especially considering the very high parasitism rates observed in some cicada populations.

Variation in host parasitism rates among study sites

Host populations at the two westernmost field sites appeared to have consistently higher parasitism rates than sites further east (Table 1, Figure 3). The biogeography of potential host cicadas might offer one explanation for this pattern. The western sites were located on the semiarid High Plains, where there are fewer species of large cicada present than on the more mesic midgrass prairies of the study sites further east. *E. erro* parasitizes other cicada species besides *T. dorsatus* (B. Stucky, in prep.), so higher parasitism rates of *T. dorsatus* on the High Plains could be a consequence of local differences in the communities of potential host species.

However, as noted in the 'Results' section, because these western sites were also sampled later in the season than the eastern sites, higher parasitism rates could have also been caused by seasonal effects rather than intrinsic differences among the sites. One might expect parasitism rates to increase throughout the season as *E. erro* populations reach their peak and host populations decline, as has been observed for several other species of dipteran parasitoids, including some acoustic parasitoids (e.g., Tamaki et al. 1983; Allen 1995; Lehmann 2008). It seems likely that this accounts for at least some of the among-site differences in parasitism rates found in this study. Furthermore, both host and parasitoid population sizes undoubtedly also play a role in determining parasitism rates. As evidenced by some of the small population sample sizes, host cicadas were uncommon and difficult to collect for some years at some field sites, which suggests that there was variation in host population sizes from year to year. Future studies that estimate host and parasitoid population sizes and sample both High Plains and central Plains sites multiple times throughout the season will be needed to fully disentangle the effects of these variables on host parasitism rates.

Superparasitism by *E. erro*

The strong, positive relationship between parasitism rate and parasitoid load (Figure 8), as well as the significant difference between the mean parasitoid load of field-collected hosts and the mean clutch size of larvipositing females (4.97 and 2.53 larvae/host, respectively), can both be explained as a consequence of superparasitism in the field. If at least some host cicadas are superparasitized in the field, then we should expect the mean parasitoid load of host cicadas to be larger than the mean clutch size of individual female flies. Furthermore, for gregarious parasitoids such as *E. erro*, superparasitism is expected to be more common when unparasitized hosts are rare, simply because female parasitoids have a harder time finding hosts that have not already been infected (Godfray 1994). Unparasitized hosts are rare when parasitism rates are high, so higher parasitism rates should correspond with increasing rates of superparasitism. Increased superparasitism would, in turn, likely result in larger parasitoid loads per host, which means that higher population parasitism rates should correspond with higher parasitoid loads. This prediction matches the pattern of the data quite well (Figure 8).

Additionally, anecdotal evidence of superparasitism was found in the relative sizes of larvae emerging from some of the most heavily parasitized hosts. In some cases, two distinct larval size classes were evident, presumably due to the smaller larvae having been deposited on the host later than the larger larvae. In other cases, though, all larvae emerging from heavily parasitized hosts were approximately the same size, suggesting that either a single female deposited all of the larvae at once, or more likely, that two (or more) female flies discovered an uninfected host at nearly the same time.

Conclusions

E. erro is a widespread, common parasitoid of the cicada *T. dorsatus* on the grasslands of the Great Plains in the central United States. Female flies locate potential hosts by eavesdropping on the acoustic mating calls of male cicadas, then use visual cues to larviposit on the host while it is in motion. Larviposition often occurs while the cicada is in flight. Parasitization by *E. erro* is always fatal for *T. dorsatus*, which seems to have few consistently effective defenses against attack. Parasitism rates for male *T. dorsatus* can exceed 70% in some host populations. Parasitoid loads of infected cicadas average about five larvae per host but can be as high as 19 larvae per host. At least some variation in parasitoid loads is likely due to superparasitism in host populations with high parasitism rates.

Even though *E. erro* is, like *E. auditrix*, an acoustically orienting parasitoid of cicadas, the close-range infection behaviors of these two species are highly divergent, and the infection behaviors of both species are very different from tachinid acoustic parasitoids of the tribe Ormiini. Indeed, the infection behavior of *E. erro* is unlike that known for any other acoustic parasitoid. There are important life history differences between *E. erro* and *E. auditrix* as well: *E. auditrix* is apparently a solitary, univoltine parasitoid with relatively low larval production per female; *E. erro* is a gregarious, multivoltine parasitoid with high larval production per female. Given the marked differences between *E. erro* and *E. auditrix*, the results of this study suggest that more work is needed to characterize the diversity of *Emblemasoma* parasitoids. An improved understanding of sarcophagid acoustic parasitoids would make it possible to more meaningfully compare sarcophagid and tachinid acoustic parasitoid lineages, and it would also allow for more robust inferences about eavesdropping parasitoids in general.

Competing interests

The author declares that he has no competing interests.

Acknowledgments

I am grateful to the private landowners who allowed me access to their land for this research. I also thank the Kansas Department of Wildlife, Parks, and Tourism, the Harvey County Parks Department, and Colorado Parks and Wildlife for allowing me access to state- and county-owned lands, as well as the employees of these agencies who provided information about the properties they manage. Erin Stucky and Bruce and Pat Stucky spent many hours assisting with field work and data collection; this study would have been much diminished without their help. Greg Dahlem provided helpful advice for how to rear sarcophagid flies. Rob Guralnick, Erin Stucky, and two anonymous reviewers gave valuable feedback on the manuscript and figures, and I also thank the Bowers Lab at CU for helpful commentary on an earlier version of the manuscript. I am very grateful for the financial support provided by research grants from the Department of Ecology and Evolutionary Biology at CU and the University of Colorado Museum of Natural History, and a Sigma Xi Grant-in-Aid of Research award.

References

Adamo SA, Robert D, Hoy RR (1995) Effects of a tachinid parasitoid, *Ormia ochracea*, on the behaviour and reproduction of its male and female field cricket hosts (*Gryllus* spp). J Insect Physiol 41:269–277, doi:10.1016/0022-1910 (94)00095-X

Agresti A, Coull BA (1998) Approximate is better than "exact" for interval estimation of binomial proportions. Am Stat 52:119, doi:10.2307/2685469

Aldrich JM (1916) Sarcophaga and allies in North America, Thomas Say Foundation Series. Murphey-Bivins Company Press, Lafayette, IN

Allen GR (1995) The biology of the phonotactic parasitoid, *Homotrixa* sp. (Diptera: Tachinidae), and its impact on the survival of male *Sciarasaga quadrata* (Orthoptera: Tettigoniidae) in the field. Ecol Entomol 20:103–110, doi:10.1111/j.1365-2311.1995.tb00435.x

Allen GR (1998) Diel calling activity and field survival of the bushcricket, *Sciarasaga quadrata* (Orthoptera: Tettigoniidae): a role for sound-locating parasitic flies? Ethology 104:645–660, doi:10.1111/j.1439-0310.1998.tb00099.x

Allen GR, Kamien D, Berry O, Byrne P, Hunt J (1999) Larviposition, host cues, and planidial behavior in the sound-locating parasitoic fly *Homotrixa alleni* (Diptera: Tachinidae). J Insect Behav 12:67–79, doi:10.1023/A:1020929231430
Audacity Team (2012) Audacity: free audio editor and recorder., http://audacity.sourceforge.net/
Beamer RH (1928) Studies on the biology of Kansas Cicadidae. Univ Kans Sci Bull 18:155–263
Beckers OM, Wagner WE (2011) Male field crickets infested by parasitoid flies express phenotypes that may benefit the parasitoids. Anim Behav 82:1151–1157, doi:10.1016/j.anbehav.2011.08.013
Bulmer MG (1979) Principles of statistics. Dover Publications, New York, NY
Cade W (1975) Acoustically orienting parasitoids: fly phonotaxis to cricket song. Science 190:1312–1313, doi:10.1126/science.190.4221.1312
Cade W (1979) The evolution of alternative male reproductive strategies in field crickets. In: Blum MS, Blum NA (eds) Sexual selection and reproductive competition in insects. Academic, New York, NY, pp 343–379
Carpenter J, Bithell J (2000) Bootstrap confidence intervals: when, which, what? A practical guide for medical statisticians. Stat Med 19:1141–1164, doi:10.1002/(SICI)1097-0258(20000515)19:9 < 1141::AID-SIM479 > 3.0.CO;2-F
Carroll RJ, Ruppert D (1996) The use and misuse of orthogonal regression in linear errors-in-variables models. Am Stat 50:1–6, doi:10.2307/2685035
Cole JA (2008) A new cryptic species of cicada resembling *Tibicen dorsatus* revealed by calling song (Hemiptera: Auchenorrhyncha: Cicadidae). Ann Entomol Soc Am 101:815–823, doi:10.1603/0013-8746(2008)101[815:ANCSOC]2.0.CO;2
Coquillett DW (1892) Report on the locust invasion of California in 1891. Bull Div Entomol US Dep Agric 27:34–57
Danyk T, Johnson DL, Mackauer M (2000) Parasitism of the grasshopper *Melanoplus sanguinipes* by a sarcophagid fly, *Blaesoxipha atlanis*: influence of solitary and gregarious development on host and parasitoid. Entomol Exp Appl 94:259–268, doi:10.1046/j.1570-7458.2000.00628.x
De Vries T, Lakes-Harlan R (2005) Phonotaxis of the female parasitoid *Emblemasoma auditrix* (Diptera, Sarcophagidae) in relation to number of larvae and age. Zoology 108:239–246, doi:10.1016/j.zool.2005.04.004
Dobson AJ, Barnett AG (2008) An introduction to generalized linear models, 3rd edn, Chapman & Hall/CRC texts in statistical science series. CRC Press, Boca Raton, FL
Efron B, Tibshirani R (1993) An introduction to the bootstrap. Monographs on statistics and applied probability. Chapman & Hall, New York, NY
Farris HE, Oshinsky ML, Forrest TG, Hoy RR (2008) Auditory sensitivity of an acoustic parasitoid (*Emblemasoma* sp., Sarcophagidae, Diptera) and the calling behavior of potential hosts. Brain Behav Evol 72:16–26, doi:10.1159/000139458
Feener DH, Brown BV (1997) Diptera as parasitoids. Annu Rev Entomol 42:73–97, doi:10.1146/annurev.ento.42.1.73
Fowler HG (1987) Field behavior of *Euphasiopteryx depleta* (Diptera: Tachinidae): phonotactically orienting parasitoids of mole crickets (Orthoptera: Gryllotalpidae: *Scapteriscus*). J N Y Entomol Soc 95:474–480
Godfray HCJ (1994) Parasitoids: behavioral and evolutionary ecology. Monographs in behavior and ecology. Princeton University Press, Princeton, NJ
Gray DA, Cade WH (1999) Sex, death, and genetic variation: natural and sexual selection on cricket song. Proc R Soc B Biol Sci 266:707–709, doi:10.1098/rspb.1999.0693
Harrison JF, Woods HA, Roberts SP (2012) Ecological and environmental physiology of insects. In: Ecological and environmental physiology series, vol 3. Oxford University Press, Oxford
Haynes KF, Yeargan KV (1999) Exploitation of intraspecific communication systems: illicit signalers and receivers. Ann Entomol Soc Am 92:960–970
Hedwig B, Robert D (2014) Auditory parasitoid flies exploiting acoustic communication of insects. In: Hedwig B (ed) Insect hearing and acoustic communication, vol 1. Springer Berlin Heidelberg, Berlin, Heidelberg, pp 45–63, http://link.springer.com/10.1007/978-3-642-40462-7_4. Accessed 26 Oct 2014
Kelly EOG (1914) A new sarcophagid parasite of grasshoppers. J Agric Res 2:435–446
Köhler U, Lakes-Harlan R (2001) Auditory behaviour of a parasitoid fly (*Emblemasoma auditrix*, Sarcophagidae, Diptera). J Comp Physiol A 187:581–587, doi:10.1007/s003590100230
Kolluru GR, Zuk M (2001) Parasitism patterns and the size-fecundity relationship in the acoustically orienting dipteran parasitoid *Ormia ochracea*. Can J Zool 79:973–979, doi:10.1139/z01-064
Lakes-Harlan R (2009) Monitoring species distribution of hearing Diptera with sound traps. Fly Times 43:10–12
Lakes-Harlan R, Lehmann GUC (2015) Parasitoid flies exploiting acoustic communication of insects - comparative aspects of independent functional adaptations. J Comp Physiol A 201:123–132, doi:10.1007/s00359-014-0958-3
Lakes-Harlan R, Stolting H, Stumpner A (1999) Convergent evolution of insect hearing organs from a preadaptive structure. Proc R Soc B Biol Sci 266:1161–1167, doi:10.1098/rspb.1999.0758
Lakes-Harlan R, Stölting H, Moore TE (2000) Phonotactic behaviour of a parasitoid fly (*Emblemasoma auditrix*, Diptera, Sarcophagidae) in response to the calling song of its host cicada (*Okanagana rimosa*, Homoptera, Cicadidae). Zoology 103:31–39
Lehmann GUC (2003) Review of biogeography, host range and evolution of acoustic hunting in Ormiini (Insecta, Diptera, Tachinidae), parasitoids of night-calling bushcrickets and crickets (Insecta, Orthoptera, Ensifera). Zool Anz J Comp Zool 242:107–120, doi:10.1078/0044-5231-00091
Lehmann GUC (2008) How different host species influence parasitism patterns and larval competition of acoustically-orienting parasitoid flies (Tachinidae: Ormiini). In: Weber EA, Krause LH (eds) Animal Behavior: New Research. Nova, New York, pp 93–132
Lehmann GUC, Lehmann AW (2006) Potential lifetime reproductive success of male bushcrickets parasitized by a phonotactic fly. Anim Behav 71:1103–1110, doi:10.1016/j.anbehav.2005.08.009
Lloyd DC (1951) A survey for grasshopper parasites in temperate South America. Can Entomol 83:213–230, doi:10.4039/Ent83213-9
Lopes HS (1981) Notes on American Sarcophagidae (Diptera). Rev Bras Biol 41:149–152
Maier CT (1982) Observations on the seventeen-year periodical cicada, *Magicicada septendecim* (Hemiptera: Homoptera: Cicadidae). Ann Entomol Soc Am 75:14–23
McFadden D (1974) Conditional logit analysis of qualitative choice behavior. In: Zarembka P (ed) Frontiers in econometrics, Economic theory and mathematical economics. Academic, New York, pp 105–142
Menard S (2000) Coefficients of determination for multiple logistic regression analysis. Am Stat 54:17–24, doi:10.1080/00031305.2000.10474502
Michelsen A, Nocke H (1974) Biophysical aspects of sound communication in insects. In: Treherne JE, Berridge MJ, Wigglesworth VB (eds) Advances in insect physiology, vol 10. Elsevier, New York, NY, pp 247–296
Middlekauff WW (1959) Some biological observations on *Sarcophaga falciformis*, a parasite of grasshoppers (Diptera: Sarcophagidae). Ann Entomol Soc Am 52:724–728
Müller P, Robert D (2002) Death comes suddenly to the unprepared: singing crickets, call fragmentation, and parasitoid flies. Behav Ecol 13:598–606, doi:10.1093/beheco/13.5.598
Pape T (1990) Revisionary notes on American Sarcophaginae (Diptera: Sarcophagidae). Tijdschr Voor Entomol 133:43–74
Pape T (1996) Catalogue of the Sarcophagidae of the world (Insecta: Diptera), Memoirs on entomology, international. Associated Publishers, Gainsville, Fla
Peterson APG (1980) Handbook of noise measurement, 9th edn. GenRad, Inc, Concord, MA, http://www.ietlabs.com/pdf/Manuals/Handbook_Noise_Measurement.pdf
Povolný D, Verves Y (1997) The flesh-flies of Central Europe (Insecta, Diptera, Sarcophagidae). Verlag Dr. F, Pfeil, München, FRG
R Core Team (2014) R: a language and environment for statistical computing, R Foundation for Statistical Computing., http://www.R-project.org
Rees NE (1973) Arthropod and nematode parasites, parasitoids, and predators of Acrididae in America north of Mexico. Tech Bull Agric Res Serv US Dep Agric 1460:1–288
Robert D (2005) In: Popper AN, Fay RR (eds) Sound source localization. vol 25. Springer, New York, pp 6–35, http://link.springer.com/10.1007/0-387-28863-5_2. Accessed 26 Oct 2014
Robert D, Miles RN, Hoy RR (1999) Tympanal hearing in the sarcophagid parasitoid fly *Emblemasoma* sp.: the biomechanics of directional hearing. J Exp Biol 202:1865–1876
Schniederkötter K, Lakes-Harlan R (2004) Infection behavior of a parasitoid fly, *Emblemasoma auditrix*, and its host cicada *Okanagana rimosa*. J Insect Sci Online 4:36
Smith HE (1915) The grasshopper outbreak in New Mexico during the summer of 1913. Bull U S Dep Agric 293:1–12
Smith RJ (2009) Use and misuse of the reduced major axis for line-fitting. Am J Phys Anthropol 140:476–486, doi:10.1002/ajpa.21090
Soper RS, Shewell GE, Tyrrell D (1976) *Colcondamyia auditrix* nov. sp. (Diptera: Sarcophagidae), a parasite which is attracted by the mating song of its host,

Okanagana rimosa (Homoptera: Cicadidae). Can Entomol 108:61–68, doi:10.4039/Ent10861-1
Spencer GJ, Buckell ER (1957) On the acridiophagous Sarcophagidae of British Columbia with records of all others taken in the province. Proc Entomol Soc Br Columbia 54:29–36
Strauß J, Lakes-Harlan R (2014) Evolutionary and phylogenetic origins of tympanal hearing organs in insects. In: Hedwig B (ed) Insect hearing and acoustic communication. vol 1. Springer Berlin Heidelberg, Berlin, Heidelberg, pp 5–26, http://link.springer.com/10.1007/978-3-642-40462-7_2. Accessed 26 Oct 2014
Tamaki G, Chauvin RL, Burditt AK (1983) Field evaluation of *Doryphorophaga doryphorae* (Diptera: Tachinidae), a parasite, and its host the Colorado potato beetle (Coleoptera: Chrysomelidae). Environ Entomol 12:386–389
Thornton PE, Running SW, White MA (1997) Generating surfaces of daily meteorological variables over large regions of complex terrain. J Hydrol 190:214–251, doi:10.1016/S0022-1694(96)03128-9
Thornton PE, Thornton MM, Mayer BW, Wilhelmi N, Wei Y, Devarakonda R, Cook RB (2014) Daymet: daily surface weather data on a 1-km grid for North America, version 2., doi:10.3334/ORNLDAAC/1219. Accessed 26 Sep 2014
US Environmental Protection Agency (2013) Level III ecoregions of the conterminous United States., http://www.epa.gov/wed/pages/ecoregions/level_iii_iv.htm. Accessed 16 Apr 2014
Vincent CM, Bertram SM (2010) Crickets groom to avoid lethal parasitoids. Anim Behav 79:51–56, doi:10.1016/j.anbehav.2009.10.001
Whitlock M, Schluter D (2009) The analysis of biological data. Roberts and Co. Publishers, Greenwood Village, Colo
Wilson EB (1927) Probable inference, the law of succession, and statistical inference. J Am Stat Assoc 22:209, doi:10.2307/2276774
Wineriter SA, Walker TJ (1990) Rearing phonotactic parasitoid flies [Diptera: Tachinidae, Ormiini, *Ormia* spp.]. Entomophaga 35:621–632, doi:10.1007/BF02375096
Yack JE (2004) The structure and function of auditory chordotonal organs in insects. Microsc Res Tech 63:315–337, doi:10.1002/jemt.20051
Yager DD (1999) Structure, development, and evolution of insect auditory systems. Microsc Res Tech 47:380–400, doi:10.1002/(SICI)1097-0029(19991215)47:6 < 380::AID-JEMT3 > 3.0.CO;2-P
Zuk M, Kolluru GR (1998) Exploitation of sexual signals by predators and parasitoids. Q Rev Biol 73:415–438

14

Life history and DNA barcode of *Oxyurella longicaudis* (Birgei, 1910) (Cladocera, Anomopoda, Chydoridae)

Maria Carolina de Almeida Castilho[1*], Maria José dos Santos Wisniewski[2], Cínthia Bruno de Abreu[2] and Tereza Cristina Orlando[2]

Abstract

Background: Cladocera is an important group of freshwater zooplankton, and the species plays an important role in energy transfer and in aquatic food webs. *Oxyurella longicaudis* is a Chydoridae species that has been recorded in North and South America. The aim of this study is to investigate the life cycle aspects of parthenogenetic females of *O. longicaudis* cultured in laboratory under controlled conditions: temperature (23°C ± 05°C), photoperiod (12 h light/12 h dark), food supply, and reconstituted water.

Results: Embryonic development duration (2.3 ± 0.5 days), post-embryonic development (5.2 ± 0.69 days), mean fecundity (two eggs female^{-1} brood^{-1}), total egg production (22.55 ± 3.98 eggs), average longevity (58 days), and body growth of the species were recorded. We also report the first DNA barcode for *O. longicaudis* isolated in Brazil, which will allow for easy identification in future zooplankton community studies. The analysis shows a genetic divergence of around 7% between our Brazilian isolate and *O. longicaudis* isolates from Mexico.

Conclusions: The time of embryonic and post-embryonic development of *O. longicaudis* was higher than that of the other species of the same family, which contributed to lower total egg production throughout its life cycle. The genetic divergence appears to be sufficient to classify the two isolates as different species.

Keywords: Conservation-priority areas; Zooplankton; Bionomics; Cryptic speciation; COI

Background

Cladocerans participate in energy transfer and aquatic food webs. In lakes and ponds, they represent a link in the food chain by consuming phytoplankton and are preyed upon by other invertebrates and fish (Sarma et al. 2005; Rocha et al. 2011). Cladocerans may be filter feeders, such as family members of Sididae, Moinidae, and Daphnidae or scrapers like the Macrothricidae and Chydoridae (Elmoor-Loureiro 2004; Castilho-Noll et al. 2010). The latter family feeds by scraping surfaces of the macrophytes or sediment.

Studies focusing on functional classification of Cladocera species are scarce, and its type of feeding is used for functional classification, so this gap in the literature needs to be addressed (Barnett et al. 2007). Studies have also shown that cladocerans' fine mesh filtering apparatus allows them to feed on small protozoa and the microbial flora of aquatic environments or feed on bacteria associated with algae (Geller and Müller 1981; Ooms-Wilms et al. 1995).

According to Frey (1980), representatives of the Chydoridae family are found in the littoral region of water bodies where they live associated with macrophytes, periphyton, and sediment. The distribution of the members of the Chydoridae family is directly related to the presence of macrophytes, most often occurring in specific association (Sacherová and Hebert 2003).

Oxyurella longicaudis is a Chydoridae species recorded in North and South America. In Brazil, they have been recorded in the northeast (Ceará, Pernambuco, Bahia, and Maranhão states), central west (Mato Grosso, Mato Grosso do Sul, and Goiás states), and the southeast (Rio de Janeiro, São Paulo, and Minas Gerais states)

* Correspondence: mariacarolcastilho@gmail.com
[1]Departamento de Zoologia, Instituto de Biociências, Universidade Estadual Paulista, CP. 510, CEP 18618-000, Botucatu, SP, Brazil
Full list of author information is available at the end of the article

(Elmoor-Loureiro 2007; Van Damme and Dumont 2010; Rocha et al. 2011; Castilho and Santos-Wisniewski 2013).

This species is rare since it was found in only 3 of 40 water bodies sampled in priority regions for conservation in southern Minas Gerais State (Castilho and Santos-Wisniewski 2013). This species has been identified in oligotrophic environments and is probably sensitive to environmental changes. *O. longicaudis* individuals are found in low densities on plankton and have restricted occurrence in coastal regions, where they play a significant role in ecosystem productivity and nutrient cycling (Kotov 2006), serving as a food source for other animals. Considering studies of the coastal region, we may assume that this species may have a wider distribution, demonstrating their role in the aquatic environment.

Studies of the Cladocera life cycle are important because they provide a deeper understanding of the biology of these animals, in addition to providing information on secondary production, population dynamics, and interactions in the food chain in aquatic environments. Ecotoxicology studies aim to control the environmental quality and the ability of zooplankton organisms to swim against a gradient of turbulence (Seuront et al. 2004; Santos-Wisniewski et al. 2002; Freitas and Rocha 2006; Castilho et al. 2012). In Brazil, studies on the life cycle of Chydoridae have been done on species such as *Chydorus dentifer* and *Acroperus harpae*, *Chydorus pubescens*, *Coronatella rectangula*, and *Alona iheringula* by Melão (1997), Santos-Wisniewski et al. (2006), Viti et al. (2013), and Silva et al. (2014), respectively. High variation was observed in their life cycle, with a duration average varying between 9 and 46 days, with the greatest longevity recorded for *A. iheringula* cultured at 25°C. Although the number of studies has been increasing recently, the knowledge of the taxonomic diversity of Cladocera remains insufficient due to morphological differentiation, phenotypic variation, and historical factors.

Nowadays, molecular identification such as the DNA barcoding has been useful for ecological studies as it allows for precise identification and determination of cryptic species. The DNA barcode includes an analysis of partial sequences of the mitochondrial gene cytochrome oxidase I (COI). Its diversity has been used to identify and detect new species in many animal groups (Hebert et al. 2003), including the Crustacea (Costa et al. 2007; Young et al. 2012). For the Chydoridae, DNA barcoding has already been used to determine some species (Sacherová and Hebert 2003; Gutiérrez et al. 2008; Gutiérrez and Valdez–Moreno 2008; Jeffery et al. 2011; Silva et al. 2014).

The aim of this study is to investigate the aspects of the life cycle of parthenogenic females of *O. longicaudis* cultured in a laboratory under controlled conditions and characterize its DNA barcode in order to allow for easy identification in future ecological studies.

Methods

Study area and sampling date

Sampling was carried out on July 7, 2010 in the Epamig Pond (21° 56′ 33″ S 45° 18′ 56″ W) situated in a priority area for conservation in Serra da Mantiqueira, Minas Gerais State, Brazil. This pond is oligotrophic, with water which has slightly acidic pH (5.7), high concentration of dissolved oxygen (9.1 $mg.L^{-1}$), and low electrical conductivity (31 $\mu s.cm^{-1}$). The pond is shallow and small in size (approximately 60 × 30 m), with an extensive macrophyte stand. The Epamig Pond is located next to the Parque Estadual de Nova Baden. The pond is comprised of a large forest fragment and located near a rice cultivation.

Sampling and acclimatization

Organisms for starting cultures were collected close to macrophyte stands in the littoral region of the Epamig Pond by horizontal hauls using a zooplankton 68-μm mesh size net. In the laboratory, parthenogenetic females of *O. longicaudis* (Figure 1) were isolated and placed in 2-L beakers containing reconstituted water. This culture medium had pH 7.6, conductivity 140 $\mu s.cm^{-1}$ and hardness of 46 $mg.L^{-1}$ $CaCO_3$. Experimental cultures were maintained in germination chambers (model 347-CDG) at a constant temperature of 23.0°C ± 0.5°C and 12 h-light/ 12 h-dark photoperiod. *O. longicaudis* was fed a suspension of the small chlorophycean *Raphidocelis subcapitata*, cultured in Chu 12 medium and cropped in the exponential phase, at a concentration of 10^5 $cells.mL^{-1}$, and 0.02 mL of a mixed suspension of yeast, and fish ration added in equal proportions (1:1) (USEPA 1994; ABNT 2009).

Life cycle

Individuals were acclimated for about ten generations (30 days). Ten females were isolated and maintained until the production of neonates. A total of 30 neonates less than 24 h old were placed in 50-mL polypropylene bottles and kept in a germination chamber with the temperature, light, and feeding conditions specified above. These organisms were observed to obtain the parameters of the life cycle. Culture media and food suspensions were completely renewed daily with a fresh suspension at the same temperature, once or twice a day. The animals were observed under a stereomicroscope to determine the number of eggs produced per brood and the longevity. The body growth of each individual was measured daily under an optical microscope, using a micrometric grid and × 50 magnification.

DNA barcode

For the DNA barcode analysis, the specimens were fixed with 95% EtOH and placed in pure water for 12 h for

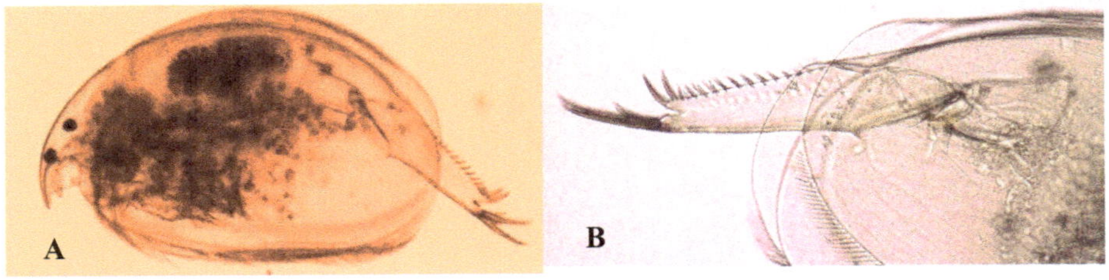

Figure 1 ***Oxyurella longicaudis*** **from Epamig Pond in Lambari, Minas Gerais state, Brazil. (A)** Female, general view (×200). **(B)** Post-abdomen (×400).

cleaning. Genomic DNA was extracted using phenol extraction and ethanol precipitation (Bucklin 2000). To amplify the mitochondrial COI gene, the universal primers, LCO 1490 and HCO2198 (Folmer et al. 1994) were used. PCR reactions had a total volume of 25 μl and were performed according to Ivanova et al. (2009) using Platinum *Taq* (Invitrogen, Carlsbad, CA, USA) as the enzyme. The PCR conditions were 94°C for 2 min as initial denaturation and 40 cycles of 94°C for 40s, 55°C for 40s, and 72°C for 1 min. DNA sequencing was done by direct sequencing of PCR amplification products, carried out in a 3130xl Genetic Analyzer automated DNA sequencer, following the manufacturer's instructions (Applied Biosystems, Foster City, CA, USA). The sequences were obtained bi-directionally two times for accurate reading.

The *O. longicaudis* COI was aligned in MEGA 6 (Tamura et al. 2013) with other COI sequences that show high sequence similarity using the BLAST tool at Genbank (http://www.ncbi.nlm.nih.gov/pubmed/). The Kimura 2-parameter (K2P) distance model (Kimura 1980) was used to calculate sequence divergences. Neighbor-joining (NJ) trees using the K2P method were generated by MEGA 6 (Tamura et al. 2013) facilities. Nonparametric bootstrapping was performed using 1,000 replicates.

Results

Life cycle

The mean duration of embryonic development was 2.3 days, and the time of post-embryonic development was 5.2 days for *O. longicaudis.* Throughout the life cycle of the species, a mean of 12 broods per female and 22 eggs female^{-1} were produced and the fecundity rate was 2 eggs female^{-1} brood^{-1}. The mean longevity was 47 days, and maximum longevity was 58 days (Table 1).

Figure 2 shows mean individual growth curve of the species as a function of time (days). The neonate of *O. longicaudis* had an average size of 504 μm and reached maturity with approximately 655 μm. On average, two juvenile instars and nine instars were recorded throughout the life cycle of the species.

DNA barcode

The sequence region of the COI gene (barcode region) was 658 bp in length and was deposited as accession number JX501501 in the Genbank. The base composition for *O. longicaudis* COI sequence was as follows: T = 41.33%, C = 13.82%, A = 23.1%, and G = 21.73%. The calculated A-T content was 64.4 %.

A genetic divergence from other *O. longicaudis* from the GenBank was found ranging from 7.0% to 7.2% (Table 2). With other Chydoridae species including *Oxyurella* sp., *Kauralona penuelasi*, *Alona* sp., and *Alona setulosa*, the genetic divergence ranged from 16.8% to 20.1%.

The NJ tree with 14 COI analyzed sequences shows *O. longicaudis* from Brazil as closely related to seven other *O. longicaudis* from Mexico (100% bootstrap support) (Figure 3). For *Oxyurella* sp., a bootstrap support of 86% was found with all *O. longicaudis* and, *Alona* sp., and *A. setulosa* were separated as another clade (Figure 3).

Table 1 Life cycle parameters of *Oxyurella longicaudis* (Cladocera: Chydoridae)

Life cycle parameters	Values
Adult mean size (μm)	883.7 ± 27.75
Maximum adult size (μm)	940
Neonate mean size (μm)	503.85 ± 52.77
Primipara mean size (μm)	654.61 ± 45.09
Minimum size of primipara (μm)	580
Number of instars between neonate and primipara	1.88 ± 0.65
Maximum number of instars in the whole life cycle	8.92 ± 1.23
Mean number of eggs in the whole life cycle	22.55 ± 3.98
Mean fecundity (eggs female^{-1} brood^{-1})	2
Maximum longevity (days)	58
Mean longevity (days)	46.96 ± 9.00
Mean embryonic development time (days)	2.30 ± 0.5
Primipara age (days)	5.20 ± 0.69

Cultured at 23.0°C ± 0.5°C in a 12-h light/dark photoperiod, fed on a mixed suspension of *Raphidocelis subcapitata* (at 10^5 cells.mL^{-1}) and yeast, and fish ration added in equal proportions.

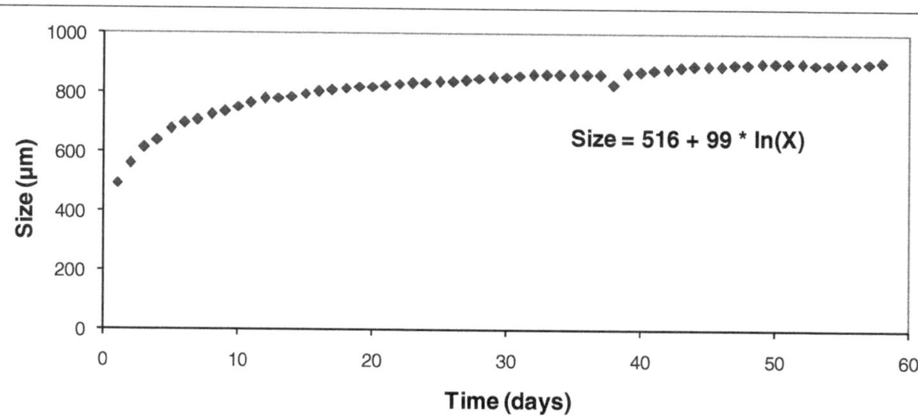

Figure 2 Mean individual growth curve of *Oxyurella longicaudis* (Cladocera, Chydoridae). Grown in the laboratory at 23.0°C, in a 12 h-light/12 h-dark photoperiod, fed on a mixed suspension of *Raphidocelis subcapitata* (at 105 cells.mL – 1) and yeast, and fish ration added in equal proportions.

Discussion

Life cycle of *O. longicaudis*

O. longicaudis occurred more abundantly in preserved environments, shallow oligotrophic water bodies, with low electrical conductivity, well-oxygenated concentration, and pH from neutral to alkaline (Rocha et al. 2011; Van Damme and Dumont 2010; Castilho and Santos-Wisniewski 2013).

Temperature is more important for determining the time of embryonic and post-embryonic development and genetic factors, such species size and size of the egg also influence these parameters. High temperatures promote quick development, while larger eggs take longer time to develop (Melão 1999). The embryonic development time observed for *O. longicaudis* at 23°C (2.3 days) is greater than that of the other Chydoridae species (Table 3): *A. iheringula* (510 μm) has an embryonic development time greater than *O. longicaudis* (940 μm) at the same temperature (23°C). In addition, *C. rectangula* (468 μm) was less than *O. longicaudis* at 25°C, which, at this temperature, reduced the time of embryonic development. *A. harpae* at a higher temperature (25°C) (Melão 1997) has a shorter embryonic development time and at a lower temperature (20°C) takes longer in embryonic development (Bottrell 1975).

The post-embryonic development of *O. longicaudis* (5.2 days) was longer than that observed for other representatives of the Chydoridae family. *A. harpae, Alonella excisa, Leydigia acanthocercoides*, and *A. iheringula* reached maturity on the third day of life (Melão 1997; Sharma and Sharma 1998; Murugan and Job 1982; Silva et al. 2014), and *C. rectangula* reaches primipara age on the second day of life (Viti et al. 2013) (Table 3). By the fifth day of life, *O. longicaudis* began to show exponential body

Table 2 K2P Genetic distance between COI sequences of *Oxyurella longicaudis* and other *Oxyurella, Kauralona* and *Alona* species

	1	2	3	4	5	6	7	8	9	10	11	12	13	14
1. *O. longicaudis* (JX501501)														
2. *O. longicaudis* (KC617722)	0.072													
3. *O. longicaudis* (KC617723)	0.070	0.002												
4. *O. longicaudis* (KC617724)	0.072	0.000	0.002											
5. *O. longicaudis* (KC617725)	0.072	0.000	0.002	0.000										
6. *Oxyurella sp.*(KC617135)	0.168	0.153	0.151	0.153	0.153									
7. *O. longicaudis* (KC617136)	0.072	0.000	0.002	0.000	0.000	0.153								
8. *O. longicaudis* (KC617138)	0.072	0.000	0.002	0.000	0.000	0.153	0.000							
9. *O. longicaudis* (KC617139)	0.070	0.002	0.000	0.002	0.002	0.151	0.002	0.002						
10. *K. penuelasi* (KC617020)	0.188	0.192	0.190	0.192	0.192	0.217	0.192	0.192	0.190					
11. *K. penuelasi* (KC617021)	0.188	0.192	0.190	0.192	0.192	0.217	0.192	0.192	0.190	0.000				
12. *K. penuelasi* (KC617022)	0.188	0.192	0.190	0.192	0.192	0.217	0.192	0.192	0.190	0.000	0.000			
13. *Alona sp.*(KC617433)	0.181	0.197	0.195	0.197	0.197	0.184	0.197	0.197	0.195	0.208	0.208	0.208		
14. *Alona setulosa* (EU701997)	0.201	0.209	0.206	0.209	0.209	0.227	0.209	0.209	0.206	0.204	0.204	0.204	0.198	

Genbank access numbers are located after each species name.

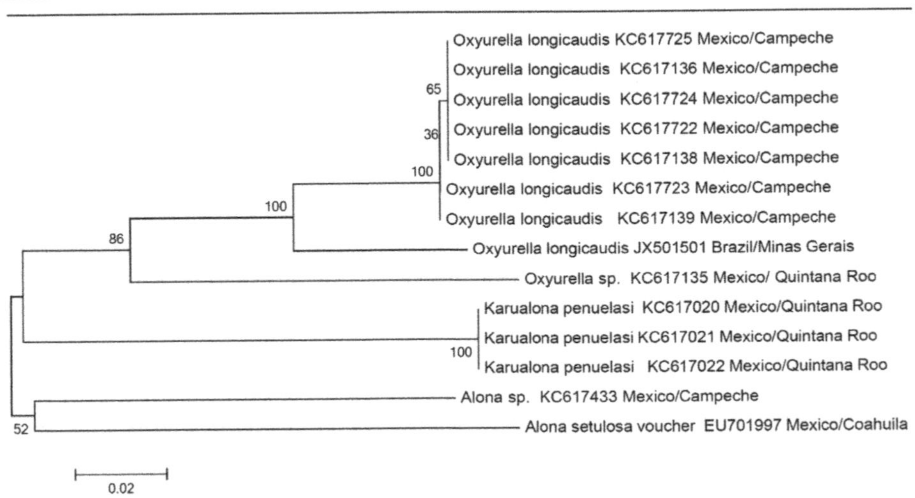

Figure 3 Identification COI tree of *Oxyurella* and other *Oxyurella, Kauralona,* and *Alona* species. Using a 658-bp COI fragment inferred by NJ. Numbers in each node designate percentage of bootstrap support (for 1,000 replicates). The bar shows the number of substitutions per site. Genbank access number and locality are located after each species name.

growth (Figure 2). Therefore, the species began to allocate energy for reproduction only from this period onward, which is indicated by the increase of *O. longicaudis*' body, being larger than most representatives of the Chydoridae family. According to Lynch (1980), larger body species invest almost all of their energy and reproduction after reaching maturity. Additionally, the time duration for post-embryonic development is higher for the Cladocera species with larger bodies under the same feeding conditions (Hardy and Duncan 1994).

The fecundity rate (two eggs brood^{-1}) found for *O. longicaudis* is common in representatives of the Chydoridae family (Bottrell 1975; Murugan and Job 1982; Robertson, 1988; Sharma and Sharma 1998; Santos-Wisniewski et al. 2006; Silva et al. 2014). This rate is the lowest fecundity rate between the families of Cladocera. The flat body of Chydoridae members prevents a higher yield of eggs per brood, as occurs with the representatives of other Cladocera families. For example, the Daphnidae *Scapholeberis armata freyi* produce up to 16 eggs brood^{-1} (Castilho et al. 2012) and the Sididae *Pseudosida ramosa* produces on average 3.4 eggs brood^{-1} when cultured at 25°C (Freitas and Rocha 2006). Moreover, the low fecundity rate of Chydoridae is related to the low levels of population growth of species (Martínez-Jerónimo and Gómez-Díaz 2011).

Among the cladocerans, small species such as *Chydorus* and *Alona* produce 20 eggs female^{-1} on average

Table 3 Comparison of life cycle parameters of Chydoridae species (data from the present study and the literature)

Species	EDD	PA	F	CF	L	T (°C)	Author
Oxyurella longicaudis	2.30	5.20	2	22	46.96	23	Present study
Chydorus pubescens	1.96	2.37	2	22.3	25.44	23.6	Santos-Wisniewski et al. 2006
Chydorus dentifer	2.20	5.73	2		11.39	25	Melão 1997
Chydorus sphaericus	3.10		2		74	20	Bottrell 1975
Acroperus harpae	1.56	3.70	1.59		9.79	25	Melão 1997
Acroperus harpae	3.18				74	20	Bottrell 1975
Pleuroxus uncinatus	3.16		2				Bottrell 1975
Alonella excisa		3.17	2	46	73.4	19-23	Sharma and Sharma 1998
Leydigia acanthocercoides		3.00	2	20	23.2	28-30	Murugan and Job 1982
Leydigia ciliata			2	50	46	28-30	Venkataraman 1990
Euryalona orientalis			2	20	23.8	28-30	Venkataraman 1990
Coronatella rectangula	1.68	2.48	1.98	27.8	28.04	23.6	Viti et al. 2013
Alona iheringula	1.79	3.24	2	47.6	46	25	Silva et al. 2014

EDD embryonic development duration (days), *PA* primiparous age (days), *F* fecundity (eggs female^{-1} brood^{-1}), *CF* cumulative fecundity (total number of eggs female^{-1}), *L* longevity (days), and *T* temperature (°C).

during their whole life cycle (Muro-Cruz et al. 2002). The total production of eggs of *O. longicaudis* over the course of its life cycle (22 eggs female^{-1}) was low compared to other Chydoridae, since it has a longer embryonic development time, and a later primipara. Similar production of eggs has been observed by Santos-Wisniewski et al. (2006) to *C. pubescens* (22.3 eggs female^{-1}), by Murugan and Job (1982) to *L. acanthocercoides* (20 eggs female^{-1}), and in *Euryalona orientalis* (20 eggs female^{-1}) by Venkataraman (1990), and longevity of these species ranged from 23 to 25 days. Longer-lived species such as *A. excisa* (73.4 days) (Sharma and Sharma 1998) and *Leydigia ciliata* (46 days) (Venkataraman 1990) produced 46 and 50 eggs female^{-1} in their whole life cycle, respectively. *A. excisa* had a shorter embryonic development time, and *L. ciliate*, grown at a higher temperature by increasing the metabolism of organisms, probably lead to a shorter embryonic development. *O. longicaudis* stopped producing eggs near the end of its life cycle, contributing to lower total fertility.

The mean size of neonate of *O. longicaudis* was about 50% less than the maximum size of adults. Smaller species tend to produce offspring with a hatching length closer to their adult size than the larger species (Lynch 1980).

DNA barcode of *O. longicaudis*

This study established the first barcode region of COI for the Cladocera species *O. longicaudis* isolated in Brazil. The percentage found for A-T (64.4%) is consistent with the data range previously described for the 60% A-T percentage for COI of Chydoridae (Sacherová and Hebert 2003; Belyaeva and Taylor 2009).

One value to consider is the 7.2% and 7.0% value of genetic divergence among O. *longicaudis* from Brazil and the other seven isolates from Mexico. For the Branchiopoda group, a genetic divergence of 3% in the COI sequence is considered a parameter for distinguishing species at the molecular level. From 3% to 5%, the species is considered provisional, and its taxonomic status should be confirmed. Above 5%, the specimens are considered different species (Jeffery et al. 2011). From this view, a genetic divergence of around 7% found between our *O. longicaudis* and the specimens from Mexico should be sufficient to classify them as different species. In order to confirm this and create a new species name, it will be necessary to perform additional morphological detailed studies combined with other molecular markers. However, among all *O. longicaudis* from Mexico, the genetic divergence ranged from 0 to 0.2% (Table 2), emphasizing that they represent the same species names.

COI analysis represents an interesting approach to new studies of taxonomy and species recognition of Brazilian isolates as new species, including cryptic species. Also, COI can be used to analyze a zooplankton community to estimate species richness of an entire zooplankton community as already proposed by Machida et al. (2009) and for further phylogeographic studies and gene flow for subpopulations as recently described for copepods (Young et al. 2014). Also, our results using COI markers strengthen the continental endemism idea for Cladocera (Forró et al. 2008; Belyaeva and Taylor 2009) and the monopolization hypothesis for aquatic organisms such as cladocerans (De Meester et al. 2002).

Conclusions

The embryonic and post-embryonic development times of *O. longicaudis* were higher than those of the other species of the same family, which contributed to lower total egg production throughout its life cycle.

For the DNA barcoding, the roughly 7% genetic divergence found between our *O. longicaudis* and the specimens from Mexico highlights the possibility of a cryptic speciation for this species and the urgent necessity to clarify the taxonomic position.

Competing interests

The authors declare that they have no competing interests.

Authors' contributions

MCAC, TCO, and MJSW conceived and designed the experiments. MCAC and TCO performed the experiments. MCAC, CBA, TCO, and MJSW analyzed the data and wrote the paper. All authors read and approved the final manuscript.

Acknowledgments

This research was funded by Eletrobrás Furnas (Programa de P&D Aneel) and FAPEMIG (Biota Minas APQ-03549-09 and Universal APQ 01518-09) grants and a graduate fellowship to C.B.A. (CAPES Program).

Author details

[1]Departamento de Zoologia, Instituto de Biociências, Universidade Estadual Paulista, CP. 510, CEP 18618-000, Botucatu, SP, Brazil. [2]Instituto de Ciências da Natureza, Universidade Federal de Alfenas-MG, CEP 37130-000, Alfenas, MG, Brazil.

References

ASSOCIAÇÃO BRASILEIRA DE NORMAS TÉCNICAS – ABNT (2009) NBR 12713: Ecotoxicologia aquática – toxicidade aguda- método de ensaio com *Daphnia* spp. (Cladocera, Crustacea). ABNT, Rio de Janeiro, p 23

Barnett AJ, Finlay K, Beisner BE (2007) Functional diversity of crustacean zooplankton communities: towards a trait-based classification. Freshw Biol 52:796–813

Belyaeva M, Taylor DJ (2009) Cryptic species within the *Chydorus sphaericus* species complex (Crustacea: Cladocera) revealed by molecular markers and sexual stage morphology. Mol Phylogenet Evol 50:534–546

Bottrell HH (1975) Generation time, length of life, instar duration and frequency of moulting, and their relationship to temperature in eight species of Cladocera from the river Thames, reading. Oecologia 19:129–140

Bucklin A (2000) Methods for population genetic analysis of zooplankton. In The ICES Zooplankton Methodology Manual, Chapter 11. International Council for the Exploration of the Sea. Academic Press,London 533–570.

Castilho MCA, Santos-Wisniewski MJ (2013) First record of *Oxyurella longicaudis* (Birgei, 1910) (Cladocera: Chydoridae) in Minas Gerais, southeastern Brazil. Check List 9(3):647–648

Castilho MCA, Wisniewski C, Santos-Wisniewski MJ (2012) Life cycle of *Scapholeberis armata freyi* Dumont & Pensaert, 1983 (Cladocera, Daphnidae). Biota Neotropica 12(4):1–5

Castilho-Noll MSM, Câmara CF, Chicone MF, Shibata EH (2010) Pelagic and littoral cladocerans (Crustacea, Anomopoda and Ctenopoda) from reservoirs of the Northwest of São Paulo State, Brazil. Biota Neotropica 10(1):21–30

Costa FO, DeWaard JR, Boutillier J, Ratnasingham S, Dooh RT, Hajibabaei M, Hebert PDN (2007) Biological identifications through DNA barcodes: the case of the Crustacea. Can J Fish Aquat Sci 64(2):272–295

Elmoor-Loureiro LMA (2004) Phylogenetic relationships among families of the order Anomopoda (Crustacea, Branchiopoda, Cladocera). Zootaxa 760:1–26

Elmoor-Loureiro MLA (2007) Phytophilous cladocerans (Crustacea, Anomopoda and Ctenopoda) from Paranã River Valley, Goiás, Brazil. Revista Brasileira de Zoologia 24(2):344–352

Folmer O, Black M, Hoen W (1994) DNA primers for amplification of mitochondrial cytochrome c oxidase subunit I from diverse metazoan invertebrates. Molec Mar Biol Biotech 3:294–299

Forró L, Korovchinski NM, Kotov A, Petrusek A (2008) Global diversity of cladocerans (Cladocera; Crustacea) in freshwater. Hydrobiologia 595:177–184

Freitas EC, Rocha O (2006) The life cycle of *Pseudosida ramosa*, Daday 1904, an endemic Neotropical cladoceran. Acta Limnologica Brasiliensia 18(34):293–303

Frey DG (1980) The non-swimming chydorid cladocera of wet forests, with descriptions of a new genus and two new species. Int Revue ges Hydrobiol 65(5):613–641

Geller W, Müller H (1981) The filtration apparatus of Cladocera: filter mesh-sizes and their implications on food selectivity. Oecologia (Berlin) 49:316–321

Gutiérrez EM, Valdez-Moreno M (2008) A new cryptic species of *Leberis* Smirnov, 1989 (Crustacea, Cladocera, Chydoridae) from the Mexican semi-desert region, highlighted by DNA barcoding. Hydrobiologia 18(1):63–74

Gutiérrez EM, Martínez Jerónimo F, Ivanova NV, Valdez-Moreno M, Hebert PDN (2008) DNA barcodes for Cladocera and Copepoda from Mexico and Guatemala, highlights and new discoveries. Zootaxa 1839:1–42

Hardy ER, Duncan A (1994) Food concentration and temperature effects on life cycle characteristics of tropical Cladocera (*Daphnia gessneri* Herbst, *Diaphanosoma sarsi* Richard, *Moina reticulata* (Daday)): I. Development time. Acta Amazon 24(1/2):119–134

Hebert PDN, Cywinska A, Ball SL, DeWaard JR (2003) Biological identifications through DNA barcodes. Proceedings of the Royal Society of London Series B-Biological Sciences 270(1512):313–321

Ivanova NV, Borisenko AV, Hebert PDN (2009) Express barcodes: racing from specimen to identification. Mol Ecol Resour 9(1):35–41

Jeffery NW, Elías-Gutiérrez M, Adamowicz SJ (2011) Species diversity and phylogeographical affinities of the Branchiopoda (Crustacea) of Churchill, Manitoba, Canada. PLoS ONE 6(5):e18364

Kimura M (1980) A simple method for estimating evolutionary rates of base substitutions through comparative studies of nucleotide sequences. J Mol Evol 16(2):111–120

Kotov AA (2006) Adaptations of Anomopoda crustaceans (Cladocera) to the benthic mode of life. Entomol Rev 86(2):210–225

Lynch M (1980) The evolution of Cladocera life history. Quaterly Review of Biology 55(1):21–42

Machida RJ, Hashiguchi Y, Nishida M, Nishida S (2009) Zooplankton diversity analysis through single-gene sequencing of a community sample. BMC Genomics 10:438

Martínez-Jerónimo F, Gómez-Díaz P (2011) Reproductive biology and life cycle of *Leydigia louisi mexicana* (Anomopoda, Chydoridae), a rare species from freshwater littotal environments. Crustaceana 84(2):187–201

De Meester, L, Gómez A, Okamura B, Schwenk K (2002) The Monopolization Hypothesis and the dispersal–gene flow paradox in aquatic organisms. Acta Oecologica 23(3)121:135

Melão MGG (1997) A comunidade planctônica (fitoplâncton e zooplâncton) e produtividade secundária do zooplâncton de um reservatório oligotrófico. Universidade Federal de São Carlos, São Carlos, Tese de doutorado

Melão, MGG (1999) Desenvolvimento e Aspectos Reprodutivos de Cladóceros e Copépodos de Águas Continentais. In: Pompêo.MLM (org) Perspectivas da Limnologia no Brasil, 1ª edição, São Luís do Maranhão, MA, Brasil: Gráfica e Editora União 1:45–57

Muro-Cruz G, Nandini S, Sarma SS (2002) Comparative life table demography and population growth of *Alona rectangula* and *Macrothrix triserialis* (Cladocera: Crustacea) in relation to algal (*Chlorella vulgaris*) food density. J Freshw Ecol 17:1–11

Murugan N, Job SV (1982) Laboratory studies on the life cycle Leydigia acanthocercoides fisher (1854) (Cladocera: Chydoridae). Hydrobiologia 89:9–16

Ooms-Wilms AL, Postema G, Gulati RD (1995) Evaluation of bacterivory of Rotifera based on measurements of in situ ingestion of fluorescent particles, including some comparisons with Cladocera. J Plankton Res 17:1057–1077

Robertson AL (1988) Life histories of some species of Chydoridae (Cladocera: Crustacea). Freshwather Biology 0:75–84

Rocha O, Santos-Wisniewski MJ, Matsumura-Tundisi T (2011) Checklist de Cladocera de água doce do Estado de São Paulo. Biota Neotropica 1ª:1–20

Sacherová V, Hebert PDN (2003) The evolutionary history of the Chydoridae (Crustacea: Cladocera). Biol J Linm Soc 79:629–643

Santos-Wisniewski MJ, Rocha O, Guntzel AM, Matsumura-Tundisi T (2002) Cladocera Chydoridae of high altitude water bodies (Serra da Mantiqueira), in Brazil. Braz J Biol 62(4A):681–687

Santos-Wisniewski MJ, Rocha O, Guntzel AM, Matsumura-Tundisi T (2006) Aspects of the life cycle of *Chydorus pubescens* Sars, 1901 (Cladocera, Chydoridae). Acta Limnologica Brasiliensia 18:315–333

Sarma SSS, Nandini S, Gulati RD (2005) Life history strategies of cladocerans: comparisons of tropical and temperate taxa. Hydrobiologia 542:315–333

Seuront L, Yamazaki H, Souissi S (2004) Hydrodynamic disturbance and zooplankton swimming behavior. Zool Stud 43(2):376–387

Sharma S, Sharma BK (1998) Observations on the longevity, instar durations, fecundity and growth in *Alonella excisa* (Fisher) (Cladocera, Chydoridae). Indian J Anim Sci 68:101–104

Silva ES, Abreu CB, Orlando TC, Wisniewski C, Santos-Wisniewski MJ (2014) *Alona iheringula* Sinev & Kotov, 2004 (Crustacea, Anomopoda, Chydoridae, Aloninae): life cycle and DNA barcode with implications for the taxonomy of the Aloninae subfamily. Plos One 9:e97050

Tamura K, Stecher G, Peterson D, Filipski A, Kumar S (2013) MEGA6: molecular evolutionary genetics analysis version 6.0. Mol Biol Evol 30:2725–2729

USEPA (1994) Methods for measuring the toxicity and bioaccumulation of sediment-associated contaminants with freshwater invertebrates. Report EPA 600/R-94/024. United States Environmental Protection Agency, Duluth, p 133

Van Damme K, Dumont HJ (2010) Cladocera of the Lençóis Maranhenses (NE - Brazil): faunal composition and a reappraisal of Sars' Method. Braz J Biol 70(3):755–779

Venkataraman K (1990) Life-history studies on some cladoceran under laboratory conditions. J And Sci Assoc 6:127–132

Viti T, Orlando TC, Santos-Wisniewski MJ (2013) Life history, biomass and production of Coronatella rectangular (Branchiopoda, Anomopoda, Chydoridae) from Minas Gerais. Iheringia, Série Zoologia 103(2):110–117

Young SS, Mei-Hui N, Min-Yun L (2012) Systematic Study of the Simocephalus Sensu Stricto Species Group (Cladocera: Daphniidae) from Taiwan by Morphometric and Molecular Analyses. Zoological Studies 51(2):222–231

Young SS, Lee YY, Liu MY (2014) Genetic variability and divergence of *Neutrodiaptomus tumidus* Kiefer 1937 (Copepoda: Calonida) among 10 subpopulations in the high mountain range of Taiwan and their phylogeographical relationships indicated by mtDNA COI gene. Zool Stud 53:22

15

Niche partitioning among three tree-climbing bird species in subtropical mountain forest sites with different human disturbance

Carlos Lara[1*], Berenice Pérez[1], Citlalli Castillo-Guevara[1] and Martín Alejandro Serrano-Meneses[2]

Abstract

Background: Closely related and ecologically similar species that overlap in ranges can coexist through resource partitioning without one pushing the others to extinction through competition. Understanding resource partitioning among species is essential to predicting how species decline can affect the functioning of communities and ecosystems. In this study, we analyzed niche overlap and resource partitioning of three tree-climbing bird species in disturbed and undisturbed forest sites at La Malinche National Park, Tlaxcala, Mexico. From January to December 2008, resource partitioning between the three species was examined through the frequency of sightings of individuals foraging in different sites in the trees of both forest types. We characterized the pattern of resource utilization by niche breadth and niche overlap. Finally, we tested if these birds divide tree space differentially according to forest type.

Results: Our results indicate that in undisturbed sites, pygmy nuthatches had a high niche breadth (foraging significantly more on the thin branches), while white-breasted nuthatches and brown creepers had similar medium niche breadth values (more often foraged in thick branches and lower trunk, respectively), causing the last two to have a high niche overlap. In contrast, in disturbed sites, niche breadth and overlap values were similar for all three tree-climbing species. All observed overlaps for both forest types were higher than expected, but expected values in disturbed sites suggest the possibility of competition on these sites. Pygmy nuthatches were more sighted than white-breasted nuthatches and brown creepers in both forest types.

Conclusions: Our study shows that differences in tree use suggest resource partitioning among the three bird species, particularly at undisturbed sites. However, activities of excessive logging and grazing in the disturbed sites resulted in niche overlap among species. This could lead to changes in competitive dynamics among them.

Keywords: Brown creeper; Pygmy nuthatch; White-breasted nuthatch; Niche overlap; Null models; Resource partitioning

Background

Numerous studies of avian communities have elucidated differences between similar species inhabiting the same habitat which lead to coexistence (MacArthur 1958, Hutchinson 1961, Alatalo and Alatalo 1979, Leyequién et al. 2007, Lara et al. 2011, Cloyed 2014). The general view, in the face of interspecific competition, is that the primary way to achieve this coexistence is for species to show a differential use of resources (Hutchinson 1961, Schoener 1974, Pyke 1982, Young et al. 2010). However, the fact that many sympatric species do not appear to differ sufficiently in their morphology to allow for niche differentiation (Wellborn and Cothran 2007) has challenged this view.

Nuthatches, woodpeckers, and brown creepers are all considered to be tree-climbing birds. They have claws that allow them to grip onto bark for insect capture. From these bird species, the Sittidae (the nuthatchers and wallcreepers *Sitta* and *Thrichoma*) are the sister group to the Certhiidae (the creepers, *Certhia* and *Salpornis*) (Sibley and Ahlquist 1990). *Certhia americana* and bird species from the *Sitta* genus are common inhabitants of North American temperate forest. These tree climbers show morphological similarities in their beaks (slender, either straight or decurved) and body size

* Correspondence: laracar@posgradouatx.com.mx
[1]Centro de Investigación en Ciencias Biológicas, Universidad Autónoma de Tlaxcala, Km. 10.5 Autopista Tlaxcala-San Martín Texmelucan, San Felipe Ixtacuixtla, Tlaxcala 90120, Mexico
Full list of author information is available at the end of the article

(ranging from 11 to 15 cm of total length). Despite this, they may show pronounced spatial segregation of primary foraging microhabitats in a tree. For example, previous studies have shown that pygmy nuthatches (*Sitta pygmaea*) forage higher up in the tree and use their branches more frequently than the trunk (Stallcup 1968, McEllin 1979). Conversely, white-breasted nuthatches (*Sitta carolinensis*) forage more often on the trunks (Stallcup 1968, McEllin 1979), as do brown creepers (*C. americana*) (Willson 1970). In this context, differences in the use of hunting tactics or substrates to find prey items should facilitate niche differentiation in sympatric tree-climbing birds (Cloyed 2014).

Forest structure has been highlighted as important for birds that feed on trees (Robinson and Holmes 1982, Whelan 2001). Constraints imposed by the structure of the vegetation and by the types and abundances of prey determine the available foraging opportunities. Such habitat parameters may affect foraging behavior to exploit a particular habitat and hence influence the patterns of bird habitat selection and ultimately the community structure (Robinson and Holmes 1982). Habitat disturbances such as grazing or fire may affect size class distribution of trees (Wimberly and Spies 2001). Thus, changes in habitat structure from habitat disturbance may affect tree use for climbing birds by altering their distribution. If this occurs, then the resource partitioning among these bird species could be affected.

Integrating these observations led us to hypothesize that due to their morphological similarities and sympatry, (1) the three tree-climbing bird species above mentioned should show a differential tree use to coexist, (2) but if the forest structure is altered, such resource partitioning can be modified. In this study, we analyzed niche overlap and resource partitioning of sympatric pygmy nuthatches, white-breasted nuthatches, and brown creepers in disturbed and undisturbed forest sites at La Malinche National Park, Tlaxcala, Mexico. We examined the frequency of sightings of individuals foraging in different sites in the trees of both forest types over an annual period. Then, we characterized the pattern of resource utilization by niche breadth and niche overlap. Finally, we tested if these birds divide tree space differentially according to forest type.

Methods

Study site

Niche partitioning among the three tree-climbing bird species was assessed from January through December 2008 in a representative area of vegetation (250 ha) at the La Malinche National Park (MNP), Tlaxcala, Mexico. About half of this area comprises undisturbed forests with ongoing management to prevent fires. The rest comprises disturbed forests that have suffered constant grazing, illegal logging, and numerous fire events during the last 20 years. The MNP forms part of the Trans-Mexican Volcanic Belt. It has an area of 46,093 ha, located between the 19°14′N and 98°14′W, with an altitude ranging from 2,300 to 4,461 m a.s.l. (López-Domínguez and Acosta 2005). Mean annual precipitation is 800 mm, the rainy season is between June and October, and mean annual temperature is 15°C. Coniferous forest is the dominant kind of vegetation. Below 3,000 m, we find primary pine forest dominated by *Pinus montezumae* Lamb., *Pinus teocote* Schl. & Chamb., and *Pinus leiophylla* Schl. & Chamb. The dominant species in the herbaceous stratum are *Lupinus aschenbornii* Schauer., *Castilleja tenuiflora* Benth., *Penstemon roseus*, *Senecio salignus* DC, and *Prunella vulgaris* L. Above 3,000 m, the forest is dominated by old-growth fir forest (*Abies religiosa* HBK.), frequently mixed with *Pinus hartwegii* Lindl. In the lower stratum are found some shrubs and trees such as *Salix mexicana* Seemen and *Buddleja parviflora* HBK. Above 3,500 m, *Pinus* forest contains pure stands of *P. hartwegii* and *A. religiosa*. The main shrubs are *Baccharis conferta* HBK and *Eupatorium glabatrum* HBK. Plants of the herbaceous layer include *Senecio platanifolius* Benth., *Muhlenbergia macroura* HBK., and *Penstemon gentianoides* HBK. (Villers et al. 2006).

We divided the study area into two forest types according to the conservation status of vegetation (undisturbed and disturbed). The undisturbed forest consisting of dense stands of fir and pine trees was 80% to 90% tree-covered, with tree heights ranging from 20 to 35 m and diameters at breast height (DBHs) of 2 to 5 m. The disturbed forests were more open (crown cover of less than 40%) and were represented by a mosaic of burned fir and pine forest (with annual fire events), pasture, and second-growth vegetation from burning of the original forest or in abandoned agricultural areas. Most of the trees in these areas are less than 20 m in height, and DBHs are smaller than 1 m (C Lara, personal observations). Both forest types are located in a range of 2,900 to 3,100 m.a.s.l and approximately 2 km apart.

Tree-climbing bird monitoring

Prior to bird monitoring and in order to determine differences between the climbing bird species in the foraging sites in a tree, we divided the fir and pine trees into three foraging sites: lower trunk (trunk below any branches), thick branches (branches with a diameter larger than 15 cm and located inside the tree crown), and thin branches (branches with a diameter smaller than 15 cm and located on the periphery of the tree crown). We measured the frequency of each foraging site for each bird species in both forest types. Four 800-m transects that were 500 m apart were established in the study area (two transects per forest type) with no previous knowledge of bird distribution. We performed the point count at an interval of 200 m along each transect in each forest type. The censuses began at 0730 hours, and we counted individuals at each point within a radius of 50 m. Binoculars (10 × 50 mm) were used, and species were identified using field guides (Sibley 2003, van Perlo 2006). The observers recorded data at each point count for 10 min and recorded sighted individuals by species and feeding site (according to categories described above). The spacing of each point count location was sufficient to ensure that one bird was not counted twice or more during a single transect survey. Thus, we used the number of sightings as proxy to bird abundance. We performed observations on each transect every 2 weeks and each transect was done a total of 24 times. The data for all four transects were compiled into a frequency matrix of tree-climbing birds sighted per counting event.

In order to identify niche partitioning among the three tree-climbing species, we first calculated the proportion of sightings in each foraging site for each bird species and forest type. According to De Cáceres et al. (2011), the notation for Rao's quadratic entropy (Rao 1982) was adopted for niche breadth: $B_D = 1/2 \, \Sigma^{24}_{j=1} \, \Sigma^{24}_{k=1} f_j f_k d^2{}_{jk}$, where f is the relative abundance of foraging sites for each counting event and d_{jk} is the distance between counting events j and k. The distance was given by difference of the foraging site values between the two counting events. The minimum niche breadth is 0; this value is obtained when all resources used are equal or when a single resource is used.

Niche overlaps were calculated between all pairs of species with the index of Pianka (1973): $O_{jk} = (\Sigma p_{ij} \, p_{ik}) \, (\Sigma p_{ij}{}^2 p_{ik}{}^2) - 1/2$, where p_{ij} and p_{ik} are proportional values of utilization of resource i by species j and k, respectively. Pianka's index is symmetrical and assumes values ranging from 0 (no resources used in common between two species) to 1 (complete overlap in resource use).

Data analysis

Niche breadth was estimated using R and the 'resniche' package (R Development Core Team 2012). Rao's quadratic entropy (Rao 1982) is a general diversity measure that has been recently recommended to assess both functional and taxonomic diversity and whose mathematical properties have been extensively studied (De Cáceres et al. 2011). This index incorporates the resemblance between resources into the calculation of resource niche metrics. This not only allows estimation of the niche breadth with greater accuracy but also makes the estimates less influenced by the way the resources are subdivided. In addition, all niche metrics can be calculated while taking into account the variation in resource availability, and confidence intervals can be obtained by bootstrapping (De Cáceres et al. 2011).

Similarity in foraging site used among the three bird species in both forest types was calculated as Bray-Curtis similarity coefficient in cluster analyses. EcoSim (Acquired Intelligence, Inc., Victoria, BC, Canada) was then used to estimate niche overlap with Pianka (1973) overlap index using Lawlor's (1980) third randomization algorithm (RA3). The RA3 has been shown to have desirable statistical properties for detecting nonrandom niche overlap patterns (Winemiller and Pianka 1990). EcoSim not only generates a niche overlap value from the observed data but also produces a statistical value based on a null distribution of randomized raw data to test the significance of the overlap value. The RA3 retains values for niche breadth but randomly reshuffles these values including any zeros among foraging site categories for each bird species. This permutation step was repeated for a total of 5,000 Monte Carlo simulations (Gotelli et al. 2004). The overlap indices from each permutation were used to generate a distribution of overlap values based on randomized data matrices that retained qualitative characteristics of the original data. We then used this null distribution of the overlap index values to test for statistical significance of the observed overlap index value. Significance levels were based on 99% confidence intervals.

To test for differences in foraging site used among the three bird species (number of sightings as response variable) in both forest types, we used a generalized linear mixed effects model (GLMM) where bird species, foraging area, and forest type were the fixed factors and month (point count event) was a random effect. Data were log transformed, and we used a Poisson distribution with log link function in the GLMM. All statistical tests were done in R (R Development Core Team 2012).

Results

Niche measurements

The three tree-climbing species were recorded foraging in different sites in the trees in both forest types throughout the study. In the undisturbed site, niche breadth of both white-breasted nuthatches and brown creepers was smaller than the niche breadth of pygmy nuthatches. However, in the disturbed site, niche breadth values were similar for the three tree-climbing species (Table 1).

Bray-Curtis coefficients of similarity in the foraging site used by the three bird species in both forest types divided the species into two major groups (Figure 1). The first group represented foraging sites where birds were sighted in the undisturbed forest. Here, white-breasted nuthatches and brown creepers had a mutual Bray-Curtis similarity in their foraging sites of about 70%. However, pygmy nuthatches were sighted in foraging sites less similar (≈30%) in relation to the other bird species. The other group represented birds from the disturbed forest. The three bird species had a mutual

Table 1 Niche breadth values (Rao's quadratic entropy) and 95% confidence intervals

Species	*n*	Niche breadth	LCI	UCI
(a)				
Brown creeper	17	0.13	0.11	0.17
Pygmy nuthatch	34	0. 28	0.19	0.35
White-breasted nuthatch	19	0.14	0.09	0.18
(b)				
Brown creeper	24	0.11	0.08	0.17
Pygmy nuthatch	50	0.12	0.09	0.19
White-breasted nuthatch	29	0.11	0.08	0.16

Niche breadth values and 95% confidence intervals for brown creepers (*Certhia americana*), pygmy (*Sitta pygmaea*), and white-breasted nuthatches (*S. carolinensis*) based on sightings and foraging use of trees in (a) disturbed and (b) undisturbed forest sites.

Table 2 Niche overlap for tree use among the three species studied

Species	Brown creeper	Pygmy nuthatch	White-breasted nuthatch
(a)			
Brown creeper	1	0.523	0.691
Pygmy nuthatch	0.523	1	0.477
White-breasted nuthatch	0.691	0.477	1
(b)			
Brown creeper	1	0.413	0.769
Pygmy nuthatch	0.413	1	0.277
White-breasted nuthatch	0.769	0.277	1

Niche overlap for tree use among brown creepers (*Certhia americana*), pygmy (*Sitta pygmaea*), and white-breasted nuthatches (*S. carolinensis*) in (a) disturbed and (b) undisturbed forest, determined by Pianka's index. After 5,000 Monte Carlo simulations, the observed mean tree use overlap in the disturbed forest (0.753) differed from the expected mean (0.614) for tree use overlap ($p < 0.01$). In the undisturbed forest, the observed mean tree use overlap (0.682) differed from the expected mean (0.192) tree use overlap ($p < 0.01$).

similarity of about 65% to 60%, showing that they were almost always sighted using the same foraging sites in this forest type.

Although the occurrence of some of the three bird species appears to be similar in both forest types, the foraging sites where they were more frequently sighted indicate a differential use of resources, which directly affects the overlap of their niches. Pianka's index of niche overlap was calculated using the frequency of birds sighted in each foraging site within a tree (Table 2). In the undisturbed forest, the higher overlap occurred between white-breasted nuthatches and brown creepers, and overlap was lowest between the nuthatch species. However, in the disturbed forest, overlap values were similar among the three bird species. All observed overlaps for both forest types were higher than expected (Table 2).

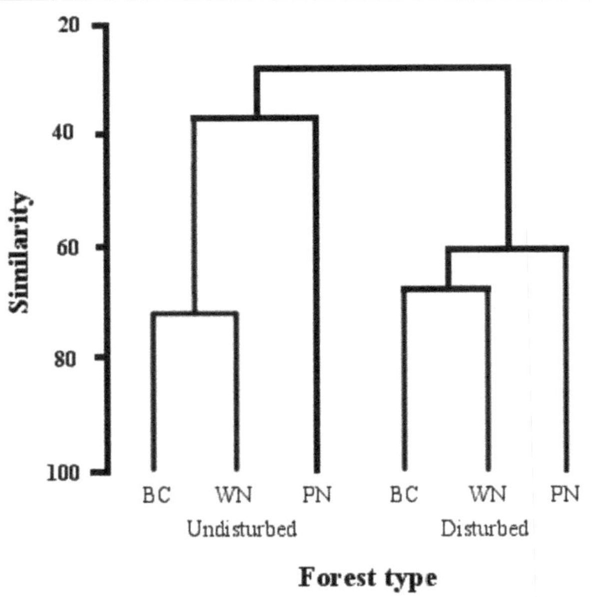

Figure 1 Cluster analysis of Bray-Curtis similarity indices. Cluster analysis of Bray-Curtis similarity indices in the foraging site used among pygmy nuthatches (PN), white-breasted nuthatches (WN), and brown creepers (BC) in both forest types.

Habitat and tree use

Pygmy nuthatches were more sighted than white-breasted nuthatches and brown creepers throughout the study. The lower number of sightings was recorded in the thick branches. Overall, the undisturbed forest had the most bird sightings (Table 3). Compared to brown creepers and white-breasted nuthatches, pygmy nuthatches foraged significantly more on the thin branches. Conversely, brown creepers more often foraged in the thick branches, and white-breasted nuthatches foraged significantly more on the lower trunk. White-breasted nuthatches and brown creepers were less sighted in disturbed forest sites (Figure 2, Table 3).

Discussion

As predicted, the three sympatric tree-climbing bird species partitioned their resources by foraging in different

Table 3 Results of GLMM

Source	*F*	*df*	*p*
Species	15.16	2	<0.001
Foraging site	17.22	2	<0.001
Forest type	13.28	1	<0.001
Species × foraging site	9.31	4	<0.001
Species × forest type	1.16	2	0.007
Foraging site × forest type	0.27	2	0.707
Species × foraging site × forest type	1.18	4	0.014

Results of GLMM testing the influence of bird species, foraging site used, forest type, and their interactions on the number of sightings (in this analysis, 'month' was included as a random effect).

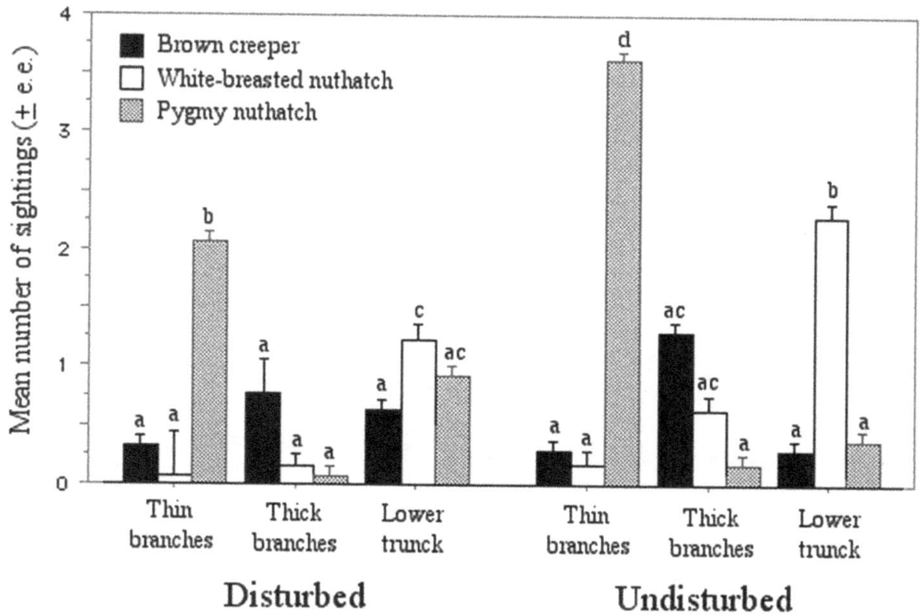

Figure 2 Mean number of sightings in the foraging site used among the three bird species in both forest types. Data with the same superscript letters are not significantly different between groups ($p < 0.05$).

parts of a tree. We found that in the undisturbed sites, pygmy nuthatches were more frequently seen foraging on the thin branches, while white-breasted nuthatches and brown creepers were seen foraging mostly in the thick branches and lower trunk, respectively. Conversely, in the disturbed sites, the three bird species were more homogeneous with respect to the site of the tree in which they foraged. This pattern caused their niche breadth and overlap values to be similar.

The differential use of substrates to find prey items is a widespread way in which ecologically similar and closely related forest-dwelling birds partition their resources (MacArthur 1958, Holmes et al. 1979, Morrison and With 1987). This niche differentiation model assumes that resource partitioning results from different fundamental niches having been occupied through past competition (Keddy 1989). Different tree use and the associated morphological attributes that make foraging in those places more efficient may represent a mechanism that minimizes interspecific competition (called the ghost of competition past, Connell 1980). Thus, resource partitioning between nuthatches and the brown creeper can be explained by the niche differentiation model. Previous studies had shown that in communities where different nuthatch species (e.g., *S. carolinensis, S. canadensis, S. pygmaea*) are sympatric with brown creepers (e.g., *C. americana, C. familiaris*), niche differentiation occurs, particularly through dividing trees spatially and by using different hunting tactics (Willson 1970, Williams and Batzli 1979, Morrison et al. 1987). This pattern has even been detected within nuthatch species. For example, in North American ponderosa pine forests, sympatric pygmy and white-breasted nuthatches partition their resources by foraging in different parts of the tree (Stallcup 1968, McEllin 1979, Cloyed 2014). Our results confirm these findings. Although prey distribution can affect habitat use among sympatric climbing bird species, additional research is needed to determine the importance of abiotic factors (e.g., light and temperature) in different seasons.

The data obtained in our undisturbed pine forest sites at La Malinche provided support to previous results of resource partitioning among tree-climbing birds. It is possible that in these forested sites, abundance asymmetry and niche breadth could potentially be acting as mechanisms for coexistence. However, changes in tree use by the studied bird species in disturbed forest sites have not been previously reported. In the only study done in this area, Cloyed (2014) examined how pygmy and white-breasted nuthatches divided their resources in two ponderosa pine forest sites, an untreated forest (mostly comprised of trees with trunks higher than 32-cm DBH), and a forest selectively thinned for small-diameter trees (less than 32 cm). He found that the areas used in the trees by the two bird species are different when foraging on large- or small-diameter trees, but partition in tree use remains in both forest conditions. From this data, the author concluded that pygmy nuthatches have a foraging niche in thinned and unthinned forests that is distinct from white-breasted nuthatches, allowing them to coexist without potential population implications. Our data, consistent with the differential tree use in nuthatches reported by Cloyed (2014) and the estimated niche overlaps (including brown creepers), suggest that changes in forest structure could lead to changes in competitive dynamics among them, as discussed below.

Preferences for larger-diameter (white-breasted nuthatches), medium-diameter (brown creepers), and small-diameter (pygmy nuthatches) tree sections have been tested in both laboratory (Pierce and Grubb 1981, Cloyed 2014) and natural conditions (Willson 1970, Williams and Batzli 1979, Morrison et al. 1987). Moreover, these preferences denote characteristics that trees should have to search for food and a change in the species or size composition of the forest that would likely alter the pattern of resource used by these bird species. For example, Morrison et al. (1987) showed that foraging behaviors used by bark foragers in the Sierra Nevada, California, were consistent with tree composition due to silvicultural management. These observations gave credence to our hypothesis that a change in forest structure must involve a change in the resource partition between sympatric species such as those studied here. Compared to undisturbed sites, the disturbed forest sites we studied showed a direct effect on the tree-climbing birds. For example, the three bird species were less abundant in these sites throughout the study, their niche breadth indices were lower, and conversely the overlap was larger among them. Thus, our results suggest that activities of excessive logging and grazing in the disturbed sites at La Malinche can affect size class distribution of trees (Peterson and Reich 2001) and that tree-climbing birds are responding to this. A possible effect of abundance and frequency of availabilities of foraging sites (lower trunk, thin branches, and thick branches) in our study site requires further investigation. Results from the null models showed that the observed overlaps were higher than expected, particularly in undisturbed forest sites. However, the expected values estimated for the disturbed sites were already high, suggesting that species have to share the habitat to a higher degree even when they distributed themselves randomly in the tree, making it difficult for them to reduce competition through niche partitioning (Gotelli and Graves 1996). This adds to the growing number of studies that demonstrate negative effects of changes in forest structure promoted by human activities for some bird species at the behavioral and population levels (Desrochers and Fortin 2000, Blake 2005, Whelan and Maina 2005, Barbaro et al. 2012).

Conclusions

In short, our study showed that tree-climbing bird species perform a differential use of space in the trees to forage, which promotes their coexistence particularly at undisturbed sites. However, activities of excessive logging and grazing in the disturbed sites resulted in niche overlap among species, and this could lead to changes in competitive dynamics among them. Future studies analyzing the mechanisms that structure tree-climbing bird communities should include the potential impact of foresting practices.

Competing interests

The authors declare that they have no competing interests.

Authors' contributions

CL conceived the study, participated in its design and coordination, performed the statistical analysis, and drafted the manuscript. BP carried out the fieldwork and participated in its design. CCG participated in the design of the study. MASM participated in the design. All authors read and approved the final manuscript.

Acknowledgements

We thank Hellen Martínez Roldán and Jheny González Méndez for their help during fieldwork. Two anonymous reviewers provided enriching comments and thorough grammar revision on a previous version. We also thank the staff of the Caseta No. 5 of the Coordinación General de Ecología del Estado de Tlaxcala for access to their facilities. This work was supported by CACyPI-UATx-2014.

Author details

[1]Centro de Investigación en Ciencias Biológicas, Universidad Autónoma de Tlaxcala, Km. 10.5 Autopista Tlaxcala-San Martín Texmelucan, San Felipe Ixtacuixtla, Tlaxcala 90120, Mexico. [2]Laboratorio de Biología Evolutiva, Centro Tlaxcala de Biología de la Conducta, Universidad Autónoma de Tlaxcala, Carretera Tlaxcala-Puebla, Km. 1.5, 90062 Tlaxcala, Mexico.

References

Alatalo RV, Alatalo RH (1979) Resource partitioning among a flycatcher guild in Finland. Oikos 33:46–54
Barbaro L, Brockerhoff EG, Giffard B, van Halder I (2012) Edge and area effects on avian assemblages and insectivory in fragmented native forests. Landsc Ecol 27:1451–63
Blake JG (2005) Effects of prescribed burning on distribution and abundance of birds in a closed-canopy oak-dominated forest, Missouri, USA. Biol Conserv 121:519–31
Cloyed CS (2014) Forest structure affects resource partitioning between pygmy and white-breasted nuthatches. Coevolution 2:26–30
Connell JH (1980) Diversity and the coevolution of competitors, or the ghost of competition past. Oikos 35:131–8
De Cáceres M, Sol D, Lapiedra O, Legendre PA (2011) Framework for estimating niche metrics using the resemblance between qualitative resources. Oikos 120:1341–50
Desrochers A, Fortin MJ (2000) Understanding avian responses to forest boundaries: a case study with chickadee winter flocks. Oikos 91:376–84
Gotelli NJ, Graves GR (1996) Null models in ecology. Press, Smithsonian Institution
Gotelli NJ, GL Entsminger. EcoSim: null models software for ecology. Version 7. Acquired Intelligence Inc. & Kesey-Bear. Jericho, VT 05465. Computer software. http://garyentsminger.com/ecosim/index.htm. (accessed on 19-04-2014); 2004.
Holmes RT, Bonney RE Jr, Pacala SW (1979) Guild structure of the Hubbard Brook bird community: a multivariate approach. Ecology 60:512–20
Hutchinson GE (1961) The paradox of the plankton. Am Nat 95:137–45
Keddy PA (1989) Competition. Chapman and Hall, London
Lara C, Martinez-Garcia V, Ortiz-Pulido R, Bravo-Cadena J, Loranca S, Cordoba-Aguilar A (2011) Temporal-spatial segregation among hummingbirds foraging on honeydew in a temperate forest in Mexico. Current Zoology 57:56–62
Lawlor LR (1980) Overlap, similarity, and competition coefficients. Ecology 61:245–51
Leyequién E, de Boer WF, Cleef A (2007) Influence of body size on coexistence of bird species. Ecol Res 22:735–41
López-Domínguez JC, Acosta R (2005) Descripción del Parque Nacional La Malinche. In: Fernández-Fernández JA, López-Domínguez JC (eds) Biodiversidad del Parque Nacional La Malinche. Coordinación General de

Ecología del Gobierno del Estado de Tlaxcala, Tlaxcala, Tlaxcala, México, pp 20–30
MacArthur RH (1958) Population ecology of some warblers of northeastern coniferous forests. Ecology 39:599–619
McEllin SM (1979) The role of insectivorous birds in forest ecosystems. In: Dickson JG, Conner RN, Fleet RR, Kroll JC, Jackson JA (eds) Population demographies, spacing, and foraging behaviors of white-breasted and pygmy nuthatches in ponderosa pine habitat. Academic Press, NY, pp 301–29
Morrison ML, With KA (1987) Interseasonal and intersexual resource partitioning in hairy and white-headed woodpeckers. Auk 104:225–33
Morrison ML, With KA, Timossi IC, Block WM, Milne KA (1987) Foraging behavior of bark-foraging birds in the Sierra Nevada. Condor 89:201–4
Peterson DW, Reich PB (2001) Prescribed fire in oak savanna: fire frequency effects on stand structure and dynamics. Ecol Appl 11:914–27
Pianka ER (1973) The structure of lizard communities. Ann Rev Ecol Syst 4:53–74
Pierce V, Grubb TC Jr (1981) Laboratory studies of foraging in four bird species of deciduous woodland. Auk 98:307–20
Pyke GH (1982) Local geographic distributions of bumblebees near Crested Butte, Colorado: competition and community structure. Ecology 63:555–73
R Development Core Team (2012) R: a language and environment for statistical computing. R Foundation for Statistical Computing, ISBN 3-900051-07-0, Vienna, Austria, Available at http://www.r-project.org/. Accessed 22 october 2012
Rao CR (1982) Diversity and dissimilarity coefficients: a unified approach. Theor Popul Biol 21:24–43
Robinson SK, Holmes RT (1982) Foraging behavior of forest birds: the relationships among search tactics, diet, and habitat structure. Ecology 63:1918–31
Schoener TW (1974) Resource partitioning in ecological communities. Science 185:27–39
Sibley D (2003) Sibley field guide to birds of eastern North America. Knopf Doubleday Publishing Group
Sibley CG, Ahlquist JE (1990) Phylogeny and classification of birds: a study in molecular evolution. Yale University Press, New Haven, Conn. Publishing Group
Stallcup PL (1968) Spatio-temporal relationships of nuthatches and woodpeckers in ponderosa pine forests of Colorado. Ecology 49:831–43
van Perlo B (2006) Birds of Mexico and Central America. Princeton University Press, Princeton
Villers RL, Rojas GF (2006) LP Tenorio. Guía Botánica del Parque Nacional Malinche Tlaxcala – Puebla. Univ. Nacional Autónoma de México, México. D. F., México
Wellborn GA, Cothran RD (2007) Niche diversity in crustacean cryptic species: complementarity in spatial distribution and predation risk. Oecologia 154:175–83
Whelan CJ (2001) Foliage structure influences foraging of insectivorous forest birds: an experimental study. Ecology 82:219–31
Whelan CJ, Maina GG (2005) Effects of season, understorey vegetation density, habitat edge and tree diameter on patch-use by bark-foraging birds. Funct Ecol 19:529–36
Williams JB, Batzli GO (1979) Competition among bark-foraging birds in central Illinois: experimental evidence. Condor 81:122–32
Willson MF (1970) Foraging behavior of some winter birds of deciduous woods. Condor 72:169–74
Wimberly MC, Spies TA (2001) Influences of environment and disturbance on forest patterns in coastal Oregon watersheds. Ecology 82:1443–59
Winemiller KO, Pianka ER (1990) Organization in natural assemblages of desert lizards and tropical fishes. Ecol Monogr 60:27–55
Young HS, Shaffer SA, McCauley DJ, Foley DG, Dirzo R, Block BA (2010) Resource partitioning by species but not sex in sympatric boobies in the central Pacific Ocean. Mar Ecol Prog Ser 403:291–301

16

The genus *Milnesium* Doyère, 1840 (Tardigrada) in South America with descriptions of two new species from Argentina and discussion of the feeding behaviour in the family Milnesiidae

Milena Roszkowska[1,2*], Marta Ostrowska[1] and Łukasz Kaczmarek[1,2]

Abstract

Background: The diversity and distribution of the tardigrades in South America are rather poor and selective, as is information about their feeding behaviour and diet. To date, only ca. 210 tardigrade taxa have been reported from the region of South America. In the present paper, we provide an update of the distribution of the genus *Milnesium* in South America and discuss some aspects of the feeding behaviour in the family Milnesiidae.

Results: In seven moss samples collected in the Argentinean province of Río Negro, 31 specimens, 4 exuviae and 32 eggs belonging to the genus *Milnesium* were found. Among them, four species were identified: *Milnesium argentinum* sp. nov., *Milnesium beatae* sp. nov., *Milnesium brachyungue* and *Milnesium granulatum*. By its dorsal sculpture, *M. argentinum* sp. nov. is most similar to *M. beatae* sp. nov., *Milnesium beasleyi* and *Milnesium berladnicorum*, but it differs from *M. beasleyi* and *M. berladnicorum* mainly by having a different claw configuration and from *M. beatae* and *M. beasleyi* by having stylet supports inserted in a more caudal position and by some other morphometric characters. In the width of its buccal tube and the claw configuration [3-3]-[3-3], *M. beatae* sp. nov. is most similar to *Milnesium bohleberi*, *M. brachyungue* and *Milnesium eurystomum*, but it differs from them mainly by having a sculptured dorsal cuticle and by some other morphometric characters.

Conclusions: The study discusses distribution and taxonomic problems of the *Milnesium* species known from South America. As of now, nine *Milnesium* taxa are known from this region (including two new species reported in this paper). Additionally, the study broadens our knowledge of tardigrades' feeding behaviour, provides some details about their diet and suggests that the type of prey chosen by some species belonging to the family Milnesiidae may be associated with the width of their buccal tube.

Keywords: Milnesiidae; *Milnesium argentinum* sp. nov; *Milnesium beatae* sp. nov; Nahuel Huapi National Park; Neotropical region; New records; Prey selection; Río Negro; Taxonomy; Water bears

Background

South America, with an area of almost 18 million km^2, is located mostly in the Southern Hemisphere, with a relatively small portion (ca. 10%) in the Northern Hemisphere. It includes 12 sovereign countries. Geographically, the western part of South America is dominated by the Andes while the eastern part contains both highland regions and large river basins such as the Amazon, Paraná and Orinoco. Most of the continent lies in the tropics and the entire South American territory belongs to Neotropical ecozone (Peel et al. 2007, Holt et al. 2013).

Argentina, located in southern part of the continent, shares land borders with Chile across the Andes to the west, Bolivia and Paraguay to the north, Brazil to the northeast, Uruguay and the South Atlantic Ocean to the east and the Drake Passage to the south. Argentina is the eighth largest country in the world and the second

* Correspondence: mil.roszkowska@gmail.com
[1]Department of Animal Taxonomy and Ecology, Faculty of Biology, Adam Mickiewicz University in Poznań, Umultowska 89, 61-614, Poznań, Poland
[2]Laboratorio de Ecología Natural y Aplicada de Invertebrados, Universidad Estatal Amazónica, Campus Principal Km 2.1/2 via a Napo (Paso Lateral) Puyo, Pastaza, Ecuador

largest in Latin America. It is subdivided into 23 provinces and one autonomous city - Buenos Aires. The exceptionally diverse climate depends on the geographic regional division and ranges from tropical in the north to subpolar in the far continental south (Edwards 2008).

Río Negro, one of the 23 provinces of Argentina, is located at the northern edge of Patagonia. The central region of the province is dominated by a series of plateaus and isolated hills with altitude ranging from 600 to 1,000 m asl. Towards the west, the foothills of the Andes are dominated by a series of low valleys. The climate of the province is temperate at low elevations and very harsh in the highest Andean peaks (Edwards 2008).

The Phylum Tardigrada consists currently of ca. 1,200 species (Guidetti and Bertolani 2005; Degma and Guidetti 2007; Vicente and Bertolani 2013; Degma et al. 2009-2014) that inhabit terrestrial and aquatic (freshwater and marine) environments throughout the world (Ramazzotti and Maucci 1983). Our knowledge of the diversity and distribution of South American water bears is poor and selective. To date, ca. 210 taxa (including seven *Milnesium* species) have been reported from this region (mostly from Argentina, Chile and Brazil) (McInnes 1994; Binda and Pilato 1994a,b, 1999a,b; Pilato and Binda 1994, 1996, 1998; Maucci 1996; Claps and Rossi 1997; Claxton 1998; Pilato and Patanè 1998; Dastych et al. 1998; Pilato et al. 1998, 2000, 2001, 2002, 2003, 2004; Dastych 1999a,b, 2000, Pilato 2000, 2007; Jerez Jaimes and Narváez Parra 2001; Nickel et al. 2001; Peluffo et al. 2002, 2007; Michalczyk and Kaczmarek 2003, 2004, 2005, 2006; Marley 2006; Moly de Peluffo et al. 2006; Degma et al. 2008; Kaczmarek and Michalczyk 2009; Rossi et al. 2009; Montoya et al. 2010; Lisi 2011; Guidetti et al. 2013; Meyer 2013; Kaczmarek et al. 2014a, b; Lisi et al. 2014; Melo et al. 2014).

The genus *Milnesium* (Doyère 1840) is known from many localities, from the Antarctic through tropical and temperate to Arctic regions (Michalczyk et al. 2012a,b). Since the genus was recently redescribed (Michalczyk et al. 2012a,b), many new records and species have been reported from various localities throughout the world (e.g. Kaczmarek et al. 2012a,b, 2014b; Meyer and Hinton 2012; Johansson et al. 2013; Hinton et al. 2013; Meyer et al. 2013, 2014; Shaw and Miller 2013; Trygvadóttir and Kristensen 2013; Bartels et al. 2014; Ciobanu et al. 2014a,b; Lisi et al. 2014; Zawierucha et al. 2014; Melo et al. 2014). Taking into consideration that some morphological characters were not mentioned in older records of the genus members, all the *Milnesium* records before 2012 should be verified (Michalczyk et al. 2012a,b). Until now in South America, only seven *Milnesium* species have been reported (see the 'Results' section). In this paper, two new species of this genus are described and illustrated. Additionally, two new records of *Milnesium* species for Argentina are given. The distribution and taxonomic problems of all *Milnesium* species known from South America are also discussed.

All *Milnesium* species, with their wide and relatively short buccal tube connected with a large pharynx without placoids, are considered carnivorous, but details of their diet are still very poorly known. They can feed on rotiferas, nematodes, other tardigrades or even on amoebas (e.g. Kinchin 1990; Suzuki 2003; Nelson et al. 2010; Miller and Williams 2012). In this paper, we discuss some details of the diet of *Milnesium* species (and more general the entire family Milnesiidae) in connection with the different constructions of the buccal tube.

Methods

Seven moss samples were collected from trees and rocks from various localities in Nahuel Huapi National Park (Patagonia, southern Argentina) in 2006 and 2012 by Dawid Diduszko, Łukasz Kaczmarek and Marta Prange (see the 'List of samples' section) and four species of the genus *Milnesium* were found. All samples were collected and examined for tardigrades using standard methods (Dastych 1980; Ramazzotti and Maucci 1983). After extraction, all specimens, exuviae and eggs were fixed and mounted on microscope slides in Hoyer's medium. Observations, measurements and photomicrographs were taken using phase contrast microscopy (PCM) (Olympus BX41 with digital camera ARTCAM-300Mi, Olympus Corporation, Shinjuku-ku, Japan). All measurements (determined with QuickPhoto Camera 2.3) are given in micrometres [μm].

Morphometric data were handled using the 'Apochela' ver. 1.1 template available from the Tardigrada Register (Michalczyk and Kaczmarek 2013). Structures were measured only if their orientations were suitable. Body length was measured from the anterior to the posterior end of the body, excluding the hind legs. All measurements followed protocols in Tumanov (2006). Buccal tube width was measured at three points as suggested by Michalczyk et al. (2012a,b). The *pt* ratio is the ratio of the length of a given structure to the length of the buccal tube, expressed as a percentage (Pilato 1981). The *pt* values are always provided in [square brackets and in *italics*]. Configuration of the number of claw points on the secondary branches ('claw configuration') is given according to Michalczyk et al. (2012a,b).

Species were identified using the keys in Michalczyk et al. (2012a,b) and other original descriptions/re-descriptions (Binda and Pilato 1990; Ramazzotti 1962; Kaczmarek et al. 2012a, b; Michalczyk et al. 2012a,b). Characteristics and measurements of the species used in the differential diagnosis are given according to the original descriptions (Ramazzotti 1962; Binda and Pilato 1990; Maucci 1991; Pilato et al. 2002; Kaczmarek et al. 2004, 2012a, b; Tumanov 2006; Kaczmarek and Michalczyk 2007;

Meyer and Hinton 2010, 2012; Michalczyk et al. 2012a, b; Bartels et al. 2014; Ciobanu et al. 2014b) or are based on direct examination of fixed specimens (holotype and paratypes of *M. beasleyi*, *M. berladnicorum*, *M. bohleberi*, *M. granulatum*, *M. katarzynae*, *M. krzysztofi* and specimens of *M. eurystomum* from Spitsbergen). Tardigrade taxonomy follows Marley et al. (2011). In the list of the species, we use Roman numerals to indicate the sample code (see the 'List of samples' and 'Results' sections) and Arabic numerals to designate the number of specimens and eggs (see the 'Results' section).

Raw data underlying the description of *Milnesium argentinum* sp. nov. and *Milnesium beatae* sp. nov. are deposited in the Tardigrada Register (Michalczyk and Kaczmarek 2013) under http://www.tardigrada.net/register/0015.htm and http://www.tardigrada.net/register/0016.htm, respectively.

List of samples

I. 41°20'S, 71°30'W, ca. 850 m asl: Río Negro, Nahuel Huapi National Park, Bariloche, Mirador Lago Mascardi, moss from rock, coll. Dawid Diduszko, 22 February 2012.

II. 41°12'S, 71°50'W, ca. 1,000 m asl: Río Negro, Nahuel Huapi National Park, Ventisquero Negro, car parking near small bar, *Nothofagus* forest, moss from rocks, coll. Łukasz Kaczmarek, 27 January 2006.

III. 41°12'S, 71°50'W, ca. 1,000 m asl: Río Negro, Nahuel Huapi National Park, Ventisquero Negro, car parking near small bar, *Nothofagus* forest, moss from rocks, coll. Łukasz Kaczmarek, 27 January 2006.

IV. 41°12'S, 71°50'W, ca. 1,000 m asl: Río Negro, Nahuel Huapi National Park, Ventisquero Negro, car parking near small bar, *Nothofagus* forest, moss from rocks, coll. Łukasz Kaczmarek, 27 January 2006.

V. 41°12'S, 71°50'W, ca. 1,000 m asl: Río Negro, Nahuel Huapi National Park, Ventisquero Negro, car parking near small bar, *Nothofagus* forest, moss from tree, coll. Łukasz Kaczmarek, 27 January 2006.

VI. 41°12'S, 71°28'W, ca. 1,400 m asl: Río Negro, Nahuel Huapi National Park, Bariloche, 1.5 km from Refugio Frey, *Nothofagus* forest, moss from stone, coll. Marta Prange, 26 January 2006.

VII. 41°13'S, 71°27'W, ca. 1,200 m asl: Río Negro, Nahuel Huapi National Park, on the trail to Refugio Frey, near to Van Titter stream, *Nothofagus* forest, moss from stone, coll. Marta Prange, 26 January 2006.

Results

Taxonomic account

Phylum: Tardigrada Doyère 1840.
Class: Eutardigrada Richters 1926.
Order: Apochela Schuster et al. 1980.
Family: Milnesiidae Ramazzotti 1962.
Genus: *Milnesium* Doyère 1840.

Milnesium argentinum sp. nov.

http://www.tardigrada.net/register/0015.htm.

Material examined: Eighteen females, all from Nahuel Huapi National Park, Río Negro, Argentina, mosses samples from rocks and stones, coll. Dawid Diduszko, Łukasz Kaczmarek and Marta Prange.

Type material: Holotype and three paratypes in sample VII.

Additional material: two specimens in sample I, ten specimens in sample III and two specimens in sample IV.

Diagnosis

Adults (Figure 1a) (measurements in micrometres, pt ratios and statistics in Table 1): The body rose before fixation and transparent afterwards, eyes present. Cuticle sculptured with pseudopores (0.4 to 0.7) not arranged in bands, sparsely distributed and not forming reticular design (Figure 2b). Six peribuccal papillae and six peribuccal lamellae around the mouth opening present. Two cephalic papillae positioned laterally. Peribuccal and cephalic papillae similar in length.

Buccal apparatus of the *Milnesium* type (Figure 2a). Buccal tube rather narrow and long (standard width on average 27% of its length) and funnel-shaped, wider anteriorly (posterior diameter on average 80% of the anterior diameter). Pharyngeal bulb elongated, pear-shaped and without placoids or septulum.

Claws of the *Milnesium* type, slender (Figure 3a,b). Primary branches on all legs with small but distinct accessory points detaching from the branch at its greatest curvature (Figure 3b). Secondary branches with rounded basal thickenings (Figure 3b). All secondary branches on all legs with three points (claw configuration: [3-3]-[3-3]). Single, long transverse, cuticular bars under claws I to III present (Figure 3a).

Eggs: Oval, smooth and deposited in exuvium as in all other known *Milnesium* species.

Males not observed.

Locus typicus: Argentina, 41°13′S, 71°27′W, ca. 1,200 m asl., Río Negro Province, Nahuel Huapi National Park.

Etymology: The new species is named after the country of Argentina, where the species was collected.

Type depositories: The holotype and 14 paratypes are deposited in the Department of Animal Taxonomy and Ecology, Adam Mickiewicz University in Poznań, Umultowska 89, Poznań, Poland; two paratypes are deposited at the Natural History Museum, University of Copenhagen Universitetsparken 15, DK-2100 Copenhagen, Denmark and one paratype is deposited at collection of

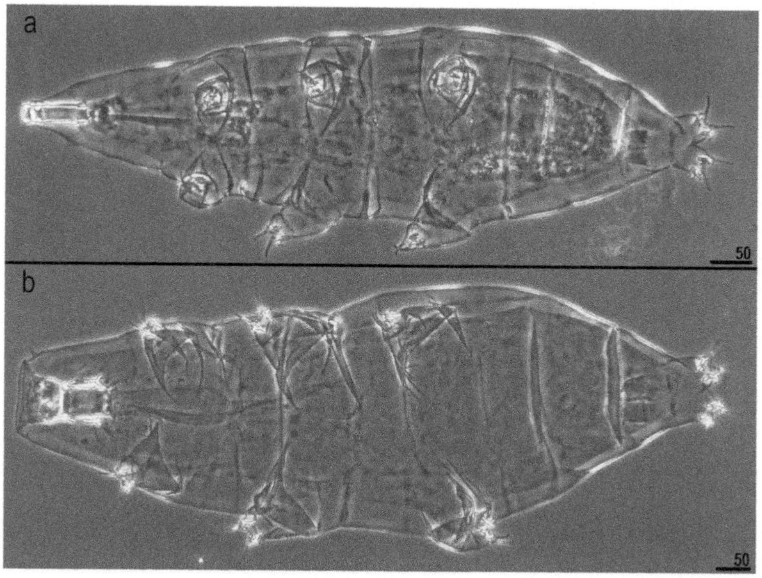

Figure 1 ***Milnesium argentinum*** **sp. nov. and** ***Milnesium beatae*** **sp. nov.** *Milnesium argentinum* sp. nov.: **(a)** habitus (ventral view) and *Milnesium beatae* sp. nov.: **(b)** habitus (ventral view) (both PCM).

Binda and Pilato, Museum of the Department of Animal Biology 'Marcello La Greca', University of Catania, Italy.

Differential diagnosis

Based on having a sculptured dorsal cuticle, *M. argentinum* sp. nov. belongs to the *granulatum* group (Michalczyk et al. 2012a,b). The new species with three points on the secondary branches of all claws (claw configuration [3-3]-[3-3]) is most similar to *M. beasleyi* Kaczmarek et al. 2012a, b, *M. beatae* sp. nov. and *M. berladnicorum* Ciobanu et al. 2014b, but it differs from the following:

1. *M. beasleyi*, known only from Turkey (Kaczmarek et al. 2012a, b) by different claw configuration ([3-3]-[3-3] in *M. argentinum* sp. nov. *vs.* [2-3]-[3-2] in *M. beasleyi*), presence of rounded basal thickenings under secondary branches of claws, stylet supports inserted in more caudal position ([*70.0-73.7*] in *M. argentinum* sp. nov. vs. [*61.6-65.6*] in *M. beasleyi*) and slightly smaller claws I to IV (see Table 1 and Table one in Kaczmarek et al. (2012a, b) for the exact differences in dimensions of claws).
2. *M. beatae* sp. nov. by stylet supports inserted in more caudal position ([*70.0-73.7*] in *M. argentinum* sp. nov. vs. [*63.8-66.7*] in *M. beatae* sp. nov.), smaller *pt* of width of buccal tube ([*30.3-39.8*], [*24.2-32.3*], [*23.1-33.6*] anterior, standard and posterior, respectively, in *M. argentinum* sp. nov. vs. [*70.3-78.9*], [*58.1-65.6*], [*49.4-56.5*] anterior, standard and posterior, respectively, in *M. beatae* sp. nov.) and smaller claws I to IV (see Tables 1 and 2 for the exact differences in dimensions of claws).
3. *M. berladnicorum*, known only from Romania (Ciobanu et al. 2014b) by different claw configuration ([3-3]-[3-3] in *M. argentinum* sp. nov. vs. [2-3]-[2-2] in *M. berladnicorum*) and smaller claws I to IV (see Table 1 and Table one in Ciobanu et al. (2014a, b) for the exact differences in dimensions of claws).

Besides the species mentioned above, *M. argentinum* sp. nov. is similar to other species of the genus *Milnesium* with the claw configuration [3-3]-[3-3] (*M. alabamae* Wallendorf and Miller 2009, *M. antarcticum* Tumanov 2006, *M. asiaticum* Tumanov 2006, *M. barbadosense* Meyer and Hinton 2012, *M. bohleberi* Bartels et al. 2014, *M. brachyungue* Binda and Pilato 1990, *M. eurystomum* Maucci 1991, *M. granulatum* Ramazzotti 1962, *M. longiungue* Tumanov 2006 and *M. zsalakoae* Meyer and Hinton 2010) or sculptured cuticle (*M. beasleyi* Kaczmarek et al. 2012a, b, *M. katarzynae* Kaczmarek et al. 2004, *M. krzysztofi* Kaczmarek and Michalczyk 2007 and *Milnesium reticulatum* Pilato et al. 2002).

The new species differs from the following:

4. *M. alabamae*, known only from the USA (Wallendorf and Miller 2009) by different dorsal sculpture (pseudopores not arranged in bands, sparsely distributed and not forming a reticular design in *M. argentinum* sp. nov. vs. pseudopores arranged in bands (especially in caudal region), densely distributed and forming a reticular design in *M. alabamae*), presence of accessory points on main branches of claws, presence of eyes and stylet

Table 1 Measurements and *pt* values of selected morphological structures of *Milnesium argentinum* sp. nov. mounted in Hoyer's medium

Character	*N*	Range						Mean		SD		Holotype	
		µm			*pt*			µm	*pt*	µm	*pt*	µm	*pt*
Body length	12	445	-	890	*921*	-	*1,290*	707	*1,093*	141	*123*	890	*1,211*
Peribuccal papillae length	10	6.7	-	12.4	*14.1*	-	*18.6*	10.3	*16.3*	1.7	*1.1*	11.6	*15.8*
Lateral papillae length	9	5.5	-	11.4	*9.4*	-	*17.6*	9.0	*14.1*	1.9	*2.3*	?	*?*
Buccal tube													
Length	12	47.6	-	74.2		-		64.3	-	8.6	-	73.5	-
Stylet support insertion point	12	35.1	-	52.3	*70.0*	-	*73.7*	46.0	*71.7*	5.8	*1.4*	52.3	*71.2*
Anterior width	12	16.0	-	29.5	*30.3*	-	*39.8*	22.3	*34.5*	4.3	*3.4*	29.1	*39.6*
Standard width	12	13.0	-	24.0	*24.2*	-	*32.3*	17.7	*27.4*	3.5	*2.8*	22.0	*29.9*
Posterior width	12	12.2	-	24.5	*23.1*	-	*33.6*	17.8	*27.6*	3.9	*3.4*	24.5	*33.3*
Standard width/length ratio	12	24%	-	32%		-		27%	-	3%	-	30%	-
Posterior/anterior width ratio	12	76%	-	87%		-		80%	-	4%	-	84%	-
Claw 1 lengths													
External primary branch	10	12.4	-	20.5	*24.3*	-	*28.6*	16.7	*26.3*	2.6	*1.5*	20.5	*27.9*
External base + secondary branch	9	10.5	-	17.8	*22.0*	-	*25.2*	14.4	*23.0*	2.4	*1.2*	17.8	*24.2*
External spur	8	3.7	-	5.8	*7.1*	-	*8.2*	4.8	*7.7*	0.8	*0.4*	?	*?*
Internal primary branch	11	12.9	-	18.7	*21.1*	-	*28.9*	15.8	*24.8*	2..1	*2.0*	17.4	*23.7*
Internal base + secondary branch	10	11.1	-	18.2	*21.1*	-	*26.1*	14.8	*23.2*	2.5	*1.5*	17.3	*23.5*
Internal spur	10	3.4	-	5.7	*7.1*	-	*8.8*	5.0	*7.8*	0.9	*0.6*	5.6	*7.6*
Claw 2 lengths													
External primary branch	10	14.3	-	20.5	*24.7*	-	*28.5*	17.8	*26.6*	2.0	*1.3*	20.5	*27.9*
External base + secondary branch	12	11.3	-	17.9	*22.1*	-	*26.3*	15.2	*23.7*	2.0	*1.3*	17.9	*24.4*
External spur	6	2.8	-	5.5	*5.9*	-	*8.5*	4.0	*6.7*	1.0	*1.0*	?	*?*
Internal primary branch	10	12.6	-	20.8	*23.1*	-	*30.5*	17.2	*26.4*	2.8	*2.5*	20.8	*28.3*
Internal base + secondary branch	11	11.6	-	16.5	*21.2*	-	*25.2*	14.8	*22.5*	1.6	*1.2*	16.1	*21.9*
Internal spur	9	3.8	-	6.3	*6.6*	-	*8.8*	5.3	*7.9*	0.9	*0.8*	6.3	*8.6*
Claw 3 lengths													
External primary branch	11	12.8	-	22.3	*23.4*	-	*30.3*	17.7	*27.3*	3.2	*2.1*	22.3	*30.3*
External base + secondary branch	10	11.9	-	17.1	*22.5*	-	*25.1*	14.9	*23.3*	2.1	*0.9*	17.1	*23.3*
External spur	7	3.3	-	5.3	*5.9*	-	*7.6*	4.5	*6.9*	0.8	*0.6*	?	*?*
Internal primary branch	11	13.9	-	21.4	*24.4*	-	*28.8*	17.5	*27.0*	2.6	*1.5*	20.7	*28.2*
Internal base + secondary branch	10	12.7	-	17.8	*20.3*	-	*25.7*	15.2	*23.1*	1.9	*1.4*	17.2	*23.4*
Internal spur	8	4.2	-	6.9	*6.8*	-	*9.9*	5.7	*8.5*	1.0	*1.1*	6.2	*8.4*
Claw 4 lengths													
Anterior primary branch	10	14.7	-	24.3	*29.9*	-	*33.8*	20.5	*31.8*	3.6	*1.3*	24.2	*32.9*
Anterior base + secondary branch	11	12.2	-	20.3	*24.4*	-	*28.8*	16.8	*26.0*	2.7	*1.4*	20.3	*27.6*
Anterior spur	11	3.8	-	7.9	*8.0*	-	*10.8*	6.3	*9.7*	1.2	*0.8*	7.3	*9.9*
Posterior primary branch	11	13.5	-	25.8	*28.4*	-	*36.4*	21.3	*33.0*	4.0	*2.7*	25.4	*34.6*
Posterior base + secondary branch	11	12.8	-	20.2	*25.0*	-	*28.3*	17.3	*26.8*	2.7	*1.0*	20.2	*27.5*
Posterior spur	8	3.3	-	6.2	*6.9*	-	*8.4*	4.9	*7.8*	1.0	*0.6*	6.2	*8.4*

Range refers to the smallest and the largest structure among all measured specimens.
The *pt* values are provided in *italics*.
N, number of specimens/structures measured; SD, standard deviation; ?, no data.

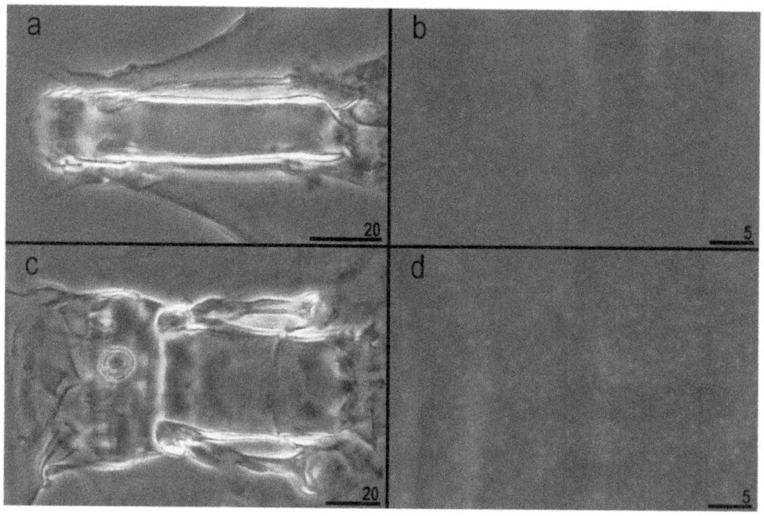

Figure 2 ***Milnesium argentinum*** **sp. nov. and** ***Milnesium beatae*** **sp. nov.** *Milnesium argentinum* sp. nov.: **(a)** buccal tube (ventral view); **(b)** dorsal cuticle with pseudopores and *Milnesium beatae* sp. nov.: **(c)** buccal tube (ventral view); **(d)** dorsal cuticle with pseudopores (all PCM).

supports inserted in more caudal position ([*70.0-73.7*] in *M. argentinum* sp. nov. vs. [*63.7*] in *M. alabamae* (in specimen 514.9 in length)).

5. *M. antarcticum*, known only from Antarctica (Tumanov 2006) by having sculptured dorsal cuticle and slightly smaller claws I to IV (see Table 1 and Table two in Tumanov (2006) for the exact differences in dimensions of claws).
6. *M. asiaticum*, known only from Kirghizia, Romania and Svalbard (Tumanov 2006; Kaczmarek et al. 2012a, b; Ciobanu et al. 2014a, b) by having sculptured dorsal cuticle and stylet supports inserted in more caudal position ([*70.0-73.7*] in *M. argentinum* sp. nov. vs. [*63.9-66.9*] in *M. asiaticum*).
7. *M. barbadosense*, known only from Barbados (Meyer and Hinton 2012) by having sculptured dorsal

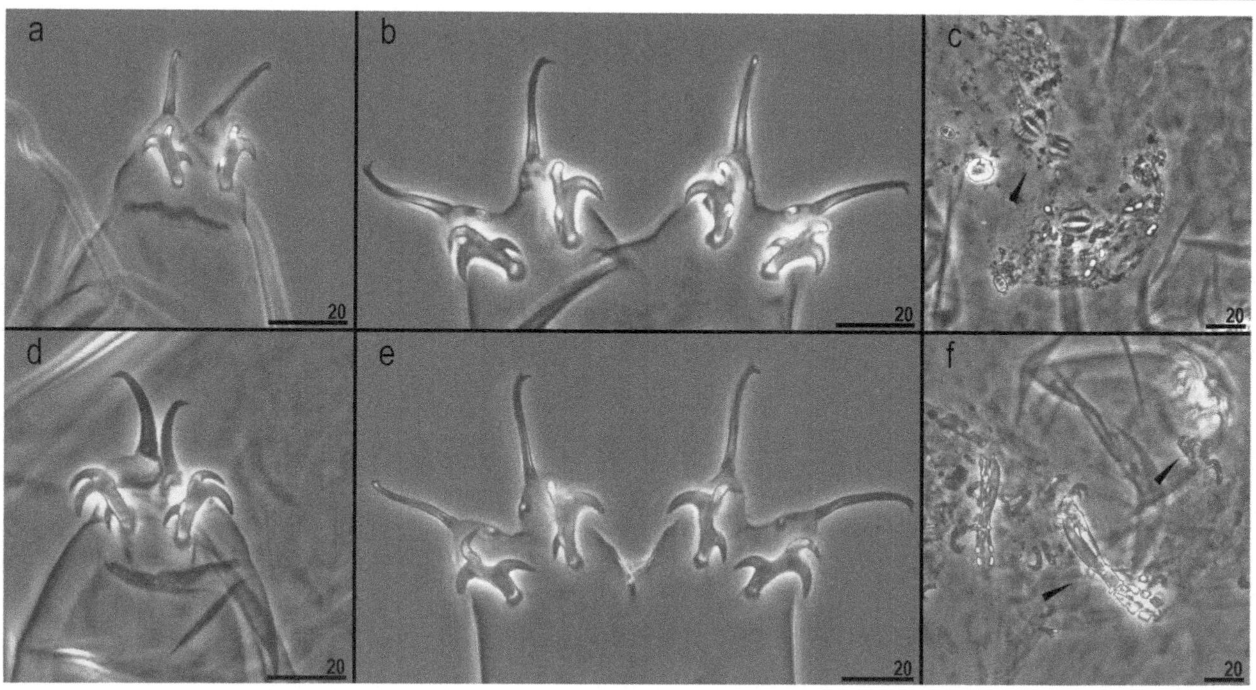

Figure 3 ***Milnesium argentinum*** **sp. nov. and** ***Milnesium beatae*** **sp. nov.** *Milnesium argentinum* sp. nov.: **(a)** claws II; **(b)** claws IV; **(c)** rotifer mastaxes in the gut (black arrowhead) and *Milnesium beatae* sp. nov.: **(d)** claws I; **(e)** claws IV; **(f)** tardigrade buccal apparatuses and claws in the gut (black arrowheads) (all PCM).

Table 2 Measurements and *pt* values of selected morphological structures of *Milnesium beatae* sp. nov. mounted in Hoyer's medium

Character	*N*	Range						Mean		SD		Holotype	
		µm			*pt*			µm	*pt*	µm	*pt*	µm	*pt*
Body length	7	662	-	1,011	*1,259*	-	*1,524*	864	*1,401*	144	*97*	979	*1,494*
Peribuccal papillae length	7	9.8	-	13.2	*18.0*	-	*23.6*	12.0	*19.7*	1.1	*1.9*	13.2	*20.2*
Lateral papillae length	6	10.0	-	14.8	*19.0*	-	*21.1*	12.8	*20.2*	1.6	*0.9*	13.2	*20.2*
Buccal tube													
Length	7	51.6	-	70.2		-		61.4	-	6.8	-	65.5	-
Stylet support insertion point	7	33.8	-	45.0	*63.8*	-	*66.7*	40.0	*65.1*	4.3	*1.1*	43.3	*66.1*
Anterior width	7	37.0	-	53.5	*70.3*	-	*78.9*	45.7	*74.3*	6.6	*3.7*	51.7	*78.9*
Standard width	7	32.0	-	42.5	*58.1*	-	*65.6*	38.0	*61.9*	4.2	*2.6*	39.2	*59.8*
Posterior width	7	25.7	-	39.0	*49.4*	-	*56.5*	33.1	*53.6*	5.2	*2.8*	37.0	*56.5*
Standard width/length ratio	7	58%	-	66%		-		62%	-	3%	-	60%	-
Posterior/anterior width ratio	7	69%	-	76%		-		72%	-	3%	-	72%	-
Claw 1 lengths													
External primary branch	7	17.9	-	26.4	*33.6*	-	*41.4*	22.0	*35.8*	3.1	*2.7*	22.0	*33.6*
External base + secondary branch	6	13.6	-	22.1	*26.4*	-	*34.6*	19.0	*30.9*	3.8	*3.0*	20.8	*31.8*
External spur	4	5.0	-	6.5	*8.0*	-	*10.3*	5.6	*9.2*	0.7	*1.0*	5.8	*8.9*
Internal primary branch	6	16.4	-	25.4	*31.2*	-	*39.8*	22.0	*34.9*	3.2	*3.0*	23.6	*36.0*
Internal base + secondary branch	6	13.9	-	22.3	*26.9*	-	*33.4*	19.1	*30.9*	3.8	*3.0*	21.5	*32.8*
Internal spur	5	4.4	-	7.4	*8.5*	-	*10.5*	5.8	*9.5*	1.2	*0.8*	?	*?*
Claw 2 lengths													
External primary branch	7	19.1	-	27.3	*36.9*	-	*42.4*	24.5	*39.8*	3.6	*2.3*	27.3	*41.7*
External base + secondary branch	6	14.8	-	22.9	*28.7*	-	*34.1*	19.4	*31.6*	3.4	*2.1*	21.1	*32.2*
External spur	5	5.5	-	7.9	*8.0*	-	*12.1*	6.3	*10.5*	1.1	*1.5*	7.9	*12.1*
Internal primary branch	6	18.6	-	25.4	*35.4*	-	*37.4*	22.2	*36.3*	2.9	*0.7*	24.5	*37.4*
Internal base + secondary branch	7	14.4	-	21.8	*27.9*	-	*33.3*	19.1	*31.0*	3.1	*2.2*	21.8	*33.3*
Internal spur	6	5.2	-	7.7	*10.1*	-	*12.3*	6.5	*10.6*	1.1	*0.9*	6.8	*10.4*
Claw 3 lengths													
External primary branch	7	18.4	-	26.7	*35.7*	-	*41.4*	23.8	*38.7*	3.5	*2.6*	26.7	*40.8*
External base + secondary branch	6	14.8	-	22.9	*28.7*	-	*36.5*	19.8	*32.5*	3.6	*3.0*	?	*?*
External spur	5	4.7	-	6.9	*8.9*	-	*10.5*	5.9	*9.7*	1.1	*0.7*	?	*?*
Internal primary branch	5	17.6	-	26.4	*34.1*	-	*41.2*	22.9	*37.8*	4.4	*3.2*	?	*?*
Internal base + secondary branch	6	15.5	-	23.1	*29.8*	-	*36.8*	20.0	*32.6*	3.4	*2.7*	21.7	*33.1*
Internal spur	5	5.8	-	8.9	*10.5*	-	*14.2*	7.3	*11.9*	1.1	*1.4*	6.9	*10.5*
Claw 4 lengths													
Anterior primary branch	6	23.5	-	31.5	*43.1*	-	*49.4*	28.7	*45.6*	2.9	*2.5*	28.2	*43.1*
Anterior base + secondary branch	6	17.7	-	25.5	*33.4*	-	*40.0*	22.9	*36.4*	2.8	*3.2*	21.9	*33.4*
Anterior spur	5	6.4	-	10.0	*12.2*	-	*14.2*	8.2	*13.0*	1.3	*0.8*	?	*?*
Posterior primary branch	6	24.0	-	32.6	*45.3*	-	*51.5*	30.0	*47.5*	3.4	*2.5*	32.6	*49.8*
Posterior base + secondary branch	6	18.2	-	26.1	*34.6*	-	*40.8*	24.0	*37.9*	2.9	*2.2*	24.1	*36.8*
Posterior spur	5	6.2	-	8.3	*11.2*	-	*11.8*	7.2	*11.5*	0.8	*0.3*	?	*?*

Range refers to the smallest and the largest structure among all measured specimens.
The *pt* values are provided in *italics*.
N, number of specimens/structures measured; SD, standard deviation; ?, no data.

cuticle, presence of eyes and slightly smaller claws I to IV (see Table 1 and Table four in Meyer and Hinton (2012) for the exact differences in dimensions of claws).

8. *M. bohleberi*, known only from the USA (Bartels et al. 2014) by having sculptured dorsal cuticle, lower *pt* of papillae length ([*14.1-18.6*], [*9.4-17.6*] peribuccal and lateral, respectively, in *M. argentinum* sp. nov. vs. [*27.2-32.3*], [*17.7-21.2*] peribuccal and lateral, respectively, in *M. bohleberi*), smaller *pt* of width of buccal tube ([*30.3-39.8*], [*24.2-32.3*], [*23.1-33.6*] anterior, standard and posterior, respectively in *M. argentinum* sp. nov. vs. [*63.4-74.7*], [*54.5-64.0*], [*52.4-62.0*] anterior, standard and posterior, respectively in *M. bohleberi*) and smaller *pt* of claws I to IV (see Table 1 and Table one in Bartels et al. (2014) for the exact differences in dimensions of claws).
9. *M. brachyungue*, known only from Argentina (present study) and Chile (Binda and Pilato 1990) by having sculptured dorsal cuticle and stylet supports inserted in more caudal position ([*70.0-73.7*] in *M. argentinum* sp. nov. vs. [*69.4*] in *M. brachyungue* (in specimen 729.0 in length)).
10. *M. eurystomum*, known from Argentina, Chile, Greenland, Mongolia (probably, see comments below), Spain, Svalbard and USA (Maucci 1991, 1996; Kaczmarek and Michalczyk 2006; Guil 2008; Kaczmarek et al. 2012a, b; Land et al. 2012; Johansson et al. 2013) by having sculptured dorsal cuticle, stylet supports inserted in more caudal position ([*70.0-73.7*] in *M. argentinum* sp. nov. vs. [*60.0-60.3*] in *M. eurystomum*) and smaller *pt* of buccal tube width ([*30.3-39.8*], [*24.2-32.3*], [*23.1-33.6*] anterior, standard and posterior, respectively, in *M. argentinum* sp. nov. vs. [*72.1-75.8*], [*61.8-64.8*], [*43.7-57.9*] anterior, standard and posterior, respectively, in *M. eurystomum*).
11. *M. granulatum*, known from Argentina (present study), Chile, Colombia, Italy, Romania and the USA (Ramazzotti 1962; Ciobanu et al. 2014a, b; Bartels et al. 2014; Melo et al. (2014)) by different dorsal sculpture (pseudopores not arranged in bands, sparsely distributed and not forming reticular design in *M. argentinum* sp. nov. vs. pseudopores densely distributed and forming a reticular design in *M. granulatum*), stylet supports inserted in more caudal position ([*70.0-73.7*] in *M. argentinum* sp. nov. vs. [*66.5*] in *M. granulatum* (in specimen 338.3 in length)), smaller *pt* of width of buccal tube ([*30.3-39.8*], [*24.2-32.3*], [*23.1-33.6*] anterior, standard and posterior, respectively, in *M. argentinum* sp. nov. vs. [*46,3*], [*39.0*], [*40.1*] anterior, standard and posterior, respectively, in *M. granulatum* (in specimen 338.3 in length)) and slightly smaller claws I to IV (see Table 1 and Table two in Michalczyk et al. (2012a,b) for the exact differences in dimensions of claws).
12. *M. katarzynae*, known from China, Colombia, Costa Rica and Taiwan (Kaczmarek et al. 2004, 2014a; Yin and Li 2011; Melo et al. (2014)) by different claw configuration ([3-3]-[3-3] in *M. argentinum* sp. nov. vs. [2-2]-[2-2] in *M. katarzynae*), different dorsal sculpture (pseudopores not arranged in bands, sparsely distributed and not forming a reticular design in *M. argentinum* sp. nov. vs. pseudopores densely distributed and forming a reticular design in *M. katarzynae*), presence of eyes and smaller summed *pt* of external primary branches of claws I to IV([*23.4-33.8*] in *M. argentinum* sp. nov. vs. [*40.0-43.8*] in *M. katarzynae*).
13. *M. krzysztofi*, known from Colombia, Costa Rica and Peru (Kaczmarek and Michalczyk 2007; Kaczmarek et al. 2014a,b; Lisi et al. 2014; Melo et al. (2014)) by different claw configuration ([3-3]-[3-3] in *M. argentinum* sp. nov. vs. [2-3]-[3-2] in *M. krzysztofi*), different dorsal sculpture (pseudopores not arranged in bands, sparsely distributed and not forming a reticular design in *M. argentinum* sp. nov. vs. pseudopores densely distributed and forming a reticular design in *M. krzysztofi*), presence of eyes, stylet supports inserted in more caudal position ([*70.0-73.7*] in *M. argentinum* sp. nov. vs. [*63.3-67.3*] in *M. krzysztofi*), slightly lower *pt* of buccal tube standard width ([*24.2-32.3*] in *M. argentinum* sp. nov. vs. [*33.1-38.4*] in *M. krzysztofi*) and slightly smaller claws I to IV (see Table 1 and Table one in Kaczmarek and Michalczyk (2007) for the exact differences in dimensions of claws).
14. *M. longiungue*, known only from India (Tumanov 2006) by having sculptured dorsal cuticle, presence of accessory points on main branches of claws, stylet supports inserted in more caudal position ([*70.0-73.7*] in *M. argentinum* sp. nov. vs. [*59.1-66.7*] in *M. longiungue*), lower *pt* of standard buccal tube width ([*24.2-32.3*] in *M. argentinum* sp. nov. vs. [*33.8-59.1*] in *M. longiungue*) and slightly smaller claws I to IV (see Table 1 and Table six in Tumanov (2006) for the exact differences in dimensions of claws).
15. *M. reticulatum*, known only from Seychelles (Pilato et al. 2002) by different claw configuration ([3-3]-[3-3] in *M. argentinum* sp. nov. vs. [2-3]-[3-2] in *M. reticulatum*), different dorsal

sculpture (pseudopores not arranged in bands, sparsely distributed and not forming a reticular design in *M. argentinum* sp. nov. vs. pseudopores arranged in nine sculptured bands, forming a reticular design in *M. reticulatum*), absence of cuticular gibbosities, having six peribuccal lamellae (four in *M. reticulatum*), stylet supports inserted in more caudal position ([*70.0-73.7*] in *M. argentinum* sp. nov. vs. [*68.5-69.8*] in *M. reticulatum*) and slightly smaller claws I to IV (see Table 1 and Table two in Pilato et al. (2002) for the exact differences in dimensions of claws).

16. *M. zsalakoae*, known only from Arizona and New Mexico, U.S.A. (Meyer and Hinton 2010) by having sculptured dorsal cuticle, presence of accessory points on main branches of claws, lower *pt* of buccal tube standard width ([*24.2-32.3*] in *M. argentinum* sp. nov. vs. [*36.8-41.9*] in *M. zsalakoae*) and smaller summed *pt* of internal claws and spurs I to IV (internal primary branch [*21.1-36.4*], internal base + secondary branch [*20.3-28.3*], internal spur [*6.6-9.9*] in *M. argentinum* sp. nov. vs. internal primary branch [*64.4-102.9*], internal base + secondary branch [*32.9-56.8*], internal spur [*11.1-23.2*] in *M. zsalakoae*).

Milnesium beatae sp. nov.

http://www.tardigrada.net/register/0016.htm.

Material examined: Seven females, all from Nahuel Huapi National Park, Río Negro, Argentina, moss sample from rocks, coll. Łukasz Kaczmarek.

Type material: Holotype and six paratypes in the sample III.

Diagnosis

Adults (Figure 1b) (measurements in micrometres, pt ratios and statistics in Table 2): The body rose before fixation and transparent afterwards, eyes present. Cuticle sculptured with pseudopores (0.3 to 0.6) not arranged in bands, sparsely distributed and not forming reticular design (Figure 2d). Six peribuccal papillae and six peribuccal lamellae around the mouth opening present. Two cephalic papillae positioned laterally. Peribuccal and cephalic papillae similar in length.

Buccal apparatus of the *Milnesium* type (Figure 2c). Buccal tube wide and short (standard width on average 62% of its length) and funnel-shaped, wider anteriorly (posterior diameter on average 72% of the anterior diameter). Pharyngeal bulb elongated, pear-shaped and without placoids or septulum.

Claws of the *Milnesium* type, slender (Figure 3d,e). Primary branches on all legs with small, but distinct accessory points detaching from the branch at its greatest curvature (Figure 3d,e). Secondary branches with rounded basal thickenings (Figure 3d,e). All secondary branches on all legs with three points (claw configuration: [3-3]-[3-3]). Single, long transverse, cuticular bars under claws I-III present (Figure 3d).

Eggs: Oval, smooth and deposited in exuvium as in all other known *Milnesium* species.

Males not observed.

Locus typicus: Argentina, 41°12′S, 71°50′W, ca. 1,000 m asl., Río Negro, Nahuel Huapi National Park.

Etymology: This species is named after first author's secondary school Biology teacher - Mrs. Beata Ostasiewicz.

Type depositories: The holotype and all paratypes are deposited in the Department of Animal Taxonomy and Ecology, Adam Mickiewicz University in Poznań, Umultowska 89, Poznań, Poland.

Differential diagnosis

By having a sculptured dorsal cuticle, *M. beatae* sp. nov. belongs to the *granulatum* group within the genus (Michalczyk et al. 2012a,b). This new species with three points on the secondary branches of all claws [3-3]-[3-3] and very short and wide buccal tube is most similar to *M. bohleberi*, *M. brachyungue* Binda and Pilato 1990 and *M. eurystomum*, but it differs from the following:

1. *M. bohleberi* by having sculptured dorsal cuticle and lower *pt* of peribuccal papillae length ([*18.0-23.6*] in *M. beatae* sp. nov. vs. [*27.2-32.3*] in *M. bohleberi*).
2. *M. brachyungue* by having sculptured dorsal cuticle, stylet supports inserted in more cephalic position ([*63.8-66.7*] in *M. beatae* sp. nov. vs. [*69.4*] in *M. brachyungue* (in specimen 729.0 in length)) and lower *pt* of claws I to IV (see Table 2 and Table one in Binda and Pilato (1990) for the exact differences in dimensions of claws).
3. *M. eurystomum* by having sculptured dorsal cuticle, stylet supports inserted in more caudal position ([*63.8-66.7*] in *M. beatae* sp. nov. vs. [*60.0-60.3*] in *M. eurystomum*) and slightly different *pt* of claws I to IV (see Table 2 and Table two in Michalczyk et al. (2012a) for the exact differences in dimensions of claws).

Besides the abovementioned the most similar species, *M. beatae* sp. nov. is similar to other species of the genus *Milnesium* with the claw configuration [3-3]-[3-3] (*M. alabamae*, *M. antarcticum*, *M. asiaticum*, *M. barbadosense*, *M. eurystomum*, *M. granulatum*, *M. longiungue* and *M. zsalakoae*) or with a sculptured cuticle (*M. beasleyi*, *M. katarzynae*, *M. krzysztofi* and *M. reticulatum*).

This new species differs from the following:

4. *M. alabamae* by different dorsal sculpture (pseudopores not arranged in bands, sparsely distributed and not forming a reticular design in *M. beatae* sp. nov. vs. pseudopores arranged in bands (especially in caudal region), densely distributed and forming a reticular design in *M. alabamae*), presence of accessory points on main branches of claws, presence of eyes and lower *pt* of standard width of buccal tube ([*58.1-65.6*] in *M. beatae* sp. nov. vs. [*29.5-44.0*] in *M. alabamae*).
5. *M. antarcticum* by having sculptured dorsal cuticle, higher *pt* of standard width of buccal tube ([*58.1-65.6*] in *M. beatae* sp. nov. vs. [*35.4-43.9*] in *M. antarcticum*) and stylet supports inserted in more cephalic position ([*63.8-66.7*] in *M. beatae* sp. nov. vs. [*70.0-73.7*] in *M. antarcticum*).
6. *M. argentinum* sp. nov. by stylet supports inserted in more cephalic position ([*63.8-66.7*] in *M. beatae* sp. nov. vs. [*70.0-73.7*] in *M. argentinum* sp. nov.), lower *pt* of width of buccal tube ([*70.3-78.9*], [*58.1-65.6*], [*49.4-56.5*] anterior, standard and posterior, respectively, in *M. beatae* sp. nov. vs. [*30.3-39.8*], [*24.2-32.3*], [*23.1-33.6*] anterior, standard and posterior, respectively, in *M. argentinum* sp. nov. and longer claws I to IV (see Tables 1 and 2 for the exact differences in dimensions of claws).
7. *M. asiaticum* by having sculptured dorsal cuticle, lower *pt* of standard width of buccal tube ([*58.1-65.6*] in *M. beatae* sp. nov. vs. [*30.0-41.6*] in *M. asiaticum*) and slightly different *pt* of claws I to IV (see Table 2 and Table four in Tumanov (2006) for the exact differences in dimensions of claws).
8. *M. barbadosense* by having sculptured dorsal cuticle, presence of eyes and lower *pt* of standard width of buccal tube ([*58.1-65.6*] in *M. beatae* sp. nov. vs. [*27.2-49.7*] in *M. barbadosense*).
9. *M. beasleyi* by different claw configuration ([3-3]-[3-3] in *M. beatae* sp. nov. vs. [2-3]-[3-2] in *M. beasleyi*), presence of rounded basal thickenings under secondary branches of claws, higher *pt* of width of buccal tube ([*70.3-78.9*], [*58.1-65.6*], [*49.4-56.5*] anterior, standard and posterior, respectively, in *M. beatae* sp. nov. vs. [*35.3-41.8*], [*31.2-39.8*], [*29.6-33.2*] anterior, standard and posterior, respectively, in *M. beaslei*,]) and slightly smaller claws I to IV (see Table 2 and Table one in Kaczmarek et al. (2012a, b) for the exact differences in dimensions of claws).
10. *M. berladnicorum* by different claw configuration ([3-3]-[3-3] in *M. beatae* sp nov vs. [2-3]-[2-2] in *M. berladnicorum*), higher *pt* of width of buccal tube ([*70.3-78.9*], [*58.1-65.6*], [*49.4-56.5*] anterior, standard and posterior, respectively, in *M. beatae* sp. nov. vs. [*35.0-47.1*], [*30.6-38.9*], [*27.7-36.0*] anterior, standard and posterior, respectively, in *M. berladnicorum*]) and smaller claws I to IV (see Table 2 and Table one in Ciobanu et al. (2014b) for exact differences in dimensions of claws).
11. *M. granulatum* by different dorsal sculpture (pseudopores not arranged in bands, sparsely distributed and not forming a reticular design in *M. beatae* sp. nov. vs. pseudopores densely distributed and forming a reticular design in *M. granulatum*), higher *pt* of width of buccal tube ([*70.3-78.9*], [*58.1-65.6*], [*49.4-56.5*] anterior, standard and posterior, respectively, in *M. beatae* sp. nov. vs. [*46,3*], [*39.0*], [*40.1*] anterior, standard and posterior, respectively, in *M. granulatum* (in specimen 338.3 in length)) and slightly larger claws I to IV (see Table 2 and Table two in Kaczmarek et al. (2012a, b) for the exact differences in dimensions of claws).
12. *M. katarzynae* by a different claw configuration ([3-3]-[3-3] in *M. beatae* sp. nov. vs. [2-2]-[2-2] in *M. katarzynae*), different dorsal sculpture (pseudopores not arranged in bands, sparsely distributed and not forming a reticular design in *M. beatae* sp. nov. vs. pseudopores densely distributed and forming a reticular design in *M. katarzynae*), presence of eyes, higher *pt* of standard width of buccal tube ([*58.1-65.6*] in *M. beatae* sp. nov. vs. [*21.7-26.6*] in *M. katarzynae* and stylet supports inserted in more cephalic position ([*63.8-66.7*] in *M. beatae* sp. nov. vs. [*73.3-78.3*] in *M. katarzynae*).
13. *M. krzysztofi* by different claw configuration ([3-3]-[3-3] in *M. beatae* sp. nov. vs. [2-3]-[3-2] in *M. krzysztofi*), different dorsal sculpture (pseudopores not arranged in bands, sparsely distributed and not forming a reticular design in *M. beatae* sp. nov. vs. pseudopores densely distributed and forming a reticular design in *M. krzysztofi*), presence of eyes, higher *pt* of buccal tube standard width ([*58.1-65.6*] in *M. beatae* sp. nov. vs. [*33.1-38.4*] in *M. krzysztofi*) and smaller claws I to IV (see Table 2 and Table one in Kaczmarek and Michalczyk (2007) for the exact differences in dimensions of claws).
14. *M. longiungue* by having sculptured dorsal cuticle, presence of accessory points on the main branches of claws, slightly higher *pt* of standard width of buccal tube ([*58.1-65.6*] in *M. beatae* sp. nov. vs. [*33.8-59.1*] in *M. longiungue*) and smaller claws I to IV (see Table 2 and Table six in Tumanov (2006) for the exact differences in dimensions of claws).
15. *M. reticulatum* by different claw configuration ([3-3]-[3-3] in *M. beatae* sp. nov. vs. [2-3]-[3-2] in *M. reticulatum*), different dorsal sculpture

(pseudopores not arranged in bands, sparsely distributed and not forming a reticular design in *M. beatae* sp. nov. vs. pseudopores arranged in nine sculptured bands, forming a reticular design in *M. reticulatum*), absence of cuticular gibbosities, having six peribuccal lamellae (four in *M. reticulatum*), stylet supports inserted in more cephalic position ([*63.8-66.7*] in *M. beatae* sp. nov. vs. [*68.5-69.8*] in *M. reticulatum*), higher *pt* of standard width of buccal tube ([*58.1-65.6*] in *M. beatae* sp. nov. vs. [*30.4-37.4*] in *M. reticulatum*) and different measurements of claws I to IV (see Table 2 and Table two in Pilato et al. (2002) for the exact differences in dimensions of claws).

16. *M. zsalakoae* by having sculptured dorsal cuticle, presence of accessory points on main branches of claws, higher *pt* of buccal tube standard width ([*58.1-65.6*] in *M. beatae* sp. nov. vs. [*36.8-41.9*] in *M. zsalakoae*), stylet supports inserted in more cephalic position ([*63.8-66.7*] in *M. beatae* sp. nov. vs. [*68.2-71.1*] in *M. zsalakoae*) and higher *pt* of internal claws I to IV (see Table 2 and the description of *M. zsalakoae* in Meyer and Hinton (2010) for the exact differences in dimensions of claws).

Milnesium brachyungue Binda and Pilato, 1990

Terra typica: Chile (South America).

Localities and number of specimens in present studies: Five females in sample VI.

Remarks: Until now this species was considered as an endemic of Chile (Binda and Pilato 1990) (Table 3). This is a second report of the species and a new record for Argentina. *Milnesium brachyungue* is relatively similar to *M. tardigradum* and it is possible that the specimens of *M. tardigradum* sensu stricto reported from South America (especially from southern regions of Chile or Argentina) in the past belong rather to *M. brachyungue*, but this hypothesis needs to be confirmed. We strongly suggest that all specimens reported as *M. tardigradum* s. s. from this region should be re-examined and determined based on the modern taxonomic characters (see also Michalczyk et al. 2012a,b).

Milnesium eurystomum Maucci, 1991

Terra typica: Denmark (Arctic, Greenland).

Localities and number of specimens in present studies: None.

Remarks: Species reported from Argentina, Chile, Mongolia (probably), Spain, Svalbard, USA and *locus typicus* on Greenland (Maucci 1991, 1996; Kaczmarek and Michalczyk 2006; Guil 2008; Kaczmarek et al. 2012a, b; Land et al. 2012; Johansson et al. 2013) (Table 3). This species is characterized by a very wide and relatively short buccal tube, claw configuration [3-3]-[3-3] and smooth cuticle. Recently, a similar species (with a short and wide buccal tube), *M. bohleberi*, was described from North America (USA, North Carolina and Tennessee) (Bartels et al. 2014). In this situation, we suggest that the examples, especially from South and North America, should be re-examined (see also Michalczyk et al. 2012a,b). In Mongolia this species was reported as *M.* cf. *eurystomum* Maucci 1991 (Kaczmarek and Michalczyk 2006) and its presence in this region needs confirmation.

Minesium granulatum Ramazzotti, 1962

Terra typica: Chile (South America).

Localities and number of specimens in present studies: Two females in sample II.

Remarks: Species reported from Argentina, Chile, Colombia, Italy, Romania and USA (Ramazzotti 1962; Michalczyk et al. 2012a,b; Ciobanu et al. 2014a; Melo et al. 2014) (Table 3). As mentioned by Bartels et al. (2014), such wide and discontinuous geographic distribution can suggest a complex of cryptic species, but testing this hypothesis requires molecular data which are not yet available. We suggest that at least the examples

Table 3 Distribution of the species of genus _Milnesium_ in South America

Species	Country									
	Argentina	Bolivia	Brazil	Chile	Colombia	Ecuador	Paraguay	Peru	Uruguay	Venezuela
M. argentinum	X									
M. beatae	X									
M. brachyungue	X			X						
M. eurystomum	X			X						
M. granulatum	X			X	X					
M. katarzynae					X					
M. krzysztofi					X			X		
M. t. tardigradum	X	X	X	X	X	X	X	X	X	X
M. t. trispinosa			X	X						

from Europe should be carefully re-examined (see also Michalczyk et al. 2012a,b).

Milnesium katarzynae Kaczmarek et al., 2004

Terra typica: China (Asia).

Localities and number of specimens in present studies: None.

Remarks: Until now, this species was known only from China (*locus typicus*) (Kaczmarek et al. 2004), Costa Rica (Kaczmarek et al. 2014a), Colombia (Melo et al. 2014) and Taiwan (Yin and Li 2011) (Table 3). Such distribution may suggest a pantropical distribution of this species.

Milnesium krzysztofi Kaczmarek and Michalczyk 2007

Terra typica: Costa Rica (Central America).

Localities and number of specimens in present studies: None.

Remarks: Until now, this species was known only from Costa Rica (*locus typicus*) (Kaczmarek and Michalczyk 2007), Colombia (Lisi et al. 2014; Melo et al. 2014) and Peru (Kaczmarek et al. 2014a,b) (Table 3). Such distribution may suggest a neotropical distribution of this species.

Milnesium tardigradum tardigradum Doyère 1840

Terra typica: Chile (South America).

Localities and number of specimens in present studies: None.

Remarks: This species has been reported from many localities in South America (McInnes 1994, see also the 'Discussion' section and Table 3). For over a century, *Milnesium* was considered to be a monotypic, highly cosmopolitan genus. However, currently there are more than 20 species within the genus and it is very likely that the zoogeographic range of *M. tardigradum* s. s. is limited to the Palaearctic (Michalczyk et al. 2012a,b). This means that all South American reports for *M. tardigradum* that occurred prior to the re-description of *M. tardigradum* s. s. (Michalczyk et al. 2012a,b) should be considered dubious and need a re-examination based on the modern taxonomy of the genus *Milnesium*. In other words, currently, there is no evidence that *M. tardigradum* s. s. inhabits South America; thus, its records may represent a number of different *Milnesium* taxa (Michalczyk et al. 2012a,b, see also remarks on other *Milnesium* species, especially on *M. brachyungue*).

Milnesium tardigradum trispinosa Rahm 1932 [T]

Terra typica: Chile (South America).

Localities and number of specimens in present studies: None.

Remarks: A subspecies reported from single localities in Chile and Brazil (Table 3). This unique subspecies is characterized by the presence of short spines on dorsal cuticle. It should be probably promoted to the species level, but the re-description based on type material (or based on specimens collected in the type locality) is necessary (Michalczyk et al. 2012a,b).

Discussion

Distribution

Milnesium species have been recorded from many localities throughout the world (Michalczyk et al. 2012a,b) and one of them, *M. swolenskyi* Bertolani and Grimaldi 2000, is also known from Cretaceous amber collected in the USA (Bertolani and Grimaldi 2000). For a long time, *M. tardigradum* Doyère 1840 was the only recognised species in the genus and it has been considered as an extremely cosmopolitan taxon (e.g. Ramazzotti and Maucci 1983). In the past, even morphologically very different specimens were still considered as only different forms or subspecies of the type *M. tardigradum* (e.g. Rahm 1932; Ramazzotti 1962; Sudzuki 1964; Horning et al. 1978; Nelson and Horning 1979; Dastych 1984). Interestingly, even that Nederström 1919 described a new *Milnesium* species, *Milnesium quadrifidum*, this new taxon was later recognised only as a specific form of *M. tardigradum* (Szymańska 1994 according to Marcus 1936) and finally excluded from the list of tardigrade taxa at all. It was not until 1990 that Binda and Pilato described a new *Milnesium* species, *M. brachyungue*, from Tierra del Fuego. Later, new species were described only occasionally mainly due to the lack of a clear diagnosis of the nominal species *M. tardigradum* s. s. (Maucci 1991; Bertolani and Grimaldi 2000; Pilato et al. 2002; Yang 2003; Kaczmarek et al. 2004). In 2006, Tumanov published the first partial revision of the genus *Milnesium* and described five new species. Since this time, many new species have been described and now 25 taxa (including the two new species described in this paper) are known in the genus (Degma et al. 2009-2014). The majority of the newly described species are known only from their type localities and Michalczyk et al. (2012a,b) hypothesised that *M. tardigradum* s. s. is not a cosmopolitan species, but previous reports from various localities should be re-examined in light of new taxonomic characters because they probably belong to different species (see also Meyer 2013). Based on the present data, most of the species from the genus *Milnesium* have truly restricted geographic ranges, as was also shown in other tardigrade genera (e.g. Pilato and Binda 2001).

As of now, nine *Milnesium* taxa are known from South America (including two new species reported in this paper) (Table 3). These are *M. argentinum* sp. nov. and *M. beatae* sp. nov. (currently restricted to Argentina), *M. brachyungue* (known from Argentina and Chile), *M. eurystomum* (known from Argentina, Chile, Greenland, Mongolia (probably), Spain, Svalbard and USA), *M.*

granulatum (known from Argentina, Chile, Colombia, Italy, Romania and USA), *M. katarzynae* (known from China, Colombia, Costa Rica and Taiwan), *M. krzysztofi* (known from Colombia, Costa Rica and Peru), *M. tardigradum* sensu lato (known from Argentina, Bolivia, Brazil, Chile, Colombia, Ecuador, Paraguay, Peru, Uruguay and Venezuela) and *M. t. trispinosa* (known from Chile and Brazil) (Richters (1911a); Murray and Wailes (1913); Heinis (1914); Rahm (1931, 1932); de Barros (1943); (du Bois-Reymond 1944); Ramazzotti (1957, 1962, 1964); Iharos (1963); Mihelčič (1967, 1972) Rossi and Claps (1980, 1989, 1991); Claps and Rossi (1981, 1984, 1988, 1997); Maucci (1988); Binda and Pilato (1990), Maucci (1991, 1996); Séméria (1993); McInnes (1994); Jerez Jaimes and Narváez Parra (2001); Nickel et al. (2001); Pilato et al.(2003); Kaczmarek et al. (2004, 2012a,b); Kaczmarek and Michalczyk (2006, 2007); Moly de Peluffo et al.(2006); Peluffo et al.(2007); Guil (2008); Rossi et al.(2009); Kaczmarek et al. (2012a,b), Land *et al.* (2012), Michalczyk *et al.* (2012a,b), Johansson *et al.* (2013); Ciobanu et al. (2014a); Lisi et al. (2014); Melo et al. (2014)).

As shown above, only *M. tardigradum* s. l. has a wide distribution in South America, however the presence of this species in this region is doubtful and needs confirmation. The present study seems to confirm this hypothesis, because although four *Milnesium* species were found in the samples studied, none of them was recognised as the nominal *M. tardigradum*. This also confirms the hypothesis proposed by Michalczyk et al. (2012a,b) that *M. tardigradum* have rather a narrower geographic range (restricted to Holarctic or Palearctic), although more extensive studies (molecular) are necessary in other regions of South America.

Similar to *M. tardigradum* s.s., other species of the genus *Milnesium* also have restricted geographic ranges (e.g. *M. alabamae, M. almatyense* Tumanov 2006, *M. antarcticum, M. barbadosense, M. beasleyi, M. jacobi* Meyer and Hinton 2010, *M. brachyungue, M. dujiangensis* Yang 2003, *M. lagniappe* Meyer et al. 2013, *M. longiungue, M. reductum* Tumanov 2006, *M. reticulatum, M. tetralamellatum* Pilato and Binda 1991 and *M. zsalakoae*), which seems to support the hypothesis that none of the *Milnesium* species is cosmopolitan (Bartels et al. 2014). However, as was also shown by Bartels et al. (2014), some of the *Milnesium* species have rather wide and discontinuous geographic ranges (*M. asiaticum, M. eurystomum, M. granulatum, M. katarzynae* and *M. krzysztofi*), which can suggest the presence of cryptic species complexes or a specific distribution of some of the taxa (e.g. pantropical for *M. katarzynae* or neotropical for *M. krzysztofi*). In this situation, only detailed morphological and molecular studies could shed light on this issue.

Remarks on feeding behaviour

Very little is known about food preferences in tardigrades, although, many different types of food sources have been reported, i.e. plant cell fluids, algae, bacteria, protozoa and small invertebrates like nematodes, rotifers and other tardigrades (for the review, see Schill et al. 2011). Previous studies have focused on the following: a) the rate of consumption, e.g. in the carnivorous *Paramacrobiotus richtersi* (Murray, 1911), b) predator-prey interactions (*P. richtersi* vs. nematodes), c) population dynamics and food availability in the herbivorous tardigrade *Apodibius confusus* Dastych, 1983, d) tardigrade food preferences in the herbivorous *Echiniscus granulatus* (Doyère, 1840), *Macrobiotus sapiens* Binda and Pilato, 1984, *M. persimilis* Binda and Pilato, 1972, *Richtersius coronifer* (Richters, 1903) and the carnivorous *P. richtersi*, e) life histories of some species and f) different aspects of tardigrade histology (e.g. Suzuki 2003; Hohberg and Traunspurger 2005, 2009; Poprawa 2005, 2011; Hohberg 2006; Horikawa et al. 2008; Hohberg et al. 2011; Lemloh et al. 2011; Schill et al. 2011; Schill 2013), give us some incomplete information on tardigrade feeding behaviour.

The species in the genus *Milnesium* are relatively large (occasionally more than 2 mm but most often between 0.5 to 1.0 mm) and inhabit mainly limno-terrestrial habitats (Guil 2008). Their buccal tube is wide, relatively short, connected with a large pharynx without placoids and associated with large buccal lamellae on a wide mouth ring, which is in agreement with the definition of the 'carnivore type' of buccal apparatus proposed by Guidetti et al. (2012). All *Milnesium* species are considered carnivorous and feed on other small invertebrates like rotifers, nematodes, tardigrades and sometimes also on amoebas (e.g. Kinchin 1990; Suzuki 2003; Nelson et al. 2010; Miller and Williams 2012).

In contrast to other predatory tardigrades from genera such as *Bertolanius* Özdikmen, 2008, *Macrobiotus* C.A.S. Schultze, 1834 or *Paramacrobiotus* Guidetti et al., 2009, *Milnesium* species suck the entire prey into the gut rather than only single cells or body fluids (McInnes et al. 2001). However, also species from the genus *Bertolanius* (i.e. *Bertolanius weglarskae* (Dastych 1972)) can suck in entire prey (see Figure 1a in Guidetti and Bertolani 2001). McInnes et al. (2001) analysed the diet of sub-Antarctic specimens of *M.* cf. *tardigradum*. They found in the guts of *M.* cf. *tardigradum* not only remnants but also entire specimens of three different species of Rotifera (Bdelloidea: *Adineta* sp., *Philodina* spp. and Monogononta: *Dicranophorus* sp.), unidentified species of Nematoda and two species of Tardigrada (*Diphascon* sp. and *Calohypsibius* cf. *ornatus*). Miller and Williams (2012) reported a Greenland tardigrade *M. eurystomum* feeding on shelled (testate amoebae) (similar to *Euglypha*).

Additionally, the same authors suggested that *M. eurystomum* has the ability to properly orient their prey, so that the amoeba can be extracted by the sucking action and later the empty shell can be expelled. However, specific differences in the feeding behaviour in connection with the structure of buccal apparatus in different *Milnesium* species have never been discussed.

In the present research, a total of 25 adult specimens of two new *Milnesium* species: *M. argentinum* (18 specimens) and *M. beatae* (7 specimens) were studied. Both species were found in the same region of Nahuel Huapi National Park (Río Negro Province, Argentina) in a few moss samples. Five specimens (two specimens of *M. argentinum* sp. nov. and three specimens of *M. beatae* sp. nov.) were found with the remnants of food (rotifer mastaxes and tardigrade buccal apparatuses and claws) in their guts. Both species were found in the moss samples (sometimes also together in the same piece of moss) that also contained rotifers, nematodes and other tardigrades (*Diphascon chilenense* Plate, 1888, *Echiniscus bigranulatus* Richters 1907, *Hebesuncus mollispinus* Pilato et al. 2012, *Macrobiotus* cf. *andinus*, *M. harmsworthi* group, *M. kazmierskii* Kaczmarek and Michalczyk 2009, *M. patagonicus* Maucci 1988, *M. szeptyckii* Kaczmarek and Michalczyk 2009, *Mopsechiniscus granulosus* Mihelčič 1967). Both new *Milnesium* species were able to choose from the same type of prey, because the prey were similar in every sample.

The main morphological differences between *M. argentinum* sp. nov. and *M. beatae* sp. nov. are visible in the structure of buccal tube. *Milnesium argentinum* sp. nov. has a relatively long and narrow buccal tube while *M. beatae* sp. nov. has a relatively short and wide buccal tube (in the specimens of similar lengths, see Tables 1 and 2). Based on gut contents, both species seem also to have a different diet: the two specimens of *M. argentinum* sp. nov. fed on undefined species of rotifers (Figure 3c), while the three specimens of *M. beatae* sp. nov. fed on other tardigrades (exclusively *M. szeptyckii*) (Figure 3f). In general, specimens of *M. szeptyckii* are larger (360 to 650 μm) and more bulky (a wider buccal tube is probably necessary to swallow them) than typical rotifers which are shorter (100 to 500 μm) and more slender (a narrower buccal tube is sufficient to swallow them). Although the numbers are small, these data suggest that the buccal tube width may affect the type of prey consumed. However, it should be also noted that McInnes et al. (2001) showed that both tardigrades and rotifers were found in the gut of the same specimen of *M.* cf. *tardigradum*. But, according the photo (Figure 1a in McInnes et al. 2001), the *Milnesium* species studied by McInnes et al. (2001) had rather long and narrow buccal tube (similarly to *M. argentinum* sp. nov.) and fed on rather small and slender prey (rotifers, nematodes and small and slender tardigrades from the genera *Diphascon* Plate, 1888 and *Calohypsibius* Thulin 1928), which is in agreement with the observations made in the present research.

Although neither McInnes et al. (2001) observations nor the present research provided a definitive answer whether the buccal tube width has a crucial role in the choice of different types of prey, these studies are the basis for further, more comprehensive studies.

Such studies are necessary also to understand the evolution of buccal apparatuses in other genera of the family Milnesiidae (*Bergtrollus* Dastych 2011, *Limmenius* Horning, Schuster and Grigarick, 1978 and *Milnesioides* Claxton 1999). However, we know almost nothing about feeding behaviour and prey choice of the members of these monophyletic genera. Only Claxton (1999) mentioned some specimens of *M. exsertum* that had remnants of rotifers in their guts. This is a very interesting observation, because this species has a very long and narrow buccal tube, similarly to the 'rotifer-feeding' *M. argentinum* sp. nov.

Based on the available data, it can be initially hypothesised that the species from the family Milnesiidae with long and narrow buccal tubes (e.g. *M. argentinum* sp. nov., Antarctic *M.* cf. *tardigradum* or *M. exsertum*) choose smaller and more slender prey (rotifers or small slender tardigrades from genera *Diphascon* or *Calohypsibius*) in contrast to the species with relatively short and wide buccal tubes (e.g. *M. beatae* sp. nov. or *M. eurystomum*) which choose more robust and larger or armoured prey (large and robust tardigrades from the genus *Macrobiotus* or testate amoebae). Certainly, such conclusions based only on the 'post-mortem' studied specimens give us only very limited knowledge on the food that had been eaten just before death (preservation) of the animal. Such information needs confirmation in further, more detailed experimental studies with molecular methods (e.g. Schill et al. 2011) or based directly on the observations in cultured animals (e.g. Suzuki 2003).

Conclusions

Milnesium argentinum and *M. beatae* are new taxa for science. As of now, nine *Milnesium* taxa are known from South America (including the two new species and two newly recorded taxa for Argentina, reported in this paper). The presence of *M. tardigradum* s. s. in South America needs confirmation and for now it should be considered as dubious. It is probable that width of buccal tube may limit a prey size and play an important role in the feeding behaviour in the family Milnesiidae. Studies on the feeding behaviour in tardigrades are in initial phase and definitive conclusions are not possible at this moment.

Competing interests

The authors declare that they have no competing interests.

Authors' contributions

MR examined and analysed the material, identified and described the species, made the measurements, figures and tables and drafted the manuscript. MO made the measurements and tables. ŁK collected part of the material, invented the research conception, analysed the examined material, identified and described the species, corrected tables and drafted the manuscript. All authors read and approved the final manuscript.

Acknowledgements

The authors want to thank Prof. Diane Nelson of East Tennessee State University for very valuable comments to the manuscript and help in improving of the English. This work was partially supported by the Polish Ministry of Science and Higher Education via the 'Iuventus Plus' programme (grant: IP2010 015570, Invertebrate biodiversity in temporary ponds of Costa Rica - verification of the 'Great American Biotic Interchange' hypothesis) and by the Prometeo Project of the Secretariat for Higher Education, Science, Technology and Innovation of the Republic of Ecuador. Studies have been conducted in the framework of activities of BARg (Biodiversity and Astrobiology Research group).

References

Bartels PJ, Nelson DR, Kaczmarek Ł, Michalczyk Ł (2014) The genus *Milnesium* (Tardigrada: Eutardigrada: Milnesiidae) in the Great Smoky Mountains National Park (North Carolina and Tennessee, USA), with the description of *Milnesium bohleberi* sp. nov. Zootaxa 3826(2):356–368

Bertolani R, Grimaldi D (2000) A new eutardigrade (Tardigrada: Milnesiidae) in amber from the Upper Cretaceous (Turonian) of New Jersey. In: Grimaldi D (ed) Studies on fossils in amber, with particular reference to the Cretaceous of New Jersey. Backhuys Publishers, Leiden The Netherlands, pp 103–110

Binda MG, Pilato G (1972) Tardigradi muscicoli di Sicilia (IV Nota). Boll Accad Gioenia Sci Nat Catania 11:47–60

Binda MG, Pilato G (1984) *Macrobiotus sapiens*, nuova specie di eutardigrado di Sicilia. Animalia 11:85–90

Binda MG, Pilato G (1990) Tardigradi di Terra del Fuoco e Magallanes, I. *Milnesium brachyungue*, nuova specie di tardigrado Milnesiidae. Animalia 17:105–110

Binda MG, Pilato G (1994a) *Echiniscus peruvianus*, nuova specie di eterotardigrado del Sudamerica. Animalia 21:57–62

Binda MG, Pilato G (1994b) *Dactylobiotus caldarellai*, nuova specie di eutardigrado della Terra del Fuoco. Animalia 21:87–91

Binda MG, Pilato G (1999a) *Dactylobiotus lombardoi* sp. n. (Eutardigrada: Macrobiotidae) from Tierra del Fuego, with a Key to the *Dactylobiotus*-species. Zool Anz 238:147–155

Binda MG, Pilato G (1999b) *Macrobiotus erminiae*, new species of eutardigrade from southern Patagonia and Tierra del Fuego. Entomol Mit Zool Mus Hamb 13:151–158

Ciobanu DA, Moglan I, Zawierucha K, Kaczmarek Ł (2014a) New records of terrestrial tardigrades (Tardigrada) from Ceahlău National Park with zoogeographical and taxonomical remarks on Romanian water bears. North-West J Zool 10:art.140301

Ciobanu DA, Zawierucha K, Moglan I, Kaczmarek Ł (2014b) *Milnesium berladnicorum* sp. n. (Eutardigrada, Apochela, Milnesiidae), a new species of water bear from Romania. ZooKeys 429:1–11

Claps MC, Rossi GC (1981) Contribucion al conocimiento de los tardigrados de Argentina. II Rev Soc Entomol Arg 40(1–4):107–114

Claps MC, Rossi GC (1984) Contribucion al conocimiento de los tardigrados de Argentina. IV. Acta Zool Lilloana 38:45–50

Claps MC, Rossi GC (1988) Contribucion al conocimiento de los tardigrados de Argentina. VI Iheringia 67:3–11

Claps MC, Rossi GC (1997) Tardigrados de Uruguay, com descripcion de dos nuevas especies (Echiniscidae, Macrobiotidae). Iheringia Sér Zool 83:17–22

Claxton SK (1998) A revision of the genus *Minibiotus* (Tardigrada: Macrobiotidae) with descriptions of eleven new species from Australia. Rec Aust Mus 50:125–160

Claxton SK (1999) *Milnesioides exsertum* gen. n. sp. n., a new tardigrade from Australia (Tardigrada: Milnesiidae). Zool Anz 238:183–190

Dastych H (1980) Niesporczaki (Tardigrada) Tatrzańskiego Parku Narodowego. Monogr Faun Pol PWN Krakow 9:1–232

Dastych H (1984) The Tardigrada from Antartica with description of several new species. Acta Zool Cracov 27:377–436

Dastych H (1999a) Redescription of the Sub-Antarctic tardigrade *Mopsechiniscus imberbis* (Richters, 1908) (Tardigrada). Mitt Hamb Zool Mus Inst 96:21–35

Dastych H (1999b) A new species of the genus *Mopsechiniscus* Du Bois-Reymond Marcus, 1944 (Tardigrada) from the Venezuelan Andes. A Biol Benrod 10:91–101

Dastych H (2000) Redescription of the Neotropical tardigrade *Mopsechiniscus granulatus* Mihelcic, 1967 (Tardigrada). Mitt Hamb Zool Mus Inst 97:45–57

Dastych H, McInnes S, Claxton SK (1998) *Oreella mollis* Murray, 1910 (Tardigrada): a redescription and revision of *Oreella*. Mitt Hamb Zool Mus Inst 95:89–113

De Barros R (1943) Tardigrados de Estado de Sao Paulo, Brasil. III. Gêneros *Hypsibius*, *Itaquascon* e *Milnesium*. Rev Brasil Biol 3:1–10

Degma P, Guidetti R (2007) Notes to the current checklist of Tardigrada. Zootaxa 1579:41–53

Degma P, Michalczyk Ł, Kaczmarek Ł (2008) *Macrobiotus derkai*, a new species of Tardigrada (Eutardigrada, Macrobiotidae, *huziori* group) from the Colombian Andes (South America). Zootaxa 1731:1–23

Degma P, Bertolani R, Guidetti R (2014) Actual checklist of Tardigrada species (2009–2014, Ver. 26:10-07-2014). http://www.tardigrada.modena.unimo.it/miscellanea/Actual%20checklist%20of%20Tardigrada.pdf

Doyère M (1840) Memoire sur les tardigrades. Ann Sci Nat Zool Paris Ser 2 (14):269–362

du Bois-Reymond ME (1944) Sobre tardígrados brasileiros. Comun Zool Mus Hist Nat Monte 1(13):1–19

Edwards TL (2008) Argentina: a global studies handbook. ABC-CLIO, Santa Barbara, CA, USA

Guidetti R, Bertolani R (2001) The tardigrades of Emilia (Italy). III. Piane di Mocogno (Northern Apennines). Zool Anz 240:377–383

Guidetti R, Bertolani R (2005) Tardigrade taxonomy: an updated check list of the taxa and a list of characters for their identification. Zootaxa 845:1–46

Guidetti R, Schill RO, Bertolani R, Dandekar T, Wolf M (2009) New molecular data for tardigrade phylogeny, with erection of *Paramacrobiotus* gen. n. J Zool Syst Evol Res 47:315–321

Guidetti R, Altiero T, Marchioro T, Sarzi Amade L, Avdonina AM, Bertolani R, Rebecchi L (2012) Form and function of the feeding apparatus in Eutardigrada (Tardigrada). Zoomorphol 131(2):127–148

Guidetti R, Peluffo JR, Rocha AM, Cesari M, Moly de Peluffo MC (2013) The morphological and molecular analyses of a new South American urban tardigrade offer new insights on the biological meaning of the *Macrobiotus hufelandi* group of species (Tardigrada: Macrobiotidae). J Nat Hist 47(37–38):2409–2426

Guil N (2008) New records and within-species variability of Iberian tardigrades (Tardigrada), with comments on the species from the *Echiniscus-blumi-canadensis* series. Zootaxa 1757:1–30

Heinis F (1914) Die Moosfauna Columbiens. Mém Soc Sci Nat Neu 5:713–724

Hinton JG, Meyer HA, Soileau BN, Dupuid AP (2013) Tardigrada of the Caribbean island of Dominica (West Indies). J Limnol 72(S1):108–112

Hohberg K (2006) Tardigrade species composition in young soils and some aspects on life history of *Macrobiotus richtersi* J. Murray, 1911. Pedobiologia 50:267–274

Hohberg K, Traunspurger W (2005) Predator–prey interaction in soil food web: functional response, size-dependent foraging efficiency, and the influence of soil texture. Biol Fert 41(6):419–427

Hohberg K, Traunspurger W (2009) Foraging theory and partial consumption in a tardigrade-nematode system. Behav Ecol 20(4):884–890

Hohberg K, Russell DJ, Elmer M (2011) Mass occurrence of algal-feeding tardigrade *Apodibius confusus*, in the young soils of a post-mining site. J Zool Syst Evol Res 49(S1):62–65

Holt BG, Lessard JP, Borregaard MK, Fritz SA, Araújo MB, Dimitrov D, Fabre PH, Graham CH, Graves GR, Jønsson KA, Bravo DN, Wang Z, Whittaker RJ, Fjeldså J, Rahbek C (2013) An update of Wallace's Zoogeographic Regions of the World. Science 339(6115):74–78

Horikawa D, Kunieda T, Abe W, Watanabe M, Nakahara Y, Yukuhiro F, Sakashita T, Hamada N, Wada S, Funayama T, Katagiri C, Kobayashi Y, Higashi S, Okuda T (2008) Establishment of a rearing system of the extremotolerant Tardigrade *Ramazzottius varieornatus*: a new model animal for astrobiology. Astrobiology 8:549–556

Horning D, Schuster R, Grigarick A (1978) Tardigrada of New Zealand. New Zeal J Zool 5:185–280

Iharos G (1963) The zoological results of Gy. Topal's collections in South Argentina, 3. Tardigrada. Ann Hist Natur Ms Nat Hung Budapest 55:293–299

Jerez Jaimes JH, Narváez Parra EX (2001) Tardigrados (Animalia, tardigrada) de la Reserva El Diviso - Santander, Colombia. Biota Colomb 2:145–151

Johansson C, Miller WR, Linder ET, Adams BJ, Boreliz-Alvarado E (2013) Tardigrades of Alaska: distribution patterns, diversity and species richness. Polar Res 32:18793

Kaczmarek Ł, Michalczyk Ł (2006) Tardigrada fauna of Mongolia (Central Asia) with a description of *Isohypsibius altai* sp. nov. (Eutardigrada: Hypsibiidae). Zool Stud 45(1):11–23

Kaczmarek Ł, Michalczyk Ł (2007) A new species of Tardigrada (Eutardigrada: Milnesiidae): *Milnesium krzysztofi* from Costa Rica (Central America). New Zeal J Zool 34:297–302

Kaczmarek Ł, Michalczyk Ł (2009) Two new species of Macrobiotidae, *Macrobiotus szeptyckii* (*harmsworthi* group) and *Macrobiotus kazmierskii* (*hufelandi* group) from Argentina. A Zool Crac 52B:87–99

Kaczmarek Ł, Michalczyk Ł, Beasley CW (2004) *Milnesium katarzynae* sp. nov., a new species of eutardigrade (Milnesiidae) from China. Zootaxa 743:1–5

Kaczmarek Ł, Jakubowska N, Michalczyk Ł (2012a) Current knowledge on Turkish tardigrades with a description of *Milnesium beasleyi* sp. nov. (Eutardigrada: Apochela: Milnesiidae, the *granulatum* group). Zootaxa 3589:49–64

Kaczmarek Ł, Zawierucha K, Smykla J, Michalczyk Ł (2012b) Tardigrada of the Revdalen (Spitsbergen) with the descriptions of two new species: *Bryodelphax parvuspolaris* (Heterotardigrada) and *Isohypsibius coulsoni* (Eutardigrada). Polar Biol 35:1013–1026

Kaczmarek Ł, Michalczyk Ł, McInnes SJ (2014a) Annotated zoogeography of non-marine Tardigrada. Part I: Central America. Zootaxa 3763(1):1–62

Kaczmarek Ł, Cytan J, Zawierucha K, Diduszko D, Michalczyk Ł (2014b) Tardigrades from Peru (South America), with descriptions of three new species of Parachela. Zootaxa 3790(2):357–379

Kinchin IM (1990) The cosmopolitan tardigrade *Milnesium tardigradum* Doyère: an observation from Northern Ireland. Microscopy 36:412–414

Land M, Musto A, Miller WR, Starkey DE, Miller JD (2012) Tardigrades of the University of Central Arkansas Campus, Conway, AR. Southeast Nat 11 (3):469–476

Lemloh ML, Brummer F, Schill RO (2011) Life-history traits of the bisexual tardigrades *Paramacrobiotus tonollii* and *Macrobiotus sapiens*. J Zool Syst Evol Res 49(S1):58–61

Lisi O (2011) Remarks on *Doryphoribius flavus* (Iharos, 1966), and description of three new species (Tardigrada, Hypsibiidae). Zootaxa 2834:17–32

Lisi O, Londoño R, Quiroga S (2014) Tardigrada from a sub-Andean forest in the Sierra Nevada de Santa Marta (Colombia) with the description of *Itaquascon pilatoi* sp. nov. Zootaxa 3841(4):551–562

Marcus E (1936) Tardigrada. Das Tierreich 66:1–340

Marley NJ (2006) A new species of Tardigrada, *Platicrista ramsayi* sp. nov. from the páramo of Volcán Chiles, Ecuador. Zootaxa 1166:35–48

Marley NJ, McInnes SJ, Sands CJ (2011) Phylum Tardigrada: a re-evaluation of the Parachela. Zootaxa 2819:51–64

Maucci W (1988) Tardigrada from Patagonia (Southern South America) with description of three new species. Rev Chilena Entomol 16:5–13

Maucci W (1991) Tre nuove specie di eutardigradi della Groenlandia meridionale. Boll Mus Civ St Nat Verona 15:279–289

Maucci W (1996) Tardigrada of the Arctic tundra with description of two new species. Zool J Linn Soc 116:185–204

McInnes SJ (1994) Zoogeographic distribution of terrestrial/freshwater tardigrades from current literature. J Nat Hist 28:257–352

McInnes SJ, Chown SL, Dartnall JG, Pugh PJA (2001) *Milnesium* cfr. *tardigradum* (Milnesiidae, Apochela, Tardigrada): a monitor high altitude meiofauna on sub-Antarctic Marion Island. Zool Anz 240:461–465

Melo JC, Beltrán-Pardo E, Bernal JE, Kaczmarek Ł (2014) New records of tardigrades from Colombia (Guatavita, Cundinamarca Department). Turk J Zool 38:1–9. doi:10.3906/zoo-1405-13

Meyer HA (2013) Terrestrial and freshwater Tardigrada of the Americas. Zootaxa 3747:1–71

Meyer HA, Hinton JG (2010) *Milnesium zsalakoae* and *M. jacobi*, two new species of Tardigrada (Eutardigrada: Apochela: Milnesiidae) from the southwestern United States. Proc Biol Soc Wash 123:113–120

Meyer HA, Hinton JG (2012) Terrestrial Tardigrada of the Island of Barbados in the West Indies, with the description of *Milnesium barbadosense* sp. n. (Eutardigrada: Apochela: Milnesiidae). Caribb J Sci 46(2–3):194–202

Meyer HA, Hinton JG, Dupré MC (2013) *Milnesium lagniappe*, a new species of water bear (Tardigrada, Eutardigrada, Apochela, Milnesiidae) from the southern United States. West N Am Naturalist 73(3):295–301

Meyer HA, Domingue MN, Hinton JG (2014) Tardigrada of the west Gulf Coastal Plain, with descriptions of two new species from Louisiana. Southeast Nat 13(5):117–130

Michalczyk Ł, Kaczmarek Ł (2003) *Minibiotus constellatus*, a new species of Tardigrada from Peru (Eutardigrada: Macrobiotidae). Genus 14:295–305

Michalczyk Ł, Kaczmarek Ł (2004) *Minibiotus eichhorni* sp. nov., a new species of eutardigrade (Eutardigrada: Macrobiotidae) from Peru. Ann Zool 54:673–676

Michalczyk Ł, Kaczmarek Ł (2005) The first record of the genus *Calohypsibius* Thulin, 1928 (Eutardigrada: Calohypsibiidae) from Chile (South America) with description of a new species *Calohypsibius maliki*. New Zeal J Zool 32:287–292

Michalczyk Ł, Kaczmarek Ł (2006) Revision of the *Echiniscus bigranulatus* group with a description of a new species *Echiniscus madonnae* (Tardigrada: Heterotardigrada: Echiniscidae) from South America. Zootaxa 1154:1–26

Michalczyk Ł, Kaczmarek Ł (2013) The Tardigrada Register: a comprehensive online data repository for tardigrade taxonomy. J Limnol 72(S1):175–181

Michalczyk Ł, Wełnicz W, Frohme M, Kaczmarek Ł (2012a) Redescriptions of three *Milnesium* Doyère, 1840 taxa (Tardigrada: Eutardigrada: Milnesiidae), including the nominal species for the genus. Zootaxa 3154:1–20

Michalczyk Ł, Wełnicz W, Frohme M, Kaczmarek Ł (2012b) Corrigenda of *Zootaxa*, 3154:1–20 Redescriptions of three *Milnesium* Doyère, 1840 taxa (Tardigrada: Eutardigrada: Milnesiidae), including the nominal species for the genus. Zootaxa 3393:66–68

Mihelčič F (1967) Ein Beitrag zur Kenntnis der Tardigraden Argentiniens. Verh Zool-Botan Ges Wien 107:43–56

Mihelčič F (1972) Ein weiterer Beitrag zur Kenntnis der Tardigraden Argentiniens. Verh Zool-Botan Ges Wien 110(111):47–52

Miller WR, Williams S (2012) A note on a Greenland tardigrade: *Milnesium eurystomum* Maucci, 1991 eating an amoeba. Trans Kans Acad Sci 115(1–2):26–27

Moly de Peluffo MC, Peluffo JR, Rocha AM, Doma IL (2006) Tardigrada distribution in a medium-sized city of the central Argentina. Hydrobiologia 558:141–150

Montoya E, Rull V, van Geel B (2010) Non-pollen palynomorphs from surface sediments along an altitudinal transect of the Venezuelan Andes. Palaeogeogr Palaeocl 297:169–183

Murray J, Wailes GH (1913) Notes on the natural history of Bolivia and Peru. Scott Oceanogr Lab, Edinburgh, pp 1–45

Nederström P (1919) Die bis jetzt aus Finland bekannten Tardigraden. A Soc Fauna Flora Fenn 46(8):1–25

Nelson DR, Horning DS (1979) Tardigrada of the Kowhai Bush, Kaikoura, New Zealand. Zesz Nauk Univ Jagiell 25:125–142

Nelson DR, Guidetti R, Rebecchi L (2010) Tardigrada. In: Thorp JH, Covich AP (eds) Ecology and Classification of North American Freshwater Invertebrates, vol 3E. Academic Press (Elsevier), San Diego, pp 455–484

Nickel K, Miller WR, Marley N (2001) Tardigrades of South America: Machu Picchu and Ollantayambo, Peru. Zool Anz 240:505–509

Özdikmen H (2008) *Bertolanius* nom. nov., a replacement name for the genus *Amphibolus* Bertolani, 1981 (Tardigrada: Parachela) with type species designation. Mun Ent Zool 3(1):330–332

Peel MC, Finlayson BL, McMahon TA (2007) Updated world map of the Köppen-Geiger climate classification. Hydrol Earth Syst Sc 11:1633–1644, http://dx.doi.org/10.5194/hess-11-1633-2007

Peluffo JR, Moly de Peluffo MC, Rocha AM (2002) Rediscovery of *Echiniscus rufoviridis* du Bois-Reymond Marcus, 1944. (Heterotardigrada, Echiniscidae). New contributions to the knowledge of its morphology, bioecology and distribution. Gayana 66:97–101

Peluffo JR, Rocha AM, Moly de Peluffo MC (2007) Species diversity and morphometrics of tardigrades in a medium-sized city in the Neotropical Region: Santa Rosa (La Pampa, Argentina). Anim Biodivers Conserv 30:43–51

Pilato G (1981) Analisi di nuovi caratteri nello studio degli Eutardigradi. Animalia 8:51–57

Pilato G (2000) *Macrobiotus centesimus*, new species of eutardigrade from the South America. Boll Accad Gioenia Sci Nat Catania 33:97–101

Pilato G (2007) *Echiniscus quitensis*, a new species of tardigrade from Ecuador (Heterotardigrada: Echiniscidae). Zootaxa 1389:55–60

Pilato G, Binda MG (1991) *Milnesium tetralamellatum*, new species of Milnesiidae from Africa (Eutardigrada). Trop Zool 4:103–106
Pilato G, Binda MG (1994) *Dactylobiotus caldarellai*, nuova specie di Eutardigrado della Terra del Fuoco. Animalia 21:87–91
Pilato G, Binda MG (1996) *Mixibius fueginus*, nuova specie di eutardigrado della Terra del Fuoco. Boll Accad Gioenia Sci Nat Catania 29(351):27–32
Pilato G, Binda MG (1998) A comparison of *Diphascon (D.) alpinum* Murray, 1906, *D. (D.) chilenense* Plate, 1889 and *D. (D.) pingue* Marcus, 1936 (Tardigrada), and description of a new species. Zool Anz 236:181–185
Pilato G, Binda MG (2001) Biogeography and limno-terrestrial tardigrades: are they truly incompatible binomials? Zool Anz 240:511–516
Pilato G, Patanè M (1998) *Macrobiotus ovostriatus*, a new species of eutardigrade from Tierra del Fuego. Boll Acc Gioenia Sci Nat Catania 30(353):263–268
Pilato G, Binda MG, Qualtieri F (1998) *Diphascon (Diphascon) mitrense*, new species of eutardigrade from Tierra del Fuego. Boll Acc Gioenia Sci Nat Catania 31:101105
Pilato G, McInnes SJ, Lisi O (2012) *Hebesuncus mollispinus* (Eutardigrada, Hypsibiidae), a new species from maritime Antarctica. Zootaxa 3446:60–68
Pilato G, Binda MG, Napolitano A, Moncada E (2000) The specific value of *Macrobiotus coronatus* De Barros 1942, and description of two new species of the *harmsworthi* group (Eutardigrada). Boll Acc Gioenia Sci Nat Catania 33:103–120
Pilato G, Binda MG, Napolitano A, Moncada E (2001) Notes on South American tardigrades with the description of two new species: *Pseudechiniscus spinerectus* and *Macrobiotus danielae*. Trop Zool 14:223–231
Pilato G, Binda MG, Lisi O (2002) Notes on tardigrades of the Seychelles with the description of two new species. Boll Sedute Accad Gioenia Sci nat Catania 35:503–517
Pilato G, Binda MG, Lisi O (2003) Remarks on some species of tardigrades from South America with description of *Minibiotus sidereus* n. sp. Zootaxa 195:1–8
Pilato G, Binda MG, Napolitano A, Moncada E (2004) Remarks on some species of tardigrades from South America with description of two new species. J Nat Hist 38:1081–1086
Poprawa I (2005) The ovary structure, previtellogenic and vitellogenic stages in parthenogenetic species *Dactylobiotus dispar* (Murray, 1907) (Tardigrada: Eutardigrada). Tissue Cell 37(5):385–392
Poprawa I (2011) Ultrastructural studies of the formation of the egg capsule in the hermaphroditic species, *Isohypsibius granulifer granulifer* Thulin, 1928 (Eutardigrada: Hypsibiidae). Zool Sci 28(1):37–40
Rahm G (1931) Tardigrada of the South of America. Rev Chil Hist Nat 35:118–141
Rahm G (1932) Freilebende Nematoden, Rotatorien und Tardigraden aus Südamerika (besondersaus Chile). Zool Anz 98:94–112 and 113–128
Ramazzotti G (1957) Due nuove specie di Tardigradi extra-europei. Atti Soc Ital Sci Nat Mus Civ Stor Nat Milano 96:188–191
Ramazzotti G (1962) Tardigradi del Cile, con descrizione di quattro nuove specie e di una nuova varieta. Atti Soc Ital Sci Nat Mus Civ Stor Nat Milano 101:275–287
Ramazzotti G (1964) Tardigradi del Cile II, con descrizione di due nuove specie e note sulla scultura degli Echiniscidae. Atti Soc Ital Sci Nat Mus Civ Stor Nat Milano 103:89–100
Ramazzotti G, Maucci W (1983) Il Phylum Tardigrada. III edizione riveduta e aggiornata. Mem Ist Ital Idrobiol 41:1–1012
Richters F (1903) Nordische Tardigraden. Zool Anz 27:168–172
Richters F (1911) Südamerikanische Tardigraden. Zool Anz 38:273–277
Richters F (1926) Tardigrada. In: By Kükenthal W, Krumbach T (eds) Handbuch der Zoologie, vol 3. Walter de Gruyter & Co, Berlin and Leipzig, pp 58–61
Rossi GC, Claps MC (1980) Contribucion al conocimiento de los tardigrados de Argentina. I. Rev Soc Ent Argent 39:243–250
Rossi GC, Claps MC (1989) Contribucion al conocimiento de los Tardigrados de la Argentina V. Rev Soc Ent Argent 47:133–142
Rossi GC, Claps MC (1991) Tardigrados dulceacuicolas de a Argentina. Faun Aqu Repub Argent 19:1–70
Rossi G, Claps M, Ardohain D (2009) Tardigrades from northwestern Patagonia (Neuquén Province, Argentina) with the description of three new species. Zootaxa 2095:21–36
Schill RO (2013) Life-history traits in the tardigrade species *Paramacrobiotus kenianus* and *Paramacrobiotus palaui*. J Limnol 72(S1):160–165
Schill RO, Jonsson KI, Pfannkuchen M, Brummer F (2011) Food of tardigrades: a case study to understand food choice, intake and digestion. J Zool Syst Evol Res 49(S1):66–70
Schultze CAS (1834) *Macrobiotus Hufelandii* animal e crustaceorum classe novum, reviviscendi post diuturnam asphyxian et ariditaten potens, etc. 8 Seiten, 1 tab. C. Curths, Berlin.
Schuster RO, Nelson DR, Grigarick AA, Christenberry D (1980) Systematic criteria of the Eutardigrada. Trans Am Microsc Soc 99:284–303
Séméria Y (1993) Description d'une nouvelle de Tardigrade du Vénézuéla, *Ramazzottius edmondabouti* n. sp. (Eutardigrada, Hypsibiidae). Bull Mens Soc Linn Lyon 62:215–216
Shaw MW, Miller WR (2013) Tardigrades of North America: a preliminary survey of New Jersey, U.S.A. Newslet New York Microscop Soc 7(27,8):36–56
Sudzuki M (1964) On the microfauna of the Antarctic Region. I. Moss-water community at Langhovde. Sci Rep Ser E 19:1–4
Suzuki AC (2003) Life history of *Milnesium tardigradum* Doyere (Tardigrada) under the rearing environment. Zool Sci 20(1):49–57
Szymańska B (1994) The Tardigrada from Axel Heiberg Island and the associate bryophyte species. Folia Entom Hung 55:359–368
Trygvadóttir BV, Kristensen RM (2013) A zoogeographic study of the limnoterrestrial tardigrade fauna on the Faroe islands. J Limnol 72(S1):113–122
Tumanov D (2006) Five new species of the genus *Milnesium* (Tardigrada, Eutardigrada, Milnesiidae). Zootaxa 1122:1–23
Vicente F, Bertolani R (2013) Considerations on the taxonomy of the Phylum Tardigrada. Zootaxa 3626(2):245–248
Wallendorf M, Miller WR (2009) Tardigrades of North America: *Milnesium alabamae* nov. sp. (Eutardigrada: Apochela: Milnesiidae) a new species from Alabama. Trans Kans Acad Sci 112:181–186
Yang T (2003) Two new species and three new records of the Tardigrada (Heterotardigrada, Echiniscidae; Eutardigrada, Milnesiidae, Macrobiotidae, Hypsibiidae). A Zootax Sin 28:235–240
Yin H, Li X (2011) New records of tardigrades (Tardigrada) from Taiwan. J Shaanxi Normal Univ Nat Sci Ed 39(4):55–59
Zawierucha K, Coulson SJ, Michalczyk L, Kaczmarek L (2014) Current knowledge of the Tardigrada of Svalbard with the first records of water bears from Nordaustlandet (High Arctic). Polar Res 32:20886, doi.org/10.3402/polar.v32i0.20886

17

Population structure of *Hirundichthys oxycephalus* in the northwestern Pacific inferred from mitochondrial cytochrome oxidase I gene

Chang-En Chou[1†], Te-Yu Liao[2†], Hsueh-Wen Chang[1*] and Shui-Kai Chang[3*]

Abstract

Background: *Hirundichthys oxycephalus* is an important flyingfish resource in eastern Taiwan and northwestern Japan. A substantial catch decline in Taiwan has caused serious concerns on stock status of the fish, prompting the government to impose a set of regulations on flyingfish egg fishery since 2008. However, the regulations were set in a precautionary manner, without considering the fundamental understanding of the population genetic structure. This study aims to investigate the population genetic structure of *H. oxycephalus* in the region based on mtDNA cytochrome oxidase I (COI) gene and to thus provide scientific information for sustainable management of the resource.

Results: Tissue samples (156) from six localities of eastern Taiwan and western Japan were collected, and 616 bp of mtDNA COI gene were sequenced. Seventy haplotypes were determined, and the haplotype diversity and nucleotide diversity were estimated as 0.93% and 0.57%, respectively. Results of various statistical analyses suggested that the genetic differentiations among the six localities were small and most variation occurred within populations, indicating a high gene flow in the region with undergoing population expansion. Although the study showed that the fishes were genetically divided into two groups, the support was low and the separation was not geologically evident.

Conclusions: The study revealed two groups of *H. oxycephalus* in the northwestern Pacific Ocean. However, due to high gene flow, an association of either group to a spatial distribution was not observed, and so the two groups may be considered as one population. Thus, the results favored the conclusion that *H. oxycephalus* from eastern Taiwan and western Japan belong to the same population and, consequently, that the management unit of the current regulations only covering eastern Taiwan does not match the spatial structure of the population. Rather, the results suggest that joint efforts from countries within the population boundary are necessary to maintain a sustainable exploitation.

Keywords: Ecological population; Flyingfish; Taiwan; Northwestern Pacific; COI

Background

The flyingfish family Exocoetidae is a group of small fish comprising 52 species and is distributed globally in the tropical and subtropical areas of three oceans (Nelson 2006; Chang et al. 2012a; Xu et al. 2013). Exocoetids are low-tropical level species (Wu et al. 2010) and are a food source for dolphinfish, tuna, and swordfish (Oxenford and Hunte 1999; Wu et al. 2006). They are known as flyingfish due to their ability to glide in the air over the water using large and elongated pectoral and asymmetric caudal fins (Davenport 1994). This uncommon flying behavior serves as an escape mechanism from predators and is uniquely present in extant fish species of the Exocoetidae group as a homoplasy of the extinct family Thoracopteridae (Davenport 1994; Kutschera 2005; Xu et al. 2013).

The flyingfish is economically important in the Caribbean and the western Pacific Ocean (Rennie 2002; Potts et al.

* Correspondence: hwchang@mail.nsysu.edu.tw; skchang@faculty.nsysu.edu.tw
†Equal contributors
[1]Department of Biological Sciences, National Sun Yat-sen University, 70, Lien-hai Road, Kaohsiung 804, Taiwan
[3]Institute of Marine Affairs, National Sun Yat-sen University, 70, Lien-hai Road, Kaohsiung 804, Taiwan
Full list of author information is available at the end of the article

2003). It is the national fish and one of the national cuisines of Barbados (Cumberbatch and Hinds 2013) and provides a commercially important harvest in Japan, especially in the southwestern regions (Masuda et al. 1984). In Taiwan, it supports the stability of the Kuroshio ecosystem as an intermediary in the bioenergetics transfer, the sustainability of artisanal fisheries as a locally consumed dish and major bait for fishing, and the continuity of aboriginal Tao culture as a respectable emblem (Chang et al. 2012a).

At least 29 species of flyingfish in seven genera are reportedly found in the waters off Taiwan and adjacent areas (Chang et al. 2012a). Among these, *Hirundichthys oxycephalus* (bony flyingfish) is one of the dominant flyingfishes in the Kuroshio Current (Chang et al. 2012b). The fish can be found from south of Taiwan to northwestern Japan (Ichimaru 2007; Chang et al. 2012b); however, it has not been found in the Philippines according to a 3-year sampling program conducted in Batanes (SK Chang and CH Lin, unpublished data). The fish was utilized as the main source of eggs for flyingfish egg fisheries in northern Taiwan and the bait of longline and troll fisheries in eastern Taiwan (Chang et al. 2012a, b). Juveniles of the fish also constituted an important target of boat seine and setnet fisheries in northwestern Japan (Ichimaru 2007). A decline of more than 60% in catches of the fish and its eggs occurred in Taiwan during the 2006 to 2007 period (Fisheries Agency 2010). This decline has caused serious concerns regarding the fish stock status and prompted the Taiwan government to implement a set of regulations on flyingfish egg fishery since 2008, including an annual catch limit. The catch limit regulation is precautionary, however, and has no scientific support regarding the optimal fish stock abundance or, more basically, on the boundary of the fish population that is crucial for effective fishery management (Garcia 2005). Therefore, the population structure of *H. oxycephalus* needs to be resolved urgently in order to confirm if the population is trans-boundary and is exploited by more than one country.

Genetic analysis is a useful tool for species identification and the determination of population structure which could provide a scientific basis for fishery management (Ward 2000). The technique has been successfully utilized for the sustainable management of many economically and ecologically important fish, such as albacore and bigeye tunas (Ward 2000). For flyingfish, the molecular phylogeny of selected species of the Exocoetidae group has been reconstructed based on the mitochondrial cytochrome *b* and the nuclear RAG2 genes (Lewallen et al. 2011). Other than this exception, research on population genetics of flyingfish is rare and previous studies all focused on *Hirundichthys affinis* (Gomes et al. 1998; Gomes et al. 1999, 2000). This is the first study that aims to investigate the population structure of *H. oxycephalus* in the northwestern Pacific Ocean inferred from mitochondrial cytochrome oxidase I (COI), and to provide a scientific basis for proper fishery management.

Methods

From 2008 to 2011, samples of *H. oxycephalus* were collected from four localities along the eastern coasts of Taiwan, including Keelung (KL), Yilan (YL), Hualien (HL), and Ludao (LD), as well as two localities along the northwestern coasts of Japan, including Nagasaki (NS) and Yakushima (YS). Tissue samples were obtained from 150 adult fish caught by drive-in net, gillnet, and setnet in the six localities, four juveniles from KL, and two eggs from YL using hand-held net (Figure 1 and Table 1). Tissue samples were stored frozen at −20°C before the extraction of DNA.

DNA sequencing

Crude DNA was extracted from the muscle tissue by proteinase K using a Wizard Genomic DNA Purification Kit (Promega, Madison, WI, USA). The mitochondrial COI gene was amplified with the following two primers (Ward et al. 2005): FishF1 (5′TCAACCAACCACAAA GACATTGGCAC3′) and FishR1 (5′TAGACTTCTGGG TGGCCAAAGAATCA3′). PCR was performed using a GeneAmp PCR System 2400 thermal cycler (Applied Biosystems, Carlsbad, CA, USA) with an initial denaturation at 95°C for 4 min, followed by 35 cycles of amplification (denaturing at 94°C for 0.5 min, annealing at 55°C for 0.5 min, and extension at 72°C for 50 seconds) and a final extension at 72°C for 10 min. Autosequencing was performed on ABI 3100 Avanter Genetic Analysis with an ABI PRISM BigDye™ Terminators v 2.0 Cycle Sequence Kit (Applied Biosystems, Carlsbad, CA, USA). The sequences were double checked with forward and reverse strands.

Sequences analysis

All sequences were checked visually and aligned with Clustal W Multiple Alignment using Bioedit 7.1.3 (Hall 1999). Genetic distances were calculated using the Kimura 2-parameter model as suggested by MEGA 5.02 (Tamura et al. 2011).

A minimum spanning network was constructed using the Arlequin 3.5.1.3 program (Excoffier and Lischer 2010). In order to measure genetic variation, haplotype diversity (h) and nucleotide diversity (π) were calculated. In order to estimate the nucleotide differences between populations, the Nm and fixation indexes (Fst) were calculated. All of the above indices were calculated by DNASP v5. To analyze if the flyingfish from Taiwan and Japan are from the same population, all samples were hierarchically clustered into Taiwan and Japan groups.

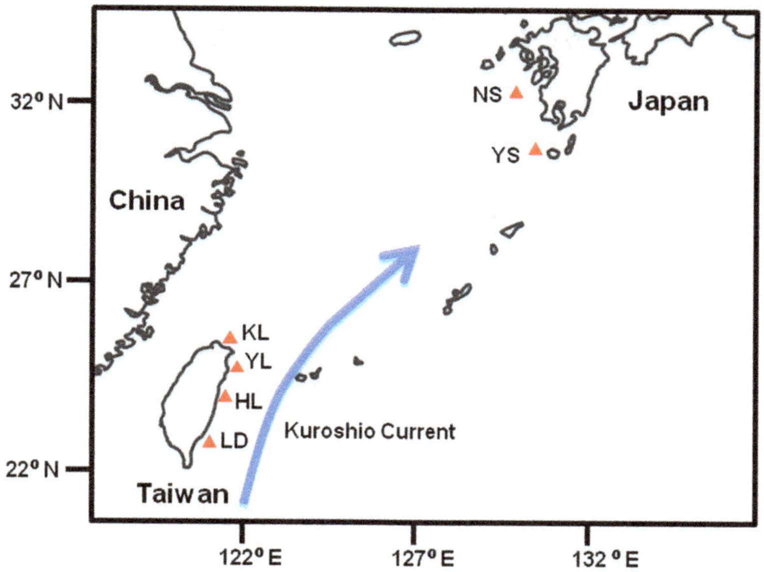

Figure 1 Sampling localities of *Hirundichthys oxycephalus* from the northwestern Pacific Ocean. NS, Nagasaki; YS, Yakushima; KL, Keelung; YL, Yilan; HL, Hualien; LD, Ludao.

Then, a nested analysis of molecular variance (AMOVA) (Excoffier et al. 1992) was performed using Arlequin to estimate population differentiation from the genetic variation of each hierarchical level. Furthermore, neutrality tests, including Tajima's *D*, a frequency-based test to identify the genes under selection (Tajima 1989), and Fu's *Fs* (Fu 1997), a test sensitive to population demographic expansion, were performed using Arlequin. Mismatch distribution was analyzed to examine the model of sudden population expansion (Harpending 1994). Demographic history and the divergence time for the *H. oxycephalus* COI sequence were further estimated using Bayesian evolutionary analysis sampling trees (BEAST) ver. 1.6.2 (Drummond and Rambaut 2007). Bayesian skyline analysis (which was calculated using BEAST) was used to infer the changes in effective population size (*Ne*) across time. The analysis was run with 50 million steps in a Markov chain Monte Carlo (MCMC) simulation under the HKY + I model, relaxed molecular clock model (uncorrelated log-normal), and 1.2% nucleotide substitution rate (Bermingham et al. 1997). Other operators were optimized automatically. BEAST under the Yule process model was also used to estimate the time to the most recent common ancestor (TMRCA). The other model parameters were the same with Bayesian skyline analysis. The results were viewed using Tracer v1.5 (Rambaut and Drummond 2003).

Exocoetus volitans was designated as an out-group for tree rooting. In addition to *H. oxycephalus*, *Cypselurus poecilopterus* was included in all molecular analyses. An NJ tree was reconstructed with a Kimura 2-parameter model using MEGA 5.02 (Tamura et al. 2011). Bayesian

Table 1 Summary of genetic diversity indices of *Hirundichthys oxycephalus* samples from the northwestern Pacific Ocean

Code	Locality	*N*	*h*	π	Tajima's *D*	Fu's *Fs*	SSD	RI
NS	Nagasaki	54	0.90	0.0054	−1.5788*	−15.5619*	0.0220	0.0450
YS	Yakushima	8	1.00	0.0056	−1.5649*	−5.0985*	0.0541	0.1859
KL	Keelung	41	0.87	0.0060	−1.4178	−14.1934*	0.0058	0.0305
YL	Yilan	7	1.00	0.0063	−1.4463	−3.6463*	0.0330	0.0192
HL	Hualien	12	0.92	0.0052	−0.7891	−2.3297	0.0075	0.0198
LD	Ludao	34	0.96	0.0055	−1.8500*	−22.9093*	0.0474*	0.1837
Group I		112	0.93	0.0039	−2.3552*	−26.9307*	0.0008	0.0434
Group II		44	0.54	0.0018	−2.4259*	−14.0031*	0.0022	0.0624
Total		156	0.93	0.0057	−2.1052*	−26.0292*	0.0076	0.0221

*$p < 0.05$.

analysis was conducted using MrBayes v3.1.2 (Ronquist et al. 2012) with the GTR + I + G model as suggested by MrModeltest (Nylander 2004) and partitions based on codon position (first, second, and third). All parameters, with the exception of topology and branch length, were allowed to vary independently with sampling for 27 million generations (two concurrent analyses, *n*runs = 2; three heated chains, *n*chains = 4; chain temperature 0.1; sample frequency 1,000; burn in = 7 million generations). Convergence was checked using the software Tracer v1.5 (Rambaut and Drummond 2003) as well as the SUMP and PLOT commands of MrBayes. Maximum parsimonious analysis was analyzed using TNT (Goloboff et al. 2008) by the heuristic algorithm with zero-length branches permitted. Reliability of NJ and MP trees were examined by bootstrapping with 1,000 replications (Felsenstein 1985) using MEGA 5.02 and TNT, respectively. Bayesian posterior probability was calculated by using the MCMC method with MrBayes v3.1.2.

Results

A total of 616 bp of mtDNA COI gene were sequenced from the 156 *H. oxycephalus* tissue samples. There are 62 variable sites, including 28 parsimoniously informative sites and 34 singletons. The mean nucleotide diversity (π) is 0.57%, with values ranging from 0.52% for HL to 0.63% for YL (Table 1). Seventy haplotypes were determined, and the mean haplotype diversity for all samples is 0.93. Two of them were found in almost every locality in Taiwan and Japan. Furthermore, they were shared by 37% of the total individuals; 13 haplotypes were shared by at least two localities and 28% of the total individuals; and the remaining 55 haplotypes were found in only one locality. The frequency is 35% of the total individuals. The overall mean genetic distance of *H. oxycephalus* is 0.58%. The minimum spanning network of all samples revealed two distinct groups. Group I comprises most samples (71%) collected from all localities, while group II was smaller (29%) and comprised of samples from all localities except YS (Figure 2A).

All 15 *F*st values for the six localities were smaller than 0.15; more than half of them (8) were 0, indicating high gene flow of *H. oxycephalus* in the northwestern Pacific Ocean. The AMOVA results (Table 2) did not show differentiation between Taiwan (HL, KL, LD, and YL) and Japan (NS and YS; $\Phi CT = -0.0306$, $p = 0.1$), and the variation was −3.0613%. Most variation was within populations (98.92%) and the ΦST value was 0.0108 ($p = 0.006$).

In addition, the mean Tajima's *D* and Fu's *Fs* values over all samples and three localities (e.g., NS, YS, and LD) were negative, indicating a deviation from neutrality with population expansion (Table 1). Mismatch distribution analysis of all samples of *H. oxycephalus* revealed a unimodal shape. The sum of squared deviation (SSD) and Harpending's raggedness index (RI) were not significant ($SSD = 0.0076$, $p = 0.349$; $RI = 0.0221$, $p = 0.514$), which indicates that the observed results fit with the growth expansion model (Figure 3). Both group I and group II showed similar results for the mismatch distribution analysis with the whole group of *H. oxycephalus* (Table 1). The Bayesian skyline plot for total *H. oxycephalus* revealed Pleistocene demographic expansion,

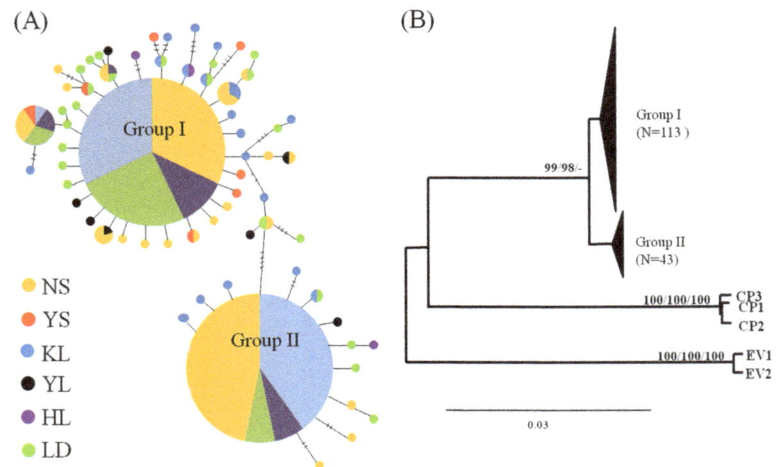

Figure 2 Minimum spanning network and phylogenetic tree of *Hirundichthys oxycephalus* from the northwestern Pacific. (A) Minimum spanning network among haplotypes. Each circle represents a unique haplotype, and the diameter is proportional to the individual number sharing that haplotype. Different colors stand for different sampling localities (NS, Nagasaki; YS, Yakushima; KL, Keelung; YL, Yilan; HL, Hualien; LD, Ludao). **(B)** Phylogenetic tree based on neighbor-joining analysis. Topologies of Bayesian and maximum parsimony analyses are similar, but group I is not recovered. Numbers above nodes are support values of neighbor-joining, Bayesian, and maximum parsimony, respectively. Only values over 50 are shown. (Number of individuals of each locality in group I: NS = 36, YS = 8, KL = 25, YL = 6, HL = 9, LD = 29; number of individuals of each locality in group II: NS = 18, KL = 16, YL = 1, HL = 3, LD = 5).

Table 2 Genetic structure of *Hirundichthys oxycephalus* population based on mtDNA COI gene

Source of variation	*df*	Sum of squares	Variance components	Percentage of variation	Fixation index, Φ
Among groups	1	0.753	−0.0535	−3.0613	−0.0306
Among populations within groups	4	12.387	0.0723	4.1397	0.0402*
Within populations	151	259.206	1.7280	98.9216	0.0108*

*$p < 0.05$.

which occurred between 40,000 and 140,000 years ago (Figure 4). TMRCA for total *H. oxycephalus* was 0.531 million years ago (HPD 95% confidence interval was 0.317 to 0.743 million years ago).

The neighbor-joining tree (Figure 2B) showed that the haplotypes of *H. oxycephalus* were divided into two groups with high bootstrap value but low genetic distance (0.9%). Despite varying in abundance, haplotypes from most localities were shared by the two groups, with the exception that those from YS were only present in group I. Bayesian and maximum parsimony trees showed similar topology, but group I was not recovered in these two analyses.

Discussion

In this study, we used a 616-bp fragment of the mtDNA COI gene to analyze the population genetic structure of *H. oxycephalus*. The mean haplotype diversity of the six localities was high, but nucleotide diversity was low (Table 1). Marine fish populations can be classified into four categories according to the different combinations of values for nucleotide diversity and haplotype diversity (Grant and Bowen 1998). In *H. oxycephalus*, the high nucleotide diversity and low haplotype diversity seem to

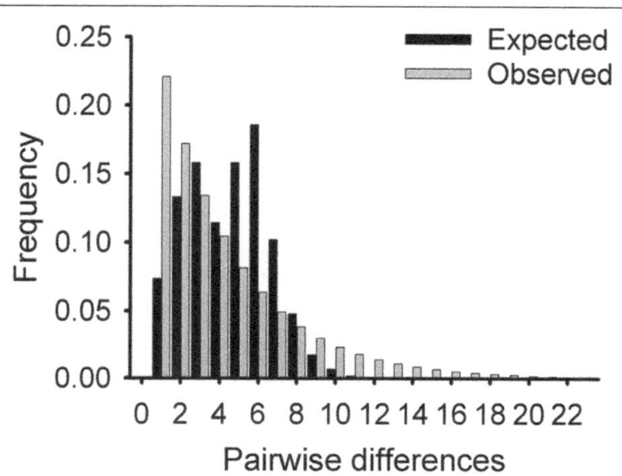

Figure 3 Mismatch distribution of *Hirundichthys oxycephalus* in the northwestern Pacific.

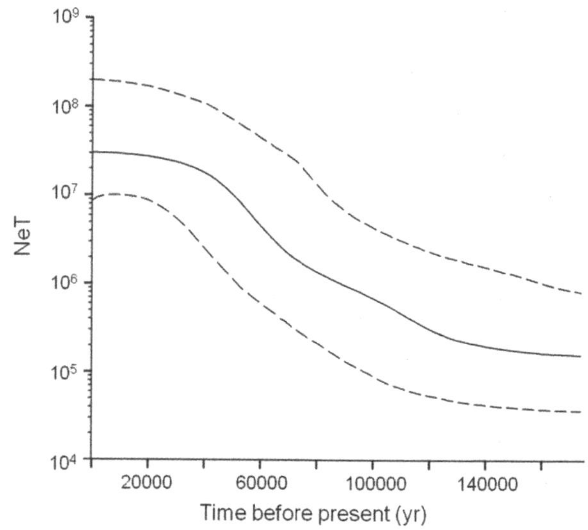

Figure 4 Bayesian skyline plot for total *Hirundichthys oxycephalus* in the northwestern Pacific. The Bayesian skyline plot is derived from mtDNA COI sequences, where the *x*-axis is time in years and the *y*-axis is the product of effective population size (*Ne*) and generation time (*T*). The black line indicates the median estimate, and dashed lines indicate the 95% highest posterior density (HPD) region.

fit the population of the second category, i.e., with a scenario of population bottleneck followed by a rapid population growth and accumulation of mutations. Negative and significant values of Tajima's *D* and Fu's *Fs* tests, as well as the unimodal shape of mismatch distribution and the Bayesian skyline plot (Figure 3; Table 1), also support the above classification. During the last glacial maximum (LGM), the sea surface temperature was maximally 3°C cooler than it is today (Stott et al. 2002). The cooler temperature could have further limited the distribution of tropical and subtropical species, such as flyingfish, and thus eventually reducing the population size. The fit to an expansion model might be due to its recent colonization from warm refuge after Pleistocene cooling had occurred (Díaz-Jaimes et al. 2006). A similar episode was also observed among dolphinfish, yellowfin tuna, and whiskered velvet shrimp (Díaz-Jaimes et al. 2006; Dammannagoda et al. 2008; Chu et al. 2012).

According to the Bayesian skyline plot for *H. oxycephalus*, demographic expansion was estimated to occur during the Pleistocene period (Figure 4). Similar observations were noted for some marine fauna, such as mitten crab, mottled spinefoot, and ice goby in the western Pacific, whose Bayesian skyline plots also indicated that the populations had experienced Pleistocene demographic expansion (Xu et al. 2009; Ravago-Gotanco and Juinio-Meñez 2010; Kokita and Nohara 2011). In addition, the TMRCA estimates for all samples of *H. oxycephalus* indicated that the expansion occurred approximately in the middle Pleistocene period, which is much later than the flathead mullet in this region which occurred in the early Pleistocene period or even in the Pliocene period (Shen et al. 2011).

The low genetic differentiation among six localities in the northwestern Pacific Ocean, along with the variation mostly found within populations rather than among them (AMOVA, Table 2), implies a high gene flow of *H. oxycephalus* among populations. The strong and constant Kuroshio Current might facilitate the gene flow of species distributed within the northwestern Pacific Ocean and result in a weak genetic structure (Hohenlohe 2004). This may account for the low genetic differentiation of *H. oxycephalus* and the other species distributed in the range of the Kuroshio Current, including albacore (Wu et al. 2009), Japanese eel (Han et al. 2010), Japanese anchovy (Liu et al. 2006), crocker (Han et al. 2008a,b), mackerel (Chiou and Lee 2004), mullet (Liu et al. 2007; Liu et al. 2009; Shen et al. 2011), goby (Iida et al. 2010; Ju et al. 2013), and whiskered velvet shrimp (Chu et al. 2012).

Historical and ecological factors may influence the demography of marine fishes and result in significant population structure (Riginos and Nachman 2001; Yu et al. 2002; Lourie and Vincent 2004; Liu et al. 2007). On the contrary, species with high dispersal potential and longer pelagic larval duration, such as bigeye tuna (Chiang et al. 2008), albacore (Wu et al. 2009), Atlantic mackerel (Nesbø et al. 2000), and billfish (Graves and McDowell 1995), tend to have high level of gene flow and a lack of population genetic structure (Palumbi 1994, 2003; Shulman 1998; Bohonak 1999; Planes et al. 2001; Purcell et al. 2006, 2009). Despite their distant dispersal and pelagic stage, *H. affinis* of the central western Atlantic can be divided into three lineages with distinct geographic distribution. Furthermore, its population genetic structure is considered to be associated with spawning behavior (Gomes et al. 1999). In the present study, phylogenetic trees and minimum spanning network of *H. oxycephalus* revealed two distinct clades with minor genetic distance. The subdivision implied that the specimens of *H. oxycephalus* were composed of two populations. However, the interrelationship of either stock to any sampling locality, the Kuroshio, or other minor currents was not detected. Similar results have also been observed in other marine fishes in the northwestern Pacific (Chiou and Lee 2004; Liu et al. 2007; Han et al. 2008a,b; Liu et al. 2009; Wu et al. 2009; Han et al. 2010; Iida et al. 2010; Shen et al. 2011; Chu et al. 2012; Ju et al. 2013), in which major ocean current patterns are not good predictors of gene flow. On the other hand, historical factors such as Pleistocene glaciations might account for the genetic pattern of *H. oxycephalus*. Further study that includes more comprehensive samplings throughout its distribution range would facilitate classification of the genetic pattern.

There are two major definitions for a biological population: the ecological definition that reflects mainly the co-occurrence and demographic interaction of individuals in time and space, and the evolutionary definition that is based on genetic structure due to reproductive interactions among individuals (Waples and Gaggiotti 2006; Reiss et al. 2009). If a significant and reproducible genetic differentiation can be detected, the populations could be considered as separate evolutionarily defined populations with demographical independence (Bentzen 1998). Accordingly, the present study suggested that genetically speaking, there were two populations of *H. oxycephalus* that existed in the studied region; however, the genetic differentiation was not 'significant' enough according to the results of the genetic analysis (Figure 2B).

On the other hand, the present study showed that the haplotypes from most localities were shared by the two groups with the exception of those from YS, which were only present in group I (Figure 2). There were only eight specimens available from YS for the present study, and the limited sample size might have inevitably introduced some uncertainties to the results of the analysis. YL, a locality with only seven specimens, also showed similar pattern, i.e., all haplotypes except one were placed in group I (Figure 2). Therefore, if temporarily ignoring samples with possible sampling bias, the results tended to suggest that the *H. oxycephalus* from eastern Taiwan and from western Japan was a single population. A separate 3-year study on the spatio-temporal distributions of fish density, body length, and gonadosomatic index (GSI) data of the fish from the same region also suggested one population in the region with two possible migration routes (SK Chang, CH Lin, S Jan, T Ichimaru, and CE Chou, unpublished data). These observations suggested that *H. oxycephalus* from the northwestern Pacific Ocean belonged to one ecologically defined population.

Fish populations are the fundamental biological unit. Ignoring the congruence of spatial scales between the population structure of fish species and management units can result in reduced productivity and local reduction of population (Kenchington et al. 2003; Waples and Gaggiotti 2006; Worm et al. 2006). Currently, the only known spawning ground of *H. oxycephalus* is north of Taiwan (Chang et al. 2012b). The spawning activities of the fish take place from May to July, which is the major spawning season based on the back-calculation performed in a daily growth increments study (Chang et al. 2012a). A reproductive biology study on *H. oxycephalus* suggested that the fish was a batch spawner with asynchronous oocyte development patterns (CH Lin, YT Tsuei, and SK Chang, unpublished data). Based on the assumption of one ecological population, there might be another unknown distinct spawning ground for the same population during the second spawning season in winter time. Currently, there are no implemented management measures on this species in Japan, and the measures implemented by Taiwan are only for flyingfish egg fishery in eastern Taiwan. Although the hypothesis of either one or two populations has not been confirmed, it is evident that the current management scheme covering only the eastern waters of Taiwan mismatches with the spatial structure of the biological population. To properly and effectively manage the resource of *H. oxycephalus*, negotiations and joint efforts between the Taiwanese and Japanese governments are needed.

Conclusions

Study of population structures is a significant step toward realizing the goal of management of fishery and conservation of the fish resources in their natural population (Abdul-Muneer 2014). This approach has been previously applied to *H. affinis* in the central western Atlantic (Gomes et al. 1999). To our knowledge, this is the first study to investigate the population genetics of flyingfish in the northwestern Pacific Ocean, particularly with a focus on an economically valuable species, *H. oxycephalus*, distributed among East Asian countries (Shao 2009). Neutrality tests and mismatch distribution analysis based on the fragment of the mtDNA COI gene of *H. oxycephalus* collected from the northwestern Pacific Ocean suggested that *H. oxycephalus* was undergoing population expansion, probably due to recent colonization from warm refuge after Pleistocene cooling (Díaz-Jaimes et al. 2006). AMOVA showed that most variation was within populations, implying that high gene flow of *H. oxycephalus* occurred among populations, which was probably facilitated by the Kuroshio Current. Phylogenetic analyses and minimum spanning network showed that the fish were genetically divided into two groups. However, the genetic divergence between the two groups was low, and the separation was not geologically evident. On the other hand, based on the genetic data of the present study, which is supported by an independent ecological study (SK Chang, CH Lin, S Jan, T Ichimaru, and CE Chou, unpublished data), *H. oxycephalus* from eastern Taiwan and from northwestern Japan might be considered as one population. Effective fishery management of *H. oxycephalus* requires joint efforts from countries within its population boundary in order to maintain sustainable exploitation.

Competing interests
The authors declare that they have no competing interests.

Authors' contributions
CEC conducted the analyses and prepared the draft. TYL provided the technical text and discussion and made contributions equal to those of CEC. HWC commented on and edited the draft. SKC provided the idea for the study and text on fisheries, management and conclusions and edited the manuscript. SKC is also the primary corresponding author of this manuscript. All authors read and approved the final manuscript.

Acknowledgements
The authors are thankful for the special efforts from Mr. Chung-Hui Lin in fish sampling and Mr. Tsung-Wei Lin in molecular analysis. Constructive comments from Dr. Tzong-Der Tzeng of the Shu-Te University on the early work and from the two anonymous reviewers on the manuscript are much appreciated. This study was financially supported by the Ministry of Science and Technology (NSC99-2611-M-110-017, NSC100-2611-M-110-016, and NSC101-2611-M-110-013) and partially by the Asia-Pacific Ocean Research Centre, National Sun Yat-sen University.

Author details
[1]Department of Biological Sciences, National Sun Yat-sen University, 70, Lien-hai Road, Kaohsiung 804, Taiwan. [2]Department of Oceanography, National Sun Yat-sen University, 70, Lien-hai Road, Kaohsiung 804, Taiwan. [3]Institute of Marine Affairs, National Sun Yat-sen University, 70, Lien-hai Road, Kaohsiung 804, Taiwan.

References

Abdul-Muneer PM (2014) Application of microsatellite markers in conservation genetics and fisheries management: recent advances in population structure analysis and conservation strategies. Genet Res Int 2014:11, doi:10.1155/2014/691759

Bentzen P (1998) Seeking evidence of local stock structure using molecular genetic methods. In: Herbing IH, Kornfield I, Tupper M, Wilson J (eds) The implications of localized fisheries stocks. Regional Agricultural Engineering Service, New York, pp 20–30

Bermingham E, McCafferty SS, Martin PP (1997) Fish biogeography and molecular clocks: perspectives from the Panamian Isthmus. In: Stepien CA (ed) Kocher TD. Molecular Systematics of Fishes, San Diego, pp 113–228

Bohonak AJ (1999) Dispersal, gene flow, and population structure. Q Rev Biol 74(1):21–45

Chang CW, Lin CH, Chen YS, Chen MH, Chang SK (2012a) Age validation, growth estimation and cohort dynamics of the bony flying fish *Hirundichthys oxycephalus* off eastern Taiwan. Aquat Biol 15(3):251–260, doi:10.3354/ab00425

Chang SK, Chang CW, Ame E (2012b) Species composition and distribution of the dominant flyingfishes (Exocoetidae) associated with the Kuroshio Current, South China Sea. Raffles Bull Zool 60(2):539–550

Chiang HC, Hsu CC, Wu GCC, Chang SK, Yang HY (2008) Population structure of bigeye tuna (*Thunnus obesus*) in the Indian Ocean inferred from mitochondrial DNA. Fish Res 90(1–3):305–312, doi:10.1016/j.fishres.2007.11.006

Chiou WD, Lee LK (2004) Migration of kawakawa *Euthynnus affinis* in the waters near Taiwan. Fish Sci 70(5):746–757, doi:10.1111/j.1444-2906.2004.00867.x

Chu TJ, Wang D, Huang HL, Lin FJ, Tzeng TD (2012) Population structure and historical demography of the whiskered velvet shrimp (*Metapenaeopsis barbata*) off China and Taiwan inferred from the mitochondrial control region. Zool Stud 51(1):99–107

Cumberbatch JA, Hinds CJ (2013) Barbadian bio-cultural heritage: an analysis of the flying fish. Int J Intangible Heritage 8:118–134

Dammannagoda ST, Hurwood DA, Mather PB (2008) Evidence for fine geographical scale heterogeneity in gene frequencies in yellowfin tuna (*Thunnus albacares*) from the north Indian Ocean around Sri Lanka. Fish Res 90(1–3):147–157, http://dx.doi.org/10.1016/j.fishres.2007.10.006

Davenport J (1994) How and why do flying fish fly? Rev Fish Biol Fish 4(2):184–214, doi:10.1007/BF00044128

Díaz-Jaimes P, Uribe-Alcocer M, Ortega-García S, Durand JD (2006) Spatial and temporal mitochondrial DNA genetic homogeneity of dolphinfish populations (*Coryphaena hippurus*) in the eastern central Pacific. Fish Res 80(2–3):333–338, doi:10.1016/j.fishres.2006.04.015

Drummond AJ, Rambaut A (2007) BEAST: Bayesian evolutionary analysis by sampling trees. BMC Evol Biol 7:214, doi:10.1186/1471-2148-7-214

Excoffier L, Lischer HEL (2010) Arlequin suite ver 3.5: a new series of programs to perform population genetics analyses under Linux and Windows. Mol Ecol Resour 10(3):564–567, doi:10.1111/j.1755-0998.2010.02847.x

Excoffier L, Smouse PE, Quattro JM (1992) Analysis of molecular variance inferred from metric distances among DNA haplotypes: application to human mitochondrial DNA restriction data. Genetics 131(2):479–491

Felsenstein J (1985) Confidence limits on phylogenies: an approach using the bootstrap. Evolution 39:783–791, doi:10.2307/2408678

Fisheries Agency (2010) Fisheries statistical yearbook - Taiwan, Kinmen and Matsu Area, 2009. Fisheries Agency, Council of Agriculture, Taipei

Fu YX (1997) Statistical tests of neutrality of mutations against population growth, hitchhiking and background selection. Genetics 147(2):915–925

Garcia SM (2005) Fisheries and aquaculture topics. Defining fishery stocks. Topics Fact Sheets. FAO Fisheries and Aquaculture Department [online]. http://www.fao.org/fishery/topic/14787/en. Accessed 23 April 2014

Goloboff PA, Farris JS, Nixon KC (2008) TNT, a free program for phylogenetic analysis. Cladistics 24(5):774–786, doi:10.1111/j.1096-0031.2008.00217.x

Gomes C, Dales RBG, Oxenford HA (1998) The application of RAPD markers in stock discrimination of the four-wing flyingfish, *Hirundichthys affinis* in the central western Atlantic. Mol Ecol 7(8):1029–1039, doi:10.1046/j.1365-294x.1998.00427.x

Gomes C, Oxenford HA, Dales RBG (1999) Mitochondrial DNA D-loop variation and implications for stock structure of the four-wing flyingfish, *Hirundichthys affinis*, in the central western Atlantic. Bull Mar Sci 64(3):485–500

Gomes C, Oxenford HA, Dales RBG (2000) Restriction site mapping of the mitochondrial DNA of the four-wing flyingfish, *Hirundichthys affinis*. Mitochondr DNA 11(3):277–280, doi:10.3109/10425170009033243

Grant W, Bowen B (1998) Shallow population histories in deep evolutionary lineages of marine fishes: insights from sardines and anchovies and lessons for conservation. J Hered 89(5):415–426, doi:10.1093/jhered/89.5.415

Graves JE, McDowell JR (1995) Inter-ocean genetic divergence of istiophorid billfishes. Mar Biol 122(2):193–203, doi:10.1007/BF00348932

Hall TA (1999) BioEdit: a user-friendly biological sequence alignment editor and analysis program for Windows 95/98/NT. Nucleic Acids Symp Ser 41:95–98, doi:citeulike-article-id:691774

Han ZQ, Gao TX, Yanagimoto T, Sakurai Y (2008a) Deep phylogeographic break among white croaker *Pennahia argentata* (Sciaenidae, Perciformes) populations in North-western Pacific. Fish Sci 74(4):770–780, doi:10.1111/j.1444-2906.2008.01588.x

Han ZQ, Gao TX, Yanagimoto T, Sakurai Y (2008b) Genetic population structure of *Nibea albiflora* in Yellow Sea and East China Sea. Fish Sci 74(3):544–552, doi:10.1111/j.1444-2906.2008.01557.x

Han YS, Hung CL, Liao YF, Tzeng WN (2010) Population genetic structure of the Japanese eel *Anguilla japonica*: panmixia at spatial and temporal scales. Mar Ecol Prog Ser 401:221–232, doi:10.3354/meps08422

Harpending HC (1994) Signature of ancient population growth in a low-resolution mitochondrial DNA mismatch distribution. Hum Biol 66(4):591–600

Hohenlohe PA (2004) Limits to gene flow in marine animals with planktonic larvae: models of Littorina species around Point Conception, California. Biol J Linn Soc 82(2):169–187, doi:10.1111/j.1095-8312.2004.00318.x

Ichimaru T (2007) The life cycle of three species of flyingfish in the north waters of Kyusyu and the recruitment of young flyingfish to the fishing ground. Bull Nagasaki Prefectural Inst Fisheries 33:7–110

Iida M, Zenimoto K, Watanabe S, Kimura S, Tsukamoto K (2010) Larval transport of the amphidromous goby *Sicyopterus japonicus* by the Kuroshio Current. Coast Mar Sci 34:42–46

Ju YM, Hsu CH, Fang LS, Lin HD, Wu JH, Han CC, Chen IS, Chiang TY (2013) Population structure and demographic history of *Sicyopterus japonicus* (Perciformes; Gobiidae) in Taiwan inferred from mitochondrial control region sequences. Genet Mol Res 12:4046–4059, doi:10.4238/2013

Kenchington E, Heino M, Nielsen EE (2003) Managing marine genetic diversity: time for action? ICES J Mar Sci: J Conseil 60(6):1172–1176, doi:10.1016/s1054-3139(03)00136-x

Kokita T, Nohara K (2011) Phylogeography and historical demography of the anadromous fish *Leucopsarion petersii* in relation to geological history and oceanography around the Japanese Archipelago. Mol Ecol 20:143–164, doi:10.1111/j.1365-294X.2010.04920.x

Kutschera U (2005) Predator-driven macroevolution in flyingfishes inferred from behavioural studies: historical controversies and a hypothesis. Ann Hist Phil Biol 10:59–77

Lewallen EA, Pitman RL, Kjartanson SL, Lovejoy NR (2011) Molecular systematics of flyingfishes (Teleostei: Exocoetidae): evolution in the epipelagic zone. Biol J Linn Soc 102(1):161–174, doi:10.1111/j.1095-8312.2010.01550.x

Liu JX, Gao TX, Zhuang ZM, Jin XS, Yokogawa K, Zhang YP (2006) Late Pleistocene divergence and subsequent population expansion of two closely related fish species, Japanese anchovy (*Engraulis japonicus*) and Australian anchovy (*Engraulis australis*). Mol Phylogenet Evol 40(3):712–723, doi:10.1016/j.ympev.2006.04.019

Liu JX, Gao TX, Wu SF, Zhang YP (2007) Pleistocene isolation in the Northwestern Pacific marginal seas and limited dispersal in a marine fish, *Chelon haematocheilus* (Temminck & Schlegel, 1845). Mol Ecol 16(2):275–288, doi:10.1111/j.1365-294X.2006.03140.x

Liu JY, Brown CL, Yang TB (2009) Population genetic structure and historical demography of grey mullet, *Mugil cephalus*, along the coast of China, inferred by analysis of the mitochondrial control region. Biochem Syst Ecol 37(5):556–566, http://dx.doi.org/10.1016/j.bse.2009.09.002

Lourie SA, Vincent ACJ (2004) A marine fish follows Wallace's Line: the phylogeography of the three-spot seahorse (*Hippocampus trimaculatus*, Syngnathidae, Teleostei) in Southeast Asia. J Biogeogr 31(12):1975–1985, doi:10.1111/j.1365-2699.2004.01153.x

Masuda H, Amaoka K, Araga C, Uyeno T, Yoshino T (1984) The fishes of the Japanese archipelago. Tokai University Press, Tokyo

Nelson JS (2006) Fishes of the world, 4th edn. John Wiley & Sons, Inc., New York

Nesbø CL, Rueness EK, Iversen SA, Skagen DW, Jakobsen KS (2000) Phylogeography and population history of Atlantic mackerel (*Scomber scombrus* L.): a genealogical approach reveals genetic structuring among the eastern Atlantic stocks. Proc R Soc Biol Sci Ser B 267(1440):281–292

Nylander JAA (2004) MrModeltest v2. Program distributed by the author. Evolutionary Biology Centre, Uppsala University. doi:citeulike-article-id:7578084

Oxenford HA, Hunte W (1999) Feeding habits of the dolphinfish (*Coryphaena hippurus*) in the eastern Caribbean. Sci Mar 63:303–315, doi:10.3989/scimar.1999.63n3-4317

Palumbi SR (1994) Genetic divergence, reproductive isolation, and marine speciation. Annu Rev Ecol Syst 25(1):547–572, doi:10.1146/annurev.es.25.110194.002555

Palumbi SR (2003) Population genetics, demographic connectivity, and the design of marine reserves. Ecol Appl 13(sp1):146–158, doi:10.1890/1051-0761(2003)013[0146:PGDCAT]2.0.CO;2

Planes S, Doherty PJ, Bernardi G (2001) Strong genetic divergence among populations of a marine fish with limited dispersal, *Acanthochromis polyacanthus*, within the Great Barrier Reef and the Coral Sea. Evolution 55(11):2263–2273, doi:10.1111/j.0014-3820.2001.tb00741.x

Potts AC, Thomas AD, Nichols E (2003) An economic and social assessment of the flying fish (pelagic) fishery of Tobago, Trinidad and Toba. Proc Gulf Caribb Fish Inst 54:635–649

Purcell JFH, Cowen RK, Hughes CR, Williams DA (2006) Weak genetic structure indicates strong dispersal limits: a tale of two coral reef fish. Proc R Soc Biol Sci Ser B 273(1593):1483–1490, doi:10.1098/rspb.2006.3470

Purcell JFH, Cowen RK, Hughes CR, Williams DA (2009) Population structure in a common Caribbean coral-reef fish: implications for larval dispersal and early life-history traits. J Fish Biol 74(2):403–417, doi:10.1111/j.1095-8649.2008.02078.x

Rambaut A, Drummond A (2003) Tracer: a program for analysing results from Bayesian MCMC programs such as BEAST & MrBayes. http://tree.bio.ed.ac.uk/software/tracer. Accessed 18 April 2013

Ravago-Gotanco RG, Juinio-Meñez MA (2010) Phylogeography of the mottled spinefoot *Siganus fuscescens*: Pleistocene divergence and limited genetic connectivity across the Philippine archipelago. Mol Ecol 19:4520–4534, doi:10.1111/j.1365-294X.2010.04803.x

Reiss H, Hoarau G, Dickey-Collas M, Wolff WJ (2009) Genetic population structure of marine fish: mismatch between biological and fisheries management units. Fish Fisheries 10(4):361–395, doi:10.1111/j.1467-2979.2008.00324.x

Rennie J (2002) Review of the social and economic status of the flyingfish fishery in Grenada. In: FAO/WECAFC (ed) Report on the second meeting of the WECAFC Ad Hoc Flyingfish Working Group of the Eastern Caribbean. Bridgetown, Barbados, 8–12 January 2001. FAO Fisheries Report No. 670, 156 pp., FAO, Rome, pp 103–157

Riginos C, Nachman MW (2001) Population subdivision in marine environments: the contributions of biogeography, geographical distance and discontinuous habitat to genetic differentiation in a blennioid fish, Axoclinus nigricaudus. Mol Ecol 10(6):1439–1453, doi:10.1046/j.1365-294X.2001.01294.x

Ronquist F, Teslenko M, van der Mark P, Ayres DL, Darling A, Höhna S, Larget B, Liu L, Suchard MA, Huelsenbeck JP (2012) MrBayes 3.2: efficient Bayesian phylogenetic inference and model choice across a large model space. Syst Biol 61:539–542, doi:10.1093/sysbio/sys029

Shao KT (2009) Taiwan fish database (web electronic publication). http://fishdb.sinica.edu.tw. http://fishdb.sinica.edu.tw. Accessed 10 Dec 2013

Shen KN, Jamandre BW, Hsu CC, Tzeng WN, Durand JD (2011) Plio-Pleistocene sea level and temperature fluctuations in the northwestern Pacific promoted speciation in the globally-distributed flathead mullet *Mugil cephalus*. BMC Evol Biol 11(1):83–83, doi:10.1186/1471-2148-11-83

Shulman MJ (1998) What can population genetics tell us about dispersal and biogeographic history of coral-reef fishes? Aust J Ecol 23(3):216–225, doi:10.1111/j.1442-9993.1998.tb00723.x

Stott L, Poulsen C, Lund S, Thunell R (2002) Super ENSO and global climate oscillations at millennial time scales. Science 297(5579):222–226, doi:10.1126/science.1071627

Tajima F (1989) Statistical method for testing the neutral mutation hypothesis by DNA polymorphism. Genetics 123(3):585–595

Tamura K, Peterson D, Peterson N, Stecher G, Nei M, Kumar S (2011) MEGA5: molecular evolutionary genetics analysis using maximum likelihood, evolutionary distance, and maximum parsimony methods. Mol Biol Evol 28:2731–2739, doi:10.1093/molbev/msr121

Waples RS, Gaggiotti O (2006) What is a population? An empirical evaluation of some genetic methods for identifying the number of gene pools and their degree of connectivity. Mol Ecol 15(6):1419–1439, doi:10.1111/j.1365-294X.2006.02890.x

Ward RD (2000) Genetics in fisheries management. Hydrobiologia 420(1):191–201, doi:10.1023/A:1003928327503

Ward RD, Zemlak TS, Innes BH, Last PR, Hebert PDN (2005) DNA barcoding Australia's fish species. Philos Trans R Soc Lond B Biol Sci 360(1462):1847–1857, doi:10.1098/rstb.2005.1716

Worm B, Barbier EB, Beaumont N, Duffy JE, Folke C, Halpern BS, Jackson JBC, Lotze HK, Micheli F, Palumbi SR, Sala E, Selkoe KA, Stachowicz JJ, Watson R (2006) Impacts of biodiversity loss on ocean ecosystem services. Science 314 (5800):787–790, doi:10.1126/science.1132294

Wu CC, Lin JC, Su WC (2006) Diet and feeding habits of dolphin fish (*Coryphaena hippurus*) in the waters off eastern Taiwan. J Taiwan Fish Res 14(1):13–27

Wu GC-C, Chiang H-C, Chen K-S, Hsu C-C, Yang H-Y (2009) Population structure of albacore (*Thunnus alalunga*) in the Northwestern Pacific Ocean inferred from mitochondrial DNA. Fish Res 95(1):125–131, doi:10.1016/j.fishres.2008.07.014

Wu GC-C, Chiang H-C, Chou Y-W, Wong Z-R, Hsu C-C, Chen C-Y, Yang H-Y (2010) Phylogeography of yellowfin tuna (*Thunnus albacares*) in the Western Pacific and the Western Indian Oceans inferred from mitochondrial DNA. Fish Res 105(3):248–253, doi:10.1016/j.fishres.2010.03.015

Xu J, Chan T, Tsang LM, Chu KH (2009) Phylogeography of the mitten crab *Eriocheir sensu stricto* in East Asia: Pleistocene isolation, population expansion and secondary contact. Mol Phylogenet Evol 52:45–56, doi:10.1016/j.ympev.2009.02.007

Xu GH, Zhao LJ, Gao KQ, Wu FX (2013) A new stem-neopterygian fish from the Middle Triassic of China shows the earliest over-water gliding strategy of the vertebrates. Proc R Soc Biol Sci Ser B 280(1750):20122261, doi:10.1098/rspb.2012.2261

Yu H-T, Lee Y-J, Huang S-W, Chiu T-S (2002) Genetic analysis of the populations of Japanese anchovy (Engraulidae, *Engraulis japonicus*) using microsatellite DNA. Mar Biotechnol 4(5):471–479, doi:10.1007/s10126-002-0035-8

18

Differences in the male mating calls of co-occurring epauletted fruit bat species (Chiroptera, Pteropodidae, *Epomophorus wahlbergi* and *Epomophorus crypturus*) in Kruger National Park, South Africa

Rick A Adams* and Emily R Snode

Abstract

Background: Almost nothing is known about the mating ecology and behavior of epauletted fruit bats (*Epomophorus* spp) of which eight species occur worldwide. Two species of epauletted fruits bats (*Epomophorus wahlbergi* and *Epomophorus crypturus*) overlap in their distributional ranges in Kruger National Park (KNP), South Africa. Morphologically, these two species are nearly identical to the human eye and field recognition is based upon the number and position of palatal rugae of captured individuals. In addition, the males of both species perform mating rituals during overlapping breeding seasons that appear quite similar and involve wing flapping and monotone vocalizations from calling stations in proximity to fruiting sycamore fig trees where females congregate to feed. The overlap in breeding seasons as well as physical and behavioral characteristics between these two species brings into question how species recognition occurs, and no research is available to understand how males are identified by the females of each species for mating purposes. We recorded vocalizations from calling males in local areas of KNP known to support both species and compared the sonograms to determine if the call structure of mating vocalizations between males of each species differs.

Results: We recorded 25 mating vocalizations from seven male epauletted fruit bats near the Shingwedzi Research Camp and 31 mating vocalizations from nine individuals along the Sabie River near Skukuza. Analysis of calls showed significant distinctiveness of male mating vocalizations between the two species in terms of mean fundamental frequency, mean high frequency, mean low frequency, mean bandwidth, and mean call slope at the two sites.

Conclusions: We hypothesize that differences in male mating vocalizations recorded at each of our study sites represent call structure differences that potentially may be used to avoid cross-mating between species.

Keywords: Epauletted fruit bats; Mating calls; Mating behavior; Kruger National Park; South Africa

Background

Two species of epauletted fruit bats (*Epomophorus crypturus* and *Epomophorus wahlbergi*) co-occur in Kruger National Park (KNP), South Africa. Although the range of *E. wahlbergi* is quite extensive throughout Sub-Saharan Africa, *E. crypturus* is near-endemic to South Africa, occurring only marginally outside of South Africa in the extreme south of Tanzania (Monadjem et al. 2010). Morphologically, the species are very similar to each other and are only told apart after capture by opening the oral cavity and counting the number of palatal ridges that reside behind the last upper molar (see 'Methods' section below); however, an analysis of museum specimens also revealed that maxillary width of the skull is significantly wider in *E. crypturus* than in *E. wahlbergi* (Taylor and Monadjem

* Correspondence: rick.adams@unco.edu
School of Biological Sciences, Ross Hall, University of Northern Colorado, 501 20 Street, Greeley, CO 80639, USA

2008). Both species show sexual dimorphism with males being larger in body size than females (Bergmans 1988), and during the mating season, males of both species call from tree perches at night to attract females for mating (McCracken and Wilkinson 2000). Both *E. crypturus* and *E. wahlbergi* have a mating system in which year-round multi-male/multi-female groups copulate away from the roost (Wickler and Seibt 1976; McCracken and Wilkinson 2000). Although *E. wahlbergi* has a bimodal breeding season (December and May) in the southern part of its range (Monadjem and Reside 2012), the breeding cycle of *E. crypturus* has not been thoroughly documented.

The mating call of male *E. wahlbergi* is described as a loud, audible, repetitive, monotone call (Wickler and Seibt 1976; Acharya 1992; Fenton et al. 1985) and appears to serve two functions: a) to attract females and b) to space out male conspecifics calling in the same area (Kingdon 1974; Wickler and Seibt 1976; Adams and Snode 2013). Although the mating call of *E. wahlbergi* has been described, herein, we provide the first quantification of the male mating call of *E. crypturus*.

The detailed distribution of either species of epauletted fruit bat in Kruger National Park (KNP) is poorly known. Monadjem et al. (2010) reports no records for *E. crypturus* from the southern areas of KNP, but Bonaccorso et al. (2014) captured both species in southern and northern KNP between 2004 and 2007, but showed skewed population numbers with *E. crypturus* in higher abundance in the north and *E. wahlbergi* in higher abundance in the south. Fenton et al. (1985) captured and radio-tagged 10 male and 10 female *E. wahlbergi* near Pafuri, about 30 km north of Shingwedzi, where our study took place, thus showing that concentrations of this species can and do occur in the northern KNP. It also seems apparent that, although these species do occur in sympatry on a regional scale, one or the other appears to dominate at the local scale (Monadjem et al. 2010).

Seasonal movement patterns of either species are not well documented. However, a study on *E. wahlbergi* in the urbanized environment of Pietermaritzburg campus, University of KwaZulu-Natal, South Africa, showed that males and females had larger home ranges in winter compared to spring and that males on average were relatively sedentary throughout the year (having relatively small home ranges). Females, on average, made longer flights and had much larger home ranges (Rollinson et al. 2013). Our observations that both species mate within the same timeframe and that males of both species set up calling perches in proximity to fruiting fig trees in KNP resulted in us proposing the hypothesis that the males of each epauletted fruit bat species should show differences in mating calls to avoid cross-mating with the other species even though, to the human ear, little to no discernible differences are evident.

Methods

Study area

Our study sites were located in proximity to Skukuza in southern KNP and near Shingwedzi in Northern KNP (Figure 1). The habitat near Skukuza is diverse and includes mixed woodland with thorn thickets and Marula woodlands, whereas Shingwedzi is dominated by stunted Mopane woodlands and open savanna. There are riverine corridors near both Skukuza and Shingwedzi that are rich with sycamore fig trees (*Ficus sycomorus*) that fruit asynchronously and provide a major food supply for many vertebrate species in KNP, including epauletted fruit bats (Grant et al. 2001).

Catching and identifying bats

We used mist nets raised into the canopy of ripe fig trees (*F. sycomorus*) to capture epauletted fruit bats in May and June of 2007, 2008, and 2009. In two cases, we captured epauletted fruit bats while mist netting over water sources for insectivorous bats. We identified species using the position of palatal ridges in relation to the last cheek tooth in the upper tooth row (Figure 2) (Fenton et al. 1985). All individuals were released immediately after identifying to species, taking measurements of body mass and forearm length, and noting reproductive condition.

Recording mating calls of male epauletted bats

We recorded the male mating calls of epauletted bats in areas where our trapping indicated only one of each of the species was locally present. We used a Pettersson D240× bat detector (Pettersson Elektronik, Uppsala, Sweden) set to a 3.5-s recording time and attached to a digital recorder (Samson H2 Zoom, Samson Technologies, Hauppauge, New York, USA) to capture male calls in the vicinity of Shingwedzi in northern KNP and in the vicinity of Skukuza in southern KNP. We drove roads throughout each area at night and when we heard a male bat calling we approached in our vehicle slowly to a position that allowed clear recordings of the calls to be taken but far enough away to not cause that individual to move from its calling perch. All recordings in each area happened on the same night, and because we could hear any given male calling from its perch as we left the recording position, we are confident that we did not record the same male more than once.

Data analysis and statistics

We analyzed call structure of male mating vocalizations using SonoBat 3.1 (Arcata, Oregon, USA) analysis software to determine the fundamental frequencies (first harmonic), high/low frequencies, bandwidths, and slope (downward or upward change in frequency from the initiation of the vocalization to the end of the vocalization) of each call. We compared means of call parameters of

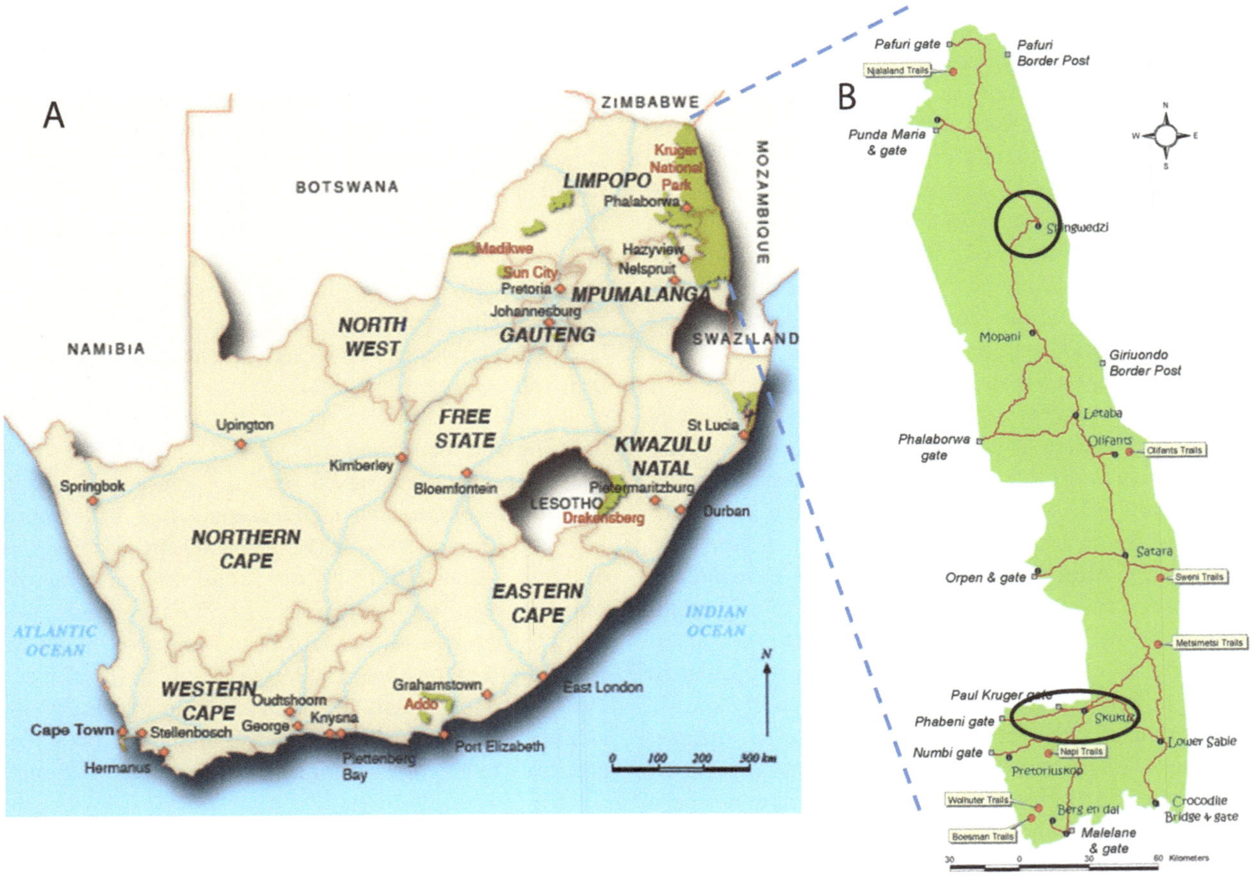

Figure 1 Map of survey areas. (A) South Africa showing the location of Kruger National Park and **(B)** Kruger National Park showing study areas where epauletted fruit bats were captured and sonar calls were recorded near Shingwedzi in the north and Skukuza in the south.

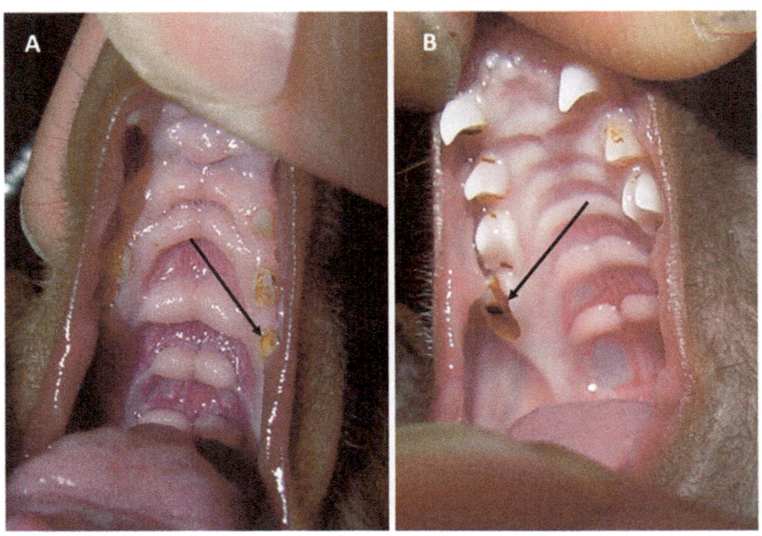

Figure 2 Photographs showing palatal ridges. (A) *E. crypturus* having two post-dental palatal ridges and **(B)** *E. wahlbergi* having one post-dental palatal ridge behind the last molar in the tooth row indicated by arrows.

individuals recorded in the Skukuza area versus those recorded in the Shingwedzi area using a two-sample *T*-test (NCSS Statistical Software, Kaysville, Utah, USA) with the null hypothesis that there are no significant differences between mating calls gathered in each area. All *P* values were adjusted using the Bonferroni correction for multiple comparisons.

Results

Capture data

All individuals captured in and around the Skukuza tourist camp during the dry season in May and June in 2007 (N = 5), 2008 (N = 9), 2009 (N = 3), and during the wet season (December) in 2011 (N = 1) (total captures = 18) were identified as *E. wahlbergi*. All captures in the area of the Shingwedzi tourist camp were during the dry season in May and June in 2007 (N = 8), 2008 (N = 5), and 2009 (N = 2) (total captures = 15) and were identified as *E. crypturus*.

Male mating calls

We recorded 25 mating vocalizations from seven male epauletted fruit bats (two to three calls per individual) near the Shingwedzi Research Camp on 22 June 2008 and 31 mating vocalizations from nine individuals (one to three calls per individual) along the Sabie River near Skukuza on 20 June 2009. If multiple calls were recorded from the same individual, we calculated mean values for each call parameter for statistical analyses. Sonograms depicted by SonoBat of male mating calls recorded near Skukuza (Figure 3A) showed distinctive attributes from those recorded near Shingwedzi (Figure 3B). All calls recorded near Skukuza (Figure 4, N = 31) showed a mean fundamental frequency at 7.88 kHz (first harmonic, standard deviation = 0.11 kHz) followed by a second harmonic having a mean frequency of 15.71 kHz (SD = 0.23 kHz), whereas all male mating calls recorded near Shingwedzi (Figure 4, N = 25) showed a mean fundamental frequency of 15.71 kHz (SD = 0.07 kHz), and in only one case were multiple harmonics recorded and these were at 16, 48, 64, 96, and 112 kHz. The number of harmonics recorded from a vocalization is likely a product of the distance between our microphone and the individual vocalizing and also the level of power (dB) in the vocalization emitted by the bat itself. There were significant differences between the male calls from each area in fundamental frequency (t = 2.04, df = 54, P = 0.0005), high frequency (t = 2.00, df = 54, P = 0.0005), low frequency (t = 2.04, df = 54, P = 0.0005), bandwidth (t = 2.00, df = 54, P = 0.0005), and total slope (t = 2.05, df = 54, P = 0.0005) (Figure 4, Table 1).

Discussion

Analysis of mating calls of males recorded near Skukuza versus those calls gathered near Shingwedzi (separated

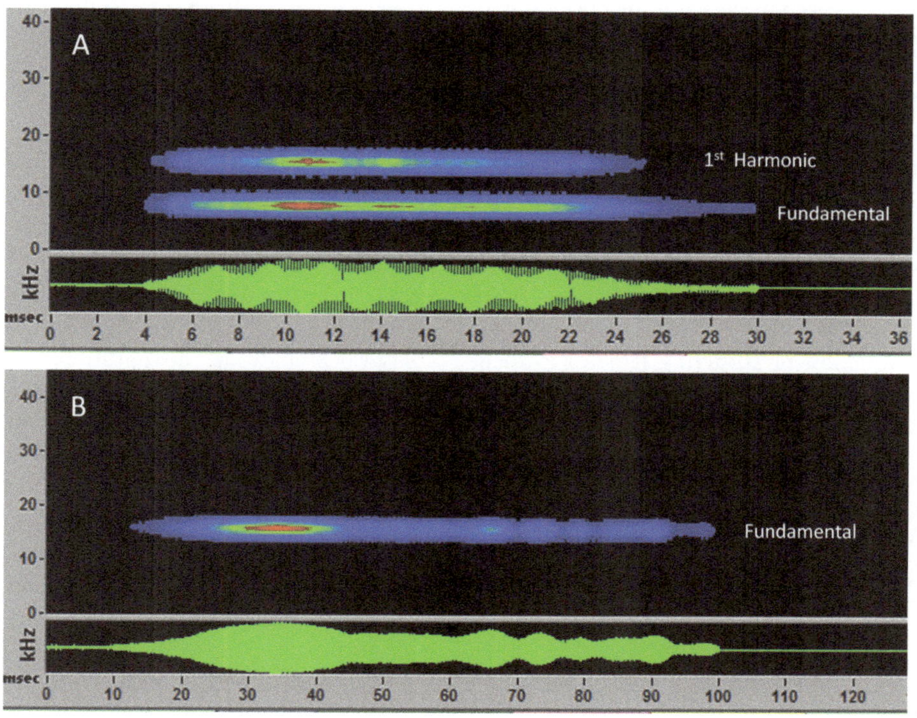

Figure 3 Sonograms of male vocalizations. (A) Near Skukuza in southern KNP and **(B)** near Shingwedzi in northern KNP. *X*-axes are time in milliseconds, *Y*-axes are frequencies displayed in kilohertz.

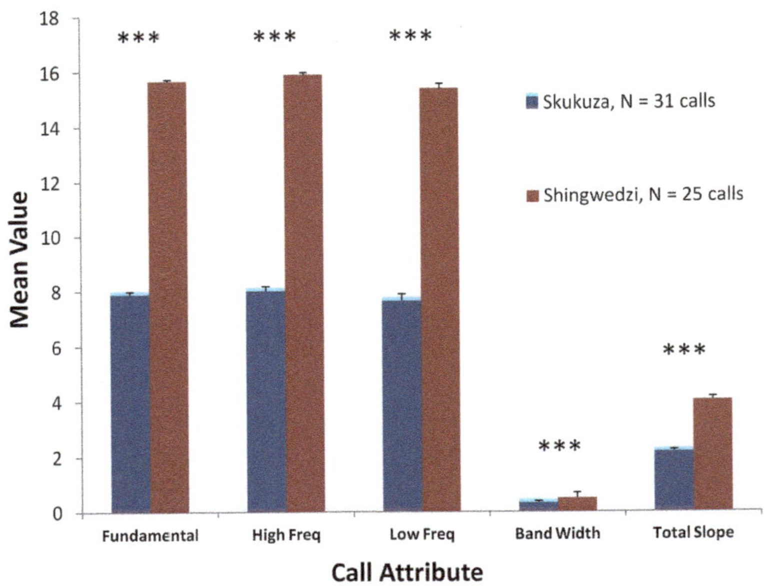

Figure 4 Comparative plot of mean values of five call attributes. Calls recorded near Skukuza (blue bars) versus calls recorded near Shingwedzi (red bars). Fundamental refers to the first harmonic where the highest energy occurred. Two-sample *T*-test determined that all four attributes differed significantly between sites after Bonferroni correction for multiple comparisons. ***$P = 0.0005$.

by approximately 270 km) showed significant differences in basic call structure and thus supported our hypothesis. We found that the call structure of males around specific fruiting sycamore fig trees were consistent with single-species presence for males at a very localized scale. However, there is also range overlap between the species, and because the male mating calls travel at least 0.5 km from the calling perch near where females are feeding on figs (Wickler and Seibt 1976, personal observations), there is potential for females of each species in the area to approach calling males of the other species if there is not a discernible difference in call structure.

From our data, we conclude that the male mating calls we recorded from each area depicted *E. wahlbergi* from Skukuza and *E. crypturus* from Shingwedzi. As mentioned, the mating call of male epauletted bats appears to serve both the function of attracting females to their perch because male call rates increase as females approach, but secondly, male calls appear to act as a deterrent to other conspecific males attempting to call from perches too close to other calling males (Wickler and Seibt 1976; Adams and Snode 2013). Although one could argue that because we could not definitively identify individuals to species from which calls were recorded, we could not make this determination. However, several factors help in corroborating our interpretation of results: a) there are unequivocal distinctions in the type of male mating calls recorded in different areas of KNP; b) the calls are consistently different in the two regions of KNP, with exceedingly small degrees of variation in each of the call parameters; c) our capture data, as well as data from other researchers (Bonaccorso et al. 2014), indicate that *E. wahlbergi* dominates the area around Skukuza, whereas *E. crypturus* dominates the area near Shingwedzi; d) capture data from other studies in KNP (Fenton et al. 1985), and other areas in Africa, show spatial separation between epauletted fruit bat species on both local (Wickler and Seibt 1976) and, in some cases, regional scales (Kingdon 1974); and e) there are no records of mixed-species feeding groups of epauletted fruit bats at the same fig trees, and radio-telemetry data indicate that female *E. wahlbergi* from a particular colony moved to the same feeding area of ripe fig trees nightly (Fenton et al. 1985).

Table 1 Means and standard deviations (SD) among five call attributes by site

	Mean attributes		Mean attributes	
	Skukuza	SD	Shingwedzi	SD
Fundamental	7.88	0.11	15.66	0.07
High frequency	8	0.18	15.89	0.074
Low frequency	7.66	0.22	15.38	0.18
Bandwidth	0.33	0.071	0.5	0.2
Total slope	0.002	0.0008	0.004	0.0012

An alternative hypothesis to ours would be that males of either *E. wahlbergi* or *E. crypturus* alter their call structure in different areas of KNP, and thus, we were recording the same species in both areas using calls that were geographically distinctive within species. Although we could not refute this hypothesis with our study, from a theoretical perspective, this seems unlikely, and we can think of no benefits of such mating behavior. In addition, the extreme lack of variation in call structure in the variables measured for this study suggests the possibility of stabilizing selection via mate choice that would not likely promote such strong regional differences in intraspecific mating calls. In fact, just the opposite would be expected (Ptacek 2000).

It should be noted that the call structure of male *E. wahlbergi* we recorded in KNP was substantially different from those recorded by Wickler and Seibt (1976) in Kenya. Our recordings contained only a fundamental and single second harmonic, whereas they found a fundamental with three harmonics (however, this could simply be the distance away from a perch where recording was taken). In addition, the fundamental frequency reported by Wickler and Seibt (1976) equated to 1.7 kHz, whereas our recordings show the fundamental at a much higher frequency of about 7 kHz. Although there are no published descriptions of *E. crypturus* calls available for comparison, it is also curious to note that *E. crypturus* calls in KNP correspond in frequency to the second harmonic of *E. wahlbergi* calls. Tonal differences in the calls of each species may be the product of differences in the slope of the calls. Although both species produce relatively flat calls, the total slope of the calls emitted in the Shingwedzi area was twice that of the calls we recorded in the Skukuza area on average. Alterations in call structure attributes presented herein represent a relatively simple measure by which males of co-occurring species of epauletted bats may use harmonic displacement to avoid call overlap with another similar species. The question of if species-specific differences in mating calls are learned or have inherent genetic components remains unanswered. However, our study raises the possibility that differences in mating calls between species of male epauletted fruit bats may serve to avoid hybridization between the two species of epauletted fruit bats.

Conclusions

There are two species of epauletted fruit bats (*E. wahlbergi* and *E. crypturus*) that co-occur in Kruger National Park, South Africa. Males of each species position themselves at perches near feeding trees where females congregate, and males use mating vocalizations, as well as wing flapping, to attract females to their perch to mate. The two species have similar mating calls to the naked ear, but through structural analysis of call structure in two areas separated by 270 km, we provide evidence that the males of each species have distinctive calls. We hypothesize that the differences in the structure of calls from each area produce distinctions that are important to the mating success of each species.

Competing interests

The authors declare that they have no competing interests.

Authors' contribution

RAA designed the study, carried out part of the data collection, and statistically analyzed and interpreted the data. ERS carried out part of the data collection and edited the manuscript. Both authors have read and approved the final manuscript.

Acknowledgements

The authors gratefully acknowledge the University of Northern Colorado for funding this study. We also greatly appreciate the hospitality and support provided by the staff of Kruger National Park in allowing us to conduct research and keeping us safe.

References

Acharya L (1992) Epomophorus wahlbergi. Mamm Species 394:1–4
Adams RA, Snode ER (2013) Unique insights into dispersion distances among males of Wahlberg's epauletted fruit bat in Kruger National Park, South Africa. Open Ecol J 6:54–60
Bergmans W (1988) Taxonomy and biogeography of African fruit bats (Mammalia, Megachiroptera). The genus *Epomophorus* Bennett, 1836. Beaufortia 38:75–146
Bonaccorso FJ, Winkelmann JR, Todd CM, Miles AC (2014) Foraging movements of epauletted fruit bats (Pteropodidae) in relation to the distribution of sycamore figs (Moraceae) in Kruger National Park, South Africa. Acta Chirop 16:41–52
Fenton MB, Brigham RM, Mills AM, Rautenbach IL (1985) The roosting and foraging areas of *Epomophorus wahlbergi* (Pteropodidae) and *Scotopholis viridis* (Vespertilionidae) in Kruger National Park, South Africa. J Mammal 66:461–468
Grant R, Thomas V, Van Gogh J (2001) Sappi tree spotting: lowveld including Kruger National Park, 2nd edn. Jacana Press, Johannesburg
Kingdon J (1974) East African mammals, vol. 2. Insectivores and bats. Academic, New York
McCracken GF, Wilkinson GS (2000) Bat mating systems. In: Chrichton EG, Krutzsch P (eds) Reproductive biology of bats. Academic, New York
Monadjem A, Reside AE (2012) Breeding seasons of *Epomophorus wahlbergi* in the lowveld of Swaziland. Afr Zool 47:321–325
Monadjem A, Taylor PJ, Cotterill FDP, Corrie-Schoeman M (2010) Bats of Southern and Central Africa. Wits University Press, Johannesburg
Ptacek PT (2000) The role of mating preferences in shaping interspecific divergence in mating signals in vertebrates. Behav Processes 51:111–134
Rollinson P, Coleman JC, Downs CT (2013) Seasonal differences in foraging dynamics, habitat use and home range sizes of Wahlberg's epauletted fruit bat in an urban environment. Afr Zool 48:340–350
Taylor PJ, Monadjem A (2008) Maxillary shape as a diagnostic tool for identifying fruit bats, *Epomophorus crypturus* and *E. wahlbergi* from museum specimens and in the field. S Afr J Wild Res 28:22–27
Wickler W, Seibt U (1976) Field studies on the African fruit bat, *Epomophorus wahlbergi* (Sundevall), with special reference to male calling. Z Tierpsychol 45:225–255

19

Anatomic and molecular characterization of the endocrine pancreas of a teleostean fish: Atlantic wolffish (*Anarhichas lupus*)

Jessica S Fortin[1*], Ariane Santamaria-Bouvier[2], Stéphane Lair[2], André D Dallaire[1] and Marie-Odile Benoit-Biancamano[1*]

Abstract

Background: The biologic attributes of the endocrine pancreas and the comparative endocrinology of islet amyloid polypeptide (IAPP) of fish are not well described in the literature. This study describes the endocrine pancreas of one teleostean fish. Ten captive Atlantic wolffish (*Anarhichas lupus*) from the Montreal Biodome were submitted for necropsy and their pancreata were collected.

Results: Grossly, all the fish pancreata examined contained 1-3 nodules of variable diameter (1-8 mm). Microscopically, the nodules were uniform, highly cellular, and composed of polygonal to elongated cells. Immunofluorescence for pancreatic hormones was performed. The nodules were immunoreactive for insulin most prominent centrally, but with IAPP and glucagon only in the periphery of the nodules. Exocrine pancreas was positive for chromogranin A. Not previously recognized in fish, IAPP immunoreactivity occurred in α, glucagon-containing, cells and did not co-localize with insulin in β cells. The islet tissues were devoid of amyloid deposits. IAPP DNA sequencing was performed to compare the sequence among teleost fish and the potency to form amyloid fibrils. In silico analysis of the amino acid sequences 19-34 revealed that it was not amyloidogenic.

Conclusions: Amyloidosis of pancreatic islets would not be expected as a spontaneous disease in the Atlantic wolffish. Our study underlines that this teleost fish is a potential candidate for pancreatic xenograft research.

Keywords: Beta cells; Endocrine pancreas; Glucagon; Insulin; Islet amyloid polypeptide (IAPP); Teleost fish

Background

The pancreas is an endodermal organ found in all vertebrates with two different functional components that are essential for the homeostatic regulation of glucose and food digestion. The exocrine pancreas is composed of acinar and duct cells involved in the production and transport of digestive enzymes into the gut, whereas the endocrine compartment regulates the production and secretion of peptide hormones into the bloodstream. In higher vertebrates, there are five endocrine cell types that cluster as the islets of Langerhans. The hormonal expression is specific to each cell type: glucagon (α cells), insulin (β cells), somatostatin (δ cells), ghrelin (ε cells), and pancreatic polypeptide (PP cells) (Ali 1985).

The pancreatic tissues of several species of teleost fish have been described and a complete separation of exocrine and endocrine parenchyma occurs (Ali 1985; Nguyen et al. 1995). The islet tissues of some teleost fish are concentrated in two principal islets (Brockmann bodies): one is found near the spleen and the other is located inside the wall of the duodenum, at the pyloric junction (Cutfield et al. 1986). These islets contain insulin, glucagon, and somatostatin, but only the pyloric Brockmann bodies secrete pancreatic polypeptide (Falkmer and Olsson 1962). The isolation and characterization of glucagon, insulin, somatostatin, and pancreatic polypeptide from islet tissues of several teleost fish were previously described (Cutfield et al. 1986; Conlon et al. 1987a; Conlon et al. 1987b; Conlon et al. 1986; Cutfield et al. 1987; Cutfield and Cutfield 1993). The amino acid

* Correspondence: Jessica.fortin.2@umontreal.ca; marie-odile.benoit-biancamano@umontreal.ca
[1]Département de pathologie et de microbiologie, Faculté de médecine vétérinaire, Université de Montréal, 3200 Sicotte, Saint-Hyacinthe, QC J2S 2M2, Canada
Full list of author information is available at the end of the article

sequence and primary structure of teleost islet hormones slightly differs from their higher vertebrate counterparts (Nguyen et al. 1995; Cutfield et al. 1986; Falkmer and Olsson 1962; Conlon et al. 1987a; Conlon et al. 1987b; Conlon et al. 1986; Cutfield et al. 1987; Cutfield and Cutfield 1993). Islet amyloid polypeptide (IAPP) is a hormone of 37 amino acids that is co-secreted with insulin by the β cells of mammals (Westermark et al. 2011). This protein, namely amylin, can generate amyloid fibril deposits found naturally in the human and cats, but pancreatic islet amyloid fibrils, in contrast to serum amyloid fibrils, are not found in rat and mouse (Westermark et al. 2011). Amyloid fibril formation contributes to the pathogenesis of diabetes mellitus in humans and cats (Westermark et al. 2011). IAPP has been described in pancreatic islet parenchyma of daddy sculpin (*Myoxocephalus scorpius*) (Westermark et al. 2002); however, the comparative endocrinology of IAPP and the propensity to form amyloid deposits remain to be elucidated in lower vertebrates.

Since the last decade, teleost species have become important as model systems in different areas of hormonal, metabolic, and immunologic research (Intine et al. 2013; McGonnell and Fowkes 2006). One advantage of using teleost fish in studies of diabetes mellitus is that the pancreatic endocrine cells are separated from the pancreatic exocrine tissue and can be easily isolated and harvested. Therefore, the Brockmann body of teleost fish, particularly the tilapia (*Oreochromis nilotica*), has been investigated as a potential xenograft tissue for patients with type 1 diabetes (Nguyen et al. 1995). The transplantation of tilapia Brockmann bodies into a diabetic nude mice model has been shown to promote long-term normoglycemia responsive to intraperitoneal glucose administration (Wright et al. 2004; Yang et al. 1997). However, islet amyloid formation is a major hindrance to pancreatic transplant resulting in non-immunologic transplanted islet graft loss in a primate transplant model (Liu et al. 2012). It is important for the xenograft success of micro-encapsulated pancreatic endocrine cells to identify which species of teleost fish could be used without the amyloidosis risk. To promote comparative understandings in endocrine pathophysiology, the pancreatic hormonal expression and localization in lower vertebrates should be further clarified.

In this study, we describe the principal islets (Brockmann bodies) of a different teleost species, the Atlantic wolffish (*Anarhichas lupus*), using immunohistochemical and immunofluorescence techniques in an attempt to better understand the physiological regulation of glucose and the potential usefulness of this teleost fish in endocrinological studies. These endocrine pancreas bodies of fish have occasionally been mistaken for endocrine tumors by pathologists; thus, we also clarify endocrine pancreas structure and function for comparative pathologists. The comparative pancreatic endocrinology of IAPP is poorly described in fish; we therefore focused our study on the isolation of the IAPP gene from the Atlantic wolfish pancreatic tissues. The amino acid sequences of the IAPP were determined, and the in silico analysis was performed to compare the amyloidogenicity of *Anarhichas* IAPP to other vertebrate species.

Methods

Atlantic wolffish (*A. lupus*)

All fish were obtained from the Montreal Biodome that maintained aquariums. Atlantic wolffish were approximately 9 to 10 years old and submitted dead on ice at the necropsy services for diagnostic purpose. Necropsy was done approximately 1 h after death, and all were well preserved. Pancreata were collected and fixed in 10% neutral buffered formalin, trimmed, and embedded in paraffin.

Routine staining, special stains, and immunohistochemistry

The paraffin sections (4 μm) were stained with hematoxylin-eosin-phloxine-saffron (HEPS) for the histologic analysis of all organs. Pancreatic sections were stained with Congo red to reveal amyloid deposits and Grimelius to identify neuroendocrine cells.

For immunohistochemistry procedure, pancreatic sections were deparaffinized and washed with PBS (pH 7.4) for 20 min. They were next permeabilized and blocked with 0.1% saponin and 3% (*w*/*v*) BSA in PBS. The tissue sections were further incubated during 1 h at 37°C. Immunofluorescence was performed with monoclonal antibodies against insulin, IAPP, glucagon, and somatostatin (Santa Cruz Biotechnology, Dallas, TX, USA). All were rabbit monoclonal antibody except for IAPP, which was a mouse monoclonal antibody. Rabbit anti-chromogranin A, anti-pancreatic polypeptide, and anti-synaptophysin antibodies were purchased from Thermo Fisher Scientific (Waltham, MA, USA). Each antibody was applied overnight at 4°C at a 1/50 dilution with a solution containing 0.1% saponin and 3% (*w*/*v*) BSA. The pancreatic sections were washed three times with PBS containing 0.05% Tween 20™ and then incubated 1 h at 37°C in the blocking buffer containing a secondary antibody. Anti-rabbit Alexa 488 and anti-mouse Alexa 594 were used as secondary antibodies at 1/1,000 dilutions (Santa Cruz Biotechnology, Dallas, TX, USA). Negative controls were performed using the serum corresponding to the type of the secondary antibody used (mouse or rabbit). After three washes, slides were mounted with anti-fade reagent (DakoCytomation, Carpinteria, CA, USA). The cellular distribution of each pancreatic hormone was assessed using a DMI 4000B reverse fluorescence microscope. Images were captured as 8-bit tagged image format files with a DFC 490 digital camera, and the images were analyzed using the Leica Application Suite Software, version 2.4.0

(Leica Microsystems Inc., Richmond Hill, ON, Canada). Data are representative of all specimens collected.

Isolation and sequencing of *Anarhichas* IAPP DNA

To sequence the IAPP gene, DNA was isolated from paraffin-embedded pancreatic tissues using a DNeasy tissue kit (Qiagen, Toronto, ON, Canada). Fast-cycling PCR DNA synthesis was performed with a Techne TC-512 thermocycler with the following PCR protocol: 5 min at 96°C; 45 cycles of 5 s at 96°C, 5 s at 53°C, and 5 s at 68°C; and a final extension of 1 min at 72°C. Each PCR reaction consisted of 500 nM of IAPP primers (forward and reverse), 300 ng of DNA, and 10 μL of fast-cycling Taq DNA polymerase master mix (Qiagen, Toronto, ON, Canada). Degenerate IAPP forward (5′-AAGTGCAACACAGCCACCTG-3′) and reverse (5′-CGTTTKCCGTAGGTRTTDCGA-3′) primers were used and covered the seven first and three last amino acids of the IAPP peptide (Westermark et al. 2002). PCR products were visualized in a 2% (*m*/*v*) agarose gel with ethidium bromide, and bands of the expected size were extracted (Zymoclean™ Gel DNA Recovery Kit, Cedarlane Labs, Burlington, NC, USA). A second PCR was performed using the same primers with the following PCR protocol: 5 min at 96°C; 45 cycles of 5 s at 96°C, 5 s at 54°C, and 5 s at 68°C; and a final extension of 1 min at 72°C. Sequencing was performed using the Mcgill gene sequencing platform. The amino acid sequence of *Anarhichas* IAPP was deducted based on DNA sequence (http://didac.free.fr/seq/dna2pro.htm).

In silico analysis of fibril formation

The tendency for β-sheet aggregation of each amino acid sequence was calculated based on the Agg parameter, obtained using in silico analysis with the TANGO program (http://tango.crg.es), which predicts protein aggregation. A higher Agg score is indicative of a peptide that can form potentially toxic aggregates. The amino acid sequence in the 19-34 segment of Atlantic salmon (*Salmo salar*), daddy sculpin (*M. scorpius*), and zebrafish (*Danio rerio*) IAPP previously published were selected to compare the Agg parameter (Westermark et al. 2002). Human and feline IAPP amino acid sequences were used as positive controls, and rodent IAPP was representative of a negative control. Control sequences were obtained from GenBank (www.ncbi.nlm.nih.gov/Genbank), with the following accession numbers: *Homo sapiens* [GenBank: M26650.1], *Felis catus* [GenBank: NM001043338.1], *Rattus norvegicus* [GenBank: NM012586.2].

Results

Macroscopic and microscopic examination

At necropsy, the pancreas was surrounded by adipose tissue. Figure 1A shows the gross appearance of the pancreas from an Atlantic wolffish, the arrows indicate the pancreatic tissue surrounded by adipose tissue. All the fish pancreata examined contained one to three nodules (Figure 1B) of variable diameter (1-8 mm) within the pancreas. Nodules were further isolated for histopathological examination (Figure 2A) revealing pancreatic islet cells in nodules sometimes surrounded by exocrine pancreatic tissue, though the specific subtopographical location of each nodule was not recorded at necropsy. The nodules consisted of uniform polygonal to elliptical cells and were surrounded by a delicate connective tissue capsule. Smaller nodules were completely embedded in exocrine tissue (Figure 2B) whereas larger nodules were not consistently completely encased in exocrine pancreas. Discrete nodules consisting of exocrine pancreatic tissue were also observed (Figure 2C).

Immunohistochemistry and special stains

Using immunofluorescence with an antibody against insulin (Figure 3A), a large number of β cells was observed. The elliptical β cells were found in most parts of the nodular islet not just on the periphery of the nodule. Other islet cells stained immunocytochemically positive for glucagon (Figure 3B). The α cells (glucagon containing) were confined to the periphery of the nodular islet encircling the β cells located in the center. Infrequently, glucagon-containing cells were noticed in the center of the nodule. Interestingly, IAPP was found at the periphery

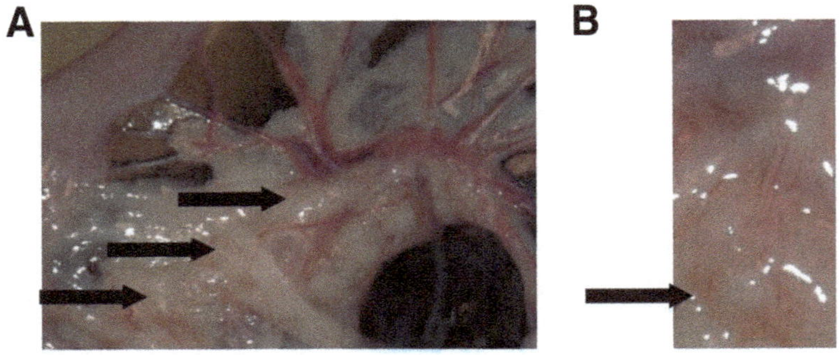

Figure 1 Brockmann bodies isolated from an Atlantic wolffish. **(A)** Pancreatic tissues are scattered in the mesentery (black arrows). **(B)** The Brockmann body is indicated with an arrow.

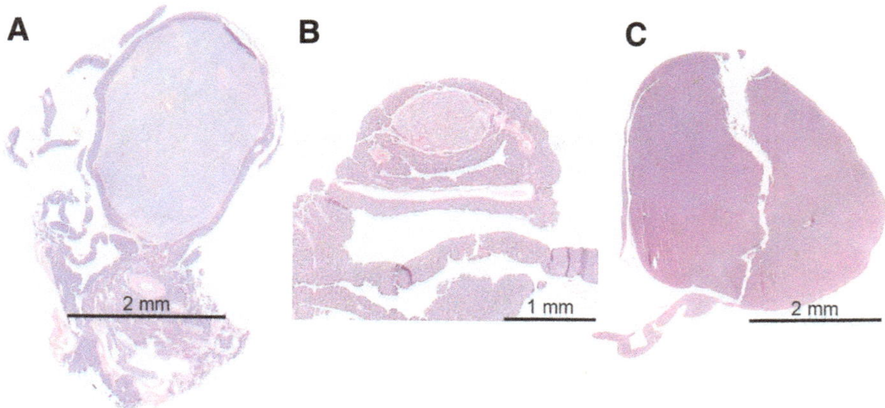

Figure 2 Microscopic features of excised nodules from the pancreas (HEPS stain). (A) Principal islet. **(B)** Smaller islets embedded in exocrine tissues. **(C)** Exocrine nodule from an Atlantic wolffish.

of the islet tissue and did not co-localize with insulin (Figure 3A). Instead, IAPP co-localized in α cells containing glucagon (Figure 3B).

Several islet nodules were surrounded by exocrine tissue. The exocrine cells were immunoreactive with chromogranin A (Figure 3C). One nodule of exocrine tissue occurred and immunoreacted positively with the antibody targeting chromogranin A (Figure 3D). None of the antibodies against somatostatin, pancreatic polypeptide, or synaptophysin reacted with islet tissue. When standard controls were carried out, negative immunofluorescence occurred consistently in all experiments. Noteworthy, Congo red and Grimelius staining of the islet cells from Atlantic wolffish were negative (Figure 4) for amyloid or neuroendocrine cells respectfully.

DNA sequence of *Anarhichas* IAPP and in silico analysis

We isolated DNA of teleost fish to amplify and sequence the IAPP gene and compared it to other species. PCR of DNA from the Atlantic wolffish resulted in a band of 120 base pairs. The amino acid sequence translated after the sequencing reaction is shown in Table 1. *Anarhichas* IAPP DNA is very different among the mammals and other teleost fish. The amino acid at position 19, 20, and 33 are highly conserved among the teleost fish. The IAPP amino acid sequences from all teleost fish are devoid of the GAILS region (amino acid positions 24-28) found in mammal IAPP amino acid sequences. GAILS represents an amino acid sequence at positions 24-28 in IAPP. The amino acid sequence GAILS contributes in part to the fibrillization process of IAPP. Two proline residues were

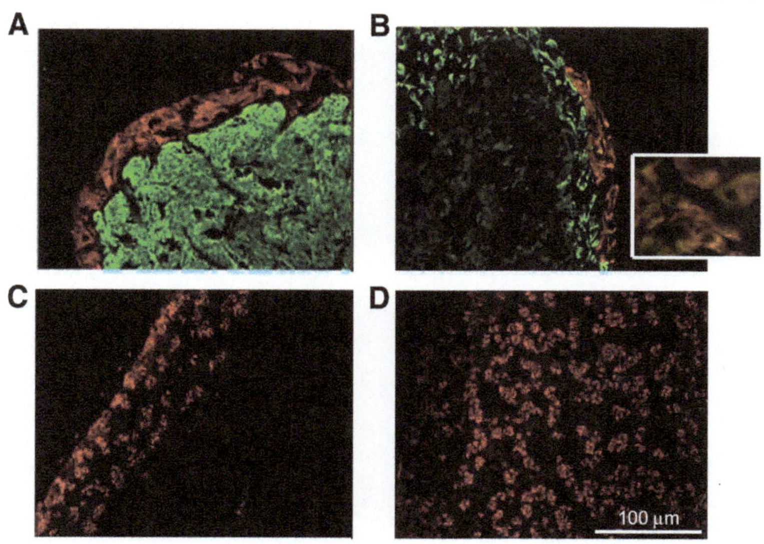

Figure 3 Histological sections of the pancreatic nodules with immunoreactivity to the pancreatic hormones of Atlantic wolffish. (A) Insulin (green) and IAPP (red) in the endocrine tissue. **(B)** Glucagon (green) and IAPP (red) in the endocrine tissue. **(C)** Chromogranin A surrounding the endocrine nodules. **(D)** The exocrine nodule with immunoreactivity to chromogranin A was isolated from an Atlantic wolffish.

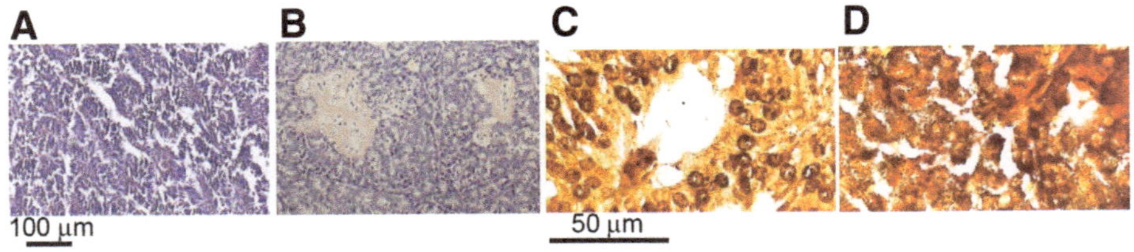

Figure 4 Congo red and Grimelius staining of the islet cells from an Atlantic wolffish. Congo red: **(A)** Atlantic wolffish. **(B)** Positive control from a feline pancreatic tissue with islet amyloidosis. Grimelius: **(C)** Atlantic wolffish. **(D)** Positive control from a canine adrenal tumor.

found in the IAPP amino acid sequences of the Atlantic wolffish and the zebrafish, but not in Atlantic salmon or daddy sculpin.

We compared the in silico potential of IAPP to form β-sheet precipitates, as expressed by the Agg parameter (Table 1). The human IAPP amyloidogenic sequence has the highest Agg parameter. Among the mammals, the rat non-amyloidogenic IAPP sequence shows the lowest Agg parameter. Interestingly, the Atlantic wolfish and the zebrafish exhibit a null Agg parameter, and the daddy sculpin Agg parameter is lower than the rat. In contrast, the Atlantic salmon IAPP sequence reveals Agg parameter higher than the cat, though spontaneous islet amyloidosis has not yet been reported.

Discussion

The pancreatic islets in the Atlantic wolffish appear as a few grossly evident nodules embedded in pancreatic parenchyma, designated Brockmann bodies in other teleost species. These are the homologs of the islets of Langerhans as has been previously described in other teleostean fish (Falkmer 1961). According to the Grimelius staining, the pancreatic tissues did not contain neuroendocrine cells, which preclude the islet nodule as the occurrence of a neuroendocrine tumor.

Demonstrated via immunofluorescence, the α cells (glucagon) occur at the periphery of the islet, and the β cells are distributed throughout the entire nodule. This distribution of islet cells is similar for other teleost species such as the daddy sculpin (*M. scorpius*) (Falkmer and Olsson 1962; Stefan and Falkmer 1980), gilthead sea bream (*Sparus auratus* L.) (Abad et al. 1986), swordtail (*Xiphophorus helleri* H.) (Klein and Lang 1977), and rosy barb (*Barbus conchonius*) (Rombout and Taverne-Thiele 1982).

Antibodies against somatostatin, pancreatic polypeptide, or synaptophysin did not react with nodular islet tissue, perhaps because the specific epitope does not occur in the Atlantic wolffish. Further research will be necessary to evaluate specific antibodies against teleostean proteins or clarify the affinity of current antibodies to each protein in teleostean fish. However, δ cells, which secrete somatostatin, are reported to be located near the β cells (Stefan and Falkmer 1980; Abad et al. 1986; Klein and Van 1978), and pancreatic polypeptide is found in the Brockmann bodies near the pylorus in other teleostean fish (Stefan and Falkmer 1980; Rombout and Taverne-Thiele 1982).

Chromogranin A was found in the exocrine pancreas and lining the connective tissue capsule of the pancreatic islet as previously reported with other teleostean fish (Falkmer and Olsson 1962; Stefan and Falkmer 1980; Abad et al. 1986; Klein and Lang 1977; Rombout

Table 1 Identification of the 37 amino acids of IAPP using PCR and sequencing

	1									10										20										30								Agg Segment 19-34
Human	K	C	N	T	A	T	C	A	T	Q	R	L	A	N	F	L	V	H	S	S	N	N	F	G	A	I	L	S	S	T	N	V	G	S	N	T	Y	23
Cat	K	C	N	T	A	T	C	A	T	Q	R	L	A	N	F	L	**I**	**R**	S	S	N	N	**L**	G	A	I	L	S	**P**	T	N	V	G	S	N	T	Y	9
Rat	K	C	N	T	A	T	C	A	T	Q	R	L	A	N	F	L	V	**R**	S	S	N	N	**L**	G	**P**	I	L	**P**	**P**	T	N	V	G	S	N	T	Y	0
Atlantic wolffish																			S	S	**P**	**S**	**R**	**S**	**G**	I	**S**	**P**	**R**	**N**	**T**	**Y**	G	**K**				0
Atlantic salmon									T	Q	R	L	A	**D**	F	L	**T**	**R**	S	S	N	**T**	**I**	G	**T**	**V**	**Y**	**A**	**P**	T	N	V	G	S				17
Daddy sculpin																			S	**N**	**S**	**T**	**I**	G	**T**	**V**	**N**	**A**	**P**	T	N	V	G	S				2
Zebrafish																	**T**	**R**	S	S	**S**	**P**	**I**	G	**T**	**V**	**N**	**A**	**P**	T	N	V	G	S				0

The amino acid sequences of teleost fish are compared to the amyloidogenic (human, cat) or non-amyloidogenic (rat) sequences. Sequences from species written in italics were obtained from a published work (Westermark et al. 2002). The Agg parameters were obtained using the Tango algorithm. The amino acids in boldface represent amino acid variations between species.

and Taverne-Thiele 1982). Nodules or discrete areas consisting of a dense and compact mass of acinar cells were observed and was chromogranin immunofluorescent, confirming an exocrine origin.

IAPP or amylin (the precursor to amyloid) was identified at the peripheral region of the islets for the Atlantic wolffish, co-localizing with glucagon. In contrast to these results in the Atlantic wolffish, IAPP has been localized in endocrine pancreas of mammals, co-localizing with insulin within the β cells and centrally, but not peripherally in the islets of the Atlantic salmon (*S. salar*) (Westermark et al. 2002). Amyloid was not found in the pancreatic tissues from the ten Atlantic wolffish.

IAPP has been previously isolated and characterized in three species of teleosts: zebrafish (*D. rerio*), Atlantic salmon (*S. salar*), and daddy sculpin (*M. scorpius*) (Westermark et al. 2002; Martinez-Alvarez et al. 2008). The amino acid sequences are different for the daddy sculpin, Atlantic salmon, zebrafish, and Atlantic wolffish (Table 1). Compared to mammals, the teleosts herein do not present the GAILS region (segment 24-28), which contributes to the amyloid formation in human and cats. The Atlantic salmon IAPP was shown to be amyloidogenic in vitro (Westermark et al. 2002). Interestingly, the Atlantic salmon Agg parameter was higher than the cat IAPP segment 19-34, which is known to be amyloidogenic in vivo. The in silico analysis performed with the Tango algorithm revealed that the *Anarhichas* IAPP segment 19-34 was not prone to form amyloid fibrils. The Atlantic wolffish and the zebrafish have two proline residues in contrast to the other teleost fish which have only one residue. Proline residues are known to protect against fibril formation (Westermark et al. 2011); thus, their presence possibly explains the lowest Agg parameter obtained from the Tango algorithm in the Atlantic wolfish and the zebrafish. However, further research is needed to evaluate the potential of fish full-length IAPP to form amyloid deposits.

Conclusions

We confirmed the species-specific differences in the hormonal tissue distribution in teleost fish and demonstrated a unique distribution of IAPP in the Atlantic wolffish. The presence of this difference in the population of cells secreting IAPP deserves further investigation. Previous phylogenetic analyses have shown that pancreatic peptides are relatively well conserved among vertebrates. However, the IAPP amino acid sequence differs among teleost fish. Our study shows that the Atlantic wolffish is devoid of amyloid in vivo and the segment 19-34 is non-amyloidogenic, according to the in silico secondary structure analysis. Amyloidosis of pancreatic islets would not be expected as a spontaneous disease in the Atlantic wolffish. The Atlantic wolffish thus could be a potential candidate to isolate the endocrine tissues for pancreatic xenografts. This study adds new insight in the characterization of fish pancreatic peptides whose physiological functions remain to be elucidated.

Competing interests

The authors declare that they have no competing interests.

Authors' contributions

JSF performed the experimental work and data analysis and drafted the manuscript. ASB, AD, and SL collected pancreatic tissues. All authors read and approved the final version of the manuscript.

Acknowledgements

The authors acknowledge the financial support of the Diabète Québec foundation. Jessica S. Fortin was also a recipient of a studentship from IRSC. We would like to thank Dr. Wong's collaboration (Biodome) for this study and also Dr. Carl Gagnon and Dr. Josée Harel for the access to their apparatus.

Author details

[1]Département de pathologie et de microbiologie, Faculté de médecine vétérinaire, Université de Montréal, 3200 Sicotte, Saint-Hyacinthe, QC J2S 2M2, Canada. [2]Département des sciences cliniques, Faculté de médecine vétérinaire, Université de Montréal, 3200 Sicotte, Saint-Hyacinthe, QC J2S 2M2, Canada.

References

Abad ME, Agulleiro B, Rombout JH (1986) An immunocytochemical and ultrastructural study of the endocrine pancreas of *Sparus auratus* L. (Teleostei). Gen Comp Endocrinol 64:1–12

Ali S (1985) Microvasculature of the principal islets in the scorpion fish, Myoxocephalus scorpius. Arch Histol Jpn 48:363–371

Conlon JM, Schmidt WE, Gallwitz B, Falkmer S, Thim L (1986) Characterization of an amidated form of pancreatic polypeptide from the daddy sculpin (*Cottus scorpius*). Regul Pept 16:261–268

Conlon JM, Davis MS, Falkmer S, Thim L (1987a) Structural characterization of peptides derived from prosomatostatins I and II isolated from the pancreatic islets of two species of teleostean fish: the daddy sculpin and the flounder. Eur J Biochem 168:647–652

Conlon JM, Falkmer S, Thim L (1987b) Primary structures of three fragments of proglucagon from the pancreatic islets of the daddy sculpin (*Cottus scorpius*). Eur J Biochem 164:117–122

Cutfield SM, Cutfield JF (1993) A second glucagon in the pancreatic islets of the daddy sculpin Cottus scorpius. Gen Comp Endocrinol 91:281–286

Cutfield JF, Cutfield SM, Carne A, Emdin SO, Falkmer S (1986) The isolation, purification and amino-acid sequence of insulin from the teleost fish *Cottus scorpius* (daddy sculpin). Eur J Biochem 158:117–123

Cutfield SM, Carne A, Cutfield JF (1987) The amino-acid sequences of sculpin islet somatostatin-28 and peptide YY. FEBS Lett 214:57–61

Falkmer S (1961) Experimental diabetes research in fish. Acta Endocrinol (Copenh) Supplementum 37(Suppl 59):1–122

Falkmer S, Olsson R (1962) Ultrastructure of the pancreatic islet tissue of normal and alloxan-treated Cottus scorpius. Acta Endocrinol (Copenh) 39:32–46

Intine RV, Olsen AS, Sarras MP Jr (2013) A zebrafish model of diabetes mellitus and metabolic memory. J Vis Exp 72:e50232

Klein C, Lang RH (1977) Principal cell types in the pancreatic islet of a teleost fish, *Xiphorphorus helleri* H. Cell Tissue Res 176:529–551

Klein C, Van NS (1978) Use of immunocytochemical staining of somatostatin for correlative light and electron microscopic investigation of D cells in the pancreatic islet of *Xiphophorus helleri* H. (Teleostei). Cell Tissue Res 194:399–404

Liu C, Koeberlein B, Feldman MD, Mueller R, Wang Z, Li Y, Lane K, Hoyt CC, Tomaszewski JE, Naji A, Rickels MR (2012) Accumulation of intrahepatic islet amyloid in a nonhuman primate transplant model. Endocrinology 153:1673–1683

Martinez-Alvarez RM, Volkoff H, Cueto JA, Delgado MJ (2008) Molecular characterization of calcitonin gene-related peptide (CGRP) related peptides

(CGRP, amylin, adrenomedullin and adrenomedullin-2/intermedin) in goldfish (Carassius auratus): cloning and distribution. Peptides 29:1534–1543

McGonnell IM, Fowkes RC (2006) Fishing for gene function—endocrine modelling in the zebrafish. J Endocrinol 189:425–439

Nguyen TM, Wright JR Jr, Nielsen PF, Conlon JM (1995) Characterization of the pancreatic hormones from the Brockmann body of the tilapia: implications for islet xenograft studies. Comp Biochem Physiol C Pharmacol Toxicol Endocrinol 111:33–44

Rombout JH, Taverne-Thiele JJ (1982) An immunocytochemical and electron-microscopical study of endocrine cells in the gut and pancreas of a stomachless teleost fish, Barbus conchonius (*Cyprinidae*). Cell Tissue Res 227:577–593

Stefan Y, Falkmer S (1980) Identification of four endocrine cell types in the pancreas of *Cottus scorpius* (Teleostei) by immunofluorescence and electron microscopy. Gen Comp Endocrinol 42:171–178

Westermark GT, Falkmer S, Steiner DF, Chan SJ, Engstrom U, Westermark P (2002) Islet amyloid polypeptide is expressed in the pancreatic islet parenchyma of the teleostean fish, *Myoxocephalus (cottus) scorpius*. Comp Biochem Physiol B Biochem Mol Biol 133:119–125

Westermark P, Andersson A, Westermark GT (2011) Islet amyloid polypeptide, islet amyloid, and diabetes mellitus. Physiol Rev 91:795–826

Wright JR Jr, Pohajdak B, Xu BY, Leventhal JR (2004) Piscine islet xenotransplantation. ILARJ 45:314–323

Yang H, Dickson BC, O'Hali W, Kearns H, Wright JR Jr (1997) Functional comparison of mouse, rat, and fish islet grafts transplanted into diabetic nude mice. Gen Comp Endocrinol 106:384–388

20

DNA barcoding supports identification of *Malacobdella* species (Nemertea: Hoplonemertea)

Jose E F Alfaya[1,2], Gregorio Bigatti[1,2], Hiroshi Kajihara[3], Malin Strand[4], Per Sundberg[5] and Annie Machordom[6*]

Abstract

Background: Nemerteans of the genus *Malacobdella* live inside of the mantle cavity of marine bivalves. The genus currently contains only six species, five of which are host-specific and usually found in a single host species, while the sixth species, *M. grossa*, has a wide host range and has been found in 27 different bivalve species to date. The main challenge of *Malacobdella* species identification resides in the similarity of the external morphology between species (terminal sucker, gut undulations number, anus position and gonad colouration), and thus, the illustrations provided in the original descriptions do not allow reliable identification. In this article, we analyse the relationships among three species of *Malacobdella*: *M. arrokeana*, *M. japonica* and *M. grossa*, adding new data for the *M. grossa* and reporting the first for *M. japonica*, analysing 658 base pairs of the mitochondrial cytochrome *c* oxidase subunit I gene (*COI*). Based on these analyses, we present and discuss the potential of DNA barcoding for *Malacobdella* species identification.

Results: Sixty-four DNA barcoding fragments of the mitochondrial *COI* gene from three different *Malacobdella* species (*M. arrokeana, M. japonica and M. grossa*) are analysed (24 of them newly sequenced for this study, along with four outgroup specimens) and used to delineate species. Divergences, measured as uncorrected differences, between the three species were *M. arrokeana-M. grossa* 11.73%, *M. arrokeana-M. japonica* 10.62% and *M. grossa-M. japonica* 10.97%. The mean intraspecific divergence within the ingroup species showed a patent gap with respect to the interspecific ones: 0.18% for *M. arrokeana*, 0.13% for *M. grossa* and 0.02% for *M. japonica* (ranges from 0 to 0.91%).

Conclusions: We conclude that there is a clear correspondence between the molecular data and distinguishing morphological characters. Our results thus indicate that some morphological characters are useful for species identification and support the potential of DNA barcoding for species identification in a taxonomic group with subtle morphological external differences.

Keywords: DNA barcoding; *COI* gene; Bivalvia; Entocommensal nemertean; *Malacobdella*

Background

The phylum Nemertea is a group of organisms whose identification and taxonomy requires specialized methods, mainly histology. Recently, molecular methods have been a useful tool for ascertaining the actual biodiversity of these worms and increasing our knowledge of several of the problematic species (Chen et al. 2010; Fernández-Álvarez and Machordom 2013; Kvist et al. 2013). The nemertean genus *Malacobdella* de Blainville 1827 originally contained 13 nominal species, of which six are currently regarded as valid (Gibson 1995; Ivanov et al. 2002). The species of the genus are entocommensal in the mantle cavity of marine bivalves, mainly from the subclass Heterodonta (Jensen and Sadeghian 2005). The phylogenetic position of the genus *Malacobdella* is controversial within the phylum, mainly because this genus is always represented by sequences belonging only to the species *M. grossa* (Thollesson and Norenburg 2003; Andrade et al. 2012), the most studied and cosmopolitan species. The *Malacobdella* species are distributed in distant locations around the world: *M. japonica*, *M. macomae* and *M. siliquae* were described in the eastern (Japan) and western (west coast of the USA) Pacific Ocean; *M. grossa* was described in the Pacific (west coast of the USA) and Atlantic Ocean (northern Europe); and *M. arrokeana* was the only southern species described in the South Atlantic Ocean (Ivanov et al. 2002

* Correspondence: annie@mncn.csic.es
[6]Museo Nacional de Ciencias Naturales (MNCN), CSIC, José Gutiérrez Abascal, 2, 28006 Madrid, Spain
Full list of author information is available at the end of the article

and references herein). The geographic distribution of the genus requires a huge sampling effort to work with all species. Identifying *Malacobdella* species is difficult because important diagnostic features were not initially recognized and thus not mentioned in earlier descriptions (Ivanov et al. 2002). In addition, the similarity of the external morphology between species (terminal sucker, number of intestinal loops (undulations), anus position and immature gonad colouration) and the illustrations provided in the original descriptions of the species do not allow reliable identification. There is little knowledge about the biology of *Malacobdella* species. The genus *Malacobdella* belongs to the hoplonemertean non-pilidiophora species (Thollesson and Norenburg 2003); this group presents direct development and non-feeding planuliform larvae (Maslakova and von Döhren 2009). However, there are no studies on its larval development, dispersion and settlement.

The host specificity of the *Malacobdella* species is generally high; five of the six known species have been reported from only one or two bivalve species: *M. arrokeana* Ivanov et al. 2002 from *Panopea abbreviata* (Heterodonta: Hiatelloidea) (Ivanov et al. 2002; Alfaya et al. 2013); *M. japonica* Takakura 1897 from *Spisula sachalinensis* (Heterodonta: Mactroidea) (Takakura 1897); *M. macomae* Kozloff 1991 from *Macoma nasuta* and *Macoma secta* (Heterodonta: Tellinoidea) (Kozloff 1991); *M. minuta* (Coe 1945) from *Yoldia cooperii* (Protobranchia: Nuculanoidea) (Coe 1945); and *M. siliquae* Kozloff 1991 from *Siliqua patula* (Heterodonta: Solenoidea) (Kozloff 1991). The sixth species, *M. grossa*, is the type specimen of the genus and was originally obtained from *Dosinia exoleta* (Linnaeus 1758) (Heterodonta: Veneroidea) (Müller 1776). However, the most complete morphological description of *M. grossa*, reported by Riepen (1933), was based on material from *Arctica islandica* (Linnaeus 1767) (Heterodonta: Arcticoidea). Nemerteans are traditionally identified and classified using morphological criteria, but the relatively low number of qualitative morphological characters, the lack of adequate fixation procedures for histological studies, vague descriptions in the original papers and the paucity of species-specific characters make species delimitation problematic, especially when comparing closely related species (Chen et al. 2010; Sundberg et al. 2010; Fernández-Álvarez and Machordom 2013; Kvist et al. 2013). The difficulty of morphological recognition of a great part of the more than 1,280 species included in the phylum Nemertea has been previously discussed (Andrade et al. 2012; Sundberg and Strand 2010; Kvist et al. 2013, among others), and different authors have advocated new tools and comprehensive studies for correctly identifying species and thus providing accurate biodiversity knowledge. The combination of molecular and morphological methods has been useful in elucidating nemertean taxonomy in other genera (Sundberg et al. 2009; Junoy et al. 2010; Puerta et al. 2010; Kajihara et al. 2011; Taboada et al. 2013). DNA barcoding accelerated the discovery ratio of new species (Wiens 2007) and identified some inconsistencies between species assignment and previously sequenced specimens (Kvist et al. 2013). Nevertheless, only a small number of nemerteans have been analysed through DNA barcoding (Sundberg et al. 2009; Chen et al. 2010; Fernández-Álvarez and Machordom, 2013; Kvist et al. 2013; Strand et al. 2014). The potential use of DNA barcoding needs to be substantiated in well-established taxonomic groups before it can be fully exploited in all nemertean genera. Currently, only 6% of the phylum Nemertea has an associated barcode sequence (Bucklin et al. 2011; Fernández-Álvarez and Machordom 2013).

Our objectives are to provide the most comprehensive molecular data set for the *Malacobdella* genus and test the usefulness of this data in the delimitation of the three most widely distributed species of the *Malacobdella* genus, and to provide for the first time *COI* sequences for *M. japonica*. We also estimate the phylogenetic relationship between 64 individuals from the 3 species based on partial sequences of *COI*.

Methods

Specimen collection and sequences used

Thirty-eight sequences of *M. arrokeana* were obtained from specimens collected at three northern Patagonian gulfs in Argentina (previously studied by Alfaya et al. 2013): San Matías Gulf ($n = 23$), San José Gulf ($n = 7$) and Nuevo Gulf ($n = 8$) (Table 1). Each of the *M. arrokeana* specimens was collected from a different host specimen of *P. abbreviata*. Specimens of *M. japonica* were collected from 15 different *S. sachalinensis* bivalves at Shinryu Beach in Akkeshi, Hokkaido, Japan. Individuals of *M. grossa* ($n = 11$) were obtained from different *A. islandica* clams in the waters of Tjärnö, Skagerak, Sweden. Additionally, two *M. grossa COI* sequences from GenBank (HQ848591 and AJ436905, from Tjärnö, Sweden, and the White Sea, Russia, respectively) were included in the analysis (Table 1). Fresh specimens were stored in absolute ethanol, and DNA was extracted from preserved tissues using a DNeasy extraction kit (Qiagen, Inc., Hilden, Germany), following the manufacturer's protocol.

To test for the monophyly of the genus, we sequenced *Ramphogordius sanguineus* specimens collected along the Argentinean coast to use as the outgroup. We also used *COI* sequences available from GenBank for nemerteans that are closely related to *Malacobdella* and have been previously used in other phylogenetic studies (Andrade et al. 2012; Mateos and Giribet 2008; Thollesson and Norenburg 2003), including two *R. sanguineus* outgroup sequences (Table 1).

Table 1 List of species used in the analysis, including the sample locality, number of specimens analysed (*N*) and GenBank accession numbers

Species	Specimen ID	Locations	Position	*N*	GenBank acc. number	References or voucher numbers[a]
Malacobdella arrokeana	Ma1	San Matías Gulf, Argentina	40°50′S/65°04′W	23	JX220596	CNP-INV 1879
	Ma2				JX220597	CNP-INV 1880
	Ma4				JX220599	CNP-INV 1881
	Ma5				JX220600	CNP-INV 1882
	Ma6				JX220601	CNP-INV 1883
	Ma7				JX220602	CNP-INV 1884
	Ma8				JX220603	CNP-INV 1885
	Ma9				JX220604	CNP-INV 1886
	Ma10				JX220605	CNP-INV 1887
	Ma11				JX220606	CNP-INV 1888
	Ma13				JX220607	CNP-INV 1889
	Ma14				JX220608	CNP-INV 1890
	Ma15				JX220609	CNP-INV 1891
	Ma16				JX220610	CNP-INV 1892
	Ma17				JX220611	CNP-INV 1893
	Ma18				JX220612	CNP-INV 1894
	Ma19				JX220613	CNP-INV 1895
	Ma20				JX220614	CNP-INV 1896
	Ma21				JX220615	CNP-INV 1897
	Ma22				JX220616	CNP-INV 1898
	Ma23				JX220617	CNP-INV 1899
	Ma24				JX220618	CNP-INV 1900
	Ma27				JX220620	CNP-INV 1901
	MaA1	Nuevo Gulf, Argentina	42°55′S/64°30′W	7	JX220621	CNP-INV 1902
	MaA2				JX220622	CNP-INV 1903
	MaA3				JX220623	CNP-INV 1904
	MaA4				JX220624	CNP-INV 1905
	MaA6				JX220625	CNP-INV 1906
	MaB1				JX220626	CNP-INV 1907
	MaB2				JX220627	CNP-INV 1908
	GSJ5A	San José Gulf, Argentina	42°20′S/64°10′W	8	JX220629	CNP-INV 1909
	GSJ5B				JX220630	CNP-INV 1910
	GSJJ5				JX220638	CNP-INV 1911
	GSJJ6				JX220639	CNP-INV 1912
	GSJJ7				JX220640	CNP-INV 1913
	GSJJ8				JX220641	CNP-INV 1914
	GSJJ9				JX220642	CNP-INV 1915
	GSJJ10				JX220643	CNP-INV 1916
M. japonica	Mja1	Hokkaido, Japan	43°3′N/144°51′E	13	KF597252	MNCN-5.02/3
	Mja2				KF597253	MNCN-5.02/4
	Mja4				KF597254	MNCN-5.02/5
	Mja5				KF597255	MNCN-5.02/6
	Mja6				KF597256	MNCN-5.02/7

Table 1 List of species used in the analysis, including the sample locality, number of specimens analysed (*N*) and GenBank accession numbers *(Continued)*

	Mja7				KF597257	MNCN-5.02/8
	Mja8				KF597258	MNCN-5.02/9
	Mja10				KF597259	MNCN-5.02/10
	Mja11				KF597260	MNCN-5.02/11
	Mja12				KF597261	MNCN-5.02/12
	Mja13				KF597262	MNCN-5.02/13
	Mja14				KF597263	MNCN-5.02/14
	Mja15				KF597264	MNCN-5.02/15
M. grossa	Mgrossa197	Tjärnö, Sweden	58°53'N/011°5'E	12	KF597241	MNCN-5.02/16
	Mgrossa198				KF597242	MNCN-5.02/17
	Mgrossa201				KF597243	MNCN-5.02/18
	Mgrossa202				KF597244	MNCN-5.02/19
	Mgrossa203				KF597245	MNCN-5.02/20
	Mgrossa204				KF597246	MNCN-5.02/21
	Mgrossa205				KF597247	MNCN-5.02/22
	Mgrossa206				KF597248	MNCN-5.02/23
	Mgrossa207				KF597249	MNCN-5.02/24
	Mgrossa208				KF597250	MNCN-5.02/25
	Mgrossa209				KF597251	MNCN-5.02/26
	Mgrossa GB 1				HQ848591	Andrade et al. (2012)
	Mgrossa GB 2	White Sea, Russia		1	AJ436905	Thollesson and Norenburg (2003)
Ramphogordius sanguineus	R.s1	Maine, USA		1	HQ848580	Andrade et al. (2012)
	R.s2	Anglesey, UK		1	AJ436938	Thollesson and Norenburg (2003)
	Li2	Nuevo Gulf, Argentina		4	KM387723	CNP-INV 1917
	Li3				KM387724	CNP-INV 1918
	Li4				KM387725	CNP-INV 1919
	Li5				KM387726	CNP-INV 1920
Amphiporus lactifloreus	A.l	Anglesey, UK	53°17'N/04°03'W	1	HQ848611	Andrade et al. (2012)
Paradrepanophorus crassus	P.c	Galicia, Spain		1	HQ848603	
Geonemertes pelaensis	G.p1	St. Davis, Bermuda		2	HQ848592	
	G.p2	Bermuda			EU255602	Mateos and Giribet (2008)

[a]Voucher numbers are given only for *Malacobdella arrokeana* specimens from the previous work of Alfaya et al. (2013), and for *M. grossa* and *M. japonica* newly sequenced here. MNCN, Museo Nacional de Ciencias Naturales (Madrid, Spain); CNP-INV, Invertebrate Collection of the Centro Nacional Patagónico (Argentina).

PCR amplification and sequencing

Partial *COI* sequences were amplified by PCR using the following primers: LCO1490 (5′-GGTCAACAAATCATAAAGATATTGG-3′) (Folmer et al. 1994) and *COI*-H (5′-TCAGGGTGACCAAAAAATCA-3′) (Machordom et al. 2003). Amplification was carried out in a 50-μl final volume reaction consisting of 5 μl buffer containing 10 × 2 mM $MgCl_2$, 1 μl dNTP mix (10 mM), 0.4 μl of *Taq* DNA polymerase (5 U/μl) (Biotools, Madrid, Spain), 1–3 μl of genomic DNA and 0.8 μl of each primer (10 μM). Thermal cycling conditions were an initial 4-min denaturation at 94°C, followed by 40 cycles of 45 s at 94°C, 1 min at 45°C and 1 min at 72°C, ending with a final 10-min extension at 72°C. The products were visualized under blue light in 0.8% agarose gels stained with SYBR Safe (Invitrogen, Carlsbad, CA, USA), with co-migrating 100-bp or 1-kb ladder molecular weight markers. The amplification products (around 700 bp) were purified by ethanol precipitation. Both strands were sequenced using the BigDye Terminator sequencing kit and an ABI PRISM 3730 DNA Sequencer (Applied Biosystems, Grand Island, NY, USA).

Data analysis

Special alignment was unnecessary as the *COI* sequences from the analysed species did not present any gaps. Phylogenetic analyses were performed with PAUP 4.0 b10 (Swofford 2000) for maximum parsimony (MP) and maximum likelihood (ML), and with MrBayes 3.2 (Huelsenbeck 2000; Huelsenbeck and Ronquist 2001) for Bayesian inference (BI). MP parameters included a heuristic search with tree bisection-reconnection (TBR) branch swapping and ten random additions. ML was also estimated through a heuristic search, with stepwise addition, applying the TIM3 + G + I (transitional) model (Posada 2003). Two runs of 5,000,000 generations were performed for BI, sampling one tree per 1,000 replicates. The model that best fits the data (TIM3 + I + G) was found with jModelTest (Posada 2008). Branch supports for MP (1,000 replicates) and ML (150 replicates) were determined by bootstrapping (Felsenstein 1985) and by posterior probabilities (after a burn-in of 20% of the obtained trees) for BI.

Results

According to original descriptions, the three studied species of the *Malacobdella* genus are very similar in their external morphology but differ in internal morphological characters. *Malacobdella arrokeana* differs from *M. japonica* and *M. grossa* in the origin of the proboscis retractor muscle. This muscle is curved dorsally and attaches to the internal body wall in *M. arrokeana*, whereas in *M. grossa*, it originates ventrally next to the terminal sucker (Table 2). In *M. japonica*, the muscle also originates ventrally but ends freely in the parenchyma (Table 2). *M. japonica* differs from *M. arrokeana* and *M. grossa* in the position of the nerve commissure. In *M. japonica*, the posterior nerve commissure is situated around the terminal sucker, while in the other two species, the nerve commissure is situated dorsally in the posterior part of the intestine just before the anus (Table 2).

M. grossa also differs from *M. arrokeana* and *M. japonica* in the position of the excretory pores, being ventro-lateral in *M. grossa* and dorso-lateral in *M. arrokeana* and *M. japonica*. The rhynchocoel occupies most of the body length in the three species, up to 80% to 90% with the proboscis occupying two thirds of the rhynchocoel. Gonad colouration varies according to species and stage of gonad development. Generally, the gonads are dark olive green in colour in mature specimens of *M. grossa*, rosy or purple in mature *M. arrokeana* and rosy in mature *M. japonica* (Figure 1, Table 2). However, gonad colouration is always white in immature specimens (Figure 1A).

The *COI* fragments sequenced for *M. arrokeana*, *M. grossa* and *M. japonica* were 658 bp in length. These sequences were deposited in GenBank under accession numbers KF597241 to KF597264. The different phylogenetic analyses resulted in the same topology, with high support within the ingroup (Figure 2). The divergences, measured as uncorrected distances (percentage of nucleotide substitutions), between the three species were as follows: *M. arrokeana-M. grossa* 11.73%, *M. arrokeana-M. japonica* 10.62% and *M. grossa-M. japonica* 10.97%. There was a clear gap between these interspecific divergences and the intraspecific ones: the mean intraspecific divergence within the ingroup species was 0.18% for *M. arrokeana* (range 0% to 0.92%), 0.13% for *M. grossa* (range 0% to 0.61%) and 0.02% for *M. japonica* (range 0% to 0.15%). It is noteworthy that within the *R. sanguineus* sequences selected as the outgroup (Table 1), the divergence between one (GenBank accession number

Table 2 Characteristic features of the analysed *Malacobdella* species

Features	*M. arrokeana* (Ivanov et al. 2002)	*M. japonica* (Takakura 1897)	*M. grossa* (Müller 1776)
Excretory pores	Dorso-lateral (40% of length)	Dorso-lateral (25% of length) (Yamaoka 1940)	Ventro-lateral
Proboscis length	Most of the rhynchocoel, 80 to 90	Two thirds of the rhynchocoel length (Yamaoka 1940)	Two thirds of the rhynchocoel length
Proboscis retractor muscle	Curved dorsally and attached to the body muscular wall	Originates ventrally and ends freely in the parenchyma (Yamaoka 1940)	Originates ventrally next to the terminal sucker (Riepen 1933)
Ovary colour (mature)	White or purple (Teso et al. 2006)	Rosy (Yamaoka 1940)	Olive green or yellowish green (Gibson 1968)
Testicle colour (mature)	Pale rose (Teso et al. 2006)	White (Yamaoka 1940)	Rosy or pinkish hue (Gibson 1968)
Nerve commissure	Dorsal in the posterior part of the intestine just before the anus	Far behind the anus along the posterior margin of the sucker (Yamaoka 1940) (posterior around the terminal sucker)	Dorsal in the posterior part of intestine, above the anus
Host specificity	High (only in *Panopea abbreviata*)	High (only in *Spisula sachalinensis*)	Low (27 species)
Geographic distribution	South Atlantic Ocean (from Uruguay to north Patagonian gulfs)	North Pacific Ocean (northern Japan)	North Atlantic Ocean (Europe and North America) North Pacific Ocean (North America)

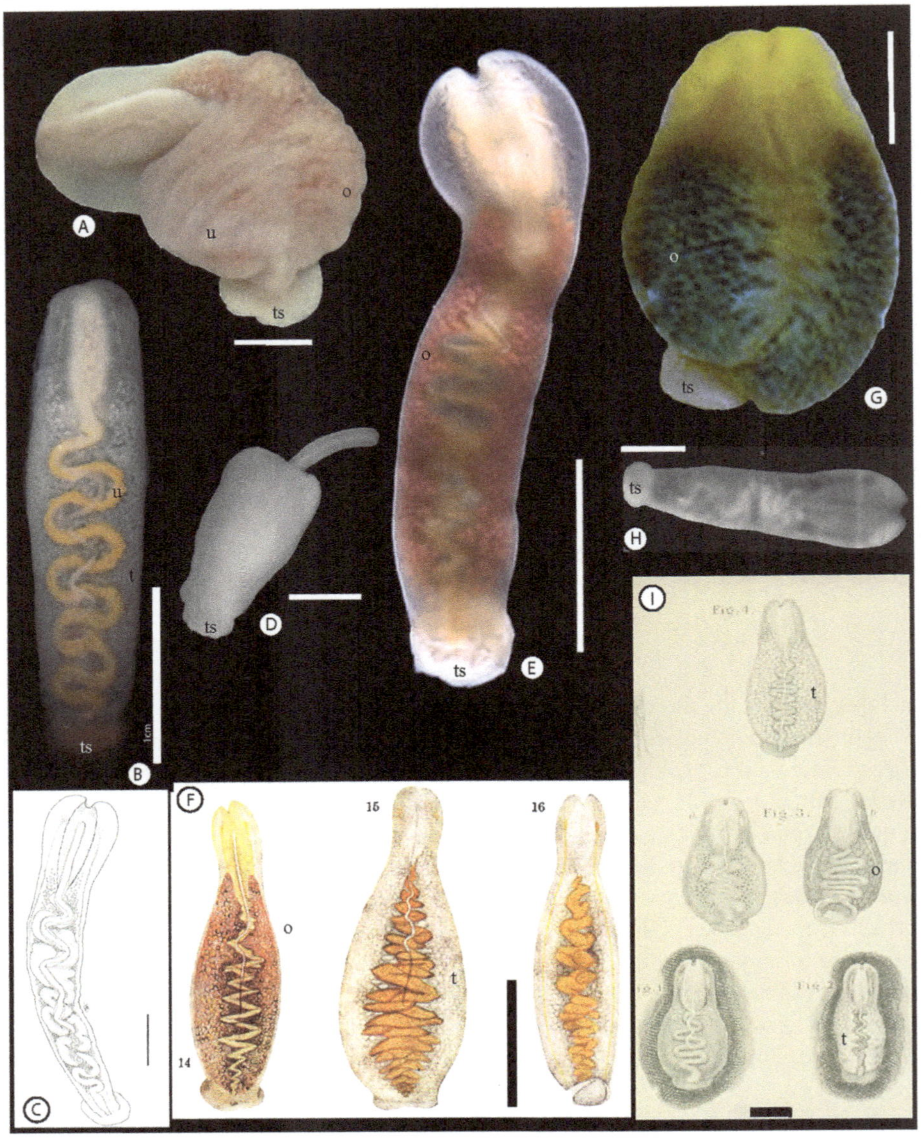

Figure 1 Specimens of the different *Malacobdella* species here analysed. *Malacobdella arrokeana*: **(A)** unrelaxed mature female, **(B)** relaxed male, **(C)** original illustration (Ivanov et al. 2002) and **(D)** immature specimen (scale bars: **A**, **B**: 10 mm; **C**: 3.5 mm, **D**: 1 mm). *Malacobdella japonica*: **(E)** relaxed mature female and **(F)** original illustration (Yamaoka 1940) (scale bars: **E**: 10 mm; **F**: 1 mm). *Malacobdella grossa*: **(G)** unrelaxed mature female, **(H)** immature specimen and **(I)** original illustration (Riepen 1933) (scale bars: **A**: 10 mm, **B**: 4 mm; **C**: 1 mm). Abbreviations: ts, terminal sucker; u, gut undulations; o, ovaries; t, testes.

HQ848580) of the sequences and the rest of the sequences of this species (GenBank AJ436938 and those collected by the authors) was more than 15% (15.75%), while the maximum divergence between the others was 100-fold lower (around 0.15%), suggesting that the *COI* sequence HQ848580 from GenBank is not actually from *R. sanguineus*. The five substitutions found among the *M. grossa* haplotypes were synonymous, as was the unique change among specimens of *M. japonica*, while of the 15 substitutions found among the haplotypes of *M. arrokeana*, three lead to amino acid changes.

Discussion

As previously mentioned, nemerteans of the *Malacobdella* genus are very difficult to identify based only on external characters. Rather, it is the internal morphology that provides the main taxonomic features used for identification, which can be differentiated only after a rigorous histological procedure. The genetic analysis performed here clearly shows that the distinguishing internal morphological differences used to separate the three species of *Malacobdella* analysed are concordant with differences in a fragment of the *COI* gene. However, previous

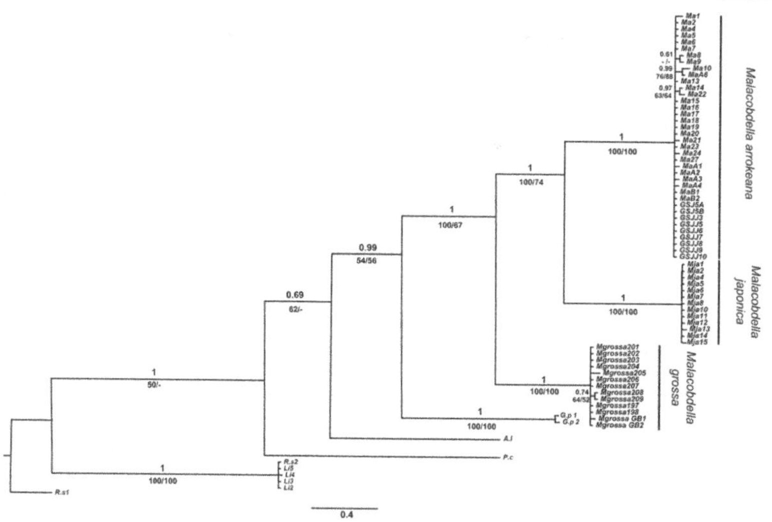

Figure 2 Haplotype phylogeny estimated by Bayesian analysis of the *Malacobdella* species studied. The numbers above the branches represent the posterior probabilities; the numbers below the branches represent the bootstrap percentages of the MP and ML analysis, respectively. When a certain node was not recovered by one of the methods, a hyphen was added. Ma, GSJ, *M. arrokeana*; Mja, *M. japonica*; *Mgrossa*, *M. grossa*; Li and R.s, *Ramphogordius sanguineus*, from Argentina and GenBank respectively; A.l, *Amphiporus lactifloreus*; P.c, *Paradrepanophorus crassus*; G.p, *Geonemertes pelaensis*.

DNA barcoding studies with nemerteans of the genera *Tetrastemma* and *Cerebratulus* revealed a lack of concordance between morphological and molecular characters (see Strand and Sundberg 2005; Sundberg et al. 2010). These authors argued that this difference could be the result of intraspecific variation or changes in the external morphology during development (Cantell 1975). They also concluded that the morphological characters used to describe *Tetrastemma* and *Cerebratulus* species are inadequate to identify evolutionary lineages. Our results showed that the divergence found between the sequences of the *Malacobdella* species (10% to 11% between the three species) are consistent with their status as distinct species. The values here found for intra- and interspecific divergences are in agreement with values found for other nemertean groups (e.g. Sundberg et al. 2010; Chen et al. 2010; Andrade et al. 2012; Kvist et al. 2013). The only exception was the divergence obtained between samples of *R. sanguineus*, 100-fold higher than other intraspecific values, suggesting that the *COI* sequence from GenBank (HQ848580) is not actually from *R. sanguineus*.

The phylogenetic analysis presented here showed that *M. arrokeana* and *M. japonica* are more closely related to each other than to *M. grossa*. Further studies using more molecular markers could clarify the phylogeny of the *Malacobdella* genus, taking into account the high host specificity of the majority of the species *versus* the cosmopolitan distribution and low host specificity of *M. grossa*. In this sense, the sequence of this last species from many hosts throughout its entire distribution would clarify whether *M. grossa* is indeed a single cosmopolitan species or a complex of species.

Based on the available literature by Takakura (1897), Yamaoka (1940), Riepen (1933), Gibson (1967, 1968, 1994), Gibson and Jennings (1969), Kozloff (1991), Ivanov et al. (2002) and the present work, the positions of the proboscis muscle and the nerve commissure appear to be good diagnostic characters for identifying species within the genus *Malacobdella* (since they are unique characteristic autapomorphies).

Several authors have concluded that nemertean species identification based only on morphological characters could be unreliable (see Strand and Sundberg 2005; Sundberg et al. 2009; Thornhill et al. 2008; Chen et al. 2010; Fernández-Álvarez and Machordom 2013; Kvist et al. 2013). More comprehensive future studies using the results presented here may strengthen species identification of other *Malacobdella* species by DNA barcoding.

Conclusions

Analysis of the *COI* (DNA barcoding) sequences presented in this work is clearly a powerful tool for species identification, at least of the three *Malacobdella* species studied. These molecular data are congruent with identifications based on internal morphological characters used in the original descriptions and re-descriptions of these three species and could be used to delineate between species in this genus with a similar external morphology.

Competing interests

The authors declare that they have no competing interests.

Authors' contributions

The work presented here was carried out in collaboration among all authors. JEFA, GB and AM defined the research theme and wrote the manuscript. JEFA, HK, MS and PS collected the samples in Argentina, Japan and Sweden, respectively, and took the photos. PS and MS provided *COI* sequences from *M. grossa*. JEFA and AM carried out the laboratory work (DNA extraction, PCR amplification, edition and sequences analysis), analysed the data and interpreted the results. JEFA prepared the tables and figures. PS, MS and HK improved the manuscript and made the English revision. All authors read and approved the final manuscript.

Acknowledgements

The authors thank CONICET, Secretaría de Ciencia Tecnología e Innovación Productiva de la Provincia de Chubut, Agencia Española de Cooperación Internacional y Desarrollo (AECID: A/023484/09 and A/032441/10) and the iBOL-CONICET project, for financial support. We are grateful also to Melinda Modrell for her conscientious revision of the English.

Author details

[1]LARBIM, IBIOMAR- Centro Nacional Patagónico (CENPAT), CONICET, Bvd. Brown 2915, U9120ACV Puerto Madryn, Chubut, Argentina. [2]Facultad de Ciencias Naturales, Universidad Nacional de la Patagonia San Juan Bosco (UNPSJB), Bvd. Brown S/N, U9120ACV Puerto Madryn, Chubut, Argentina. [3]Faculty of Science, Hokkaido University, Kita 10, Nishi 8, Kita-ku, Sapporo 060-0810, Japan. [4]The Swedish Species Information Centre, SLU, Bäcklösavägen 10, Box 7007, SE-750 07 Uppsala, Sweden. [5]Department of Biological and Environmental Sciences, University of Gothenburg, Medicinaregatan 18, PO Box 463, SE-40530 Gothenburg, Sweden. [6]Museo Nacional de Ciencias Naturales (MNCN), CSIC, José Gutiérrez Abascal, 2, 28006 Madrid, Spain.

References

Alfaya JEF, Bigatti G, Machordom A (2013) Mitochondrial and nuclear markers reveal a lack of genetic structure in the entocommensal nemertean *Malacobdella arrokeana* in the Patagonian gulfs. Helgoland Mar Res 67(2):407–412

Andrade SCS, Strand M, Schwartz M, Chen H, Kajihara H, von Döhren J, Sun S, Junoy J, Thiel M, Norenburg JL, Turbeville JM, Giribet G, Sundberg P (2012) Disentangling ribbon worm relationships: multi-locus analysis supports traditional classification of the phylum Nemertea. Cladistics 28:141–159

Bucklin A, Steinke D, Blanco-Bercial L (2011) DNA barcoding of marine metazoa. Annu Rev Mar Sci 3:471–508

Cantell CE (1975) Anatomy, taxonomy, and biology of some Scandinavian heteronemertines of the genera *Lineus*, *Micrura*, and *Cerebratulus*. Sarsia 58:89–122

Chen H, Strand M, Norenburg JL, Sun S, Kajihara H, Chernyshev AV, Maslakova SA, Sundberg P (2010) Statistical parsimony networks and species assemblages in Cephalotrichid nemerteans (Nemertea). PLoS ONE 5(9):e12885. doi:10.1371/journal.pone.0012885

Coe WR (1945) *Malacobdella minuta*, a new commensal nemertean. J Wash Acad Sci 35:65–67

de Blainville HM (1827) Sangsue. In: Levrault EG (ed) Dictionnaire des Sciences Naturelles, 47. Normant, Paris, pp 270–271

Felsenstein J (1985) Confidence limits on phylogenies: an approach using the bootstrap. Evolution 39:783–791

Fernández-Álvarez FÁ, Machordom A (2013) DNA barcoding reveals a cryptic nemertean invasion in Atlantic and Mediterranean waters. Helgoland Mar Res 67(3):599–605

Folmer O, Black M, Hoeh W, Lutz R, Vrijenhoek R (1994) DNA primers for amplification of mitochondrial cytochrome *c* oxidase subunit I from diverse metazoan invertebrates. Mol Mar Biol Biotech 3:294–299

Gibson R (1967) Occurrence of the entocommensal rhynchocoelan, *Malacobdella grossa*, in the oval piddock, *Zirfaea crispata*, on the Yorkshire coast. J Mar Biol Assoc UK 49:301–317

Gibson R (1968) Studies on the biology of the entocommensal rhynchocoelan *Malacobdella grossa*. J Mar Biol Assoc UK 48:637–656

Gibson R (1994) Nemerteans. Synopses of the British fauna (new series, no. 24), 2nd edn. The Dorset Press, Dorchester, UK

Gibson R (1995) Nemertean genera and species of the world: an annotated checklist of original names and description citations, synonyms, current taxonomic status, habitats and recorded zoogeographic distribution. J Nat Hist 29:1–562

Gibson R, Jennings JB (1969) Observations on the diet, feeding mechanisms, digestion and food reserves of the entocommensal rhynchocoelan *Malacobdella grossa*. J Mar Biol Assoc UK 49:17–32

Huelsenbeck JP (2000) MrBayes: Bayesian inference of phylogeny. University of Rochester, Rochester, NY, Department of Biology

Huelsenbeck JP, Ronquist F (2001) MrBayes: Bayesian inference of phylogeny. Bioinformatics 17:754–755

Ivanov VA, Bigatti G, Penchaszadeh PE, Norenburg JL (2002) *Malacobdella arrokeana* (Nemertea: Bdellonemertea), a new species of nemertean from the southwestern Atlantic Ocean entocommensal in *Panopea abbreviata* (Bivalvia, Heterodonta, Hiatellidae) in Argentina. P Biol Soc Wash 115:359–367

Jensen K, Sadeghian PS (2005) Nemertea. In: Rohde K (ed) Marine parasitology. CSIRO press, Australia, pp 205–210

Junoy J, Andrade SCS, Giribet G (2010) Phylogenetic placement of a new hoplonemertean species commensal on ascidians. Invertebr Syst 24:616–629

Kajihara H, Olympia M, Kobayashi N, Katoh T, Chen HX, Strand M, Sundberg P (2011) Systematics and phylogeny of the hoplonemertean genus *Diplomma* (Nemertea) based on molecular and morphological evidence. Zool J Linn Soc 161:695–722

Kozloff EN (1991) *Malacobdella siliquae* sp. nov. and *Malacobdella macomae* sp. nov., commensal nemerteans from bivalve mollusks on the Pacific coast of North America. Can J Zool 66:1612–1618

Kvist S, Laumer C, Junoy J, Giribet G (2013) New insights into the phylogeny, systematics and DNA barcoding of Nemertea. Invertebr Syst 28:287–308

Linnaeus C (1758) Systema naturae per regna tria naturae: secundum classes, ordines, genera, species, cum characteribus, differentiis, synonymis, locis, vol I, 10th edn. Laurentius Salvius, Stockholm, pp 824

Linnaeus C (1767) Systema naturae per regna tria naturae, secundum classes, ordines, genera, species, cum characteribus & differentiis. Vol. II. Salvius, Holmiae, p 735

Machordom A, Araujo R, Erpenbeck D, Ramos MA (2003) Phylogeography and conservation genetics of endangered European Margaritiferidae (Bivalvia: Unionoidea). Biol J Linn Soc 78:235–252

Maslakova SA, von Döhren J (2009) Larval development with transitory epidermis in *Paranemertes peregrina* and other hoplonemerteans. Biol Bull 216:273–292

Mateos E, Giribet G (2008) Exploring the molecular diversity of terrestrial nemerteans (Hoplonemertea, Monostilifera, Acteonemertidae) in a continental landmass. Zool Scr 37:235–243

Müller OF (1776) Zoologiae Danicae prodromus, seu Animalium Daniae et Norvegiae indigenarum characters, nomina, et synonyma imprimis popularium. Typis Hallageriis, Havniae, Copenhagen

Posada D (2003) Using Modeltest and PAUP* to select a model of nucleotide substitution. In: Baxevanis AD, Davison DB, Page RDM, Petsko GA, Stein LD, Stormo GD (eds) Current protocols in bioinformatics. John Wiley & Sons, New York, pp 6.5.1–6.5.14

Posada D (2008) jModelTest: phylogenetic model averaging. Mol Biol Evol 25:1253–1256

Puerta P, Andrade SCS, Junoy J (2010) Redescription of *Lineus acutifrons* Southern, 1913 (Nemertea) and comments on its phylogenetic position. J Nat Hist 44:37–40

Riepen O (1933) Anatomie und histologic von *Malacobdella grossa* (Müll.). Z Wiss Zool Abt A 143:323–496

Strand M, Sundberg P (2005) Delimiting species in the hoplonemertean genus *Tetrastemma* (phylum Nemertea): morphology is not concordant with phylogeny as evidenced from mtDNA sequences. Biol J Linn Soc 86:201–212

Strand M, Hererra-Bachiller A, Nygren A, Kånneby T (2014) A new nemertean species: what are the useful characters for ribbon worm descriptions? J Mar Biol Assoc UK 94:317–330

Sundberg P, Strand M (2010) Nemertean taxonomy - time to change lane? J Zool Syst Evol Res 48:283–284

Sundberg P, Thuróczy Vodoti E, Zhou H, Strand M (2009) Polymorphism hides cryptic species in *Oerstedia dorsalis* (Nemertea, Hoplonemertea). Biol J Linn Soc 98:556–567

Sundberg P, Vodoti E, Strand M (2010) DNA barcoding should accompany taxonomy the case of *Cerebratulus* spp. (Nemertea). Mol Ecol Res 10:274–281
Swofford DL (2000) *PAUP**. Phylogenetic analysis using parsimony (*and other methods), Version 4. Sinauer Associates, Sunderland, Massachusetts, USA
Taboada S, Junoy J, Andrade SCS, Giribet G, Cristobo J, Avila C (2013) On the identity of two Antarctic brooding nemerteans: redescription of *Antarctonemertes valida* (Burger, 1893) and description of a new species in the genus *Antarctonemertes* Friedrich, 1955 (Nemertea, Hoplonemertea). Polar Biol 36:1415–1430
Takakura U (1897) On a new species of *Malacobdella* (*M. japonica*). Annot Zool Japon 1:105–112
Teso SV, Bigatti G, Bazterrica MC, Ciocco NF, Penchaszadeh PE (2006) The reproductive cycle of the entocommensal nemertean *Malacobdella arrokeana* and its symbiosis with the geoduck *Panopea abbreviata*. Invertebr Biol 125(4):314–324
Thollesson M, Norenburg JL (2003) Ribbon worm relationships: a phylogeny of the phylum Nemertea. P Roy Soc Lond B Biol 270:407–415
Thornhill DJ, Mahon AR, Norenburg JL, Halanych KM (2008) Open-ocean barriers to dispersal: a test case with the Antarctic Polar Front and the ribbon worm *Parborlasia corrugatus* (Nemertea: Lineidae). Mol Ecol 17:5104–5117
Wiens JJ (2007) Species delimitation: new approaches for discovering diversity. Syst Biol 56:875–878
Yamaoka T (1940) The fauna of Akkeshi Bay. IX. Nemertini. J Fac Sci, U Hokkaido 6(7):205–261

Feeding interactions in an assemblage of terrestrial carnivores in central Mexico

Yuriana Gómez-Ortiz[1], Octavio Monroy-Vilchis[1*] and Germán D Mendoza-Martínez[2]

Abstract

Background: We evaluated the strategies of resources partitioning among species, dietary overlap and niche breadth in an assemblage of carnivores integrated by top predators (*Puma concolor* and *Panthera onca*) and mesopredators (*Leopardus pardalis, Leopardus wiedii, Puma yagouaroundi, Nasua narica* and *Urocyon cinereoargenteus*). The aim of this study was to investigate the mechanisms explaining the coexistence among species at a temperate zone in central Mexico.

Results: We collected 259 scats of carnivores and identified 45 food items. The analysis showed the common consumption of mammals in the assemblage, and the correspondence analysis showed three guilds: 1) top predators associated with the use of medium-sized mammals and the exclusive consumption of large mammals, 2) carnivore mesopredators associated with the consumption of small mammals and birds and 3) omnivore mesopredators associated with the consumption of arthropods and plants. The dietary overlap analysis indicated a low overlap between guilds and a high overlap within them. Top predators were specialist foragers, whilst the carnivore mesopredators showed generalist consumption.

Conclusions: The coexistence in this carnivore assemblage seems to be related to body size, morphology and prey segregation because such characteristics suggest the presence of three guilds. We observed high dietary overlap within guilds and resource partitioning between guilds.

Keywords: Coexistence; Community; Niche breadth; Niche overlap; Trophic interactions

Background

The evaluation of feeding relationships between predators, prey and their environment is used to identify factors structuring communities (Carvalho and Gomes 2004). To reach this aim, it is necessary to analyse trophic interactions between species that use similar resources, namely guilds (Root 1967; Jaksic 1981). The guild approach allows to explain the association among species, the ecological structure and community dynamics (Carvalho and Gomes 2004; Zapata et al. 2007). Ecological interactions and the mechanisms involved in resource partitioning and species coexistence are studied in relation to food, space or time segregation (MacArthur and Levins 1967; Schoener 1974). The knowledge of trophic structure in carnivore communities improves their management and conservation, particularly in scarcely studied communities such as temperate zones (Ray and Sunquist 2001; Guerrero et al. 2002; Carvalho and Gomes 2004).

The predation theory predicts a convergence of trophic niches when resources are abundant (Schoener 1982), such that the dietary overlap might suggest high abundance of resources in the environment and moderate competition (Silva-Pereira et al. 2011). Because carnivore communities are structured by intraguild predation and competition, all members play important ecological roles in the ecosystem and show different strategies to coexist (Bianchi et al. 2011). Likewise, the presence of generalist predators in the ecosystem induces the exploitation of a wide range of prey by other carnivores (Pianka 1973). Thus, the opportunistic ability of some species represents an advantage in the use of resources and favours coexistence on the guild (Jaksic et al. 1996; Guerrero et al. 2002).

We considered that the body size of predators influences the community structure and resource sharing between species (Simberloff and Dayan 1991; Hayward

* Correspondence: tavomonroyvilchis@gmail.com
[1]Estación Biológica Sierra Nanchititla, Facultad de Ciencias, Universidad Autónoma del Estado de México, Instituto Literario 100, Colonia Centro, Toluca, CP 50000, México
Full list of author information is available at the end of the article

et al. 2006). We defined the top predators as species occupying the apex trophic position in a community (large-bodied) and mesopredators as species occupying trophic positions below them (medium-bodied, Ritchie and Johnson 2009). We investigated the feeding strategies that may explain the coexistence among members of a carnivore assemblage at a temperate zone of central Mexico in relation to morphological similarity. As the body size of predators is thought to shape community structure, we expected high dietary overlap between *Puma concolor* and *Panthera onca* (top predators), between *Leopardus pardalis*, *Leopardus wiedii* and *Puma yagouaroundi* (carnivore mesopredators) and between *Nasua narica* and *Urocyon cinereoargenteus* (omnivore mesopredators).

We obtained data regarding diet composition, niche overlap and niche breadth to identify resource partitioning strategies. The study was conducted through the analysis of field-collected scats. The most effective identification of scats has been performed using molecular techniques such as bile acids identification using thin-layer chromatography (TLC) or DNA extraction (Ernest et al. 2000; Cazón and Sühring 1999). TLC has been suggested as an effective means of identification of carnivore scats, since the profiles of faecal bile acids have been shown to be species-specific. Species often display a unique bile acid profile, so TLC uses specific differences in bile acid profiles to distinguish between species (Major et al. 1980; Fernández et al. 1997; Taber et al. 1997; Ray and Sunquist 2001). This is an inexpensive and non-invasive technique that has been effective with carnivore species, although some studies have described a number of limitations when applied in omnivores due to the presence of vegetal pigments (Cazón and Sühring 1999). Our study is the first to describe the diet of small felids (*L. pardalis*, *L. wiedii* and *P. yagouaroundi*) using TLC, which improves species assignment and allows us to analyse the association strategies among guilds.

Methods

Study area

Sierra Nanchititla Natural Park (SNNP) is located in central Mexico, between extreme coordinates 100° 36′ 49′′-100° 16′ 03′′ west longitude and 18° 45′ 13′′-19° 04′ 22′′ north latitude. Altitude ranges from 410 to 2,080 m a.s.l. and covers 663.87 km^2 (Figure 1). The vegetation types are oak forest (30%), induced grasslands (30%), deciduous tropical forest (18%), pine-oak forest (17%) and cultivations (4%, Monroy-Vilchis et al. 2008). Overall, 53 species of mammals have been recorded, belonging to 6 orders, including 17 families and 38 genera; of these, 14 species are carnivores, 11 are endemic and 9 are endangered or considered rare (Monroy-Vilchis et al. 2011).

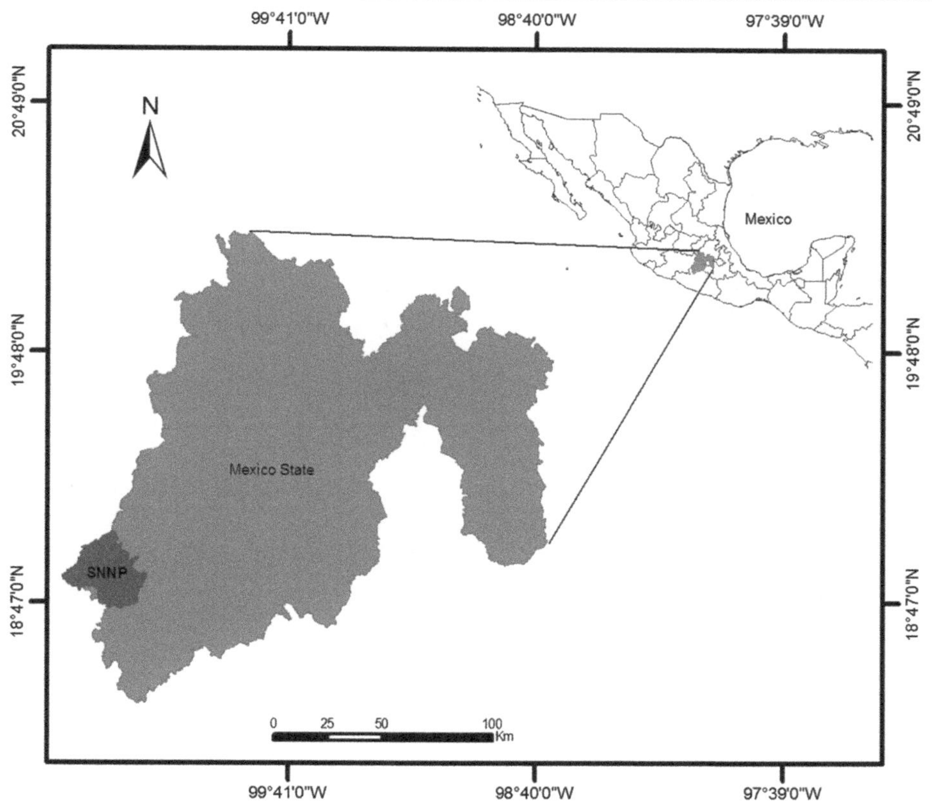

Figure 1 Geographic position of Sierra Nanchititla Natural Park (SNNP) in central Mexico.

Scat identification

From September 2009 to March 2012, carnivore scats were collected over different paths distributed throughout the SNNP. The scats were identified considering the diameter, associated traces (tracks or scrapes) and extraction of bile acid using TLC (Salame-Méndez et al. 2012). We collected two known-origin reference scats from individual *P. concolor* (three individuals) and *P. onca* (three individuals) and one known-origin reference scat from individual *L. pardalis* (six individuals), *L. wiedii* (eight individuals) and *P. yagouaroundi* (six individuals). The reference bile acid profiles were obtained through the following methods. We used a gram approximately of each scat collected and preserved it in ethanol; the solution was mixed and 1 μl of the supernatants was used to extract bile acids. We used two chromatography techniques to separate all bile acids by polarity, each one with a different mobile phase: 1) chloroform:methanol:acetic acid (80:12:0.5 *v*/*v*) and 2) chloroform:ethanol:acetic acid (80:45:3 *v*/*v*).

Samples were run against a known mixture of eight bile acids (cholic acid, chenodeoxycholic acid, deoxycholic acid, lithocholic acid, taurodeoxycholic acid, taurocholic acid, glycolic acid and dehydrocholic acid) and cholesterol, from Sigma-Aldrich® (St. Louis, MO, USA). The bile acid spots of the mixtures were identified and RF values were calculated as the ratio of the distance moved by the solute (i.e. bile acid) and the distance moved by the solvent front along the TLC plate. Finally, each bile acid profile obtained from field-collected scats was compared against reference bile acid profiles extracted from known-origin reference scats.

Diet composition

The scat analysis was based on the standard methodology reported by Monroy-Vilchis et al. (2009a) and Gómez-Ortiz et al. (2011). Food items were identified considering remains in scats (hair, bones, claws, teeth, seeds, feathers etc.). Mammalian identifications were carried out through hair identification (Monroy-Vilchis and Rubio-Rodríguez 2003). Birds, reptiles, arthropods and fruits were identified by comparing them with specimen collections.

The contribution of each food item was presented as follows: the frequency of occurrence (FO), defined as the percentage of presences of a given food item in the total number of scats, and the percentage of occurrence (PO), obtained as the number of occurrences of a given food item divided by the total food items identified.

The resource partitioning among species was evaluated using an ordination procedure that permits the arrangement of similar species (Ray and Sunquist 2001). We applied a correspondence analysis using the frequency of occurrence of each food item. A matrix was constructed by grouping food items; mammals (commonly consumed in the assemblage) were separated into three groups according to body size: large mammals (>6.13 kg, maximum biomass ingested by puma/day, Monroy-Vilchis et al. 2013), medium-sized mammals (>1 and <6.13 kg) and small mammals (<1 kg), birds, reptiles, arthropods and plants.

The estimates of niche were calculated through frequency of occurrence; niche breadth was calculated using Levins' index (B'), which ranges from 0 (specialist foraging) to 1 (generalist foraging). Dietary overlap was measured using Pianka's index; this ranges from 0 (total separation) to 1 (total overlap). Pairwise niche overlap values were calculated in the software EcoSim 7.0 (Gotelli and Entsminger 2005). We compared the observed overlap in the diet of this assemblage with null models of dietary overlap generated using the niche overlap module in this software. To determine this probability, 10,000 interactions were randomly generated with a level of significance of 0.05. The RA3 algorithm was used because it preserves the specialisation of each species, but allows for the potential use of other resources (Winemiller and Pianka 1990). We assumed that all food items were equally available to all species, due to the lack of data regarding their availability.

Results

In total, 263 carnivore scats were identified. The largest number of scats was from *U. cinereoargenteus* (90), whilst the rest were from *P. concolor* (54), *P. yagouaroundi* (42), *L. pardalis* (21), *N. narica* (20), *L. wiedii* (16) and *P. onca* (16). All scats, excluding four belonging to *Canis latrans* (1) and *Procyon lotor* (3), were used in the dietary analysis.

We found 45 food items in a total of 259 carnivore scats examined. We identified 51.10% of food items belonging to mammals, 24.40% to plants, 12% to arthropods, 8.10% to birds and 4.40% to reptiles. One mammal, one bird and two plants were not identified.

Trophic structure of carnivore assemblage

It was found that all members of this carnivore assemblage consumed mammals, with differences in the body size and species consumed, although these were more frequent in diet of felids (>50%). Particularly, felids showed a key association with *Dasypus novemcinctus*. Overall, reptiles and birds were rarely consumed, and the occasional consumption of lagomorphs was evident in this assemblage.

The correspondence analysis generated two axes that explain approximately 90% of the total variation in diet. Axis I (68.27% of variance) showed a clear separation of top predators (*P. onca* and *P. concolor*), characterised by the consumption of large- and medium-sized mammals. Axis II (19.81%) clearly separates mesopredators into

Table 1 Diet of carnivore assemblage analysed through percentage of occurrence and frequency of occurrence, central Mexico

Species	*Puma concolor*		*Panthera onca*		*Leopardus wiedii*		*Leopardus pardalis*		*Puma yagouaroundi*		*Nasua narica*		*Urocyon cinereoargenteus*	
	(*n* = 54)		(*n* = 16)		(*n* = 16)		(*n* = 21)		(*n* = 42)		(*n* = 20)		(*n* = 90)	
	PO	FO	PO	FO	PO	FO	PO	FO	PO	FO	PO	FO	PO	FO
Mammals														
Baiomys musculus	-	-	-	-	-	-	7.41	9.52	-	-	-	-	-	-
Bassariscus astutus	-	-	-	-	6.67	6.25	-	-	1.78	2.38	-	-	-	-
Capra hircus	3.33	3.7	13.04	18.75	-	-	-	-	-	-	-	-	-	-
Conepatus leuconotus	-	-	4.35	6.25	-	-	-	-	-	-	-	-	-	-
Dasypus novemcinctus	38.33	42.59	39.13	56.25	6.67	6.25	7.41	9.52	5.36	7.14	-	-	-	-
Didelphis virginiana	1.67	1.85	-	-	6.67	6.25	3.7	4.76	-	-	-	-	-	-
Geomyidae	-	-	-	-	-	-	-	-	1.79	2.38	-	-	-	-
Liomys irroratus	1.67	1.85	-	-	33.33	31.25	22.22	28.57	16.07	21.43	11.11	20	8.47	16.67
Nasua narica	25	27.78	13.04	18.75	-	-	-	-	-	-	-	-	-	-
Odocoileus virginianus	5	5.56	4.35	6.25	-	-	-	-	5.36	7.14	-	-	-	-
Peromyscus maniculatus	-	-	-	-	-	-	-	-	1.79	2.38	-	-	0.56	1.11
Peromyscus melanotis	-	-	-	-	-	-	-	-	1.79	2.38	-	-	-	-
Peromyscus aztecus	-	-	-	-	-	-	-	-	3.57	4.76	-	-	0.56	1.11
Peromyscus melanophrys	-	-	-	-	-	-	-	-	3.57	4.76	2.78	5	0.56	1.11
Peromyscus sp.	-	-	-	-	-	-	7.41	9.52	-	-	-	-	1.69	3.33
Procyon lotor	1.67	1.85	4.35	6.25	-	-	-	-	5.36	7.14	2.78	5	-	-
Sciurus aureogaster	-	-	-	-	13.33	12.5	-	-	1.79	2.38	-	-	0.56	1.11
Sylvilagus cunicularius	6.67	7.41	-	-	6.67	6.25	3.7	4.76	1.79	2.38	-	-	1.13	2.22
Sylvilagus floridanus	6.67	7.41	13.04	18.75	-	-	3.7	4.76	8.93	11.9	2.78	5	1.69	3.33
Sylvilagus spp.	3.33	3.7	-	-	6.67	6.25	3.7	4.76	12.5	16.67	-	-	1.13	2.22
Spermophilus variegatus	-	-	-	-	-	-	-	-	1.79	2.38	-	-	1.13	2.22
Reitrodonthomys sp.	-	-	-	-	-	-	-	-	1.79	2.38	-	-	-	-
Unidentified mammal	5	5.56	-	-	-	-	-	-	1.79	2.38	-	-	-	-
Reptiles														
Ctenosaura pectinata	1.67	1.85	4.35	6.25	-	-	-	-	-	-	-	-	0.56	1.11
Sceloporus sp.	-	-	-	-	-	-	-	-	-	-	-	-	0.56	1.11
Arthropods														
Coleoptera	-	-	-	-	-	-	-	-	3.57	4.76	13.89	25	5.65	11.11
Crustacea	-	-	-	-	-	-	-	-	-	-	-	-	0.56	1.11
Formicidae	-	-	-	-	-	-	3.7	4.76	1.79	2.38	11.11	20	3.39	6.67
Hemiptera	-	-	-	-	-	-	-	-	1.79	2.38	-	-	-	-
Orthoptera	-	-	-	-	13.33	12.5	18.52	23.81	3.57	4.76	36.11	65	24.86	48.89
Plants														
Byrsonima crassifolia	-	-	-	-	-	-	3.7	4.76	1.79	2.38	2.78	5	1.13	2.22
Capsicum sp.	-	-	-	-	-	-	-	-	-	-	2.78	5	-	-
Crescienta sp.	-	-	-	-	-	-	-	-	-	-	-	-	5.08	10
Ficus sp.	-	-	-	-	-	-	3.7	4.76	-	-	2.78	5	5.08	10
Lysiloma sp.	-	-	-	-	-	-	-	-	-	-	-	-	1.13	2.22
Myrmeleon sp.	-	-	-	-	6.67	6.25	-	-	-	-	-	-	-	-
Physalis sp.	-	-	-	-	-	-	3.7	4.76	-	-	-	-	0.56	1.11

Table 1 Diet of carnivore assemblage analysed through percentage of occurrence and frequency of occurrence, central Mexico *(Continued)*

Psidium sp.	-	-	-	-	-	-	3.7	4.76	1.79	3.7	11.11	20	23.73	46.67
Zea mays	-	-	-	-	-	-	-	-	-	-	-	-	3.39	6.67
Unidentified plant 1 (seeds)	-	-	-	-	-	-	-	-	-	-	-	-	1.69	3.33
Unidentified plant 2 (shells)	-	-	-	-	-	-	-	-	-	-	-	-	1.13	2.22
Birds														
Buteogallus antharacincus	-	-	-	-	-	-	-	-	1.79	2.38	-	-	-	-
Seiurus sp.	-	-	-	-	-	-	3.7	4.76	-	-	-	-	-	-
Strigidae	-	-	-	-	-	-	-	-	1.79	2.38	-	-	-	-
Unidentified bird	-	-	4.35	6.25	-	-	-	-	7.14	9.52	-	-	3.95	7.78

n, scats number.

two guilds: 1) carnivore mesopredators (*P. yagouaroundi*, *L. wiedii* and *L. pardalis*), associated with the consumption of small mammals and birds, and 2) omnivore mesopredators (*N. narica* and *U. cinereoargenteus*), associated with the consumption of arthropods and plants (fruits, Table 1 and Figure 2).

The dietary niche breadth calculated in this assemblage describes top predators (*P. concolor* and *P. onca*) as a specialist group and carnivore mesopredators (*L. wiedii*, *L. pardalis* and *P. yagouaroundi*) as a generalist group. Finally, omnivore mesopredators (*N. narica* and *U. cinereoargenteus*) had an intermediate niche breadth between both guilds. Specifically, *P. concolor* showed the most specialist foraging strategies and *L. wiedii* the most generalist (Table 2).

The mean dietary overlap was 0.54, indicating an intermediate overlap in the assemblage. Comparison with null models indicated that the observed overlap was significantly greater than the mean simulated overlap (0.37, $p = 0.002$), which means that it was not random. The highest overlap (0.86 to 0.99) occurred within generated guilds. In contrast, the lowest overlap was among guilds, particularly when top predators and omnivore mesopredators were compared (0.05 to 0.09), which is consistent with the correspondence analysis results (Table 2).

Discussion

Trophic structure of carnivore assemblage

Species in a community tend to organise according to body size, which leads them to consume prey with specific characteristics. Therefore, body size differences of predators may influence their ability to hunt large prey, favouring top predators, and restricts the mesopredators to small prey (Konecny 1989; Hayward et al. 2006; Bianchi

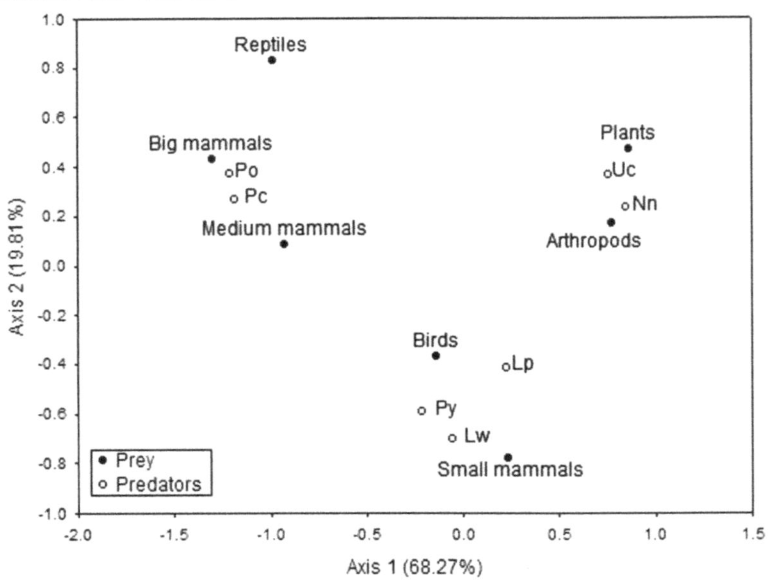

Figure 2 Trophic guild structure of carnivore assemblage from SNNP. Predators: Po, *Panthera onca*; Pc, *Puma concolor*, Lw, *Leopardus wiedii*; Lp, *Leopardus pardalis*; Py, *Puma yagouaroundi*; Nn, *Nasua narica*; and Uc, *Urocyon cinereoargenteus*. Food item: big mammals (>6.13 kg), medium mammals (1 kg < 6.13 kg), small mammals (<1 kg), arthropods, birds, plants and reptiles.

Table 2 Dietary breadth and overlap niche in the diet of a carnivore assemblage in central Mexico

Overlap	*Puma concolor*	*Panthera onca*	*Leopardus wiedii*	*Leopardus pardalis*	*Puma yagouaroundi*	*Nasua narica*	*Urocyon cinereoargenteus*
P. concolor	1	0.99	0.56	0.45	0.62	0.05	0.09
P. onca		1	0.53	0.43	0.61	0.04	0.09
L. wiedii			1	0.94	0.96	0.47	0.42
L. pardalis				1	0.92	0.67	0.69
P. yagoaroundi					1	0.39	0.41
N. narica						1	0.86
U. cinereoargenteus							1
Breadth	0.1	0.24	0.86	0.71	0.54	0.42	0.42

et al. 2011; Silva-Pereira et al. 2011). Our results suggest that the organisation of species in this assemblage should be explained by body size, because this parameter separates top predators from mesopredators, favours the differential use of feeding resources and potentially reduces competition among guilds (Davies et al. 2007). Top predators (*P. onca* and *P. concolor*) were associated with the consumption of medium and large mammals and were similar to reports from other sites of sympatry of these species (Núñez et al. 2000; Scognamillo et al. 2003; Harmsen et al. 2009).

The similarity in body size between species increases the likelihood to hunt the same prey, but our study shows that the coexistence of mesopredators can also be explained by ecological foraging segregation. This is supported by the fact that opportunism and plasticity in feeding of mesopredators reduces convergence in the use of similar resources (Fedriani et al. 1999; Guerrero et al. 2002; Carvalho and Gomes 2004). The diet of carnivore mesopredators was no different to other records where small prey were consumed (Guerrero et al. 2002; Wang 2002; Abreu et al. 2008; Bianchi et al. 2011). Moreover, omnivore mesopredators have been characterised by the consumption of fruits and arthropods (Valenzuela 1998; Fedriani et al. 1999; Guerrero et al. 2002).

Dietary breadth and overlap

Niche breadth has been used as a parameter to reference the dominance of species, so that the dominant species chooses times or sites in which their main prey can be found. Instead, subordinate species are confined to times or spaces in which the dominant species is absent, becoming a generalist forager in both niche dimensions. Resource segregation can also be modelled by evolutionary forces when the dominant species are sympatric leading to a reduction in their niche (Caro and Stoner 2003; Hayward and Slotow 2009). The niche breadth patterns in the assemblage we studied support the coexistence of predators and the community structure, placing top predators as specialist, carnivore mesopredators as generalist and omnivore mesopredators with intermediate breadth.

The high dietary overlap between species with similar morphology and diet are reported in most of the studies where these species are sympatric (Núñez et al. 2000; Guerrero et al. 2002; Scognamillo et al. 2003; Silva-Pereira et al. 2011; Gómez-Ortiz and Monroy-Vilchis 2013), because throughout their distribution, there is a similar prey base. In SNNP, competition between top predators should be taken with caution due to the low abundance of jaguar in this site and different activity patterns (Monroy-Vilchis et al. 2009b; Soria-Díaz et al. 2010); also, there are insufficient data to corroborate competition, so it is necessary to carefully evaluate resource availability and the requirements of each predator. The absence or low density of a top predator lets competitive release of subordinate predators and the expansion of their niches by changing predation patterns (Moreno et al. 2006). In this case, such release is not observed, possibly due to the low diversity of large prey (*Odocoileus virginianus*) and the existence of a high diversity of small and medium prey (Monroy-Vilchis et al. 2011).

In carnivore mesopredators, there is a high overlap that can be supported by an adequate availability of small rodents. In SNNP, rodents are represented by a high species richness (Muridae: 13 spp., Heteromidae: 1 sp., Monroy-Vilchis et al. 2011), which may favour the coexistence, as reported in Brazil (Silva-Pereira et al. 2011). Omnivore mesopredators also have a high overlap; however, *N. narica* consumed more orthopterans than fruits (*Psidium* sp.) in contrast to *U. cinereoargenteus*.

The occasional consumption of birds, reptiles and some domestic species has been reported in the diet of mesopredators (Emmons 1987; Wang 2002; Abreu et al. 2008). Moreover, our data suggest an occasional consumption on domestic goat (*Capra hircus*) and corn (*Zea mays*). In SNNR, human-wildlife conflicts have been recorded and individuals of *P. concolor* have been killed (Zarco-González et al. 2012), as well as the unjustified hunting of mesopredators (*P. yagouaroundi*, *N. narica* and *U. cinereoargenteus*) due to their supposed relation with predation on poultry and corn crops (Romero-Balderas et al. 2006).

Additional data on resource availability could provide more accurate answers regarding competition as a structuring mechanism in the assemblage we studied. In a previous study of prey selection in SNNP, we found that *P. concolor* selects its prey (Gómez-Ortiz and Monroy-Vilchis 2013). It is therefore necessary to focus future studies on assessing the availability of resources and identifying the role of key species, since the management and conservation of biodiversity must be maintained for the persistence of carnivores and those species that have a key role, such as *D. novemcinctus* in this study.

Conclusions

The analysis allowed us to identify the main feeding strategies of a carnivore assemblage in central Mexico. The coexistence in this carnivore assemblage appears to be related to body size, morphology and prey segregation, because such characteristics suggest the presence of three guilds. We observed high dietary overlap within guilds and resource partitioning between guilds.

Competing interests
The authors declare that they have no competing interests.

Authors' contributions
OM-V and YG-O conceived and designed the surveys and obtained the data. YG-O, OM-V and GDM-M analysed and interpreted the data. YG-O, OM-V and GDM-M wrote the manuscript. All authors read and approved the final manuscript.

Acknowledgements
We thank the Mexicans and the 'Consejo Nacional de Ciencia y Tecnología' for the economic support for this study through project funding (105254) and scholarship (255868). We also thank 'Programa de Mejoramiento del Profesorado' for funding the project 103.5/10/0942. We thank all students and rangers of Sierra Nanchititla Biological Station (Universidad Autónoma del Estado de México) for their support in the field and the 'Comisión Estatal de Parques Naturales y de la Fauna' for providing us with carnivore scats for identification by bile acids. We express thanks to A. Salame-Méndez and H. Domínguez-Vega for their support in this study.

Author details
[1]Estación Biológica Sierra Nanchititla, Facultad de Ciencias, Universidad Autónoma del Estado de México, Instituto Literario 100, Colonia Centro, Toluca, CP 50000, México. [2]Departamento de Producción Agrícola y Animal, Universidad Autónoma Metropolitana, Xochimilco, Calzada del Hueso 1100, Distrito Federal 04960, México.

References
Abreu KC, Moro-Rios F, Silva-Pereira JE, Miranda JMD, Jablonski EFJ, Passos FC (2008) Feeding habits of ocelot (*Leopardus pardalis*) in Southern Brazil. Mamm Biol 73:407–411
Bianchi RC, Rosa AF, Gatti A, Mendez SL (2011) Diet of margay, *Leopardus wiedii*, and jaguarondi, *Puma yagouarondi*, (Carnivora: Felidae) in Atlantic Rainforest, Brazil. Zoologia 28:127–132
Caro TM, Stoner CJ (2003) The potential for interspecific competition among African carnivores. Biol Conserv 110:67–75
Carvalho JC, Gomes P (2004) Feeding resources partitioning among four sympatric carnivores in the Peneda-Gerés National Park (Portugal). J Zool 263:275–283
Cazón AV, Sühring S (1999) A technique for extraction and thin layer chromatography visualization of fecal acids applied to Neotropical felid cats. Rev Biol Trop 47:245–249
Davies TJ, Meiri S, Barraclough TG, Gittleman JL (2007) Species co-existence and character divergence across carnivores. Ecol Lett 10:146–152
Emmons LH (1987) Comparative feeding ecology of felids in a tropical rainforest. Behav Ecol Sociobiol 20:271–283
Ernest H, Penedo M, May B, Sywanen S, Boyce W (2000) Molecular tracking of mountain lions in the Yosemite Valley region in California: genetic analysis using microsatellites and fecal DNA. Molecular Ecology 9:433–441
Fedriani JM, Palomares F, Delibes M (1999) Niche relations among three sympatric Mediterranean carnivores. Oecologia 121:138–148
Fernández GJ, Corley JC, Capurro AF (1997) Identification of cougar and jaguar feces through bile acid chromatography. J Wildl Manage 61:506–510
Gómez-Ortiz Y, Monroy-Vilchis O (2013) Feeding ecology of puma *Puma concolor* in Mexican montane forests with comments about jaguar *Panthera onca*. Wildl Biol 19:179–187
Gómez-Ortiz Y, Monroy-Vilchis O, Fajardo V, Mendoza GD, Urios V (2011) Is food quality important for carnivores? The case of *Puma concolor*. Anim Biol 61:277–288
Gotelli NJ, Entsminger GL (2005) EcoSim: null models software for ecology, vol 7. Acquired Intelligence Inc. & Kesey-Bear, Burlington
Guerrero S, Badii MH, Zalapa SS, Flores AE (2002) Dieta y nicho de alimentación del coyote, zorra gris, mapache y jaguarundi en un bosque tropical caducifolio de la costa sur del estado de Jalisco, México. Acta Zool Mex 86:119–137
Harmsen BJ, Foster RJ, Silver SC, Ostro LET, Doncaster P (2009) Spatial and temporal interactions of sympatric jaguars (*Panthera onca*) and (*Puma concolor*) in a Neotropical forest. J Mammal 90:612–620
Hayward MW, Slotow R (2009) Temporal partitioning of activity in large African carnivores: tests of multiple hypotheses. S Afr J Wildl Res 39:109–125
Hayward MW, Henschel P, O'Brien J, Hofmeyer M, Balme G, Kerley GIH (2006) Prey preferences of the leopard (*Panthera pardus*). J Zool 270:298–313
Jaksic FM (1981) Abuse and misuse of the term "guild" in ecological studies. Oikos 37:397–400
Jaksic FM, Feinsinger P, Jiménez JE (1996) Ecological redundancy and long-term dynamics of vertebrates predators in semiarid Chile. Conserv Biol 10:252–262
Konecny MJ (1989) Movement patterns and food habitats of four sympatric carnivore species in Belize, Central America. In: Redford KH, Eisenberg JF (eds) Advances in Neotropical mammalogy. Sadhill Crane Press, Florida, pp 243–264
MacArthur RH, Levins R (1967) The limiting similarity, convergence, and divergence of coexisting species. Am Natur 101:377–385
Major M, Johnson MK, Davis WS, Kellog TF (1980) Identifying scats by recovery of bile acids. J Wildl Manage 44:290–293
Monroy-Vilchis O, Rubio-Rodríguez R (2003) Guía de identificación de mamíferos terrestres del Estado de México, a través del pelo de guardia. Universidad Autónoma del Estado de México, México
Monroy-Vilchis O, Zarco-González M, Rodríguez-Soto C, Suárez P, Urios V (2008) Uso tradicional de vertebrados silvestres en la Sierra Nanchititla, México. Interciencia 33:308–313
Monroy-Vilchis O, Gómez Y, Janczur M, Urios V (2009a) Food niche of *Puma concolor* in Central Mexico. Wildl Biol 15:97–105
Monroy-Vilchis O, Rodríguez-Soto C, Zarco-González M, Urios V (2009b) Cougar and jaguar habitat use and activity patterns in central Mexico. Anim Biol 59:145–157
Monroy-Vilchis O, Zarco-González MM, Ramírez-Pulido J, Aguilera-Reyes U (2011) Diversidad de mamíferos de la Reserva Natural Sierra Nanchititla, México. Rev Mex Biodiv 82:237–248
Monroy-Vilchis O, Gómez-Ortiz Y, Urios V (2013) Expulsion rate of *Puma concolor* (Carnivora: Felidae) in captivity. Mamm Stud 38:299–302
Moreno RS, Kays RW, Samudio R (2006) Competitive release in diets of ocelot (*Leopardus pardalis*) and puma (*Puma concolor*) after jaguar (*Panthera onca*) decline. J Mammal 87:808–816
Núñez R, Miller B, Lindzey F (2000) Food habits of jaguars and pumas in Jalisco, Mexico. J Zool 252:373–379

Pianka ER (1973) The structure of lizard communities. Annu Rev Ecol Syst 4:53–74
Ray JC, Sunquist ME (2001) Trophic relations in a community of African rainforest carnivores. Oecologia 127:395–408
Ritchie EG, Johnson CN (2009) Predator interactions, mesopredator release and biodiversity conservation. Ecol Lett 12:982–998
Romero-Balderas KG, Naranjo EJ, Morales HH, Nigh RB (2006) Daños ocasionados por vertebrados silvestres al cultivo de maíz en la selva lacandona, Chiapas, México. Interciencia 31:276–283
Root RB (1967) The niche exploitation pattern of the blue-gray gnatcatcher. Ecol Monogr 37:317–350
Salame-Méndez A, Andrade-Herrera M, Zamora-Torres L, Serrano H, Soto-Mendoza S, Castro-Campillo A, Ramírez-Pulido J, Haro-Castellanos J (2012) Método optimizado para evaluar ácidos biliares de muestras fecales secas o preservadas en etanol como herramienta para identificar carnívoros silvestres. Acta Zool Mex 28:305–320
Schoener TW (1974) Resource partitioning in ecological communities. Science 185:27–39
Schoener TW (1982) The controversy over interspecific competition. Am Sci 70:27–39
Scognamillo D, Maxit I, Sunquist M, Polisar J (2003) Coexistence of jaguar (*Panthera onca*) and puma (*Puma concolor*) in a mosaic landscape in the Venezuelan llanos. J Zool 259:269–279
Silva-Pereira JE, Moro-Rios RF, Bilski DR, Passos FC (2011) Diets of three sympatric Neotropical small cats: food niche overlap and interspecies differences in prey consumption. Mamm Biol 76:308–312
Simberloff D, Dayan T (1991) The guild concept and the structure of ecological communities. Ann Rev Ecolog Syst 22:115–143
Soria-Díaz L, Monroy-Vilchis O, Rodríguez-Soto C, Zarco-González MM, Urios V (2010) Variation of abundance and density of *Puma concolor* in zones of high and low concentration of camera traps in central Mexico. Anim Biol 60:361–371
Taber A, Novaro A, Neris N, Colman F (1997) The food habits of sympatric jaguar and puma in the Paraguayan Chaco. Biotropica 29:204–213
Valenzuela D (1998) Natural history of the white-nosed coati, *Nasua narica*, in a tropical dry forest of western México. RMM 3:26–44
Wang E (2002) Diets of ocelots (*Leopardus pardalis*), margays (*L. wiedii*), and oncillas (*L. tigrinus*) in the Atlantic rainforest in southeast Brazil. Stud Neotrop Fauna E 37:207–212
Winemiller KO, Pianka ER (1990) Organization in natural assemblages of desert lizards and tropical fishes. Ecol Monogr 60:27–55
Zapata SC, Travaini A, Ferreras P, Delibes M (2007) Analysis of trophic structure of two carnivore assemblages by means of guild identification. Eur J Wildl Res 53:276–286
Zarco-González MM, Monroy-Vilchis O, Rodríguez-Soto C, Urios V (2012) Spatial factors and management associated with livestock predations by *Puma concolor* in Central Mexico. Hum Ecol 40:631–638

22

Establishment of a microsatellite set for noninvasive paternity testing in free-ranging *Macaca mulatta tcheliensis* in Mount Taihangshan area, Jiyuan, China

Bai-Shi Wang[1], Zhen-Long Wang[1], Jun-Dong Tian[1,2], Zhen-Wei Cui[1] and Ji-Qi Lu[1*]

Abstract

Background: Within multi-male and multi-female mammalian societies, paternity assignment is crucial for evaluating male reproductive success, dominance hierarchy, and inbreeding avoidance. It is, however, difficult to determine paternity because of female promiscuity during reproduction. Noninvasive molecular techniques (e.g., fecal DNA) make it possible to match the genetic father to his offspring. In the current study, a troop of free-ranging Taihangshan macaques (*Macaca mulatta tcheliensis*) in Mt. Taihangshan area, Jiyuan, China, was selected for studying the paternity. We successfully screened a set of microsatellite loci from fecal DNA and evaluated the efficiency of these loci for paternity testing using clearly recorded data of maternity.

Results: The results showed that: 1) ten loci out of 18 candidate microsatellite loci were amplified successfully in the fecal samples of Taihangshan macaques. The error probability in maternity assignments and paternity testing was very low as indicated by their power of discrimination (0.70 to 0.95), power of exclusion (0.43 to 0.84), and the values of polymorphic information content ranging from 0.52 to 0.82; 2) the combined probability of exclusion in paternity testing for ten qualified loci was as high as 99.999%, and the combined probability of exclusion reached 99.99% when the seven most polymorphic loci were adopted; 3) the offspring were assigned to their biological mother correctly and also matched with their genetic father.

Conclusions: We concluded that the ten polymorphic microsatellite loci, especially a core set of seven most polymorphic loci, provided an effective and reliable tool for noninvasive paternity testing in free-ranging rhesus macaques.

Keywords: Rhesus macaque; Microsatellite loci; Noninvasive; Fecal DNA; Paternity testing; Maternity

Background

Within multi-male and multi-female mammalian societies, maternal rather than paternal care plays a crucial role in offspring development (Rosenblatt 2003). Although rare, paternal care (e.g., male-infant affiliation and the selective support of offspring in agonistic interactions) in many cercopithecine species have been documented (Buchan et al. 2003, Fernandez-Duque et al. 2009, Langos et al. 2013). Multiple-mating by promiscuous females, however, masks paternity certainty (Soulsbury 2010), which consequently could obscure the true paternal care (Buchan et al. 2003, Widdig 2007). Paternal care is expected to result in fitness benefits for sires (Buchan et al. 2003), and known paternity is thus critical for evaluating male reproduction success in primates with a promiscuous mating system (Smith 1993, Berard et al. 1994, Bercovitch 1997). Therefore, paternity identification serves as the cornerstone of understanding male reproductive ecology. Moreover, paternity testing could clarify many important issues in the field of male reproductive ecology, including male reproductive skew (Pereira et al. 2000) and the effect of dominance hierarchy on male reproductive success

* Correspondence: lujq@zzu.edu.cn
[1]Institute of Biodiversity and Ecology, Zhengzhou University, Kexue Dadao 100, Zhengzhou 450001, PR China
Full list of author information is available at the end of the article

(Bergman et al. 2008). However, it is debatable whether the relationship between male non-human primates and their genetic offspring and unrelated individuals can be determined through epigenetic and/or transgenerational effects (Widdig 2007, Fernandez-Duque et al. 2009). As female mating promiscuously leads to paternity confusion, it is hard for researchers to determine the consanguinity between the father and his offspring through morphological and/or behavioral clues in the natural environment (Inoue et al. 1993). Fortunately, advancing molecular techniques (e.g., noninvasive genotyping) help to overcome many limitations in the field and broaden the scope of studies of free-ranging non-human primates, including paternity determination (Woodruff 1993, Constable et al. 2001, Taberlet et al. 1999).

With the development of non-invasive sampling techniques, DNA sources such as fecal DNA samples have been used in examining paternity (Constable et al. 2001). Reports of primate studies utilizing feces have become more and more prevalent in recent years (Buchan et al. 2003, Liu et al. 2008, Bergman et al. 2008, Langos et al. 2013, Liu et al. 2013). Among various molecular markers, microsatellites have been demonstrated to be ideal for analyses of genotyping, parentage, and pedigree (Estoup et al. 1998, Constable et al. 2001, Balloux and Lugon-Moulin 2002, Luna-González et al. 2012, Charpentier et al. 2012, Yang et al. 2014). A growing body of studies has examined the validity of microsatellite in parentage assessment (Vigilant et al. 2001, Reis et al. 2008, Luna-González et al. 2012, Wojtas et al. 2013) and pedigree analysis (Sefc et al. 1998, Wang et al. 2009, Li et al. 2010). Given a sufficient number of alleles, sound results for individual identification and paternity testing could be obtained with relatively small number of microsatellite loci (less than seven) (Bernatchez and Duchesne 2000). For instance, using only five microsatellite loci, the success rates of parentage identification in Plateau pike (*Ochotona curzoniae*) reached over 99.99% (Li et al. 2010). In captive rhesus macaques (*Macaca mulatta*), the probability of paternity assignment success was 99.53% when six markers were adopted (Xu et al. 2013). Along with the number of microsatellite loci, the validity of the statistics analyses also relies on the polymorphism of loci and sample size (Cooper et al. 1997). Highly polymorphic loci and fresh samples are crucial for reducing the ratio of genotype mismatch because fecal samples generally provide a low DNA concentration, and these DNA could be highly degraded due to numerous inhibitive factors (Constable et al. 2001, Liu et al. 2008). In rhesus macaques, many studies on identifying polymorphism of microsatellite loci have been reported. However, none of the markers were assured to be highly polymorphic in all rhesus macaque populations including subspecies distributed in various regions (Xu et al. 2013). For instance, different patterns of allele frequency distribution for some loci (e.g., D6S291, D6S276, and D6S2741) were documented between Indian- and Chinese-origin rhesus macaques (Penedo et al. 2005).

Taihangshan macaques, *M. mulatta tcheliensis*, are currently confined to the southern part of Mt. Taihangshan and considered as the northernmost natural population of rhesus macaques in the world (Zhang et al. 1989, Lu et al. 2007). Although studies on reproductive biology (Song and Qu 1996, Hou et al. 1998, Tian et al. 2013), behavioral ethogram and social structure (Tian et al. 2011a, 2011b) of Taihangshan macaques have been carried out, little is known about the variation in reproductive contribution among adult males in a given group. In the present study, by using noninvasive genotyping with fecal DNA, we aimed to: 1) screen for highly polymorphic and reliable microsatellite markers, 2) evaluate the effectiveness of these microsatellite markers through matching the maternity assignment to observed maternity, and then 3) determine the paternity of offspring.

Methods

Target troop and sampling

A free-ranging troop of Taihangshan macaque (Troop Wangwu 1 (WW-1)), occurring in Taihangshan Macaque National Nature Reserve (TMNNR) (35°05′ to 35°42′ N, 112°02′ to 112°45′ E) in Mt. Taihangshan area, Jiyuan, China, was assigned as our target troop. During the study period, there were six adult males (HB, BB, ZM, PZ, XBL, and SQ) and three matrilineal units (HL unit, CM unit, and BD unit; Figure 1), and HB held the highest dominance rank in WW-1 (Tian et al. 2011b). We determined mother-offspring relationship by identifying infant-to-mother attachment (e.g., infant breastfed by adult female). The majority of parental cares for infants within troop WW-1 were performed by mothers. During mating season (mainly from October to December), (adult) males (single or 'bachelor troop') from outside of WW-1 but surrounding the troop were commonly observed, and the adult males of them occasionally were observed to mate with adult females within troop WW-1.

During March 2010 and June 2011, fresh feces were collected from individually identified macaques shortly after defecation (in general, less than 5 min), which including adult males, adult females, and offspring born in 2009 (CX, JS, GX, XXC, LM, CL, and TT) and 2010 (ZY). The outer layer of the feces were harvested and then shifted into 50-mL conical tubes, one in each, filled with 95% ethanol (sample: ethanol = 1:3). The samples were kept at −20°C for further testing. In total, 136 fecal samples were collected from six adult males (mean ± SE, 6.0 ± 1.0 samples per individual), eight adult females (mean ± SE, 8.0 ± 1.5 samples per individual), and eight offspring (mean ± SE, 4.5 ± 0.5 samples per individual).

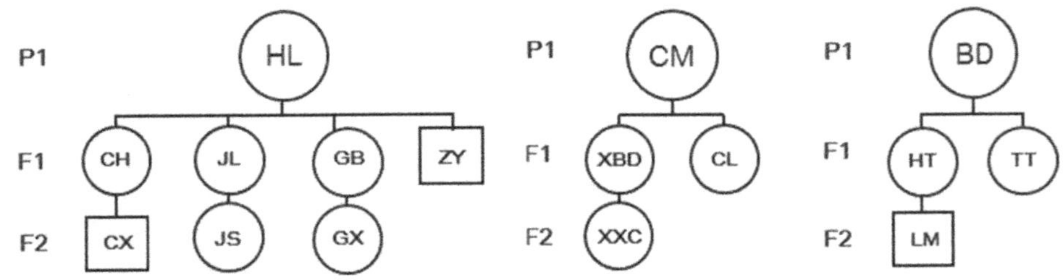

Figure 1 The kinship among individuals of three matrilineal units within troop WW-1. The circles represent females and squares represent males. Matrilineal units: HL, honglian unit; CM, changmao unit; BD, bandian unit.

The maternity within WW-1, serving as the ruler of evaluating the effectiveness of screened microsatellite markers (see results), were identified (e.g., infant breast-fed by adult female) and recorded based on long-term field observation.

DNA extractions

QIAamp® DNA Stool Kit (Qiagen, GmbH, Germany) was employed in extracting DNA from fecal samples following the manual with some modifications. To remove impurities dissolved in water and ethanol, and homogenize cells in samples, fecal samples were firstly centrifuged at 10,000 *g*, 4°C for 10 min, and the supernatant was abandoned. The residual samples were centrifuged with 97% ethanol and sterile water (sample: liquid = 1:3) three times at 10,000 *g* (4°C for 15 min each time), respectively, and the supernatants were abandoned each time. Two negative extractions (no sample) controls were processed along with each set of eight samples (in total 17 sets). All extracted fecal DNA per sample was quantified via spectrophotometry (DNA concentration ranging from 0 to 12.30 ng/μL), while samples with less than 5.00 ng/μL were not adopted because they rarely worked (Constable et al. 2001).

Microsatellite loci screening and sequence analysis

In total, 18 microsatellite markers (D12S372, D9S934, D16S403, D1S548, D3S1768, D5S820, D6S311, D5S1457, D6S2741, D14S306, D3S3045, D21S1246, D6S2419, D7S513, D4S1645, D15S644, D12S67, and D18S536) were chosen as the candidate loci based on previous genetic studies of macaques, which amplified these loci from blood or tissue samples (Kanthaswamy and Smith 1998, Kanthaswamy et al. 2006, Li et al. 2009) (Table 1).

PCR amplifications were carried out in 20 μL reaction mixtures, consisted of 1 μL template DNA, 50 mM KCl, 20 mM Tris-HCl (pH 8.3), 4.0 mM $MgCl_2$, 0.4 mM of dNTP, 0.3 μL of each forward and reverse primer (10 μM), and 0.5 U of Hotstart Taq DNA polymerase (Qiagen GMBH, Germany). Amplification conditions were as follows. Initial denaturing of 95°C for 5 min, followed by 45 cycles of 1 min at 95°C, 55°C to 60°C for 30 s, and 72°C for 1 min followed by a final 7 min extension at 72°C (see Table 2 for the optimum annealing temperature). Two negative PCR controls were employed in each set of PCRs. We repeated the amplification for two to three times.

The 5'-end of the forward primer was fluorescently labeled, and successfully amplified products were added to 3.0-μL methanamid and run in an ABI PRISM 3730 Genetic Analyser (Applied Biosystems Inc., Foster City, CA, USA). The analysis of allele sizes against the internal size standard and genotyped individuals was performed using GeneScan™-500LIZ™ Size Standard and GeneMapper™ V4.0 software.

Data analysis

Paternity analysis was mainly performed with CERVUS program version 3.0, which was also used to estimate allele frequencies, observed and expected heterozygosities (Marshall et al. 1998). This program calculates the paternity inference likelihood ratio and generates a statistic variable, delta (Δ), defined as the difference in positive log likelihood ratios (LODs) between the top two candidate fathers. If only one candidate father with a positive LOD score exists, his delta score equals his LOD score. CERVUS uses a simulation based on the observed allele frequencies, which takes typing error rates and incomplete sampling into account, to determine the statistical significance of the delta value generated for each paternity. CERVUS simulation parameters include number of offspring = 10,000, number of candidate fathers = 3, prop. sample = 1, proportion of loci typed depends on allele frequency analysis, rate of typing error = 0.01, relaxed confidence level = 80%, strict confidence level = 95%.

Moreover, following the instructions of programs, numeric errors and null alleles at 95% confidence interval were checked by the allelic size data using Micro-Checker (van Oosterhout et al. 2004). GENEPOP version 3.4 was adopted to test the Hardy-Weinberg equilibrium (Raymond and Rousset 1995). Power of exclusion (PE) and combined power of exclusion (CPE) of the ten loci

Table 1 Characteristics of 18 microsatellite loci used in this study

Locus	Primer sequence	DNA source	Species	Topic	References
D12S372	TGGACCACAGGGTATCATCT TCCAATGGAAAGAAATGGAG	Blood and tissue	Rhesus macaques	Loci screened	Rogers et al. 2005
D9S934	TTTCCTAGTAGCTCAAGTAAAGAG AGACTTGGACTGAATTACACTGC	Blood and tissue	Rhesus macaques	Loci screened	Rogers et al. 2005
D16S403	GTTTTCTCCCTGGGACATTT TATTCATTTGTGTGGGCATG	Blood and tissue	Rhesus macaques	Loci screened	Rogers et al. 2005
D1S548	GAACTCATTGGCAAAAGGAA GCCTCTTTGTTGCAGTGATT	Blood	Pig-tailed macaques	Paternity testing	Perwitasari-Farajallah 2007
D3S1768	GGTTGCTGCCAAAGATTAGA CACTGTGATTTGCTGTTGGA	Blood	Rhesus macaques	Paternity testing	Kanthaswamy and Smith 1998
D5S820	ATTGCATGGCAACTCTTCTC GTTCTTCAGGGAAACAGAACC	Blood	Rhesus macaques	Loci screened	Kayser et al. 1996
D6S311	ATGTCCTCATTGGTGTTGTG GATTCAGAGCCCAGGAAGAT	Blood	Rhesus macaques	Paternity testing	Xu et al. 2013
D5S1457	TAGGTTCTGGGCATGTCTGT TGCTTGGCACACTTCAGG	Blood	Rhesus macaques	Paternity testing	Kanthaswamy and Smith 1998
D6S2741	CTGCACTTGGCTATCTCAAC AGACTAGATGTAGGGCTAGC	Blood	Rhesus macaques	Genotyping	Penedo et al. 2005
D14S306	AAAGCTACATCCAAATTAGGTAGG TGACAAAGAAACTAAAATGTCCC	Blood	Rhesus macaques	Paternity testing	Xu et al. 2013
D3S3045	ACCAAATGAGACAGTGGCAT ATGAGGACGGTTGACATCTG	Blood and tissue	Rhesus macaques	Loci screened	Rogers et al. 2005
D21S1246	GATAAAGTAGACAGGTAAACA GGATTATAATTCAAGATGAGAT	Blood and tissue	Rhesus macaques	Loci screened	Rogers et al. 2005
D6S2419	CTAATTTGTAGATTTAAGCCTTTGC ATTCTGGTTTCAGAAACTATAGGC	Blood and tissue	Rhesus macaques	Loci screened	Rogers et al. 2005
D7S513	AGTGTTTTGAAGGTTGTAGGTTAAT ATATCTTTCAGGGGAGCAGG	Blood	Rhesus macaques	Polymorphism analyzing	Li et al. 2009
D4S1645	CAACTTTCTTCAATAAATTTGGC CAGACACTGTCCTGTTGTGG	Blood and tissue	Rhesus macaques	Loci screened	Rogers et al. 2005
D15S644	CCTTCATTGGCAGACTCACT GCAGACACCAAGATGATAACG	Blood	Rhesus macaques	Kinship estimating	Smith et al. 2000
D12S67	GCAACAGTTTATGCTAAAGC GCCTATGCAGTTCAAATCTA	Blood	Rhesus macaques	Genotyping	Kayser et al. 1995
D18S536	ATTATCACTGGTGTTAGTCCTCTG CACAGTTGTGTGAGCCAGTC	Blood	Rhesus macaques	Kinship estimating	Smith et al. 2000

were calculated from allele frequencies using PowerStats V1.2 (Promega Corporation, Madison, WI, USA). Detailed information is listed within Table 3.

Ethics statement

The present study was conducted with noninvasive method which was aimed to reduce the anthropogenic impact. Rare direct contact between the observers and the study subjects occurred, though the observers were able to view the macaques from a distance as short as 0.5 m. All protocols for the field work portion of this study complied with protocols approved by the Life Sciences Ethics Committee of Zhengzhou University and adhered to the legal requirements for animal research in China and the American Society of Primatologists principles for the ethical treatment of primates.

Table 2 Locus, amplification condition, and results of amplification success microsatellite markers tested in Taihangshan macaques

Locus	Sequence repeats	Repeat motif	Allele sizes (bp)	Annealing temperature (°C)	Success rate (%)
D12S372[a]	GATA	Tetra	156 to 199	58	62.92
D9S934[a]	AGAT	Tetra	206 to 230	56	78.33
D16S403	CA	Di	150 to 175	59	35.63
D1S548	TATC	Tetra	148 to 172	60	76.81
D3S1768[a]	TAGA	Tetra	186 to 206	57	51.32
D5S820[a]	GATA	Tetra	190 to 218	58	85.71
D6S311	CA	Di	229 to 276	58	48.72
D6S2741[a]	ATAG	Tetra	97 to 127	59	81.14
D5S1457[a]	AG	Di	199 to 299	57	66.43
D14S306[a]	AGAT	Tetra	190 to 210	58	68.79

Note: [a]Most polymorphic loci.

Results

Variation of microsatellites

In this study, we extracted DNA samples from fresh feces of 22 individuals belonging to a free-ranging rhesus macaque population. Within 136 fecal samples, 78 of them (DNA concentration >5.00 ng/μL) were adopted (mean ± SE, 3.0 ± 0.5 samples per individual). We tried to amplify 18 candidate microsatellite loci compiled from previous publications on captive macaques but derived from various samples (e.g., blood and tissues). Ten out of these 18 candidate microsatellite loci were successfully amplified, while the remaining 8 loci were discarded because of their poor amplification. Within this set of ten loci, the average success rate of amplification with fecal DNA of Taihangshan macaques was 64.20% (763 attempts) and the success rate ranged from 35.63% for locus D16S403 (31/87 attempts) to 85.71% (84/98 attempts) for locus D5S820 (Table 2).

For ten microsatellite loci, the number of alleles per locus ranged from 4 to 9 (mean 6.8 alleles per locus), and 9 loci had ≥5 alleles. Of the target monkeys, 99.55% was genotyped. The observed heterozygosity (H_o) ranged from 0.14 to 0.86 (mean 0.66), while the expected heterozygosity (H_e) ranged from 0.63 to 0.86 (mean 0.76). The distribution of alleles at D16S403 was found to deviate from the Hardy-Weinberg expectations ($P < 0.001$), while all other loci conformed to expectations (Table 4). The power of exclusion (PE) for individual loci was from 0.43 to 0.84 (mean 0.66), and the polymorphism information content (PIC) ranged from 0.52 to 0.82 (mean 0.70). The PIC was an ideal indicator for evaluating the polymorphism of fragment. The loci could be determined with high polymorphism when PIC > 0.50. For the seven most polymorphic loci (D12S372, D9S934, D3S1768, D5S820, D6S2741, D5S1457, and D14S306), the observed heterozygosity (H_o) ranged from 0.62 to 0.86 (mean 0.77), while the expected heterozygosity (H_e) ranged from 0.64 to 0.86 (mean 0.79). The PE for individual locus was from 0.43 to 0.84 (mean 0.64), and the PIC ranged from 0.55 to 0.82 (mean 0.73) (Table 4). The combined probability of exclusion (CPE) in paternity testing for ten reliable loci in this study was as high as 99.999%. Moreover, the CPE could reach 99.99% when

Table 3 Software cited for parameter analyzing in this study

Parameters	Abbreviation	Software	References
Numeric errors and null alleles	-	Micro-Checker	van Oosterhout et al. 2004
Hardy-Weinberg equilibrium	HWE	GENEPOP version 3.4	Raymond and Rousset 1995
Logarithm of odds	LOD	Cervus Software	Marshall et al. 1998; Hsu et al. 2010; Li et al. 2010; Luna-González et al. 2012
Observed heterozygotes	H_o		
Expected heterozygotes	H_e		
Power of discrimination	PD		
Polymorphism information content	PIC		
Average non-exclusion probability for one candidate parent	NE-1P		
Average non-exclusion probability for one candidate parent given the genotype of a known parent of the opposite sex	NE-2P		
Average non-exclusion probability for a candidate parent pair	NE-PP		
Estimated null allele frequency	F(null)		
Average non-exclusion probability for identity of two unrelated individuals	NE-I		
Average non-exclusion probability for identity of two siblings	NE-SI		
Power of exclusion	PE	PowerStats V1.2 (Promega Corporation, USA)	Brenner and Morris, 1990; Li et al. 2010; Stevanovic et al. 2010
Combined probability of exclusion	CPE		

Table 4 Parameters of the ten microsatellite loci

Locus	Parameters													
	K	N	H_o	H_e	PIC	NE-1P	NE-2P	NE-PP	F(null)	NE-I	NE-SI	HWE	PD	PE	CPE
D12S372[b]	6	21	0.619	0.772	0.713	0.653	0.475	0.295	0.106	0.102	0.398	NS	0.951	0.842	0.8418
D9S934[b]	8	22	0.818	0.805	0.761	0.580	0.400	0.211	−0.029	0.071	0.374	NS	0.900	0.720	0.9557673
D16S403	5	22	0.136	0.629	0.551	0.796	0.645	0.477	0.065	0.212	0.496	a	0.897	0.715	0.9873848
D1S548	4	13	0.538	0.625	0.519	0.813	0.683	0.532	0.060	0.242	0.510	NS	0.897	0.714	0.9963883
D3S1768[b]	7	20	0.800	0.826	0.779	0.558	0.381	0.199	−0.013	0.065	0.364	NS	0.881	0.691	0.9988836
D5S820[b]	7	22	0.864	0.797	0.748	0.600	0.421	0.234	−0.057	0.080	0.380	NS	0.878	0.682	0.9996451
D6S311	9	22	0.545	0.810	0.765	0.573	0.395	0.207	0.185	0.070	0.372	NS	0.889	0.681	0.9998869
D5S1457[b]	7	21	0.810	0.807	0.759	0.585	0.406	0.220	−0.039	0.074	0.375	NS	0.852	0.681	0.9999575
D6S2741[b]	6	21	0.762	0.641	0.555	0.786	0.642	0.476	−0.106	0.211	0.490	NS	0.737	0.442	0.9999765
D14S306[b]	9	22	0.727	0.863	0.823	0.484	0.316	0.146	0.071	0.044	0.340	NS	0.698	0.429	0.9999865

Note: K, number of alleles at the locus; N, the number of individuals typed at the alleles; NS, not significant; [a]Significant, lower than the 1% level. [b]Most polymorphic loci.

we used the core set of seven most polymorphic loci (Table 4), and the CPE increased slowly when added the remaining loci one by one (Figure 2).

Paternity assignment

To evaluate the effectiveness and reliability of paternity assignment, we firstly matched the maternity assigned according to the screened ten microsatellite loci with recorded maternity and found that all the offspring was assigned correctly to the observed mother. The reliability of the assignment was estimated by the LOD ranged from 2.38 to 5.01 (mean ± SE, 3.56 ± 0.32), while there were two mismatched loci, D6S311 (between ZY and HL) and D16S403 (between LM and HT).

Using the evaluated ten microsatellite loci, we assigned the paternity within the studied offspring. The LOD values for candidate paternity varied from −13.1 to 9.66 (mean ± SE, 1.94 ± 2.08), which suggested that mismatch occurred between offspring and 'genetic father' because of negative values (Table 5). We found that the estimation of only one paternity assignment (between JS and ZM) was negative (−13.1), therefore all the other offspring were assigned to their genetic fathers (Table 5). In addition, we detected a paternity relationship between two adult males that the alpha adult male (HB) was the genetic father of another adult male (BB) with 3.73 of LOD value (Table 5), but the later did not sire in the studied offspring or other target adult males (Table 5).

Discussion

From our study, ten microsatellite loci, screened with fecal DNA, exhibited higher polymorphism and higher success rate of amplification and stability. By evaluating the amplified microsatellite loci based on recorded mother-offspring pairs, we could assign all the offspring but JS to their genetic fathers. The screened markers provide sound molecular support for the study of reproductive success and parental care in free-ranging rhesus macaques. Due to the high level of polymorphism, microsatellite markers have been used for paternity

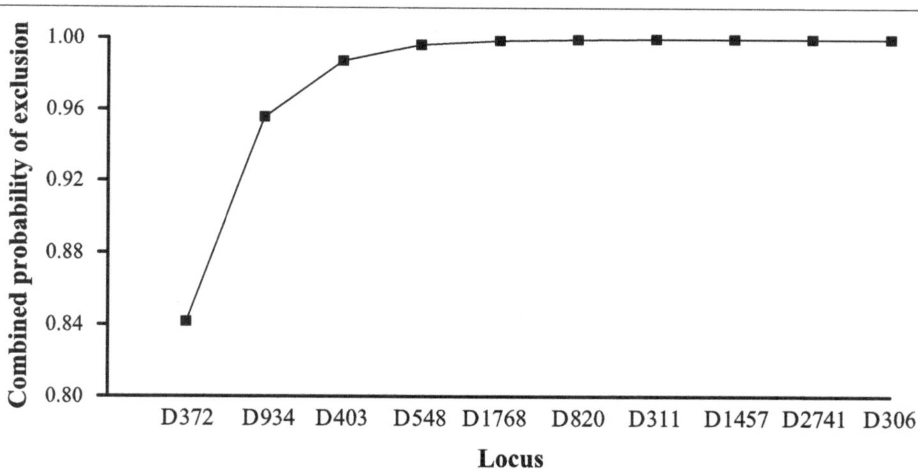

Figure 2 Combined probability of exclusion increasing with number of loci.

Table 5 Parentage assignment by ten microsatellite loci

Offspring	Candidate mother/father	Pair loci number	Pair loci mismatching	Pair LOD score	Pair Delta	Pair confidence
CX	CH^F	10	0	3.37E+00	3.37E+00	a
JS	JL^F	9	0	2.38E+00	2.38E+00	a
GX	GB^F	10	0	4.13E+00	4.13E+00	a
ZY	HL^F	10	1	3.11E+00	3.11E+00	a
XXC	XBD^F	9	0	4.37E+00	4.37E+00	a
CL	CM^F	10	0	5.01E+00	5.01E+00	a
LM	HT^F	10	1	3.56E+00	3.56E+00	a
TT	BD^F	10	0	2.54E+00	2.54E+00	a
CX	SQ^M	10	0	3.72E+00	3.72E+00	a
JS	ZM^M	8	4	−1.31E+01	0.00E+00	
GX	XBL^M	10	0	3.53E+00	3.53E+00	a
ZY	SQ^M	9	0	2.34E+00	2.34E+00	a
XXC	PZ^M	9	0	2.96E−01	2.96E−01	b
CL	XBL^M	10	0	9.66E+00	9.66E+00	a
LM	HB^M	10	0	5.53E+00	5.53E+00	a
TT	XBL^M	10	0	1.77E+00	1.77E+00	a
BB	HB^M	9	0	3.73E+00	3.73E+00	a

Note: [a]For strict confidence level. [b]For relaxed confidence level. If the candidate parent is not the most likely, this column will be blank. Pair LOD score: log-likelihood ratio for a parent-offspring relationship between the first candidate parent and the offspring. The superscripts on the adults represent the sex of individuals, ^{F}, female; ^{M}, male.

testing and population genetics studies of non-human primates successfully (Chambers et al. 2004, Bradley et al. 2007, Xu et al. 2013). In addition, the relatively new methods of noninvasive genotyping for studying free-ranging primates offer possibilities to characterize the patterns of reproduction and migration in organisms with a poor knowledge of their reproductive biology (Taberlet et al. 1999, Bergman et al. 2008, Langos et al. 2013).

Screening of microsatellite loci with fecal DNA

In published works, tri- or tetranucleotide repeat markers have been proved to reduce occurrence rate of stutter bands and provide convenience in consistent allele identification by using automated techniques (Edwards et al. 1991, Bradley et al. 2000, Liu et al. 2008). In the current study, six of the seven most polymorphic loci were tetranucleotide repeat loci (D12S372, D9S934, D3S1768, D5S820, D6S2741, and D14S306). For the other three loci, two of them (D16S403, D6S311) were dinucleotide repeats (Table 2). Considering the effectiveness of DNA extraction from blood and other tissues, microsatellite genotyping has been increasingly used in paternity testing and population genetic analysis (Domingo-Roura et al. 1997, Li et al. 2010, Luna-González et al. 2012, Wojtas et al. 2013). Fecal samples generally serve as poor DNA resource (e.g., low DNA concentration and highly degraded/fragmented DNA) with repressive factors for amplification, which results in a low success rate and a high allelic dropout rate (Gerloff et al. 1995, Kohn and Wayne 1997, Lathuillière et al. 2001, Liu et al. 2008). In this study, the incidence of allelic dropout was low and had no effect on paternity testing results. Previous studies have indicated that the incidence of allelic dropout would decrease with increased temple concentration (Taberlet et al. 1996). Morin et al. (2001) indicated that, for fecal samples of wild chimpanzees (*Pan troglodytes verus*), when amounts of template DNA being >121 pg, only two replications were needed to score with high confidence homozygous loci amplified and just 5.20% (34/656) incidence of allelic dropout occurred. In our study, the template of which the concentration was higher than 5.00 ng/μL (extracted DNA samples >250 pg/μL in 20 μL PCR reaction mixtures) was accepted for amplification. Moreover, a multiple-tube approach to genotyping was recommended for reducing the influences of allele dropout on results (Taberlet et al. 1996). However, in practical situation, few investigators working with low concentration DNA from noninvasive samples follow this approach to the recommended extent (Gerloff et al. 1999, Kohn et al. 1999). Because the multiple-tube approach would exhaust finite sample material while also add significantly the time and expense of a project. Instead, non-independent assessments of reliability are often performed, such as comparisons of mother-offspring genotypes or comparison of the results from multiple extracts (Morin et al. 2001). We found that all genotypes used in these analyses were the results of using multiple DNA extracts for each individual. For eight mother-offspring pairs studied, the sharing of alleles did not reveal any deviations from the expected Mendelian inheritance pattern.

We successfully obtained ten microsatellite loci from fecal samples with rates of 35.63% to 85.71%, and the most polymorphic seven of ten loci reached >60% success rates (Table 2). We hereafter suggested that loci with moderate or higher amplification rates can be used in future noninvasive genetic research on free-ranging rhesus macaque populations. Nonetheless, two dinucleotide repeated loci, D16S403 and D6S311, were not recommended for selection because of their low success rates (Table 2).

Previous studies have shown that reliability of paternity testing basing on the number of loci and their probability of exclusion is highly correlated with the number of alleles (Estoup et al. 1998, Bernatchez and Duchesne 2000). In this study, the average number of alleles per locus (6.80, 10 loci out of 18 loci) approximated to the study of white-handed gibbons (*Hylobates lar*) (7.00 alleles per locus, 8 loci out of 47 loci with fecal samples) (Chambers et al. 2004) but was relatively higher than 4.54 alleles per locus (13 loci out of 34 loci with fecal samples) in Yunnan snub-nosed monkeys (*Rhinopithecus bieti*) (Liu et al. 2008). Expected heterozygosity values of 10 microsatellite loci (0.76) were slightly higher than that of 11 loci used in Yugoslav pied cattle (0.75) (Stevanovic et al. 2010) and 15 loci used in Marajoara horses (0.75) and Puruca ponies (0.70) (Reis et al. 2008). We investigated seven most polymorphic loci and found that our estimate (H_o = 0.77, H_e = 0.79) showed a slightly higher level of diversity than that of the result (H_o = 0.72, H_e = 0.72) as reported by Newman et al. (2002). It indicated that the free-ranging Taihangshan macaques had the higher genetic variation coefficient. Furthermore, in our study, in several cases the observed heterozygosity exceeded the expected heterozygosity, and one locus significantly deviated from the Hardy-Weinberg equilibrium (HWE). This deviation and the variations in heterozygosity levels may be explained by the facts that macaques from a structured population were sampled and artificially treated as a randomly mating population. In our study, the PIC of 10 screened loci ranged from 0.52 to 0.82, with an average of 0.70 for 10 highly polymorphic loci and an average of 0.73 for 7 most polymorphic loci, which approximate the estimate of PIC (0.74) across 15 loci from blood samples (Kanthaswamy et al. 2006). Because the frequency of heterozygotes at a particular locus is strongly correlated with the estimate of its PIC value, microsatellites with higher PIC value are especially useful for genetically characterizing a given individuals (Ude et al. 2003).

Those ten screened microsatellite loci exhibited polymorphism and stability, and the combined probability of exclusion achieved 99.99% when the seven most polymorphic loci were employed for paternity testing (Figure 1). To evaluate the effectiveness of a paternity testing system, PE value or CPE value are the major factors in determining the genetic sire of an offspring (Sherman et al. 2004). The degree of relatedness of the candidate sire is also reflected by PE value. Therefore, PE/CPE value reflects the ability of an assay to exclude a candidate sire which is influenced by the number of loci genotyped and heterozygosity of each locus (Xu et al. 2013). The combined probability of exclusion for 10 microsatellite loci exceeded 99.99% (Table 4), which was slightly higher than that (99.62%) of 11 microsatellite loci used in paternity testing of Yugoslav pied cattle (Stevanovic et al. 2010) and relatively higher than that of 8 loci (96.60%) in white-handed gibbon (*Hylobates lar*) (Chambers et al. 2004). Comparison of our study showed that our selected microsatellite markers had greater power of exclusion, and we could reach a high level of exclusion with only seven loci.

Effectiveness of screened microsatellite loci

We amplified ten microsatellite loci which could be employed to identify the parentage and also found that seven loci of them displayed the most polymorphic. We evaluated the effectiveness of these screened microsatellite loci via matching the maternity assigned following the current study to the maternity identified by our long-term records in the field. All the studied offspring were successfully assigned to their biological mothers, which was consistent with our field observations, except for two mismatched loci between ZY and HL and between LM and HT. The mismatching was probably due to a null allele or a mutation in loci D6S311 and D16S403.

Given the positive effectiveness of our amplified loci, we could assign all the offspring and an adult male to their genetic fathers via the current methodology; however, we failed to find the genetic father for one offspring (JS). To ensure the reliability of testing, we analyzed the data of JS five times, but four mismatched loci in the paternity-offspring pair were found (LOD values < 0). This mismatch plausibly resulted from the disappearance of fecal samples of the genetic father of JS, which could be possible because of male dispersal (Southwick et al. 1996, Thierry 2004). This was true that we did observe adult males, single or from 'bachelor troop' but surrounding the studied troop, mating with adult females within WW-1 during copulating season (from October to December). Unfortunately, we failed to collect their fecal samples because these individuals were difficult to observe and follow.

Conclusions

We screened ten microsatellite markers, especially a core set of the most polymorphic seven of ten loci with fecal samples from free-ranging rhesus macaques. We verified the screened microsatellite loci by matching the maternity assignment via the current methodology to recorded maternity and assigned the paternity of our target subjects. We concluded that screened markers offer potential to characterize reproductive success and mating efforts of adult males in free-ranging rhesus macaques.

Competing interests
The authors declare that they have no competing interests.

Authors' contributions
JQL, ZLW, and BSW designed the study. BSW and JDT collected fecal samples, and BSW carried out the laboratory work. ZLW and ZWC analyzed the data and prepared figures and tables. BSW interpreted the data and drafted the manuscript. JQL and JDT made intellectual contributions to prepare the manuscript and revised it critically. All authors read and approved the final manuscript.

Acknowledgements
We thank Hou JF and Hou XQ for their sampling assistance and logistic support. We are grateful to the Jiyuan Administration of Taihangshan Macaque National Nature Reserve for their permission to our research. This work was financially supported by the Natural Science Foundation of China (30970378, 31170503) and the Graduate Programme Fund of Zhengzhou University (10L10402).

Author details
[1]Institute of Biodiversity and Ecology, Zhengzhou University, Kexue Dadao 100, Zhengzhou 450001, PR China. [2]Current address: Department of Wildlife Diseases, Leibniz Institute for Zoo and Wildlife Research, Alfred-Kowalke-Straße 17, Berlin 10315, Germany.

References
Balloux F, Lugon-Moulin N (2002) The estimation of population differentiation with microsatellite markers. Mol Ecol 11:155–165

Berard JD, Nürnberg P, Epplen JT, Schmidtke J (1994) Alternative reproductive tactics and reproductive success in male rhesus macaques. Behaviour 129:177–201

Bercovitch FB (1997) Reproductive strategies of rhesus macaques. Primates 38:247–263

Bergman TJ, Phillips-Conroy JE, Jolly CJ (2008) Behavioral variation and reproductive success of male baboons (*Papio anubis* × *Papio hamadryas*) in a hybrid social group. Am J Primatol 70:136–147

Bernatchez L, Duchesne P (2000) Individual-based genotype analysis in studies of parentage and population assignment: how many loci, how many alleles? Can J Fish Aquat Sci 57:1–12

Bradley BJ, Boesch C, Vigilant L (2000) Identification and redesign of human microsatellite markers for genotyping wild chimpanzee (*Pan troglodytes verus*) and gorilla (*Gorilla gorilla gorilla*) DNA from faeces. Conserv Genet 1:289–292

Bradley BJ, Doran-Sheehy DM, Vigilant L (2007) Potential for female kin associations in wild western gorillas despite female dispersal. Proc R Soc B Biol Sci 274:2179–2185

Brenner C, Morris JW (1990) Paternity index calculations in single locus hypervariable DNA probes: validation and other studies. In: Proceedings for the International Symposium on Human Identification. Promega Corporation, Madison, USA, pp 21–59

Buchan JC, Alberts SC, Silk JB, Altmann J (2003) True paternal care in a multi-male primate society. Nature 425:179–181

Chambers KE, Reichard UH, Möller A, Nowak K, Vigilant L (2004) Cross-species amplification of human microsatellite markers using noninvasive samples from white-handed gibbons (*Hylobates lar*). Am J Primatol 64:19–27

Charpentier MJE, Fontaine MC, Cherel E, Renoult JP, Jenkins T, Benoit L, Barthès N, Alberts SC, Tung J (2012) Genetic structure in a dynamic baboon hybrid zone corroborates behavioural observations in a hybrid population. Mol Ecol 21:715–731

Constable JL, Ashley MV, Goodall J, Pusey AE (2001) Noninvasive paternity assignment in Gombe chimpanzees. Mol Ecol 10:1279–1300

Cooper SJB, Bull CM, Gardner MG (1997) Characterization of microsatellite loci from the socially monogamous lizard *Tiliqua rugosa* using a PCR-based isolation technique. Mol Ecol 6:793–795

Domingo-Roura X, López-Giráldez T, Shinohara M, Takenaka O (1997) Hypervariable microsatellite loci in the Japanese macaque (*Macaca fuscata*) conserved in related species. Am J Primatol 43:357–360

Edwards A, Civitello A, Hammond HA, Caskey CT (1991) DNA typing and genetic mapping with trimeric and tetrameric tandem repeats. Am J Hum Genet 49:746–756

Estoup A, Gharbi K, SanCristobal M, Chevalet C, Haffray P, Guyomard R (1998) Parentage assignment using microsatellites in turbot (*Scophthalmus maximus*) and rainbow trout (*Oncorhynchus mykiss*) hatchery populations. Can J Fish Aquat Sci 55:715–723

Fernandez-Duque E, Valeggia CR, Mendoza SP (2009) The biology of paternal care in human and nonhuman primates. Annu Rev Anthropol 38:115–130

Gerloff U, Schlötterer C, Rassmann K, Rambold I, Hohmahh G, Fruth B, Tautz D (1995) Amplification of hypervariale simple sequence repeats (microsatellite) from excremental DNA of wild living bonobos (*Pan paniscus*). Mol Ecol 4:515–518

Gerloff U, Hartung B, Fruth B, Hohmann G, Tautz D (1999) Intracommunity relationships, dispersal pattern and paternity success in a wild living community of Bonobos (*Pan paniscus*) determined from DNA analysis of faecal samples. Proc R Soc Lond Ser B Biol Sci 266:1189–1195

Hou JH, Qu WY, Chen L, Zhang HX (1998) Study of the reproduction eco-behavior of *Macaca mulatta* in Taihang Mountains. Chin J Ecol 17:22–25

Hsu YC, Huang CC, Severinghaus LL (2010) Microsatellites for the hoopoes (*Upupa epops*) and their application to the parentage analysis. Conserv Genet Resour 2:187–189

Inoue M, Mitsunaga F, Nozaki M, Ohsawa H, Takenaka A, Sugiyama Y, Shimizu K, Takenaka O (1993) Male dominance rank and reproductive success in an enclosed group of Japanese macaques: with special reference to post-conception mating. Primates 34:503–511

Kanthaswamy S, Smith DG (1998) Use of microsatellite polymorphisms for paternity exclusion in rhesus macaques (*Macaca mulatta*). Primates 39:135–145

Kanthaswamy S, von Dollen A, Kurushima JD, Alminas O, Rogers J, Ferguson B, Lerche NW, Allen PC, Smith DG (2006) Microsatellite markers for standardized genetic management of captive colonies of rhesus macaques (*Macaca mulatta*). Am J Primatol 68:73–95

Kayser M, Nürnberg P, Bercovitch F, Nagy M, Roewer L (1995) Increased microsatellite variability in *Macaca mulatta* compared to humans due to a large scale deletion/ insertion event during primate evolution. Electrophoresis 16:1607–1611

Kayser M, Ritter H, Bercovitch FB, Mrug M, Roewer L, Nürnberg P (1996) Identification of highly polymorphic microsatellites in the rhesus macaque *Macaca mulatta* by cross-species amplification. Mol Ecol 5:157–159

Kohn MH, Wayne RK (1997) Facts from feces revisited. Trends Ecol Evol 12:223–227

Kohn MH, York EC, Kamradt DA, Haught G, Sauvajot RM, Wayne RK (1999) Estimating population size by genotyping faeces. Proc R Soc Lond Ser B Biol Sci 266:657–663

Langos D, Kulik L, Mundry R, Widdig A (2013) The impact of paternity on male-infant association in a primate with low paternity certainty. Mol Ecol 22:3638–3651

Lathuillière M, Ménard N, Gautier-Hion A, Crouau-Roy B (2001) Testing the reliability of noninvasive genetic sampling by comparing analyses of blood and fecal samples in Barbary macaques (*Macaca sylvanus*). Am J Primatol 55:151–158

Li RS, Zhao S, Gao R, Hu ZM (2009) Analysis of microsatellites DNA polymorphism in *Macaca mulatta* and *Macaca irus*. Acta Lab Anim Sci Sin 17:57–60

Li KX, Geng JN, Qu JP, Zhang YM, Hu SN (2010) Effectiveness of 10 polymorphic microsatellite markers for parentage and pedigree analysis in plateau pika (*Ochotona curzoniae*). BMC Genet 11:101

Liu ZJ, Ren BP, Hao YL, Zhang HR, Wei FW, Li M (2008) Identification of 13 human microsatellite markers via cross-species amplification of fecal samples from *Rhinopithecus bieti*. Int J Primatol 29:265–272

Liu ZJ, Huang CM, Zhou QH, Li YB, Wang YF, Li M, Takenaka O, Takenaka A (2013) Genetic analysis of group composition and relatedness in white-headed langurs. Integr Zool 8:410–416

Lu JQ, Hou JH, Wang HF, Qu WY (2007) Current status of *Macaca mulatta* in Taihangshan mountains area, Jiyuan, Henan, China. Int J Primatol 28:1085–1091

Luna-González A, Hernández-Arteaga S, Sánchez-Garza M, López-Revilla R, Medina-Esparza L, Cruz-Vázquez C (2012) Microsatellite loci and paternity analysis in Nubia and Boer goats. Arch Med Vet 44:123–127

Marshall TC, Slate J, Kruuk LEB, Pemberton JM (1998) Statistical confidence for likelihood-based paternity inference in natural populations. Mol Ecol 7:639–655

Morin PA, Chambers KE, Boesch C, Vigilant L (2001) Quantitative polymerase chain reaction analysis of DNA from noninvasive samples for accurate microsatellite genotyping of wild chimpanzees (Pan troglodytes verus). Mol Ecol 10:1835–1844

Newman TK, Fairbanks LA, Pollack D, Rogers J (2002) Effectiveness of human microsatellite loci for assessing paternity in a captive colony of vervets (*Chlorocebus aethiops sabaeus*). Am J Primatol 56:237–243

Penedo MCT, Bontrop RE, Heijmans CMC, Otting N, Noort R, Rouweler AJM, de Groot N, Ward T (2005) Microsatellite typing of the rhesus macaque MHC region. Immunogenetics 57:198–209

Pereira ME, Clutton-Brock TH, Kappeler PM (2000) Understanding male primates. In: Kappeler PM (ed) Primate males: causes and consequences of variation in group composition. University Press, Cambridge, pp 271–277

Perwitasari-Farajallah D (2007) Human short tandem repeat (STR) markers for paternity testing in pig-tailed macaques. HAYATI J Biosci 14:39–43

Raymond M, Rousset F (1995) GENEPOP (version 1.2): population genetics software for exact tests and ecumenicism. J Hered 86:248–249

Reis SP, Gonçalves EC, Silva A, Schneider MPC (2008) Genetic variability and efficiency of DNA microsatellite markers for paternity testing in horse breeds from the Brazilian Marajó archipelago. Genet Mol Biol 31:68–72

Rogers J, Bergstrom M, IV RG, Kaplan J, Arya A, Novakowski L, Johnson Z, Vinson A, Shelledy W (2005) A panel of 20 highly variable microsatellite polymorphisms in rhesus macaques selected for pedigree or population genetic analysis. Am J Primatol 67:377–383

Rosenblatt JS (2003) Outline of the evolution of behavioral and nonbehavioral patterns of parental care among the vertebrates: critical characteristics of mammalian and avian parental behavior. Scand J Psychol 44:265–271

Sefc KM, Steinkellner H, Glössl J, Kampfer S, Regner F (1998) Reconstruction of a grapevine pedigree by microsatellite analysis. Theor Appl Genet 97:227–231

Sherman G, Kachman S, Hungerford L, Rupp GP, Fox CP, Brown MD, Feuz BM, Holm TR (2004) Impact of candidate sire number and sire relatedness on DNA polymorphism-based measures of exclusion probability and probability of unambiguous parentage. Anim Genet 35:220–226

Smith DG (1993) A 15-year study of the association between dominance rank and reproductive success of male rhesus macaques. Primates 34:471–480

Smith DG, Kanthaswamy S, Viray J, Cody L (2000) Additional highly polymorphic microsatellite (STR) loci for estimating kinship in rhesus macaques (*Macaca mulatta*). Am J Primatol 50:1–7

Song CS, Qu WY (1996) Scientific survey of Taihangshan Macaque Nature Reserve. China Forestry Publishing House, Beijing, pp 56–114

Soulsbury CD (2010) Genetic patterns of paternity and testes size in mammals. PLoS ONE 5:e9581

Southwick CH, Zhang YZ, Jiang HS, Liu ZH, Qu WY (1996) Population ecology of rhesus macaques in tropical and temperate habitats in China. In: Fa JE, Lindburg DG (ed) Evolution and ecology of macaque societies. Cambridge University Press, Cambridge, pp 95–105

Stevanovic J, Stanimirovic Z, Dimitrijevic V, Maletic M (2010) Evaluation of 11 microsatellite loci for their use in paternity testing in Yugoslav Pied cattle (YU Simmental cattle). Czech J Anim Sci 55:221–226

Taberlet P, Griffin S, Goossens B, Questiau S, Manceau V, Escaravage N, Waits LP, Bouvet J (1996) Reliable genotyping of samples with very low DNA quantities using PCR. Nucleic Acids Res 16:3189–3194

Taberlet P, Waits LP, Luikart G (1999) Noninvasive genetic sampling: look before you leap. Trends in Ecology and Evolution 14:323–327

Thierry B (2004) Social epigenesist. In: Thierry B, Singh M, Kaumanns W (ed) Macaque societies: a model for the study of social organization. Cambridge University Press, Cambridge, pp 267–294

Tian JD, Wang ZL, Lu JQ, Guo XB, Liu JD (2011a) PAE coding system-based ethogram of Taihangshan macaques (*Macaca mulatta tcheliensis*), Jiyuan, Henan Province, China. Acta Theriol Sin 31:125–140

Tian JD, Wang ZL, Lu JQ, Guo XB, Wang BS (2011b) Social structure of wild *Macaca mulatta* in the Mt. Taihangshan area, Jiyuan, China. Acta Anthropol Sin 30:425–436

Tian JD, Wang ZL, Lu JQ, Wang BS, Chen JR (2013) Reproductive parameters of temperate-dwelling female *Macaca mulatta tcheliensis*. Am J Primatol 75:605–612

Ude GN, Kenworthy WJ, Costa JM, Cregan PB, Alvernaz J (2003) Genetic diversity of soybean cultivars from China, Japan, North America and North American ancestral lines determined by amplified fragment length polymorphism. Crop Sci 43:1858–1869

van Oosterhout C, Hutchinson WF, Wills DPM, Shipley P (2004) MICRO-CHECKER: software for identifying and correcting genotyping errors in microsatellite data. Mol Ecol Notes 4:535–538

Vigilant L, Hofreiter M, Siedel H, Boesch C (2001) Paternity and relatedness in wild chimpanzee communities. Proc Natl Acad Sci 98:12890–12895

Wang J, Liu CS, Zhang LP, Wang ZG, Yu FQ, Zhang GX, Song XZ, Han X, Wei YC (2009) Individual identification and paternity testing of bulls using microsatellite. Hereditas 31:285–289

Widdig A (2007) Paternal kin discrimination: the evidence and likely mechanisms. Biol Rev 82:319–334

Wojtas M, Piniewska D, Polańska N, Stawowiak A, Sanak M (2013) Mutations of microsatellite autosomal loci in paternity investigations of the Southern Poland population. Forensic Sci Int Genet 7:389–391

Woodruff DS (1993) Non-invasive genotyping of primates. Primates 34:333–346

Xu YR, Li JH, Zhu Y, Sun BH (2013) Development of a microsatellite set for paternity assignment of captive rhesus macaques (*Macaca mulatta*) from China. Russ J Genet 49:730–736

Yang BH, Ren BP, Xiang ZF, Yang JY, Yao H, Garber PA, Li M (2014) Major histocompatibility complex and mate choice in the polygynous primate: the Sichuan snub-nosed monkey (*Rhinopithecus roxellana*). Integr Zool 9:598–612

Zhang YZ, Quan GQ, Lin YL, Southwick CH (1989) Extinction of rhesus monkeys (*Macaca mulatta*) in Xinglung, North China. Int J Primatol 10:375–381

23

Diet adaptability by a generalist herbivore: the case of brown hare in a Mediterranean agroecosystem

Christos Sokos[1,2*], Konstantinos Andreadis[1] and Nikolaos Papageorgiou[1]

Abstract

Background: Brown hares (*Lepus europaeus*) were collected before and after autumn rains from a mixed farmland and scrubland area. The age and sex of each specimen were determined, and microhistological technique was applied to analyze the stomach contents.

Results: Hares consumed a higher number of plant species in comparison with other studies in continental European farmlands. A different pattern in diet of hare was found, where from a partial herbivory, frugivory, and granivory during the dry period, hares turn to primarily herbivory during the wet period. An expansion of diet breadth and an increase in food consumption was found in the dry season. Farming contributes to the enrichment of diet especially during the dry season. Diet composition was differed between ages, but no significant difference was found between the two sexes.

Conclusions: Hare is a facultative generalist herbivore that adapts its diet to the seasonal vegetation changes. In Mediterranean ecosystems, the seeds, fruits, and grapes are important additions to the diet. Results suggest that during the dry period juveniles cannot exploit all the available food resources, such as fruits and seeds, as effectively as adults.

Keywords: Herbivore; Feeding ecology; Dietary breadth; Forage quality; Dietary specialization

Background

The main characteristics of Mediterranean ecosystems that are important for wildlife are the prolonged dry, hot period and high biodiversity (Blondel et al. 2010; Sokos et al. 2012). Flexibility in foraging is a typical response of herbivores to cope with seasonal differences in food availability and quality in the Mediterranean (Rueda et al. 2008). Mammals such as the brown hare (hereafter hare) have higher genetic diversity (Antoniou et al. 2013), a longer reproductive period (Antoniou et al. 2008), and a more diverse diet (Kontsiotis et al. 2011) in the Mediterranean than in more northern ecosystems of its distribution.

Studies about hare diet have found that factors which cause variability are the size of the studied area (Karmiris and Nastis 2010, Kontsiotis et al. 2011), the season (Chapuis 1990), and the vegetation of the area studied (Hulbert and Iason 1996; Kontsiotis et al. 2011). Moreover, when stomach contents are analyzed, high individual variability in the diet has been recorded (Katona et al. 2010). However, this variability has been poorly studied, even though differences in diet between sexes and ages are possible. For example, a greater propensity to disperse has been found in young male hares (Avril et al. 2011); males are more active in their movements and with larger home ranges than females (Zaccaroni et al. 2013), and microhabitat use may differ (Litvaitis 1990). The different physiological status between sexes and ages can also influence diet variability (Iason and Waterman 1988).

Additionally, in Mediterranean lowland farmland areas, the diet ecology of the hare is almost unknown. However, hare is an important species for hunting economy (Sokos et al. 2003) and has fundamental roles in natural processes such as seed dispersal (Izhaki and Ne'eman 1997)

* Correspondence: sokos@vet.uth.gr
[1]Laboratory of Wildlife and Freshwater Fisheries, School of Forestry and Natural Environment, Aristotle University of Thessaloniki, Thessaloniki GR-54124, Hellas (Greece)
[2]Research Division, Hunting Federation of Macedonia and Thrace, Ethnikis, Antistasis 173-175, GR-55134 Thessaloniki, Hellas

and being prey for threatened predator species (Delibes-Mateos et al. 2009). Information on hare diet would assist evaluations on the suitability of different ecotopes for hare and to design appropriate habitat improvement actions and agri-environmental measures (Sokos et al. 2012).

The hypothesis was made that hares adjust their diet in the wet and dry seasons and that adults and males consume more plant species or fewer, depending on differentiation in movements, experience, nutritional needs, and preferred plant availability. Therefore, this study aimed to (1) identify the plant taxa consumed by hare and (2) investigate any differences in diet composition and stomach content weight in relation to season, sex, and age. Moreover, the high plant diversity of the study area provides the opportunity to test how generalist the hare is in its diet.

Methods

Study area

The study took place in SW Hellas, 65 km southwest of the city of Patras. The investigation concentrated on a 12-km^2 area of Kyllini Peninsula (37.884812°, 21.138229°). The area comprises an isolated hilly landscape with altitudes ranging from 20 to 130 m. The climate is classified as Mediterranean with dry summers and mild and rainy winters, without snowfall. The mean annual precipitation is 817 mm, and the mean annual air temperature is 16.8°C. In this lowland Mediterranean area, the growing season for most plant species begins after the autumn rains.

The Kyllini Peninsula belongs to the vegetation subzone Quercion ilicis (Dafis 1973) and the vegetation comprises shrub-dominated maquis. Farmland covers 68.5% and shrubland 31.5%. Agricultural fields of about 0.5 ha and shrublands create a diverse mosaic. The main cultivations are olive groves 60%, water melons 10%, citrus 8%, winter cereals 6%, maize 2%, tomatoes 2%, vineyards 1%, other annual crops 2.5%, plowed fields 3.5%, and set-aside 5%, whereas livestock grazing is limited, and the only wild herbivore present in the area is the hare (Sokos and Andreadis 2000).

Reference slide collection

A reference collection of slides of plant epidermis (leaves, stems, fruits, seeds) was prepared using Hertwig solution for discolouration and glycerin for preservation (Baumgartner and Martin 1939; Holechek 1982). Diagrams were drawn of all slides, and photographs were taken. A total of 162 plant species was collected (37 woody plants, 107 herbs, and 18 cultivated plant species) after vegetation survey of the area using sampling plots, thus the most plant species were recorded during the studied period (Sokos and Andreadis 2000).

Hare collection

Using a cynegetic index abundance, the population density was estimated at five hares/km^2 (Sokos and Andreadis 2000). Hares were harvested by legally licensed volunteer hunters. The authors declare that this study is based on legal hunting methods according the Hellenic Hunting Law (68719/3105/15-07-1997, Decision of the Ministry of Agriculture). A total of $n = 48$ hares have been obtained by morning hunting during three hunting seasons 1997 to 1998, 1998 to 1999, and 1999 to 2000. Each hare was dissected to obtain one eye and the stomach. The organs were preserved for further analysis in 10% formalin. Hares were divided into two age classes: juveniles with dried eye lens weights <235 mgr and adults with dried eye lens ≥235 mgr (Suchentrunk et al. 1991).

Micro-histological analysis of stomach contents

Each hare stomach was dissected, and small quantities of its contents were removed systematically from five points along the stomach axis. This method was applied as stomach content includes both newly grazed and re-ingested material that differentiates from the entrance of the stomach to the exit (pylorus). This is caused mainly due to caecotrophy during the day. As this study aimed to determine hare diet composition during all the feeding period, it was thought more appropriate to acquire samples from five different positions rather than one mixed sample (Sokos and Andreadis 2000; Kontsiotis et al. 2011).

Each content sample was washed through a sieve with pore size 0.5 mm (Brüll 1976; Frylestam 1986). The sample was then spread evenly on a glass slide, heat was applied together with Hertwig's solution, and finally, a coverslip was placed over the sample (Frylestam 1986). In the area of the coverslip (22 × 22 mm), 20 systematically chosen fields were examined. Each field was defined by the microscope using × 100 magnifications. The plant fragments observed in each field were used to identify the plant species (Holechek and Gross 1982b; Homolka 1987). Unidentified fragments were recorded according to their characteristics.

The method was tested using four caged rabbits, which had previously been fed known quantities of two or three different herb species and about 5 h after these were euthanized. Moreover, the first three hare stomachs were reanalyzed to ensure that the researcher's (first author) experience did not affect the results (Holechek and Gross 1982a).

Relatively large pieces of food originating from fruits (such as grapes and *Pyrus amygdaliformis*) and seeds could not be placed under the coverslip. For this reason, these were separated from the stomach contents, dried, weighed, and expressed as a percentage of the total dry weight of the stomach content. Sparks and Malechek

(1968) and Holechek and Gross (1982b) refer that for microhistological analyses, percentage frequency of occurrence can be expressed as percentage weight and vice versa. Thus, the percentage frequency of fruit or seed occurrence was estimated using the equation $Fi = (fi * w1/100 + wi) * 100/(w1 + wi)$, where Fi: the total frequency of occurrence of fruit or seed, fi: the frequency of occurrence of fruit or seed found by microhistological analysis, w1: the dry weight of quantity analyzed, and wi: the dry weight of the fruit or seed quantity that was separated.

Ecological and statistical analyses

Diversity in diet was estimated using the Shannon-Wiener function (H′ Log Base 10), as it is more sensitive to changes of rare species (Krebs 1999). Food niche breadth (B) was estimated using Levins' index (Krebs 1999). The standardized Levins' index (BA) was also used with the formula: $BA = (B - 1)/(n - 1)$, where B is the Levins' index, and n is the total number of plant taxa (lowest niche breadth = 0; highest niche breadth = 1). This estimate is more useful for comparing diets among different areas and seasons that may vary in the number of available palatable plant species. Additionally, some plant species are more abundant than others; therefore, the Hurlbert's niche breadth (B′) was calculated only for herbs, incorporating the availability of herb species found as median of cover and frequency of occurrence in frame quadrats applied in the study area (Sokos and Andreadis 2000).

Following the Bray-Curtis method (Krebs 1999), a dendrogram of diet data was made using Biodiversity Pro software, version 2 (McAleece et al. 1997). To test for overall differences in diet composition, statistical comparisons were made using ANOSIM two-way crossed analyses on Bray-Curtis dissimilarity matrices (program Primer-e) using 999 permutations (Clarke and Warwick 1994). Where significant differences occurred, the plant taxa most responsible for the difference were determined using a two-way similarity of percentages analysis, SIMPER (Primer-e, Clarke and Warwick 1994).

Diet parameters were tested by either one or two factors between subjects ANOVA (Zar 1996). The non-parametric test Mann-Whitney U was used when the assumptions of using the parametric tests were not met. Statistical analyses were performed using IBM SPSS 20.0 data analysis software. The level of significance was set to be $\alpha = 0.05$.

Results

In total, 48 hares were collected during the period of 3 years; differences did not happen in natural vegetation and cultivations of the studied area, thus the hares of different years were grouped for further analyses. The hares were classified according the stage of vegetation, thus 26 hares were classified in the transitional period of vegetation from summer to winter (named dry season: 20 September to 9 November) where 1) winter plant species have not yet grown sufficiently, taking into account each studied year's plant phenology, and 2) summer plant species and fruits have not yet died down or desiccated completely. The remaining 22 hares were collected in the growing period for winter herbs and many shrub species (named wet season: 19 November to 6 January).

Identification of all examined fragments to species level was not possible (caecotrophy during the day is a reason). Moreover, morphological characteristics of some plant species which belong to the same genus or family are very similar and identification is difficult (Westoby et al. 1976; Homolka 1986). To lower the risk of incorrect identifications, species with high similarity, such as sprouting winter grasses, were grouped.

Microhistological analysis recognized 36 plant taxa (recognizable in the level of species, genus or family), of which 26 were dicotyledons and 10 were monocotyledons. Careful comparisons with the reference collection permitted the estimation that hares consumed about 63 plant species (58 plant species of the 162 examined microhistologically, and 5 more species which were not in the reference slide collection). The leaves and stems mostly eaten were from winter grasses (*Avena sativa, Avena* sp., *Dasypyrum villosum, Bromus sterilis, Dactylis glomerata, Hordeum vulgare, Triticum aestivum, Lolium multiflorum*), summer grasses (*Echinochloa crus-galli, Eleusine indica, Paspalum distichum*), and other grasses and forbs (Table 1). Significant quantities of *Pyrus amygdaliformis* fruits and seeds of cereals and other grasses were also observed (Table 1).

The most common plant species found between hares (above 25% of specimens) were winter grasses, *Cynodon dactylon, Convolvulus arvensis*, summer grasses, *Allium roseum*, and *Amaranthus* sp. Overall diversity in the diet was higher in the dry season (29 taxa and H′ = 1.247) than in the wet season (27 taxa and H′ = 0.87). Similarly, the overall dietary breadth indices were higher in the dry season (Table 1).

The number of plant taxa consumed by each individual animal does not differ with season, age or sex ($p > 0.176$). However, diversity in diet H′ and dietary breadth indices B and BA were higher in dry than wet seasons ($p < 0.046$), higher in juveniles than adults in the dry season ($p < 0.014$), and higher in adults than juveniles in the wet season ($p < 0.05$). Hurlbert's niche breadth (B′) was used only for herbs (monocotyledons and dicotyledons) and did not reveal any significant differences concerning season, age, or sex ($p > 0.101$).

In the dry season, monocotyledons and dicotyledons have similar abundances in diets followed by fruits, seeds,

Table 1 Mean frequency of occurrence (F%) and percentage of hare stomachs containing a certain plant taxa (*n*%)

Species or group of species	Dry season (*n* = 26)		Wet season (*n* = 22)		Total (*n* = 48)	
	F%	*n*%	F%	*n*%	F%	*n*%
Monocotyledons	36.34	100	50.57	100	42.86	100
Grasses	32.38	100	49.52	100	40.24	100
Winter grasses	17.31	84.62	43.44	100	29.29	91.67
Summer grasses	6.63	61.54	0.12	18.18	3.65	41.67
Cynodon dactylon	2.12	53.85	3.49	81.82	2.75	66.67
Brachypodium sp.	3.02	15.38	1.18	9.09	2.18	12.50
Setaria sp.	1.94	15.38			1.05	8.33
Cynosurus echinatus	0.35	15.38	1.03	18.18	0.66	16.67
Sorgum halepense	0.19	23.08			0.10	12.50
Unidentified grasses	0.82	46.15	0.26	18.18	0.56	33.33
Other monocotyledons	3.96	38.46	1.05	27.27	2.62	33.30
Cyperus rotundus	2.67	7.69			1.45	4.17
Allium roseum	0.48	23.08	0.95	36.36	0.69	29.16
Unidentified monocotyledon	0.81	7.69	0.1	9.09	0.48	8.33
Dicotyledons	36.29	100	43.4	90.91	39.55	95.80
Medicago sp.			20.24	45.45	9.28	20.83
Convolvulus arvensis	5.27	46.15	7.96	36.36	6.50	41.67
Amaranthus sp.	9.30	46.15			5.04	25.00
Urospermum picroides	2.76	7.69	0.92	9.09	1.92	8.33
Brassicaceae			2.77	36.36	1.27	16.67
Malva sp.	0.63	7.69	0.82	18.18	0.72	12.50
Cichoriaceae	1.37	23.08			0.74	12.50
Sonhcus sp.			1.49	9.09	0.68	4.17
Solanum nigrum	0.66	3.84			0.36	4.17
Verbascum sp.	0.50	7.69	0.219	18.18	0.37	
Senecio vulgaris	0.33	3.84	1.05	27.27	0.66	
Trifolium sp.	0.33	7.69	0.05	9.09	0.20	8.33
Vicia sp.	0.08	7.69	0.14	9.09	0.11	8.33
Ornithopus compressus	0.22	7.69			0.12	4.17
Anthemis arvensis	0.16	7.69			0.09	4.17
Daucus carota			0.18	9.09	0.08	4.17
Calendula arvensis			0.10	4.54	0.05	4.17
Oxalis pes-caprae			0.06	4.54	0.03	4.17
Unidentified dicotyledon 1	0.11	3.84	0.05	4.54	0.08	8.33
Unidentified dicotyledon 2	6.74	15.38	0.28	18.18	3.78	16.67
Unidentified dicotyledon 3	1.07	23.07			0.58	12.50
Unidentified dicotyledon 4			0.96	4.54	0.44	4.17
Unidentified dicotyledon 5	0.75	23.07	0.33	4.54	0.55	16.67
Unidentified dicotyledons	6.01	76.92	5.78	90.91	5.90	83.33
Fruits, seeds, and barks	25.91	92.31	5.54	81.82	16.57	87.50
Pyrus amygdaliformis	4.56	15.38	1.72	18.18	3.26	16.67
Vitis vinifera	2.70	7.69			1.46	4.17
Olea europaea	0.27	7.69			0.15	4.17

Table 1 Mean frequency of occurrence (F%) and percentage of hare stomachs containing a certain plant taxa (*n*%) *(Continued)*

Unidentified dicotyledons	8.47	76.92	3.30	63.64	6.10	70.83
Grasses	9.91	38.46			5.37	20.83
Unidentified	1.47	69.23	0.49	18.18	1.02	45.83
Taxa number	29		27		36	
Mean ± SE of taxa number	7.07 ± 0.4		6.54 ± 0.54		6.83 ± 0.33	
H′	1.247		0.86		1.188	
Mean ± SE of H′	0.68 ± 0.023		0.56 ± 0.048		0.63 ± 0.027	
B	12.84		4.1		8.46	
Mean ± SE of B	3.76 ± 0.19		3.03 ± 0.31		3.42 ± 0.18	
BA	0.36		0.106		0.186	
Mean ± SE of BA	0.34 ± 0.017		0.27 ± 0.027		0.31 ± 0.016	
B′	0.177		0.076		-[a]	
Mean ± SE of B′	0.136 ± 0.032		0.114 ± 0.025		0.126 ± 0.021	

Mean frequency of occurrence (F%) and percentage of hare stomachs containing a certain plant taxa (*n*%) in the dry and wet seasons. Diversity and food niche breadth indices were calculated according to the mean frequencies of occurrence. Hurlbert's niche breadth (B′) was applied only to herbs.
[a]B′ was not calculated for the pooled data, due to the difference of availability of herb species between the two seasons.

and barks (Table 2). In the wet season, hares ate more monocotyledons. In the dry season, juveniles ate more dicotyledons than adults ($F1,25 = 11.9$, $p = 0.002$), and the adults ate more fruits and seeds ($F1,25 = 11.9$, $p = 0.0001$). Stomach content dry weight was higher in the dry season, in adults and females, but the difference was significant only for season (U test, $p = 0.021$).

The results of the Bray-Curtis method for the diet of the hares according to the seasons, ages, and sexes can be seen in Figure 1. Juveniles of the dry season had the lowest similarity values in diet when compared to the other groups. Adult females of the dry season begin the consumption of winter grasses and cereals earlier and thus increase the similarity with wet season groups.

Table 2 Mean ± SE of frequency of occurrence (F%) for the main food categories

Hare groups	Monocotyledons	Dicotyledons	Fruits, seeds, and barks	Stomach content dry weight
Dry				
Sex				
Males	35.37 ± 4.1	46.5 ± 8.5	15.56 ± 7.6	13.8 ± 3.03
Females	36.97 ± 5.4	29.9 ± 5	32.36 ± 4	19.1 ± 2.3
Age				
Juveniles	38.2 ± 4.9	49.3 ± 5.6[c]	10.2 ± 3.08[e]	14.8 ± 2.03*
Adults	34.2 ± 4.1	21.1 ± 5[d]	44.21 ± 3.55[f]	19.6 ± 2.75
Total	36.36 ± 3.2[a]	36.27 ± 4.7	25.9 ± 4.1[g]	17.8 ± 1.93[i]
Wet				
Sex				
Males	36.5 ± 7.7	59.52 ± 6.8	2.4 ± 0.065	9.75 ± 0.43
Females	56.3 ± 5.4	37.8 ± 5.7	5.6 ± 1.13	12.78 ± 0.92
Age				
Juveniles	52.6 ± 10.8	41.9 ± 11.7	5.51 ± 1.5	10.25 ± 1.29
Adults	49.8 ± 5.7	44.6 ± 5.85	4.73 ± 1.3	13.36 ± 0.9
Total	50.85 ± 5.2[b]	43.6 ± 5.5	5 ± 0.96[h]	12.23 ± 0.8[j]

Mean ± SE of frequency of occurrence (F%) for the main food categories in the diet of 48 examined hares, and stomach content dry weight (g) per season, age, and sex.
*Two leverets were excluded from the stomach content dry weight analysis as their stomachs contained milk.
In each column, the paired letters (ab, cd, ef, gh, ij) indicate significant differences at the 0.05 level.

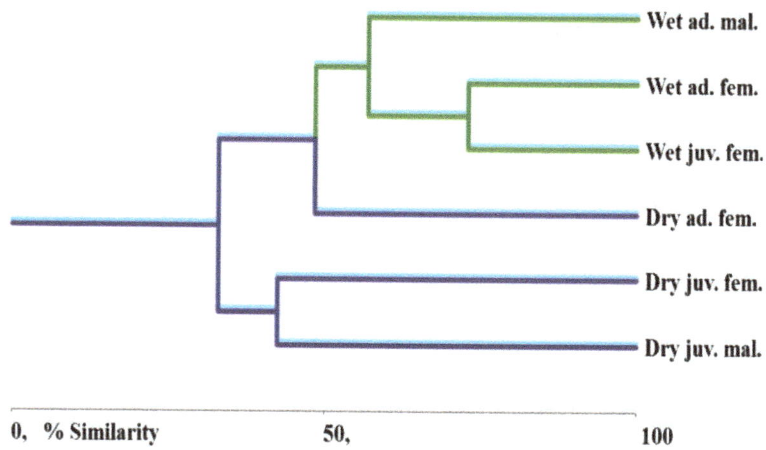

Figure 1 Dendrogram of Bray-Curtis percentage similarities for the diet composition (categories Wet juv. mal. and Dry ad. mal. were not included in sample). Juveniles of dry season had the lowest similarity values.

The ANOSIM two-way crossed analysis found a significant difference in the dietary composition between seasons (Global R = 0.299, $p = 0.002$) and a marginal difference between ages (Global R = 0.158, $p = 0.054$). The dietary composition between sexes was not significantly different (Global R = 0.183, $p = 0.098$). The SIMPER analysis showed that the dissimilarity in diet between dry and wet seasons was a result of an increased prevalence of winter grasses and *Medicago* sp. in wet season and a decreased prevalence of seeds of grasses and cereals, *C. arvensis*, *Amaranthus* sp., and fruits of *Pyrus amygdaliformis* in comparison with the dry season. In the case of ages, the dissimilarity was caused by the higher consumption by adults of winter grasses, seeds, and fruits of *Pyrus amygdaliformis* and a lower consumption of *Amaranthus* sp. and summer grasses in comparison with juveniles.

Discussion

According to Holechek (1982), the microhistological analysis gives satisfactory results. However, one weakness is the differential fragmentation and digestion between plant species which influences the frequencies of occurrence. Thus, it is possible to observe an overestimation of grasses (McInnis et al. 1983). This was one of the reasons that stomach analysis was preferred over fecal analysis, as it is considered a more accurate method of determining the diet of herbivores, although Homolka (1986) did not find significant differences. However, if hares consume a considerable amount of seed and fruits, then diet is better studied using stomach contents (Green et al. 2013).

In many stomachs, a considerable amount of large food pieces by fruits and seeds were observed. Separating and dry weighting fruit and seed content permitted a more accurate estimation of their consumption. This may explain the reason why a higher percentage of fruit and seed in diet of hare in the dry season was found compared to other studies in Mediterranean (Sfougaris et al. 2006; Paupério and Alves 2008; Kontsiotis et al. 2011).

Diet analysis revealed that hares in the study area consume mainly wild plant species (except cereals and grapes). Many studies have revealed that hares that feed in farmlands prefer wild herbs over cultivated plants (Brüll 1976; Frylestam 1986; Reichlin et al. 2006). The mosaic landscape permits hares to access many wild plant species, so any impact to agricultural production is limited at this period. Moreover, farming contributes to the enrichment of a hare's diet with plants such as *C. arvensis* and *Amaranthus* spp. These plant species can be found growing within the irrigated crops such as watermelon. Grapes from the vineyards were also consumed. In one pregnant adult female's stomach, large pieces of grapes (6 g dry weight) were found, and these comprised 20.7% of the total dry weight of the stomach content.

The number of identified plant taxa per stomach was 6.8 and was higher than in a study in Denmark where the mean number of plant taxa found in each stomach was 3 (Hansen 1996). This information is not provided by other authors. Although we examined a smaller number of hare stomachs in comparison with other studies in more northern European farmlands (Frylestam 1986; Hansen 1996; Reichlin et al. 2006; Katona et al. 2010), we found a higher number of plant species eaten by hares. In comparison with studies carried out in the Mediterranean, we found more plant taxa than Karmiris and Nastis (2010) in their small shrubland study area (300 ha). On the contrary, Sfougaris et al. (2006) and Kontsiotis et al. (2011) found a remarkably higher number of identified plant taxa when examining the autumn and winter diet of 318 and 217 hares, respectively. However, these hares were collected from broad geographical, mainly upland, study areas of thousands of km^2. The larger number of plant taxa they recorded is probably caused by the larger number of hares examined and the larger study areas, as the standardized Levins' niche dietary breadth index (BA) calculated by Kontsiotis et al. (2011) was lower than that of this study. The Shannon-Wiener function (H′) calculated in this study was approximately two times higher than that of studies conducted in continental climates (Puig et al. 2007; Katona et al. 2010), whereas it was similar with that of other studies conducted in the Mediterranean (Karmiris and Nastis 2010; Kontsiotis et al. 2011).

Grasses (Poaceae) constituted the bulk of the diet, and similar results were found in continental Europe (e.g., Katona et al. 2010) and mountainous and shrubland areas in the Mediterranean. In our farmland study area, grasses (leaves and stems) accounted for 40.24% of the diet and seeds of grasses accounted for 5.37%. Sfougaris et al. (2006) refer a total percentage of 36% of grasses in examined stomachs. Kontsiotis et al. (2011) refer that in examined stomachs, leaves and stems accounted for 24% of the diet and seeds accounted for 2.1%. In two studies where feces were examined (Paupério and Alves 2008; Karmiris and Nastis 2010), a considerably higher total percentage of grasses was found, about 65%.

In this study, a higher percentage of grass seeds (5.37%) was found to be eaten during the dry period. This is probably due to the cereal stubble fields present in the study area and the applied method of separating large food pieces. Indeed, one hare stomach contained a huge piece of starch (16 g dry weight) by grass seeds that comprised 40.5% of the total dry weight of the stomach content. A high percentage of grass seeds (9%) was also found in stomachs examined by Katona et al. (2010).

Hares adapt their diet according to vegetation in each season. ANOSIM analysis indicated that the diet composition between the two seasons differs significantly, and SIMPER analysis showed that seeds, fruits, *C. arvensis*, and *Amaranthus* spp. consumed in the dry season were replaced by winter grasses and *Medicago* spp. in the wet season. Higher consumption of grasses in winter has been found in continental Europe (e.g., Reichlin et al. 2006), whereas in year-round studies in Mediterranean mountainous and shrubland areas, the consumption of grasses was also found to increase slightly in winter (Paupério and Alves 2008; Karmiris and Nastis 2010).

Foraging theory refers an expansion of diet breadth (Pianka 2000) and an increase in food consumption (Sinclair et al. 1982; Rogers and Sinclair 1997) when the availability of nutritious food decreases. Similarly, diet breadth may be dependent on the distribution of preferred plant species in space. When few preferred plant species are available in sufficient quantity and distributed evenly in space, herbivores do not need to search and to consume more plant species (Stephens and Krebs 1986; Shipley et al. 2009). In this study, an expansion of diet breadth in the dry season was found, mainly due to the consumption of seeds and fruits. Moreover, stomach content dry weight was higher in the dry than in the wet season, indicating an increase in food consumption. These two events are indications of the nutritional reduction of herbaceous plants at the end of the dry season. Indeed, most herbs are in a mature stage in the end of the dry season, something that decreases their forage quality. Simultaneously, fruits of woody plants ripen, and seeds of many grasses and cereals provide nutritious food.

Moreover, in our study area, hare food resources are patchier during the dry season than in the wet season where grasses, the main food source, grow abundantly throughout the study area after the rains. In the dry season, cereal fields, fruit trees, vineyards, and irrigated crops are all potential feeding grounds and each contains different plant species. It is assumed that each individual hare cannot visit all these feeding grounds in one night. Therefore, dietary niche breadth expands further in the dry season in comparison with the wet season but only when overall values are compared and not the mean values of individual hares (70.5% overall diet expansion vs. 20.6% diet expansion of mean individuals, according to BA in Table 1).

The expansion of diet breadth to include woody plants has previously been found for hare in colder climates, when the ground is covered by snow and herbaceous vegetation availability is decreased (Rödel et al. 2004; Green et al. 2013). We found evidence that a similar diet breadth expansion occurs in Mediterranean climates during the warm and dry seasons, but in this case, the limiting factor is the drought that decreases the foraging quality of herbs. Indeed, the improved diet of hares following the rainy season may be reflected in reproduction rates. Until the end of October, none of the females collected were pregnant, whereas at the end of autumn, after the rains, 50% of the collected females were pregnant and during the winter this number increased to 85%. A similar pattern was found in Crete by Antoniou et al. (2008).

Comparison of dietary parameters between age and sex classes is possible due to the relatively small study area considered, the similar landscape mosaic and vegetation across the study area, and the ability of hares to disperse more than 500 m each night to find their preferred food (Homolka 1985; Kuijper and Bakker 2008) and to cover a nocturnal distance ranging between 1.39 and 6.02 km within their home range (Pepin and Cargnelutti 1994). A higher diversity in diet H′ and dietary breadth was found in juveniles than in adults in the dry season. Juveniles ate more dicotyledons than adults, and the adults ate more fruits and seeds. This indicates that juveniles cannot exploit all available food resources such as fruits and seeds. A reason may be is that fruits and seeds have more limited and patchy distributions. This is expected as juveniles have less knowledge of resource distribution (Gillingham and Bunnell 1989; Dahl 2005; Shipley et al. 2009) and, in some cases, are excluded by dominant hares (Lindlof 1978; Monaghan and Metcalfe 1985). The opposite seems to hold in the wet season when juveniles consume mainly nutritious and abundant winter grasses and have a narrower niche breadth than adults.

A second reason for fruit avoidance by juveniles may be because of the high concentration of secondary compounds contained in fruits (Jordano 2000). Iason and Waterman (1988) found that reproductively active females and leverets of *Lepus timidus* ate smaller proportions of heather *Calluna vulgaris*, than non-reproducing females and males, due to secondary compounds contained in this plant. However, in this study no differences were found between the sexes. Similarly, Homolka (1987) found no significant difference in diet composition between males and females.

Demeter and Matrai (1988) did not find differences in stomach content weights according to sex and age in Hungary, although Eley (1970) did find higher weights of female cape hares (*Lepus capensis*) than males. In this study, higher stomach content weights of adults and females were found, but they were not statistically significant. More female specimens collected in the wet season were pregnant and their higher energy requirements may force them to higher consumption (Hackländer et al. 2002).

The high plant diversity recorded in the area studied allows the examination of how specialist or generalist in diet the hare can be. Mammalian herbivores are classified as specialists if at least 60% of their diet consists of a single, distinctive plant genus or family (Shipley et al. 2009). In this study, the highest percentage of one single family was 49.52%, the Poaceae, during the winter. Therefore, the grasses, even when their availability and palatability is high, do not render the hare as a specialist. However, in other studies, the hare behaves as a specialist on grasses consuming more than 60% (Hansen 1996; Katona et al. 2010). According to (Shipley et al. 2009), the hare could be characterized as a facultative generalist as it is able to survive with a restricted diet breadth but prefers a broad dietary niche when the plant diversity is available.

Conclusions

The hare found to be a facultative generalist herbivore that adapts its diet to the seasonal vegetation changes. From a partial herbivory, frugivory, and granivory diet during the dry period, hares turn to primary herbivory during the wet period thus seeming to confirm the foraging theory for expansion of diet breadth and increase in food consumption when the availability of nutritious food decreases in the dry season due to the dying down of herbs. Therefore, in Mediterranean ecosystems, the seeds, fruits, and grapes are important additions to the diet. Stomach diet analysis seems to be more accurate in cases of frugivory and granivory than feces analysis, but additional studies are required to confirm this. During the wet season, hares probably feed on higher quality forage, and this factor appears to regulate reproduction rates; again, future studies should confirm this in Mediterranean lowland areas. Farming contributes to the enrichment of hares' diet, with cereals and grapes and other plant species that can be found growing within irrigated summer crops. Results suggest that during the dry period juveniles cannot exploit all the available food resources, such as fruits and seeds, as effectively as adults.

Competing interests

The authors declare that they have no competing interests.

Authors' contributions

CS and NP wrote the manuscript. CS carried out the statistical analyses. CS and KA carried out the laboratory analyses. CS participated in the sample collection. All authors read and approved the final manuscript.

References

Antoniou A, Kotoulas G, Magoulas A, Alves PC (2008) Evidence of autumn reproduction in female European hares (*Lepus europaeus*) from southern Europe. Eur J Wildl Res 54:581–587

Antoniou A, Magoulas A, Platis P, Kotoulas G (2013) Assessing the genetic landscape of a contact zone: the case of European hare in northeastern Greece. Genetica 141:1–18

Avril A, Leonard Y, Letty J, Peroux R, Guitton J-S, Pontier D (2011) Natal dispersal of European hare in a high-density population. Mam Biol 76:148–156

Baumgartner LL, Martin AC (1939) Plant histology as an aid in squirrel food habits studies. J Wildl Manage 3:266–268

Blondel J, Aronson J, Bodiou JY, Boeuf G (2010) The Mediterranean Region, Biological diversity in space and time. Oxford University Press, Oxford

Brüll V (1976) Nahrungsbiologishe Studien am Feldhasen in Schleswig - Holstein. In: Proceedings of the Poznan Symposium on the Ecology and Management of European hare populations. Ein Beitrag zu Äsungsverbesserung, Poland, p 1974

Chapuis JL (1990) Comparison of the diets of two sympatric lagomorphs, *Lepus europaeus* (Pallas) and *Oryctolagus cuniculus* (L.) in an agroecosystem of the Ile-de-France. Z Säugetierkd 55:176–185

Clarke KR, Warwick RM (1994) Change in Marine Communities: an Approach to Statistical Analysis and Interpretation. Plymouth Marine Laboratory, Plymouth, UK

Dafis S (1973) Classification of vegetation in Greece, School of Forestry and Natural Environment. Aristotle University of Thessaloniki, Thessaloniki

Dahl F (2005) Distinct seasonal habitat selection by annually sedentary mountain hares (*Lepus timidus*) in the boreal forest of Sweden. Eur J Wildl Res 51:163–169

Delibes-Mateos M, Farfan MA, Olivero J, Marquez AL, Vargas JM (2009) Long-term changes in game species over a long period of transformation in the Iberian Mediterranean landscape. Environ Manage 43:1256–1268

Demeter E, Matrai K (1988) Foods of the brown hare in a large scale cultivated plain area during November. Vadbiologia 2:85–90

Eley J (1970) Stomach contents, weights and volumes of cape hare. E Afr Wildl J 8:270

Frylestam B (1986) Agricultural land effects on the winter diet of brown hares (*L. europaeus* Pallas) in Southern Sweden. Mammal Rev 16:157–161

Gillingham MP, Bunnell FL (1989) Effects of learning on food selection and searching behaviour of deer. Can J Zool 67:24–32

Green K, Davis NE, Robinson WA, McAuliffe J, Good RB (2013) Diet selection by European hares (*Lepus europaeus*) in the alpine zone of the Snowy Mountains, Australia. Eur J Wildl Res 59:693–703

Hackländer K, Tataruch F, Ruf T (2002) The effect of diet quality on lactation energetics in the European hare (*Lepus europaeus*). Physiol Biochem Zool 75:19–28

Hansen K (1996) Impact of modern farming on food supply of hares (*Lepus Europaeus,* Pallas) during the summer period. In: Proceedings of the I European Congress of Mammalogy. Lisboa: Museu Nacional de Historia Natural, Portugal

Holechek JL, Gross BD (1982a) Evaluation of different calculation procedures for microhistological analysis. J Range Manage 35:721–723

Holechek JL, Gross BD (1982b) Training needed for quantifying simulated diets from fragmented range plants. J Range Manage 35:644–647

Holechek JL (1982) Sample preparation techniques for microhistological analysis. J Range Manage 35:267–268

Homolka M (1985) Spatial activity of hares (*Lepus europaeus*). Fol Zool 34:217–226

Homolka M (1986) Comparison of two methods applied to study the food of *Lepus europaeus*. Fol Zool 35:199–206

Homolka M (1987) Problems associated with investigations into the diet of the European hare. Fol Zool 36:193–202

Hulbert IA, Iason GR (1996) The possible effects of landscape change on diet composition and body weight of mountain hares *Lepus timidus*. Wildl Biol 2:269–273

Iason GR, Waterman PG (1988) Avoidance of plant phenolics by juvenile and reproducing female mountain hares in summer. Funct Ecol 2:433–440

Izhaki I, Ne'eman G (1997) Hares (*Lepus* spp.) as seed dispersers of *Retama raetam* (Fabaceae) in a sandy landscape. J Arid Environ 37:343–354

Jordano P (2000) Fruits and frugivory. In: Fenner M (ed) Seeds: the Ecology of Regeneration in Natural Plant Communities. Commonwealth Agric. Bur. Intl, Wallingford, UK

Karmiris IE, Nastis AS (2010) Diet overlap between small ruminants and the European hare in a Mediterranean shrubland. Cent Eur J Biol 5:729–737

Katona K, Biro ZS, Szemethy L, Demeš T, Nyeste M (2010) Spatial, temporal and individual variability in the autumn diet of European hare (*Lepus europaeus*). Acta Zool Acad Sci Hung 56:89–101

Kontsiotis VJ, Tsiompanoudis A, Bakaloudis DE (2011) The influence of habitat structure on the European brown hare *Lepus europaeus* food habits in mountainous areas of Northern Greece. Mammalia 75:389–394

Krebs CJ (1999) Ecological Methodology, 2nd edn. Addison-Welsey Educational Publishers Inc., Menlo Park, CA

Kuijper DPJ, Bakker JP (2008) Unpreferred plants affect patch choice and spatial distribution of European brown hares. Acta Oecol 34:339–344

Lindlof B (1978) Aggressive dominance rank in relation to feeding by European hare. Viltrevy 10:146–157

Litvaitis JA (1990) Differential habitat use by sexes of snowshoe hares (*Lepus americanus*). J Mammal 71:520–523

McAleece N, Gage JDG, Lambshead PJD, Paterson GLJ (1997) BioDiversity Professional Statistics Analysis Software. Jointly developed by the Scottish Association for Marine Science and the Natural History. Museum, London

McInnis ML, Vavra M, Krueger WC (1983) A comparison of four methods used to determine the diets of large herbivores. J Range Manage 36:302–306

Monaghan P, Metcalfe NB (1985) Group foraging in wild brown hares: effects of resource distribution and social status. Anim Behav 33:993–999

Paupério J, Alves PC (2008) Diet of the Iberian hare (*Lepus granatensis*) in a mountain ecosystem. Eur J Wildl Res 54:571–579

Pepin D, Cargnelutti B (1994) Individual variations of daily activity patterns in radiotracked European hares during winter. Acta Theriol 39:399–409

Pianka ER (2000) Evolutionary Ecology, 6th edn. Benjamin-Cummings, Addison-Wesley-Longman, San Francisco

Puig S, Videla F, Cona MI, Monge SA (2007) Diet of the brown hare (*Lepus europaeus*) and food availability in northern Patagonia (Mendoza, Argentina). Mamm Biol 72:240–250

Reichlin T, Klansek E, Hackländer K (2006) Diet selection by hares (*Lepus europaeus*) in arable land and its implications for habitat management. Eur J Wildl Res 52:109–118

Rödel HG, Völkl W, Kilias H (2004) Winter browsing of brown hares: evidence for diet breadth expansion. Mamm Biol 69:410–419

Rogers AR, Sinclair ARE (1997) Diet choice and nutrition of captive snowshoe hares (*Lepus americanus*): interactions of energy, protein, and plant secondary compounds. Ecoscience 4:163–169

Rueda M, Rebollo S, Gálvez-Bravo L, Escudero A (2008) Habitat use by large and small herbivores in a fluctuating Mediterranean ecosystem: implications of seasonal changes. J Arid Environ 72:1698–1708

Sfougaris A, Toulia S, Giannakopoulos A (2006) Botanical composition of the diet of the hare (*Lepus Europaeus*) in the Thessaly, Greece. In: Proceedings of the 4th Pan-Hellenic Rangelands Conference. Hellenic Society for Range and Pasture Management, Volos, Hellas, 10–12 November 2004

Shipley LA, Forbey JS, Moore BD (2009) Revisiting the dietary niche: when is a mammalian herbivore a specialist? Integr Comp Biol 49:274–290

Sinclair ARE, Krebs CJ, Smith JNM (1982) Diet quality and food limitation in herbivores: the case of the snowshoe hare. Can J Zool 60:889–897

Sokos C, Andreadis K (2000) Elements of Ecology and Biology of the Brown Hare (*Lepus Europaeus*) in NW Peloponnese with Emphasis on the Impact of Farming on the Composition of Food. Dissertation, School of Forestry and Natural Environment, Aristotle University of Thessaloniki, Greece

Sokos C, Mamolos AP, Kalburtji KL, Birtsas PK (2012) Farming and wildlife in Mediterranean agroecosystems. J Nat Conserv 21:81–92

Sokos C, Skordas K, Birtsas P (2003) Valuation of hunting and management of brown hare (*Lepus Europaeus*) in rangelands. In: Proceedings of the 3rd Pan-Hellenic Rangelands Conference. Hellenic Society for Range and Pasture Management, Karpenissi, Hellas, 4–6 September 2002

Sparks DR, Malechek JC (1968) Estimating percentage dry weight in diets using a microscopic technique. J Range Manage 21:264–265

Stephens DW, Krebs JR (1986) Foraging Theory. Princeton University Press, Princeton, NJ

Suchentrunk F, Willing R, Hartl GB (1991) On eye lens weights and other age criteria of the brown hare (*Lepus europaeus* Pallas 1778). Z Säugetierkd 56:365–374

Westoby M, Rest GR, Web JA (1976) Problems with estimating herbivore diets by microscopically identifying plant fragments from stomachs. J Mammal 57:167–172

Zaccaroni M, Biliotti N, Buccianti A, Calieri S, Ferretti M, Genghini M, Riga F, Trocchi V, Dessì-Fulgheri F (2013) Winter locomotor activity patterns of European hares (*Lepus europaeus*). Mammal Biol 78(6):482–485

Zar JH (1996) Biostatistical Analysis. Prentice Hall, Englewood Cliffs, New Jersey

24

Variation in the ilium of central European water frogs *Pelophylax* (Amphibia, Ranidae) and its implications for species-level identification of fragmentary anuran fossils

Hugues-Alexandre Blain[1,2*], Iván Lózano-Fernández[1,2] and Gottfried Böhme[3]

Abstract

Background: In the opinion of all field herpetologists and archaeozoologists, green frogs (living and sub-fossil specimens) are among the most difficult European amphibians to identify. Here, we examine the intra- and interspecific variation in both continuous and discrete iliac characters for refining the criteria for species-level identification among European green frogs *Pelophylax ridibundus*, *Pelophylax lessonae* and their associated klepton *Pelophylax* kl. *P. esculentus* (Amphibia, Ranidae).

Results: A total of 454 ilia of modern green frogs from the former German Democratic Republic have been studied: 168 for *P. lessonae* (86 of males and 82 of females), 118 for *P. ridibundus* (44 of males and 74 of females) and 168 for *P.* kl. *P. esculentus* (86 of males and 82 of females). The origin, sex, population structure and phenotype are known for each of the studied specimens. Eight variables have been taken (one angle and seven measurements), mainly on the distal part of the element in order to be able to apply them to fragmentary fossil ilia. Interspecific variations, sexual dimorphism and population structure have been investigated. Results suggest that the secure determination of a single fossil ilium at species level is quite impossible, but that at population level, it may be possible to distinguish between a 'pure' species or a 'pure' species plus its cohabiting hybrid form, as some minor differences have been evidenced in particular in the angle of the *tuber superior* in relation to iliac's main axis (character 2) and the width of the *pars ascendens* (character 8); two parameters significantly non size-dependent. No sexual dimorphism has been detected, except for *P. lessonae*.

Conclusions: We propose here a new biometrical method in order to differentiate between the fossil ilia of central European water frogs (genus *Pelophylax*) at population level.

Keywords: Ilium; *Pelophylax*; Osteology; Biometry; Klepton; Central Europe

Background

In the opinion of all field herpetologists, green frogs of genus *Pelophylax* ('palaearctic water frogs', formerly included in genus *Rana* until Frost et al. 2006) are among the most difficult European amphibians to identify (e.g. Lescure and de Massary 2012; Thirion and Evrard 2012). In parts of northern and central Europe, the green frog complex consists of two species: the marsh frog (*Pelophylax ridibundus*) and the pool frog (*Pelophylax lessonae*), as well as the edible frog (*Pelophylax* kl. *P. esculentus*), a hybridogenetic form between these two species (Berger 1968; Tunner 1974). Genetic studies have also revealed the existence of various other species in the Mediterranean Basin (Beerli 1994; Beerli et al. 1994), for a total of seven species and two taxa with a hybrid origin in Europe (Sillero et al. 2014).

Throughout Europe, the situation is complicated by the existence of different systems of mixed populations. Due to the hybridogenetic system, *P.* kl. *P. esculentus*

* Correspondence: hablain@iphes.cat
[1]IPHES, Institut Català de Paleoecologia Humana i Evolució Social, c/ Marcel·lí Domingo s/n (Edifici W3), Campus Sescelades, 43007 Tarragona, Spain
[2]Area de Prehistoria, Universitat Rovira i Virgili (URV), Avinguda de Catalunya 35, 43002 Tarragona, Spain
Full list of author information is available at the end of the article

can live and reproduce in the presence of only one parental species. Their occurrences depend mainly on ecological predilections of the single parental species. *P. lessonae* is more demanding in this respect, needing sunlit waters with rich vegetation (Zahn 1997). *P. ridibundus* prefers larger water bodies and hibernates in the water, in contrast to *P. lessonae*. *P.* kl. *P. esculentus* is more flexible than its parental species and poses fewer demands on its habitat (Günther 1990; Plötner 2005). The most common system occurs in central and western Europe where the hybrid *P.* kl. *P. esculentus* lives in sympatry with the parental species *P. lessonae* (LE system). Massive differences in the genetic structures of different populations are possible in LE systems, with hybrids making up between 5 and 95% of the population (Berger 1990). Other common water frog systems consist of lineage of *P.* kl. *P. esculentus* breeds with *P. ridibundus* forming populations of the RE system, which also feature high structural genetic variability (Plötner 2005). There are also populations consisting of both parental species and the hybrid (RLE system); however, in these cases, it is unclear whether *P.* kl. *P. esculentus* stems from hybridogenetic reproduction or primary hybridization between the parental species (Günther 1991; Plötner 2005). Finally, there is also all-hybrid populations (EE system) that have become reproductively independent of the parental forms owing to polyploidization, i.e. via triploid individuals (Günther 1975; Uzzell and Berger 1975; Ebendal and Uzzell 1982; Regnier and Neveu 1986; Pruvost et al. 2013) and which mainly occur in the northern region of the distribution range (Ebendal 1979; Eikhorst 1987).

While *P. lessonae* and *P. ridibundus* are significantly morphologically different, hybrids show a more or less intermediate appearance between those of their putative parents. The differentiation between these morphological types lies in several characteristics such as the size, the shape of the nose, the length of the hind legs, the extent of the webbing of the toes, the metatarsal profile tubers, the layout and size of dark spots on the sides and back, the form of the veins and the background colour of the posterior thigh, the shape and arrangement of vomerine denticles, the colour of vocal sacs and nuptial pads (on adult males), or biometric ratios (e.g. leg length, body length, tibia length, length of the first toe, height and length of the metatarsal tubercle and distance between the eye and the eardrum) (Berger 1966; Gubányi and Korsós 1992; Mayer et al. 2013). However, for living populations, the genetic approach is the only completely reliable methodology (Lescure and de Massary 2012).

According to the genetic data, the natural distribution of green frogs and hybrids is due to the differentiation of genetically closely related species that, from the Pleistocene to the Holocene, became isolated separately in refugia during glacial periods and were put into contact again during warmer periods. Secondary contact of diverged genetic entities can lead to hybridization when it happens before effective premating barriers have developed (Pruvost et al. 2013). After the last Pleistocene glaciation, the warming period that followed during the Holocene probably allowed green frogs to extend their range to the north of Europe, naturally creating hybrid zones (Thirion and Evrard 2012). However, the glacial/interglacial dynamics allow this type of hybridization throughout the Pleistocene, although the lack of reliable criteria for such determination has hardly helped provide paleontological data to support these different theories.

In the fossil and sub-fossil record, isolated anuran ilia are among the most commonly recovered anuran bones from fossil microvertebrate sites thanks to their relatively robust build and distinctive structure. They are widely considered to be the single best element upon which to base fossil identifications when isolated bones are all that is available. According to Sanchiz (1998), approximately 40% of extinct frog species' denominations are based on disarticulated material with the ilium designated as the holotype. Consequently, the proposal of this study is to examine the intra- and interspecific variation in both continuous and discrete iliac characters for refining the criteria for species-level identification among European green frogs *P. ridibundus*, *P. lessonae* and their associated klepton *P.* kl. *P. esculentus* (Amphibia, Ranidae).

Methods

The studied sample includes complete dry disarticulated skeletons from East Germany prepared in 1976 by one of us (see Böhme 1977 and Böhme and Günther 1979 for more details). This material is stored in the collections of the Museum für Naturkunde (Berlin, Germany). From this sample, we selected the isolated ilia (both right and left) of some 227 specimens of green frogs: 59 *P. ridibundus* (22 males and 37 females), 84 *P.* kl. *P. esculentus* (43 males, 41 females) and 84 *P. lessonae* (43 males, 41 females). Measurements have been taken through camera lucida-scaled drawings. Osteological nomenclature mainly follows Sanchiz (1998). For the nomenclatural treatment of klepton, we follow the proposals of Dubois and Günther (1982) and Dubois (1991).

We selected as many metric characters as possible for ilium, taking into account that they must be reproducible on fragmentary fossils (Figure 1; Table 1). Measurements have been done as follows: character 1, projected length of ilium on the line that goes from the upper part of the intersection between the acetabulum and the posterior margin and that is tangential to the highest point of the iliac's lower margin (Sanchiz et al. 1993; measurement 3 or IL); character 2, angle between the *tuber superior* and the iliac's main axis (Sanchiz et al. 1993; measurement 2 called ANG); character 3, dorsal crest height (measurement 1 or CDH in Sanchiz et al. 1993 and

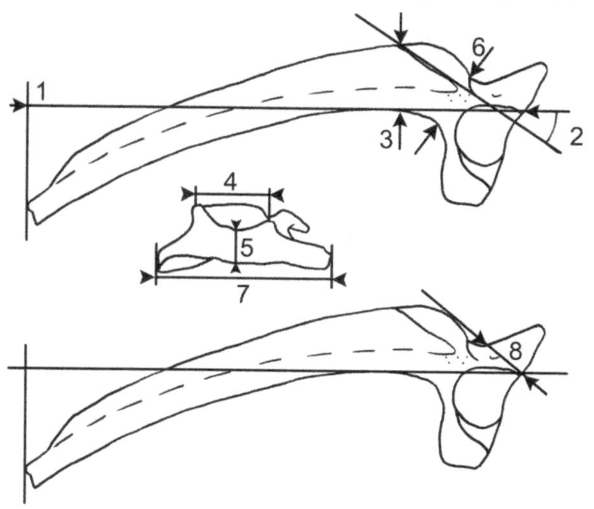

Figure 1 Measurements used on green frog's ilium. 1, length of ilium; 2, angle between *tuber superior* orientation and main iliac axis; 3, dorsal crest height; 4, acetabular diameter; 5, corpus thickness at the centre of the acetabulum; 6, 'iliac neck' (=smallest thickness of the crest on the corpus); 7, maximum width of the junctura ilioischiadica; 8, width of the *pars ascendens*.

measurement d in Böhme and Günther 1979); character 4, acetabular diameter (measurement a in Esteban and Sanchiz 1985, measurement b in Böhme and Günther 1979 and measurement d in Gleed-Owen 1998, 2000); character 5, corpus thickness at the centre of the acetabulum (measurement a in Böhme and Günther 1979 and measurement t in Gleed-Owen 1998, 2000); character 6, 'iliac neck' or smallest thickness of the crest on the corpus (measurement e in Böhme and Günther 1979); character 7, maximum width of the junctura ilioischiadica; and character 8, width of the *pars ascendens* on the imaginary line that is tangential to the *tuber superior*'s dorsal margin and that crosses the point at the intersection between the upper part of the acetabulum and the posterior edge of the ilia.

The standard non-parametric Mann-Whitney *U* test has been used to compare the measurements between species, population structure and sex and the Kendall's tau (τ) coefficient, and Spearman's rank correlation coefficient has been used to measure the association between iliac measurements and size on a personal computer with the Paleontological Statistics (PAST) programme (Hammer et al. 2001). These two tests are non-parametric hypotheses which establish whether two variables may be regarded as statistically dependent. Using the PAST programme, we can also obtain the probability (p) that the two variables are not correlated (uncorr).

To compare measurements, we used projection and boxplot sketches (Hammer et al. 2001). These boxplots display summary statistics for the distribution and plot the median, the 25th percentile, the 75th percentile and

Table 1 Statistics of iliac measurements of *Pelophylax ridibundus*, *Pelophylax lessonae* and their hybrid klepton *Pelophylax* kl. *P. esculentus*

Characters	1	2	3	4	5	6	7	8
P. ridibundus (male + female)								
N	115	116	116	113	113	116	113	116
Mean	27.60	36.00	3.65	4.20	1.70	3.40	10.00	2.50
SD	4.72	3.46	0.61	0.76	0.38	0.52	1.77	0.43
Min	17.60	28.00	2.40	2.80	1.00	2.20	6.40	1.60
Max	36.00	45.00	5.00	5.60	2.40	4.50	13.00	3.40
P. kl. *P. esculentus* (male + female)								
N	165	168	168	165	165	168	165	168
Mean	21.20	43.00	3.40	3.40	1.40	2.80	7.80	1.80
SD	2.91	4.05	0.46	0.48	0.24	0.37	1.10	0.32
Min	10.60	28.00	1.80	1.50	0.60	1.60	3.40	1.00
Max	28.10	53.00	4.40	4.40	2.00	3.60	10.80	3.00
P. lessonae (male + female)								
N	167	168	168	161	161	168	161	168
Mean	18.40	46.00	3.00	3.00	1.20	2.40	6.60	1.60
SD	2.28	4.32	0.40	0.33	0.20	0.27	0.79	0.21
Min	11.00	35.00	2.00	1.80	0.70	1.70	3.60	0.80
Max	24.20	56.00	4.40	3.80	1.80	3.20	8.40	2.00
P. kl. *P. esculentus* (RE)								
N	36	38	38	38	38	38	38	38
Mean	21.20	41.00	3.40	3.40	1.40	2.85	7.85	2.00
SD	2.09	2.12	0.43	0.42	0.20	0.34	1.00	0.31
Min	17.20	37.00	2.80	2.80	1.10	2.40	6.40	1.60
Max	26.60	46.00	4.20	4.20	2.00	3.60	10.80	3.00
P. kl. *P. esculentus* (LE)								
N	123	124	124	121	121	124	121	124
Mean	21.20	44.00	3.30	3.40	1.40	2.80	7.60	1.80
SD	3.16	4.37	0.47	0.50	0.24	0.37	1.12	0.31
Min	10.60	28.00	1.80	1.50	0.60	1.60	3.40	1.00
Max	28.10	53.00	4.40	4.40	2.00	3.60	9.60	2.60
P. ridibundus (male)								
N	42	42	42	39	39	42	39	44
Mean	28.40	36.50	3.80	4.40	1.80	3.40	10.60	2.60
SD	4.28	4.12	0.62	0.71	0.38	0.50	1.63	0.42
Min	19.20	28.00	2.40	2.80	1.10	2.40	6.80	1.80
Max	33.20	45.00	5.00	5.20	2.40	4.10	12.00	3.40
P. ridibundus (female)								
N	73	74	74	74	74	74	74	74
Mean	26.20	36.00	3.50	4.00	1.60	3.20	9.80	2.50
SD	4.98	3.06	0.61	0.79	0.38	0.54	1.84	0.45
Min	17.60	28.00	2.40	2.80	1.00	2.20	6.40	1.60
Max	36.00	42.00	4.70	5.60	2.40	4.50	13.00	3.20

Table 1 Statistics of iliac measurements of *Pelophylax ridibundus*, *Pelophylax lessonae* and their hybrid klepton *Pelophylax* kl. *P. esculentus* *(Continued)*

P. kl. *P. esculentus* (male)								
N	83	86	86	85	85	86	85	86
Mean	21.00	43.00	3.40	3.40	1.40	2.80	7.80	1.80
SD	2.60	3.59	0.47	0.47	0.23	0.36	1.09	0.31
Min	10.60	33.00	1.80	1.50	0.60	1.60	3.40	1.00
Max	26.60	53.00	4.20	4.20	2.00	3.60	10.80	3.00
P. kl. *P. esculentus* (female)								
N	82	82	82	80	80	82	80	82
Mean	21.50	42.00	3.30	3.40	1.40	2.80	7.80	1.90
SD	3.17	4.36	0.46	0.49	0.25	0.38	1.11	0.32
Min	12.60	28.00	2.20	2.00	0.80	1.70	4.40	1.20
Max	28.10	53.00	4.40	4.40	2.00	3.60	9.60	2.60
P. lessonae (male)								
N	86	86	86	81	81	86	81	86
Mean	18.00	45.00	3.00	3.00	1.20	2.40	6.40	1.60
SD	2.31	4.22	0.42	0.35	0.20	0.30	0.85	0.22
Min	11.00	37.00	2.00	1.80	0.70	1.70	3.60	0.80
Max	24.00	55.00	3.80	3.60	1.60	3.00	8.20	1.90
P. lessonae (female)								
N	81	82	82	80	80	82	80	82
Mean	19.60	46.50	3.00	3.00	1.20	2.40	6.70	1.60
SD	1.86	4.30	0.36	0.28	0.19	0.23	0.67	0.18
Min	16.40	35.00	2.60	2.40	0.90	2.10	5.90	1.20
Max	24.20	56.00	4.40	3.80	1.80	3.20	8.40	2.00

Abbreviations: N number of elements, *SD* standard deviation, *Min* minimum, *Max* maximum, *RE ridibundus-esculentus* population structure, *LE lessonae-esculentus* population structure. For measurement explanations, see Figure 1.

values more than 1.5 box lengths from 75th percentile (outliers).

Results

Interspecific variations

Mean, standard deviation and range have been computed for all variables (Table 1). According to Figure 2A,B, none of the studied measurements shows a complete separation (when taking into account minimum and maximum values) for none of the studied parental species and klepton. However, when taking into account the mean value and the standard deviation (95% of the values), some distinctions are visible. The results of the Mann-Whitney *U* test (Table 2) suggest that the difference between species and klepton are highly significant for all characters but for character 3 between *P. ridibundus* and *P.* kl. *P. esculentus* that is only significant. According to the projection of character's values (Figure 2A,B), those that present the highest differences between species are characters 1, 2, 7 and 8. Ilium length (character 1) is assumed by Sanchiz et al. (1993) as a good estimator of individual absolute size. Thus, in the studied sample, it appears that ilia from *P. ridibundus* (27.6 ± 4.7 mm) are larger than those of *P.* kl. *P. esculentus* (21.2 ± 2.9 mm) and than those of *P. lessonae* (18.4 ± 2.3 mm). Such a difference in size has already been reported by field herpetologists (e.g. Günther 1990; Plötner 2005). The angle between the orientation of the *tuber superior* and main iliac axis (character 2) shows also some differences between *P. ridibundus* (36.0 ± 3.5°) and *P.* kl. *P. esculentus* + *P. lessonae* (respectively 43.0 ± 4.0° and 46.0 ± 4.3°). For all the other characters, the ordination of values seems to follow that of the size (*P. ridibundus* > *P.* kl. *P. esculentus* > *P. lessonae*) like for character 7 (maximum width of the junctura ilioischiadica) larger in *P. ridibundus* (10.0 ± 1.8 mm) than in *P.* kl. *P. esculentus* (7.8 ± 1.1 mm) and *P. lessonae* (6.6 ± 0.8 mm), and character 8 (width of the *pars ascendens*) where *P. ridibundus* (2.5 ± 0.4 mm) is larger than *P.* kl. *P. esculentus* (1.8 ± 0.3 mm) and *P. lessonae* (1.6 ± 0.2 mm).

In order to remove from the analysis size-linked differences, we perform a statistical analysis using Spearman's rank and Kendall's Tau tests to search characters that are not size-dependent. As said above, character 1 has been taken as reference for the size of the whole individual. Among characters, only characters 2 and 8 are demonstrated to be size-independent (Table 3). Among other characters, character 7 is the one that shows the highest dependence with size with a Spearman's rank value of 0.97 followed by character 4 with 0.94 and character 6 with 0.93.

Variation according to population structure

Differences in iliac morphology in relation with the population structure (as defined in the introduction) of the hybrid klepton *P.* kl. *P. esculentus* have been investigated (Figure 2C). Characters for the LE system and RE system have been studied separately. The few triploid individuals have been removed from the analysis. The results of the Mann-Whitney *U* test (Table 2) suggest that the difference between RE and LE systems are highly significant for characters 2 and 8 and significant for character 5. There is no difference for character 1 (estimate of absolute size) between the two systems with respectively 21.20 ± 2.09 mm for RE and 21.20 ± 3.16 mm for LE systems. Character 2 (*tuber superior*'s orientation) is the one that seems to show the largest difference between the two systems (lowest value for two-tailed probability *p*), with 41.00 ± 2.12° for RE and 44.00 ± 4.37° for LE. Character 8 (width of the *pars ascendens*) although highly significant shows some tiny difference between RE (2.00 ± 0.31 mm) and LE (1.80 ± 0.31 mm) systems, as well as character 5 (acetabulum thickness) that although significant shows the same mean for RE (1.40 ± 0.20 mm) than for LE (1.40 ± 0.24 mm).

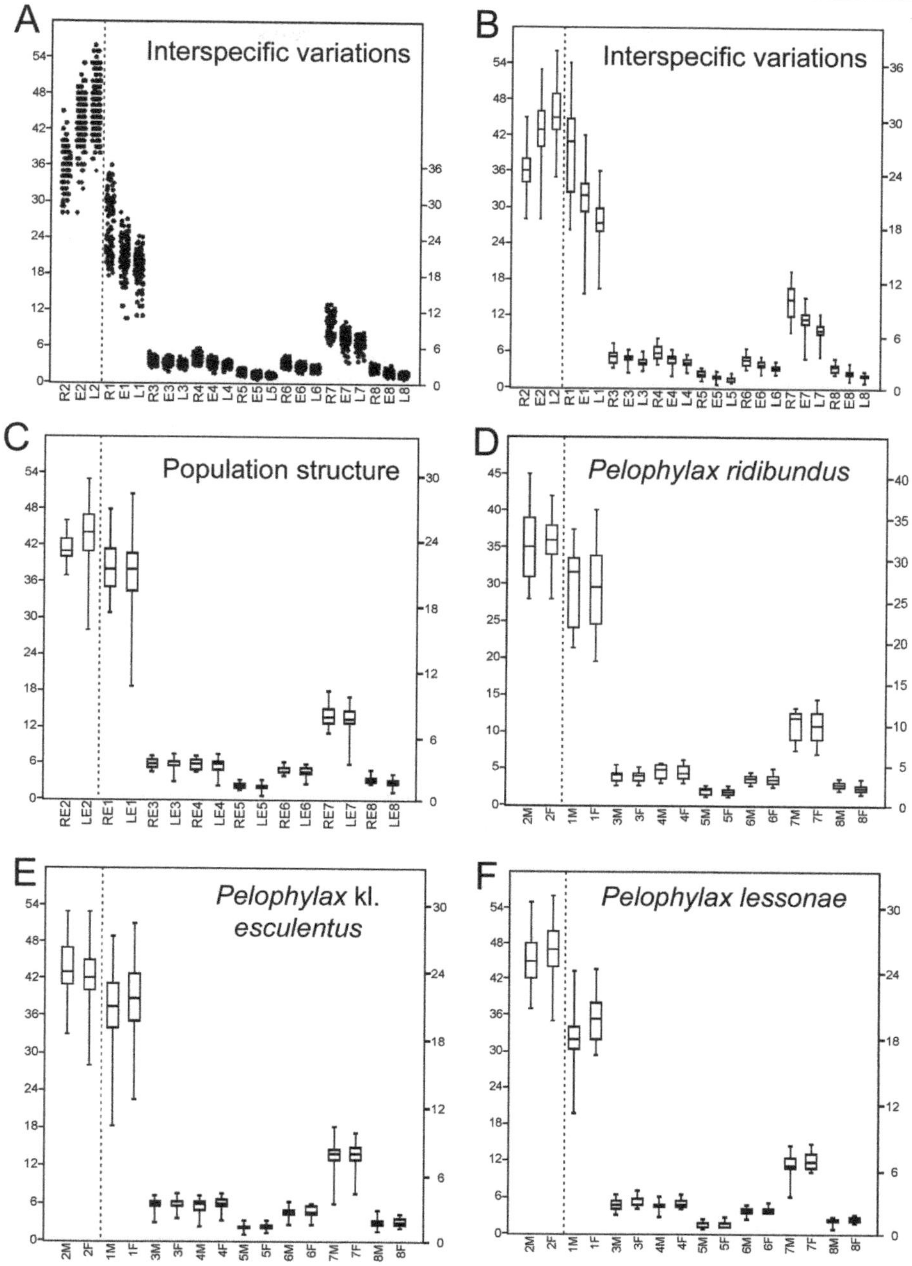

Figure 2 Projection, differences in iliac morphology and sexual dimorphism. Projection and boxplot of all characters **(A)**, **(B)** and according to their population structure **(C)** and sexual dimorphism **(D)**, **(E)**, **(F)**. Numbers in abscissa correspond to character's number. Abbreviations: E, *P.* kl. *esculentus*; L, *P. lessonae*; R, *P. ridibundus*; LE, *P.* kl. *esculentus* within the *lessonae-esculentus* system; RE, *P.* kl. *esculentus* within the *ridibundus-esculentus* system; M, male; F, female. Measurements (right of the dotted line) and angle (on the left) are respectively in millimetres and in sexigesimal degree.

Sexual dimorphism

Finally, sexual dimorphism has also been investigated (Figure 2D,E,F). According to osteological observations, all the specimens from our samples correspond to adult individuals (ossification of the pubis, etc.). The sex of the individuals was already annotated on the specimens and has been checked using the humerus (presence/ absence of crests). For character 1 (ilium length as estimate of absolute size), quite no significant differences have been found between male and female of *P. ridibundus* with the ilia of males (28.40 ± 4.28 mm) slightly larger than the ilia of females (26.20 ± 4.98 mm), as well as in *P.* kl. *P. esculentus* (male = 21.00 ± 2.60 mm; female = 21.50 ± 3.17 mm). Nevertheless, the difference between males and females in *P. lessonae* is highly significant (Table 2) with the ilia of males (18.00 ± 2.31 mm) slightly smaller than the ilia of females (19.60 ± 1.86 mm). There is an accepted hypothesis that explains the status and direction of sexual

Table 2 Results of Mann-Whitney *U* test comparing the differences for each character

	1	2	3	4	5	6	7	8
Interspecific variations								
P. ridibundus × *P.* kl. *P. esculentus*	H	H	S	H	H	H	H	H
P. ridibundus × *P. lessonae*	H	H	H	H	H	H	H	H
P. lessonae × *P.* kl. *P. esculentus*	H	H	H	H	H	H	H	H
Population structure								
RE × LE	N	H	N	N	S	N	N	H
Sexual dimorphism								
P. ridibundus	N	N	N	N	N	N	N	N
P. kl. *P. esculentus*	N	S	N	N	N	N	N	N
P. lessonae	H	S	H	S	S	H	S	H

N = two-tailed probability $p > 0.05$; S = $0.05 > p > 0.01$; H = $p < 0.01$.

size dimorphism in anurans, where males are usually smaller than females as a result of sexual selection (Monnet and Cherry 2002). In 90% of the anuran species, the females are larger than males (Shine 1979). Water frogs do not make exception to the rule (Günther 1990; Plötner 2005). These results may indicate that there may have had biases (human size selection) during the capture of the water frogs, especially for *P. ridibundus* and *P.* kl. *P. esculentus*.

The results of the Mann-Whitney *U* test (Table 2) suggest that the difference between males and females is also highly significant in *P. lessonae* for characters 3, 6 and 8 and significant for the other ones (Table 2). For other taxa, differences between male and female are not significant for all the studied characters, unless character 2 in *P.* kl. *P. esculentus*.

Biometrical differentiation between European water frogs

Characters 2 and 8 have been demonstrated to be the best characters for separation between the three studied water frog taxa, independently of the size. When considering the projection of characters 2 and 8 (Figure 3A,B), *P. lessonae* clearly separates from *P. ridibundus*. *P.* kl. *P. esculentus* has been separated according to its population structure (systems RE and LE). RE and LE largely overlaps between each others and part of the distribution of the parental species. RE mainly overlaps the distribution of *P. ridibundus*, whereas LE overlaps quite all the distribution of *P. lessonae*. In consequence, the specific attribution of a single ilium appears to be highly hypothetic, unless in some case for *P. ridibundus* that shows the largest difference with other species and systems. Nevertheless, with a minimum number of ilia, the distribution of their measurements (characters 2 and 8) within the variability of each species and system would permit to differentiate between pure species and possibly mixed populations. Only the pure populations of *P. lessonae* seem to be impossible to securely differentiate from the LE system.

Discussion

According to Sanchiz's review (1998) of the Salentia fossil record, '*Rana esculenta*' has rarely been detected as a fossil, and in older literature, references to '*R. esculenta*' should be now considered *Pelophylax* (*ridibundus*) sp. The only reliable fossil mention according to Sanchiz (1998) has been made for the Late Pleistocene and Holocene of Pisede (Germany; Böhme 1983a, b), whereas according to Martín and Sanchiz (2014) database, *P. lessonae* is mentioned since the Late Pliocene of Korotoyak-Belogor'e (Voronezh, Russia; Ratnikov 2001a) and *P. ridibundus* probably since the Early Pliocene (MN14) of Nurnus (Kotayk, Armenia; Vasilyan 2009).

Since Ecker (1889) examined the osteology of *P.* kl. *P. esculentus* in detail (=group *lessonae* + *esculentus* + *ridibundus*), very little literature has dealt with the osteology of individual species on a comparative basis. Böhme (1977) summed up the history of such work and highlighted the problems of uncertainty over identification, particularly of the green frogs. Böhme (1977) studied at least 20 specimens of each green frog species and, yet, concluded that they could not be separated using the ilium. The range of variation found within them is likely to be gradational, due to their shared genetic character. Some morphological traits are distinctive in the parent species, but these appear to be blurred in the different forms of *P.* kl. *P. esculentus*. Böhme and Günther (1979) carried out a more extensive study on a modern and sub-fossil material, using biometric methods, which permitted some specific diagnosis. Whereas frontoparietal was found to show significant differences between *P. lessonae*, *P. ridibundus* and *P.* kl. *P. esculentus*, ilium was shown to have a few characteristics permitting to separate *P.*

Table 3 Statistical analyses of the correlation between measurements and iliac length (character 1)

	2	3	4	5	6	7	8
Kendall's Tau	*−0.3298*	0.702	0.844	0.7813	0.8225	0.8794	*0.2528*
p (uncorr)	*2.08E-25*	5.72E-109	1.93E-152	6.69E-131	7.26E-145	2.51E-165	*3.09E-16*
Spearman's rank	*−0.4738*	0.8527	0.9464	0.9058	0.9388	0.9726	*0.3179*
p (uncorr)	*2.14E-26*	1.45E-127	1.10E-214	1.41E-163	1.83E-202	1.25E-276	*1.13E-11*

Abbreviations: *p (uncorr)* probability that the variables are uncorrelated. Uncorrelated variables are in italics.

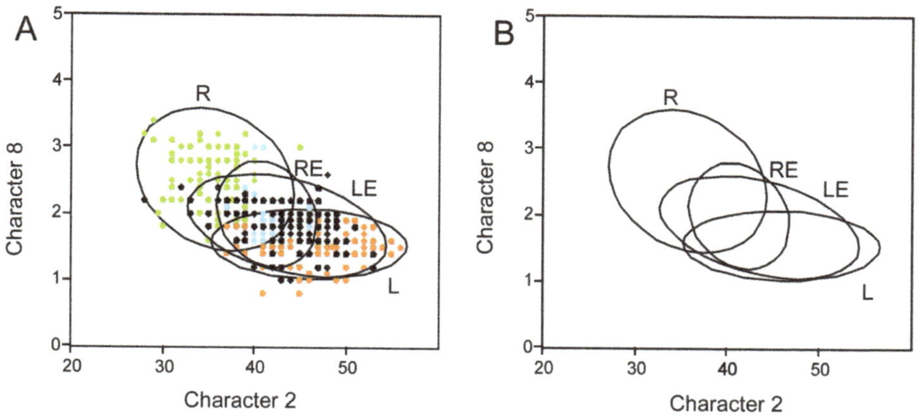

Figure 3 Biometrical separation between central European water frogs. Projections of characters 2 and 8 **(A)**, **(B)**.Abbreviations: L, *P. lessonae*; R, *P. ridibundus*; LE, *P.* kl. *P. esculentus* within the *lessonae-esculentus* system; RE, *P.* kl. *P. esculentus* within the *ridibundus-esculentus* system. Measurements (in ordinate) and angle (in abscissa) are respectively in millimetres and in sexigesimal degree.

lessonae from *P. ridibundus* + *P.* kl. *P. esculentus*. The morphological differences mainly lie in the development of the *tuber superior*, in particular in *P. lessonae* where this structure is very strong, laterally prominent and relatively steep, developing over the anterior border of the acetabulum (Böhme and Günther 1979). The same conclusions have been raised by Gleed-Owen (1998, 2000) from the study of 20 individuals each of *P. ridibundus* and *P.* kl. *P. esculentus* and 3 individuals of *P. lessonae* all from Britain, but not from a biometrical way. According to these differences, Gleed-Owen (2000) proposed that one fossil fragmentary ilium may cautiously be referred to the pool frog *P. lessonae* in a Middle Saxon (Chopdike Drove, c. 600 to 950 AD) site in Lincolnshire and so that *P. lessonae* was native from Britain. Although such an attribution has been posteriorly contested (e.g. Langton and Burton 2005), further phylogenetic studies demonstrate that *P. lessonae* was effectively native from Great Britain (Snell et al. 2005). Finally, Ratnikov (2001b) also noticed some differences in the morphology of the *tuber superior* between *P. ridibundus* and *P. lessonae* using Russian individuals, but does not include *P.* kl. *P. esculentus* in his study.

From a biometrical point of view, Böhme and Günther (1979) showed that the angle between the main iliac axis and the posterior border of the ilium was distinctive, measuring less than 55° (min/max = 43° to 55°) in *P.* kl. *P. esculentus* and *P. ridibundus*, while in *P. lessonae*, it is mainly over 55° (min/max = 52° to 58°). Unfortunately, the complete bone is needed for the measurement of this angle as proposed by Böhme and Günther (1979) and, consequently, not applicable on fragmentary fossils and so has been disregarded in our study. Our character 2 has been taken from Sanchiz et al. (1993) that studied five individuals of each *P. ridibundus*, *P.* kl. *P. esculentus* and *P. lessonae* from the collections of the Charles University in Prague (Czech Republic) and of the Museo Nacional de Ciencias Naturales de Madrid (Spain). According to Sanchiz et al. (1993) biometric analyses, the angle between the *tuber superior* orientation and the main iliac axis (ANG = our character 2) was demonstrated to be independent of the size of the element and, although the sample was reduced, permitted to distinguish *P. lessonae* (44.40 ± 7.83°; min/max = 32.00° to 52.00°) from *P. ridibundus* (37.30 ± 4.49°; min/max = 34.00° to 45.00°) and *P.* kl. *P. esculentus* (37.80 ± 6.94°; min/max = 32.00° to 48.00°). The results obtained from our larger sample (84 *P. lessonae*, 59 *P. ridibundus* and 84 *P.* kl. *P. esculentus*) are quite similar to the values obtained by Sanchiz et al. (1993) for *P. ridibundus* (36.0 ± 3.5°) and *P. lessonae* (46.0 ± 4.3°) but strongly differs for *P.* kl. *P. esculentus* (43.0 ± 4.0°). Even the separation according to the population structure does not explain such difference with 41.00 ± 2.12° for RE system and 44.00 ± 4.37° for LE system. Finally, we note that minimum (28.00°) and maximum (53.00°) values for character 2 in *P.* kl. *P. esculentus* are comprised within the minima maxima given by Sanchiz et al. (1993). Such differences may stress that in future studies, a minimum number of individuals (at least higher than five) will be required in order to separate the fragmentary fossil ilia of central European water frogs at species level.

Conclusions

A total of 454 ilia of modern central European water frogs have been studied from a biometrical point of view in order to propose criterions that will permit the differentiation between the different species and klepton *P. lessonae*, *P. ridibundus* and *P.* kl. *P. esculentus* from fossil fragmentary ilia recovered in paleontological and archaeological localities. Eight variables have been taken, mainly located on the distal part of the element in order to be able to apply them to fragmentary fossil ilia. Among our measurements, two parameters (characters 2 and 8) have been demonstrated to be significantly size-independent. Character 2 (*tuber superior*'s orientation = ANG in Sanchiz et al. 1993) and character 8 (width of the *pars ascendens*) are also the ones that show the largest difference between species and between the two systems of *P.* kl. *P. esculentus*. Character 2 is smaller for *P. ridibundus* (36.0 ± 3.5°) than for *P.* kl. *P. esculentus* + *P. lessonae* (respectively 43.0 ± 4.0° and 46.0 ± 4.3°). Character 8 is larger for *P. ridibundus* (2.5 ± 0.4 mm) than for *P.* kl. *P. esculentus* (1.8 ± 0.3 mm) and *P. lessonae* (1.6 ± 0.2 mm). As sexual dimorphism is taken into account, no major differences on the ilium have been put in evidence even if *P. lessonae* shows some significant differences in the measurements between males and females. All these results suggest that the secure determination of a single fossil ilium at species level is quite impossible, but that at population level, it must be possible to distinguish between a 'pure' species or a 'pure' species plus its cohabiting hybrid form (i.e. population structure LE or RE) according to the dispersion of the values of characters 2 and 8.

Competing interests

The authors declare that they have no competing interests.

Authors' contributions

GB prepared the osteological collections. HAB and ILF performed the statistical analyses and interpreted the data. HAB conceived the study, performed the data collection and wrote the manuscript. All authors read and approved the final manuscript.

Acknowledgements

This study is the result from the SYNTHESYS grant DE-TAF-3244 'Osteology of European green frogs (Pelophylax, Amphibia)' at the Museum für Naturkunde (Berlin, Germany). HAB wants to acknowledge Jean-Claude Rage (MNHN, Paris) and Zbyněk Roček (Laboratory of Paleobiology, Geological Institute, Prague) for their unconditional support and encouragements to this project. We are grateful to the DE-TAF team in Berlin especially to the administrative staff Manja Voss and to Florian Witzmann for their help during the realization of this work. Authors also thank Professor Sin-Che Lee managing editor, Massimo Delfino (University of Torino, Italy) and an anonymous reviewer for their comments on the manuscript. This study is also part of the research project CGL2012-38358 of the Spanish Ministry of Economy and Competitiveness and SGR2014-901 of the Government of Catalonia.

Author details

[1]IPHES, Institut Català de Paleoecologia Humana i Evolució Social, c/ Marcel·lí Domingo s/n (Edifici W3), Campus Sescelades, 43007 Tarragona, Spain. [2]Area de Prehistoria, Universitat Rovira i Virgili (URV), Avinguda de Catalunya 35, 43002 Tarragona, Spain. [3]Museum für Naturkunde, Leibniz-Institut für Evolutions- und Biodiversitätsforschung an der Humboldt-Universität zu Berlin, Invalidenstrasse 43, 10115 Berlin, Germany.

References

Beerli P (1994) Genetic Isolation and Calibration of an Average Protein Clock in Western Palearctic Water Frogs of the Aegean Region. University of Zurich, PhD Dissertation

Beerli P, Hotz H, Tunner HG, Heppich S, Uzzell T (1994) Two new water frog species from the Aegean islands Crete and Karpathos (Amphibia, Salentia, Ranidae). Notulae Naturae, Academy of Natural Sciences of Philadelphia 470:1–9

Berger L (1966) Biometrical studies on the population of green frogs from the environs of Poznan. Ann Zool 23(11):303–324

Berger L (1968) Morphology of the F1 generation of various crosses within *Rana esculenta* complex. Acta zoologica cracoviensia 13(13):301–324

Berger L (1990) On the origin of genetic systems in European water frog hybrids. Zoologica Poloniae 35:5–27

Böhme G (1977) Zur Bestimmung quartärer Anuren Europas an Hand von Skelettelementen. Wissenschaftliche Zeitschrift der Humbolt-Universität zu Berlin. Matematisch-Naturwissenschaftliche Reihe 26(3):283–300

Böhme G (1983a) Paläontologische Belege für die Batardierungen bei Raniden (Amphibia, Salientia). Schriftenreihe Geologischen Wissenschaft 19–20:31–37

Böhme G (1983b) Skelettreste von Amphibien (Urodela, Salentia) aus dem fossilen Tierbautensystem von Pisede bei Malchin. Teil 1: Taxonomie und Biostrationomie. Wissenschaftliche Zeitschrift der Humboldt-Universität zu Berlin, Mathematisch-Naturwissenschaftliche Reihe 32:657–680

Böhme G, Günther R (1979) Osteological studies in the European Water Frogs *Rana ridibunda*, *Rana lessonae* and *Rana "esculenta"* (Anura, Ranidae). Mitteilungen aus dem zoologischen Museum in Berlin 55(1):203–215

Dubois A (1991) Nomenclature of parthenogenetic, gynogenetic and "hybridogenetic" vertebrate taxons: new proposals. Alytes 8:61–74

Dubois A, Günther R (1982) Klepton and synklepton: two new evolutionary systematics categories in zoology. Zoologische Jahrbücher/Abteilung für Systematik, Ökologie und Geographie der Tiere 109:290–305

Ebendal T (1979) Distribution, morphology and taxonomy of the Swedish green frogs (*Rana esculenta* complex). Mitteilungen aus dem Zoologischen Museum in Berlin 55:143–152

Ebendal T, Uzzell T (1982) Ploidy and immunological distance in Swedish water frogs (*Rana esculenta* complex). Amphibia-Reptilia 3:125–133

Ecker A (1889) The Anatomy of the Frog (Translated by G. Haslam). Clarendon Press, Oxford

Eikhorst R (1987) Der Laich des Teichfrosches *Rana esculenta* Linnaeus, 1758 in einer reinen Bastardpopulation (Anura: Ranidae). Salamandra 23:122–131

Esteban M, Sanchiz B (1985) Herpetofauna de Erralla. Munibe (Antropología y Arqueología) 37:81–86

Frost DR, Grant T, Faivovich J, Bain RH, Haas A, Haddad CFB, De Sa RO, Channing A, Wilkinson M, Donnellan SC, Raxworthy CJ, Campbell JA, Blotto BL, Moler P, Drewes RC, Nussbaum RA, Lynch JD, Green DM, Wheeler WC (2006) The amphibian tree of life. Bull Am Museum Nat Hist 297:8–370

Gleed-Owen CP (1998) Quaternary Herpetofaunas of the British Isles: Taxonomic Descriptions, Palaeoenvironmental Reconstructions, and Biostratigraphic Implications. Coventry University, PhD Dissertation

Gleed-Owen CP (2000) Subfossil records of *Rana* cf. *lessonae*, *Rana arvalis* and *Rana* cf. *dalmatina* from Middle Saxon (c. 600–950 AD) deposits in eastern England: evidence for native status. Amphibia-Reptilia 21(1):57–65

Gubányi A, Korsós Z (1992) Morphological analysis of two Hungarian water frog (*Rana lessonae-esculenta*) populations. Amphibia-Reptilia 13(3):235–242

Günther R (1975) Zum natürlichen Vorkommen und zur Morphologie triploider Teichfrösche, « *Rana esculenta* », L., in der DDR (Anura, Ranidae). Mitteilungen aus dem zoologischen Museum in Berlin 51(1):145–158

Günther R (1990) Die Wasserfrösche Europas. A. Ziemsen Verlag, Wittenberg Lutherstadt, pp 72–193

Günther R (1991) Europäische Wasserfrösche (Anura, Ranidae) und biologisches Artkonzept. Mitteilungen aus dem Zoologischen Museum Berlin 67:39–53

Hammer O, Harper DAT, Ryan PD (2001) PAST: Paleontological Statistics. University of Oslo, Oslo

Langton TES, Burton JA (2005) Pool Frog *Rana Lessonae* in Britain: Past Nativeness Unresolved. On the Verification of Sub-fossil Bone Identification. Reports of the Green Frog Information Programme (GRIP). Herpetofauna Consultants International, Bramfield

Lescure J, de Massary J-C (2012) Atlas des Amphibiens et Reptiles de France. Biotope, Mèze; Muséum national d'Histoire naturelle. Collection Inventaires & Biodiversité, Paris

Martín C, Sanchiz B (2014) Lisanfos KMS. Version 1.2. Museo Nacional de Ciencias Naturales, MNCN-CSIC, Madrid, Spain, Online reference accessible at http://www.lisanfos.mncn.csic.es/

Mayer M, Hawlitschek O, Zahn A, Glaw F (2013) Composition of twenty green frog populations (*Pelophylax*) across Bavaria, Germany. Salamandra 49(1):31–44

Monnet J-M, Cherry MI (2002) Sexual size dimorphism in anurans. Proc Royal Soc London B Biol Sci 269(1507):2301–2307

Plötner J (2005) Die westpaläarktischen Wasserfrösche. Laurenti-Verlag, Bielefeld
Pruvost NBM, Hoffmann A, Reyer H-U (2013) Gamete production patterns, ploidy, and population genetics reveal evolutionary significant units in hybrid water frogs (*Pelophylax esculentus*). Ecol Evol 3(9):2933–2946
Ratnikov VY (2001a) Pliocene anurans of East-European Platform. Russ J Herpetol 8(3):171–178
Ratnikov VY (2001b) Osteology of Russian toads and frogs for paleontological researches. Acta zoological cracoviensia 44:1–23
Regnier V, Neveu A (1986) Structures spécifiques des peuplements en grenouilles du complexe *Rana esculenta* de divers milieux de l'Ouest de la France. Acta Oecologica 7(1):3–26
Sanchiz B (1998) Salientia. Handbuch der Paläoherpetologie, part 4. Verlag Dr Friedrich Pfeil, Stuttgart
Sanchiz B, Schleich HH, Esteban M (1993) Water frogs (Ranidae) from the Oligocene of Germany. J Herpetol 27(4):486–489
Shine R (1979) Sexual selection and sexual dimorphism in the Amphibia. Copeia 1979(2):297–306
Sillero N, Campos J, Bonardi A, Corti C, Creemers R, Crochet PA, Isailović JC, Denoël M, Ficetola GF, Gonçalves J, Kuzmin S, Lymberakis P, de Pous P, Rodríguez A, Sindaco R, Speybroeck J, Toxopeus B, Vieites DR, Vences M (2014) Updated distribution and biogeography of amphibians and reptiles of Europe based on a compilation of countrywide mapping studies. Amphibia-Reptilia 35:1–31
Snell C, Tetteh J, Evans IH (2005) Phylogeography of the pool frog (*Rana lessonae* Camerano) in Europe: evidence for native status in Great Britain and for an unusual postglacial colonization route. Biol J Linn Soc 85:41–51
Thirion J-M, Evrard P (2012) Guide des Reptiles et Amphibiens de France. Editions Belin, Paris
Tunner HG (1974) Die klonale structur eine Wasserfroshpopulation. Zeitschrift für zoologische Systematik und Evolutionforschung 12:309–314
Uzzell T, Berger L (1975) Electrophoretic phenotypes of *Rana ridibunda*, *Rana lessonae*, and their hybridogenetic associate, *Rana esculenta*. Proc Acad Nat Sci Philadelphia 127(2):13–34
Vasilyan D (2009) Lower Vertebrates of the Late Paleogene and Neogene of Armenia. PhD Dissertation. Scientific Center of Zoology and Hydroecology. National Academy of Sciences of the Republic of Armenia, Yerevan
Zahn A (1997) Untersuchungen zum *Rana* kl. *esculenta–lessonae*-Komplex in Oberbayern. Salamandra 33:79–88

Zelinkaderes yong sp. nov. from Korea - the first recording of *Zelinkaderes* (Kinorhyncha: Cyclorhagida) in Asia

Andreas Altenburger[1*], Hyun Soo Rho[2], Cheon Young Chang[3] and Martin Vinther Sørensen[1]

Abstract

Background: A new kinorhynch species, *Zelinkaderes yong* sp. nov., is described from Korea.

Results: *Zelinkaderes yong* sp. nov. is described from coastal, sandy habitats in Korea by means of light and scanning electron microscopic techniques. The new species is characterized by the presence of cuspidate spines in lateroventral positions on segments 2 and 9, ventrolateral positions on segment 5, and lateral accessory positions on segment 8; flexible tiny acicular spines in lateroventral positions on segment 2, more regular-sized lateroventral acicular spines on segment 8, and middorsal spines on segments 4, 6, 8, 9, and 11. Females furthermore have acicular spines in middorsal and midlateral positions on segment 10, whereas males have crenulated spines on this segment. The absence of acicular spines in the lateral series of segment 9 makes it easy to distinguish the new species from all previously described congeners. The new species differs most from *Zelinkaderes submersus*, whereas it is morphologically closest to *Zelinkaderes klepali*. In regard to the spine patterns, the new species only differs from *Z. klepali* by its lack of lateroventral acicular spines on segment 9.

Conclusions: The finding of a new species of *Zelinkaderes* in East Asia extends the distributional range of the genus, which suggests that the genus basically could be present anywhere in the world and could be considered as cosmopolitan.

Keywords: Biodiversity; Meiofauna; Morphology; Taxonomy; Zelinkaderidae

Background

Kinorhynchs are small, worm-like, invertebrate animals that live among the marine meiobenthos. They have been known since 1851, but knowledge about their diversity is still incomplete (Dujardin 1851; Neuhaus 2013). The current number of 217 valid described species (based on description of adult specimens) is expected to increase substantially by intensified research efforts (Sørensen 2013). South Korea is one of the countries where kinorhynch diversity is relatively well known with more than ten new species discovered within the last 4 years (Sørensen et al. 2010a; Sørensen et al. 2010b, c; Lundbye et al. 2011; Sørensen et al. 2012; Sørensen et al. 2013).

The family Zelinkaderidae and genus *Zelinkaderes* were described in 1990 by R. P. Higgins (1990). Members of the genus are characterized by the presence of 16 indistinct placids on the neck that are basally fused with the first trunk segment. The first and second trunk segments consist of complete cuticular rings, whereas the third to tenth segments are composed of single plates with midventral articulations. Oral styles are bipartite; middorsal spines are present on segments 4, 6, and 8 to 11; cuspidate spines are found on several segments; and a midterminal spine is present. Currently, Zelinkaderidae accommodates two genera: *Triodontoderes* (see Sørensen and Rho 2009) with one species and *Zelinkaderes* (see Higgins 1990) with four previously described species. The first *Zelinkaderes* species to be discovered was actually described as *Cateria submersa* Gerlach 1969. This species was found in medium coarse sand near Helgoland in Northwest Germany (Gerlach 1969). However, this species was later redescribed by Higgins (1990) and assigned to the new genus, *Zelinkaderes*, under the name *Zelinkaderes submersus* (Gerlach 1969). In the same contribution, Higgins (1990) reported a second

* Correspondence: aaltenburger@snm.ku.dk
[1]Natural History Museum of Denmark, University of Copenhagen, Universitetsparken 15, 2100 Copenhagen, Denmark
Full list of author information is available at the end of the article

species of the genus, *Zelinkaderes floridensis* Higgins 1990, that was collected from localities with muddy sand, around 20 miles off Fort Pierce at the Atlantic Coast of Florida. Only 6 years later, Bauer-Nebelsick (1995) described a third species, *Zelinkaderes klepali* Bauer-Nebelsick, 1995, from coral sand at the Egyptian Red Sea coast. The fourth, and until now most recently discovered species, *Zelinkaderes brightae* Sørensen et al., 2007, was also described from Fort Pierce, Florida but from much cleaner calcareous sand than *Z. floridensis* and also closer to the coast. The type locality of *Z. brightae* is situated 6 miles off Fort Pierce (Sørensen et al. 2007), but subsequently, Herranz et al. (2013) have recorded the species at localities only 3 and 4 miles from the coast. Generally, the taxonomy of *Zelinkaderes* appears to be relatively uncomplicated, and the four known species can be distinguished by their lateral spine patterns and in particular by the position of their cuspidate spines.

In the present contribution, we describe the fifth species of *Zelinkaderes*. This species is the first intertidal species known from the genus, and it also represents the first Asian recording of the genus.

Methods

Specimens of *Zelinkaderes yong* sp. nov. were collected at two localities. The holotype and some paratypes were collected by scuba diving on 7 September 2001 from subtidal sand on 3- to 6-m depth at Guryongpo, near Pohang (Figure 1A) on the east coast of the Korean Peninsula (position: 36° 00′ 06″N, 129° 34′ 17″E). Additional paratypes were collected on 15 May 2010 from intertidal sand around Shinchang (Figure 1B) on the west coast of Jeju Island in South Korea (33° 20′ 42″N, 126° 10′ 12″E). At this second locality, the water retracts completely from the area at low tide. The substrate was fine sand with lots

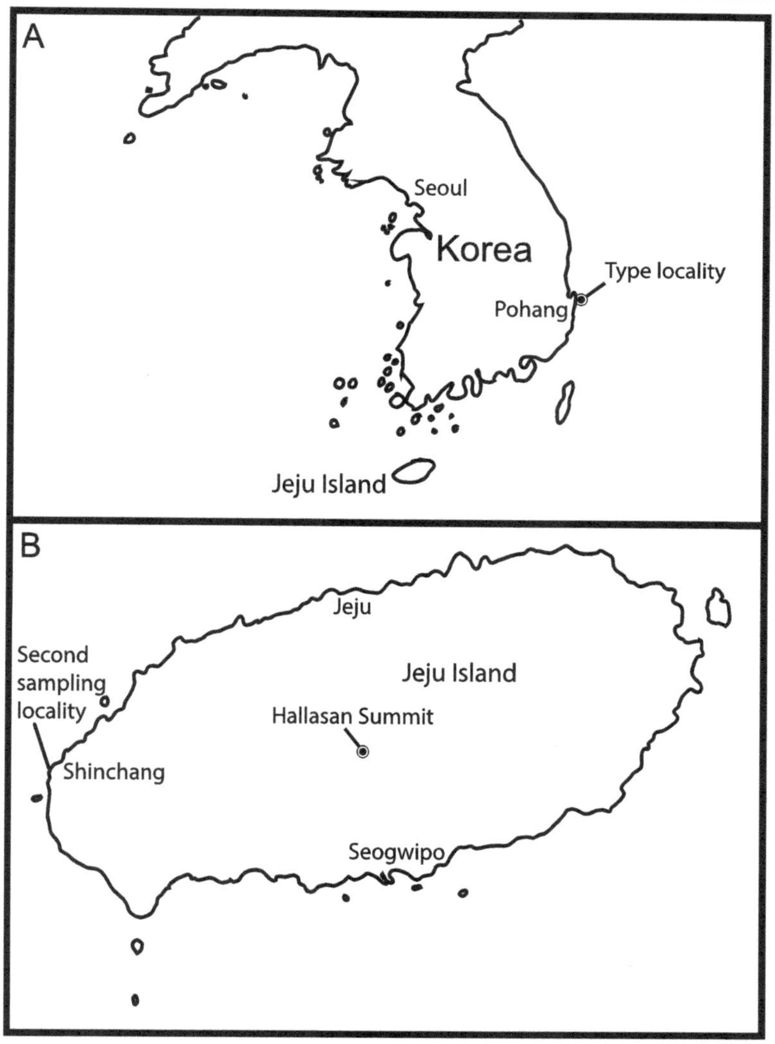

Figure 1 Maps showing the collection areas. A. Korean Peninsula showing Jeju Island and the type locality of *Zelinkaderes yong* sp. nov. at Guryongpo near Pohang. **B**. Jeju Island with the second sampling locality marked.

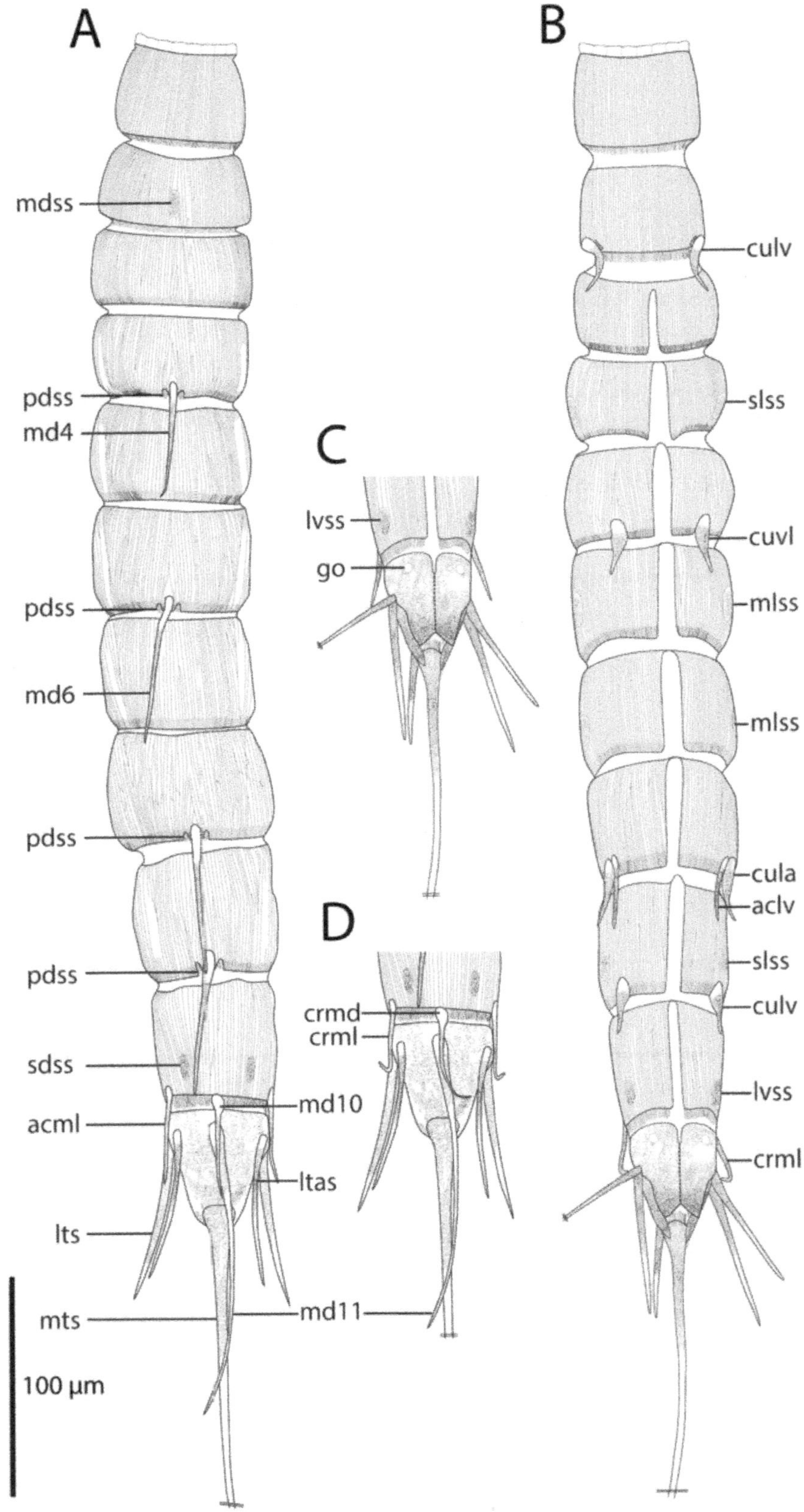

Figure 2 (See legend on next page.)

(See figure on previous page.)
Figure 2 Line art illustration of *Zelinkaderes yong* sp. nov. A. Female dorsal view. **B**. Male ventral view. **C**. Female posterior part ventral view. **D**. Male posterior part dorsal view. Drawings are based on SEM micrographs. Trunk length of the female specimen is approximately 500 μm, and trunk length of the male specimen is approximately 520 μm. Abbreviations: *aclv* acicular lateroventral spine, *acml* acicular midlateral spine, *crmd* crenulated middorsal spine, *crml* crenulated midlateral spine, *cula* cuspidate lateral accessory spine, *culv* cuspidate lateroventral spine, *cuvl* cuspidate ventrolateral spine, *go* gonopore, *ltas* lateral terminal accessory spine, *lts* lateral terminal spine, *lvss* lateroventral sensory spot, *md4-11* acicular middorsal spines on segments 4 to 11, *mdss* middorsal sensory spot, *mlss* midlateral sensory spot, *mts* midterminal spine, *pdss* paradorsal sensory spot, *sdss* subdorsal sensory spot, *slss* sublateral sensory spot.

of crustacean tubes (*Callianassa japonica* (Ortmann 1891)). Specimens at this locality were collected by digging a small pit into the sand and sieving the accumulating water with a net of 100-μm mesh size. The specimens occurred in so high numbers that no further extraction was required.

All collected specimens were fixed in 4% formalin buffered with sea water. For light microscopy, fixed specimens were transferred to distilled water, dehydrated through a graded series of glycerin, and mounted in Fluoromount G® on glass slides. They were examined using an Olympus BX51 microscope equipped with an Olympus DP20 camera and Normarski differential interference contrast optics. Measurements were made with Cell^D software version 2.7. (Olympus A/S, Ballerup, Denmark).

For scanning electron microscopy (SEM), specimens were dehydrated using a graded series of ethanol, transferred to acetone, and finally critical point dried. Dried specimens were mounted on aluminum stubs, sputter coated with a platinum/palladium mix, and examined with a JEOL JSM-6335 F field emission SEM (JEOL GmbH, Eching, Germany).

Type material is stored at the Natural History Museum of Denmark.

Results

Systematics

Order CYCLORHAGIDA (Zelinka, 1896) Higgins, 1964
Family ZELINKADERIDAE Higgins 1990
Genus ZELINKADERES Higgins 1990
Type species *Zelinkaderes yong* sp. nov.
Etymology: Like several other mud dragon species, this one is named after a dragon. The species name 'yong' refers to a dragon in Korean mythology.

Type material

Holotype: Adult male from Guryongpo, near Pohang on the East coast of the Korean Peninsula; 7th September 2001: 36° 00′ 06″N, 129° 34′ 17″E, in subtidal sand from 3- to 6-m depth, mounted in Fluoromount G®, deposited at the Natural History Museum of Denmark (NHMD) under accession number: ZMUC KIN-000818.

Paratypes: two adult females, three adult males, and one juvenile, collected at same time and locality as the holotype, mounted in Fluoromount G®, and deposited at the NHMD under accession numbers: ZMUC KIN-000819 to KIN-000824. Additional paratypes include seven adult females and two adult males, collected from an intertidal sand flat at Shinchang, Jeju Island, South Korea; 15th May 2010: 33° 20′ 42″N, 126° 10′ 12″E, mounted on an SEM stub and deposited at the NHMD under accession number: ZMUC KIN-000825.

Synonymy

Zelinkaderes sp. in Park et al. 2006; *Zelinkaderes* sp. JKP-2005: 18S rRNA sequence (accession number AY746985) in GenBank (NCBI).

Diagnosis

Zelinkaderes yong sp. nov. has middorsal spines on segments 4, 6, and 8 to 11 (middorsal spine on segment 10 crenulated in males), lateroventral cuspidate spines on segments 2 and 9, lateral accessory cuspidate spines on segment 7, ventrolateral cuspidate spines on segment 5, lateroventral acicular spines on segment 8, and midlateral spines on segment 10, as regular acicular spines in females and crenulated spines in males. Lateral terminal spines, lateral terminal accessory spines, and midterminal spine are present on segment 11. Large sensory spots are present middorsally on segment 2, midlaterally on segments 2, 6, and 7, sublaterally on segments 4 and 9, subdorsally and lateroventrally on segment 10; smaller sensory spots are present in paradorsal positions around dorsal spines and in ventrolateral positions on segment 11. Trunk segments 1 to 10 have cuticular hairs arranged in a longitudinally striated pattern with distinct hairless patches midlaterally on segments 3 to 7 and 9. Segment 11 has cuticular hairs not arranged in any particular pattern.

Description

Adult specimens possess a head, neck, and 11 trunk segments (Figures 2A,B and 3). The trunk appears long, thin, and vermiform with a semicircular cross section and a slight thickening of the trunk in the middle (Figure 4A). Trunk segments 1 and 2 are composed of closed cuticular rings and segments 3 to 11 of a single plate with a midventral joint (Figure 2B). The midventral joint appears as a distinct cuticular surface marking

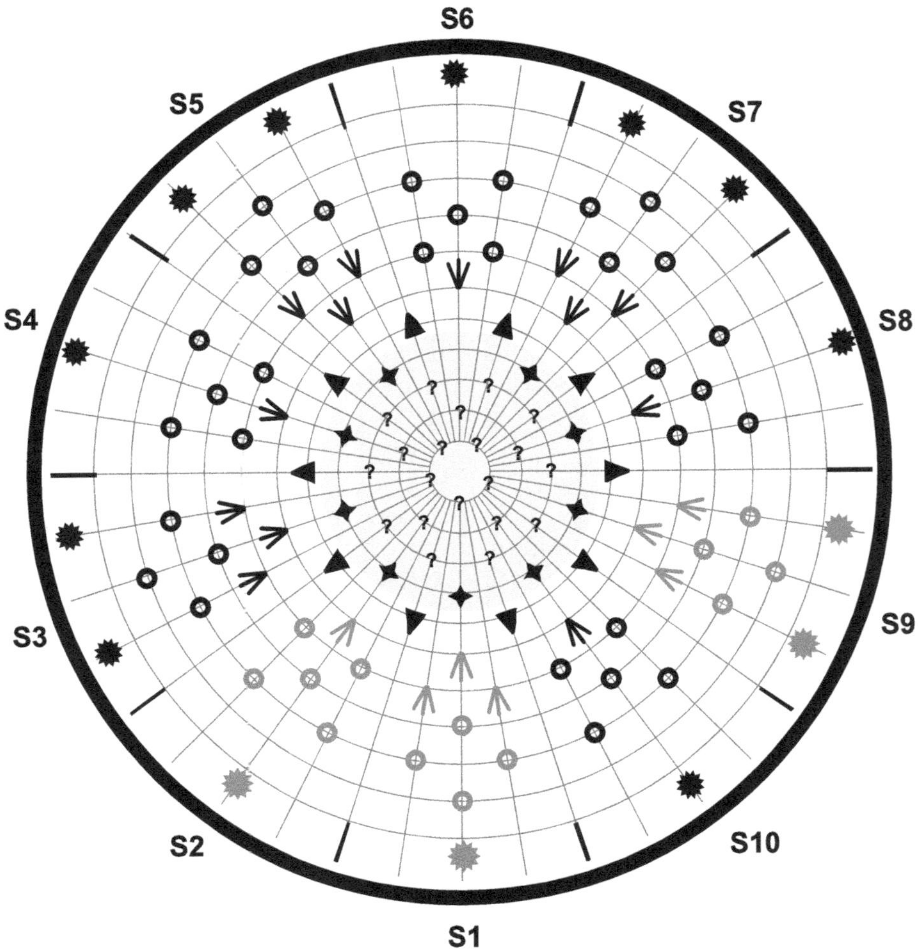

Scalid and style arrangement

Ring/Sector	1	2	3	4	5	6	7	8	9	10	Total
00 outer oral styles ✦	1	1	1	1	1	0	1	1	1	1	9
01 primary spinoscalids ▼	1	1	1	1	1	1	1	1	1	1	10
02 fringe	1	1	1	1	1	1	1	1	1	1	10
03 spinoscalids/fringe ○	2	2	2	2	2	2	2	2	2	2	20
04 spinoscalids ○	1	1	1	1	1	1	1	1	1	1	10
05 spinoscalids ○	2	2	2	2	2	2	2	2	2	2	20
06 spinoscalids ○	1	0	1	0	1	0	1	0	1	0	5
07 spinoscalids ○	0	0	0	0	0	0	0	0	0	0	0
08 trichoscalids	1	1	2	1	2	1	2	1	2	1	14
Total scalids	8	7	9	7	9	7	9	7	9	7	79

Figure 3 Mouth cone and introvert diagram. Diagram of mouth cone (gray area) and introvert in *Zelinkaderes yong* sp. nov., showing the distribution of outer oral styles, scalids, and trichcscalids. The table shows scalid arrangement by sector. Not all introvert sectors could be examined, but assuming that the species displays the usual symmetry patterns, the scalid distributions for the non-examined sectors have been predicted. Scalid numbers and distributions based on predictions are marked with gray; numbers and distributions based on actual observations are marked with black.

on segments 3 to 10, whereas it is only an intracuticular line on segment 11. The anterior border of all segments is straight. An extremely weak pectinate fringe is present on the posterior edges of segments 1 to 10. The cuticle is patterned by longitudinal furrows populated with cuticular hairs (Figure 4B). The midterminal spine can be longer than the trunk of the animal. For a complete overview of measures and dimensions, see Table 1. The distribution of cuticular structures such as sensory spots and acicular and

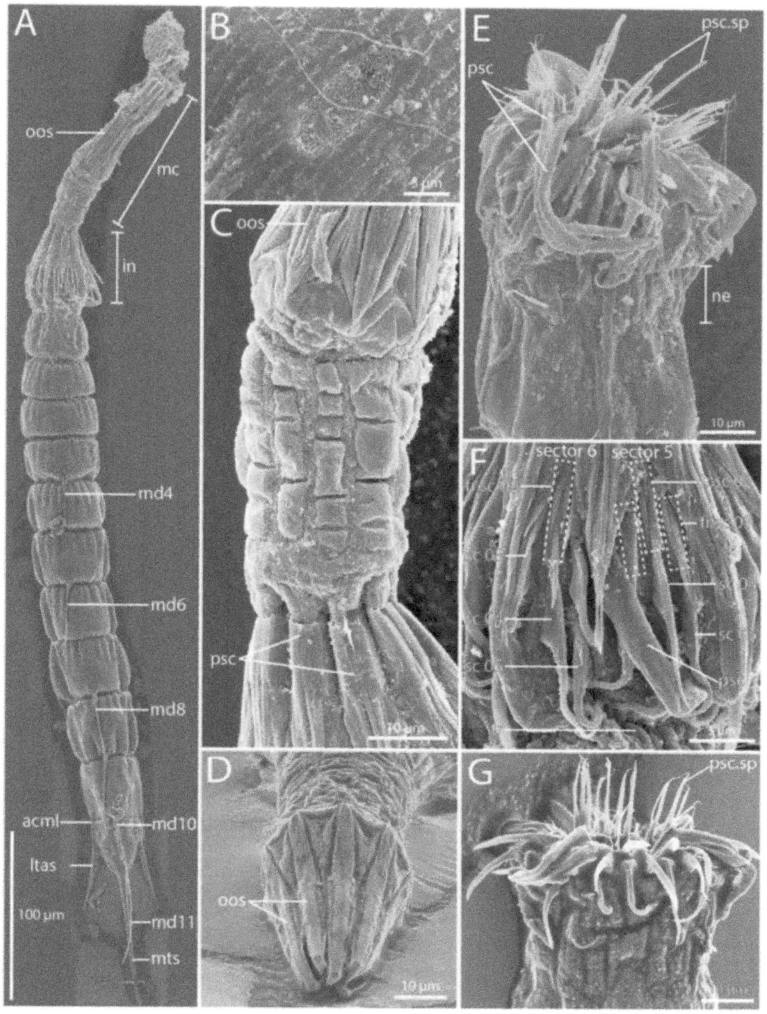

Figure 4 Overviews and details of head and trunk morphology. Scanning electron micrographs showing overviews and details of head and trunk morphology of *Zelinkaderes yong* sp. nov. paratypes (KIN-000825). **A**. Dorsal overview of female specimen. The midterminal (mt) spine is not shown entirely. **B**. Detail of the cuticle with the midlateral sensory spot on segment 6, showing the general structure of the cuticle as patterned by longitudinal furrows from where cuticular hairs emerge. **C**. Detail of the mouth cone showing the elongated proximal part of the mouth cone. **D**. Detail of the mouth cone in frontal view. **E**. Detail showing the spines at the bases of the primary spinoscalids that point in an anterior direction when the introvert is partly everted. **F**. The spines on the primary scalids as seen when the introvert is fully everted. **G**. Scalid spines pointing anteriorly as seen when the introvert is partly retracted. Abbreviations: *acml* acicular midlateral spine, *fr.sc* fringe replacing spinoscalid followed by ring number, *in* introvert, *ltas* lateral terminal accessory spine, *mc* mouth cone, *md4-11* acicular middorsal spines on segments 4 to 11, *mts* midterminal spine, *ne* neck region, *oos* outer oral styles, *psc* primary spinoscalids, *psc.sp* primary spinoscalid spines, *sc* spinoscalid followed by ring number, *tr* trichoscalid.

cuspidate spines is summarized in Table 2. Sensory spots belong to type 1.

Head

The head consists of a long retractable mouth cone and an introvert with five rings of scalids (Figures 3 and 4C-G). The exact position of the inner oral styles could not be examined, but their presence was observed in several specimens. The mouth cone is surrounded by nine anteriorly directed outer oral styles of equal size, composed of two units (Figure 4C,D). The outer oral styles alternate in length and are arranged in a way so that each style is located centrally and anterior to each introvert sector (Figure 4C), except sector 6 where the outer oral style is missing. The anterior parts of the styles point inwards (Figure 4D). The base of each style is characterized by a fringe of small anteriorly directed spines (Figure 4D). Triangular cuticular thickenings are present in between the bases of the oral styles (Figure 4D). The proximal part of the mouth cone, posterior to the outer oral styles, forms a long tube. The tube appears to have longitudinal areas with thickened

Table 1 Measurements from light microscopy of adult *Zelinkaderes yong* sp. nov. (in μm)

Character	*n*	Range	Mean	SD
TL	6	480 to 575	514.6	33.5
S1	6	38 to 55	48.0	6.7
S2	6	28 to 41	35.4	5.1
S3	6	32 to 44	36.2	4.6
S4	6	32 to 48	38.6	5.6
S5	6	38 to 49	41.9	4.2
S6	6	43 to 50	47.1	3.1
S7	6	44 to 57	49.2	4.5
S8	6	45 to 54	49.3	3.6
S9	6	47 to 58	53.2	4.4
S10	6	49 to 57	52.4	2.8
S11	6	30 to 45	37.2	5.1
MD 4 (ac)	6	16 to 30	23.2	5.4
MD 6 (ac)	6	22 to 42	33.1	7.9
MD 8 (ac)	6	31 to 45	41.0	5.2
MD 9 (ac)	6	35 to 47	41.5	4.8
MD 10 (ac)	6	35 to 53	42.6	7.1
MD 11 (ac)	6	84 to 114	95.4	11.8
LV 2 (cu)	12	10 to 20	16.4	2.6
VL 5 (cu)	12	15 to 22	20.7	2.1
LA 8 (cu)	11	21 to 30	25.1	2.7
LV 8 (ac)	12	32 to 41	36.3	3.3
LV 9 (cu)	12	20 to 27	24.5	2.0
ML 10 (ac)	4	31 to 33	31.8	0.8
ML 10 (cr)	7	28 to 44	35.5	6.0
LTS	11	58 to 76	65.6	6.2
LTAS	11	55 to 65	58.4	3.7
LTS/LTAS	10	1.0 to 1.3	1.1	0.1
MTS	4	410 to 561	489.9	81.4
MTS/TL	4	0.8 to 1.1	0.9	0.1

Abbreviations: *ac* acicular spine, *cr* crenulated spine, *cu* cuspidate spine, *LA* lateral accessory, *LTAS* lateral terminal accessory spine, *LTS* lateral terminal spine, *LV* lateroventral, *MD* middorsal, *ML* midlateral, *MTS* midterminal spine, *n* number of measured specimens, *S* segment lengths, *SD* standard deviation, *TL* trunk length, *VL* ventrolateral.

cuticle and a smooth surface, which alternate with thinner and probably more flexible cuticular areas.

The heads were everted in several paratypes mounted for SEM, but it was often difficult to examine the precise arrangement of scalids; hence, information cannot be provided for all introvert sectors. However, since the kinorhynch introvert usually follows certain symmetry patterns, and information only was missing for three sectors (sectors 1, 2, and 9), we have tried to predict the expected scalid arrangements in these sectors as well. It should be stressed though that the following description is based on incomplete data. The complete polar projection, showing mouth cone and introvert appendages, is shown in Figure 3. Structures in black are based on actual observations, whereas structures in gray are predictions.

The introvert shows spinoscalids in rings 01 and 03 to 06 and trichoscalids in ring 08, whereas ring 02 shows groups of short fringes instead of spinoscalids and ring 03 has both fringes and spinoscalids. The ten primary spinoscalids in ring 01 have three stiff spines on the base, oriented in a way that makes them stick out from the head opening once the introvert is everted (Figure 4E,G). The spines lay flat on the base of the spinoscalids once the introvert is fully everted (Figure 4F). Ring 02 has no scalids. Instead, a group of short fringes, resembling the fringes that often attach at the basal part of a spinoscalid, is present in each introvert sector (Figure 4F). Ring 03 shows both fully developed spinoscalids (two in each even-numbered sector) and the same kind of fringes as in the preceding ring (two in each odd-numbered sector) (Figure 4F). Ring 04 and the following two rings have only fully developed spinoscalids (Figure 3). Ten spinoscalids, one in each sector, are located in ring 04, whereas ring 05 has 20 spinoscalids, two in each sector, and ring 06 has a single spinoscalid in each odd-numbered sector only, giving a total of five spinoscalids in this ring.

A total of presumably 14 rather small and pointed trichoscalids are present near the neck region (Figures 3 and 4F). Placids are not clearly distinguishable on the neck (Figure 4G).

Trunk

Trunk segment 1 shows no prominent cuticular structures (Figures 4A and 5C). Anteriorly, it fuses directly with the head cuticle, and the neck region is only weakly defined as some cuticular thickenings that could be interpret as strongly reduced placids. The reduced placids show no sign of distal tripartition.

Segment 2 has one large middorsal and a pair of large midlateral sensory spots (Figures 2A, 4A, and 6A). A pair of cuspidate spines and a pair of very minute, flexible acicular spines (length ca. 5 μm from SEM), is located in lateroventral positions near the posterior end of the segment (Figures 2B, 5D, and 6A).

Segment 3 without spines or sensory spots, but hairless patches are present that correspond in size and location to the midlateral sensory spots on segment 2 (Figures 4A, 5E, and 6A).

Segment 4 has a middorsal spine that extends to the posterior edge of the following segment (Figure 7A). The spine is flanked by a pair of tiny, paradorsal sensory spots. Considerably larger sensory spots are present in sublateral positions (Figure 6B). Elongated hairless patches stretching

Table 2 Summary of nature and location of sensory spots and spines

Position segment	MD	PD	SD	ML	SL	LA	LV	VL
1								
2	ss			ss			cu, ac	
3								
4	ac	ss			ss			
5								cu
6	ac	ss		ss				
7				ss				
8	ac	ss				cu	ac	
9	ac	ss			ss		cu	
10	ac (f),cr (m)	?	ss	ac (f),cr (m)			ss	
11	ac, mts					ltas	lts	ss

Abbreviations: *LA* lateral accessory, *LV* lateroventral, *MD* middorsal, *ML* midlateral, *PD* paradorsal, *SD* subdorsal, *SL* sublateral, *VL* ventrolateral, *ac* acicular spine, *cu* cuspidate spine, *cr* crenulated spine, *(f)* female condition of sexual dimorphic character, *ltas* lateral terminal accessory spine, *lts* lateral terminal spine, *(m)* male condition of sexual dimorphic character, *mts* midterminal spine, *ss* sensory spot, *?* sensory spot expected but not observed.

over the anterior two thirds of the segment are located in mid- or laterodorsal positions (Figures 2A, 6B, and 7A).

Segment 5 has cuspidate spines in ventrolateral positions (Figure 6B). Hairless patches, similar to those on the preceding segment, are present (Figures 6B and 7A), but otherwise the segment has no conspicuous cuticular structures.

Segment 6 has middorsal spine, flanked by a pair of tiny, paradorsal sensory spots. The middorsal spine extends until the anterior part of segment 8 (Figure 7B). Large sensory spots are present in midlateral positions, elongate hairless patches are slightly dorsal to the sensory spots (Figure 6C). The widths of the hairless patches correspond to the width of the sensory spots but their areas extend almost along the entire segment (Figure 6C).

Segment 7 does not have any significant cuticular structures dorsally (Figure 4A). Large sensory spots are present midlaterally (Figure 6C), and hairless patches are similar in size and shape to those on segment 6, slightly dorsal to the sensory spots (Figure 6C).

Segment 8 has a middorsal spine, flanked by a pair of tiny, paradorsal sensory spots. The middorsal spine extends until the posterior edge of segment 9 (Figures 2A and 7C). Acicular spines are present in lateroventral positions and cuspidate spines in lateral accessory positions (Figures 2B, 6D, and 7D). A small, spiny protrusion is present on the cuticular edge of the insertion point of each cuspidate spine (Figure 6D). Lateral sensory spots are not present, whereas hairless patches appear as on the preceding segment.

Segment 9 has a middorsal spine, flanked by a pair of tiny, paradorsal sensory spots. The middorsal spine extends until the posterior edge of segment 10 (Figure 7E). Cuspidate spines are present in lateroventral positions near the posterior edge of the segment (Figure 2A). Large sensory spots are found sublaterally starting in the center of the segment and extending posteriorly, approximately one quarter of the segment (Figure 6D). Hairless patches are present midlaterally extending over 50% of the segment longitudinally (Figure 6D).

Segment 10 tapers slightly towards the posterior and has a middorsal spine that attaches close to the posterior margin of the segment (Figures 2A, 4A, and 7F). It was not possible to determine whether the spine is flanked by paradorsal sensory spots. Additional spines are present in midlateral positions (Figures 2A-B, 4A, 6E, and 7F). Middorsal and midlateral spines appear as regular acicular spines in females, whereas males have crenulated spines. Large sensory spots are present in subdorsal and lateroventral positions (Figures 2A-B and 6E).

Segment 11 tapers strongly towards the midterminal spine at its posterior end (Figures 2A-B, 4A, 5A-B, 6E, and 7G). Additional spines include a middorsal spine and a pair of lateral terminal and lateral terminal accessory spines (Figures 2A-B, 4A, 5B, and 6E). No paradorsal sensory spots are present around the middorsal or midterminal spines, but a pair of tiny sensory spots is present in ventrolateral positions, almost on the posterior edge of the tegumental plate. Female specimens have a pair of rounded gonopores on the ventral side of the segment (Figure 6E). Each gonopore is a round plate with an anterior slit-like opening and a conspicuous cuticular spike attached near the outer lateral edge of the plate (Figure 6E). The cuticular sculpturing of this segment is different from the other segments. Instead of hairs arranged in longitudinal rows, the cuticular hairs are distributed randomly on the surface (Figures 2A-B, 4A, 6E, and 7F).

Discussion

Notes on diagnostic features

Through its spine pattern, the new species can easily be distinguished from its four congeners. The single

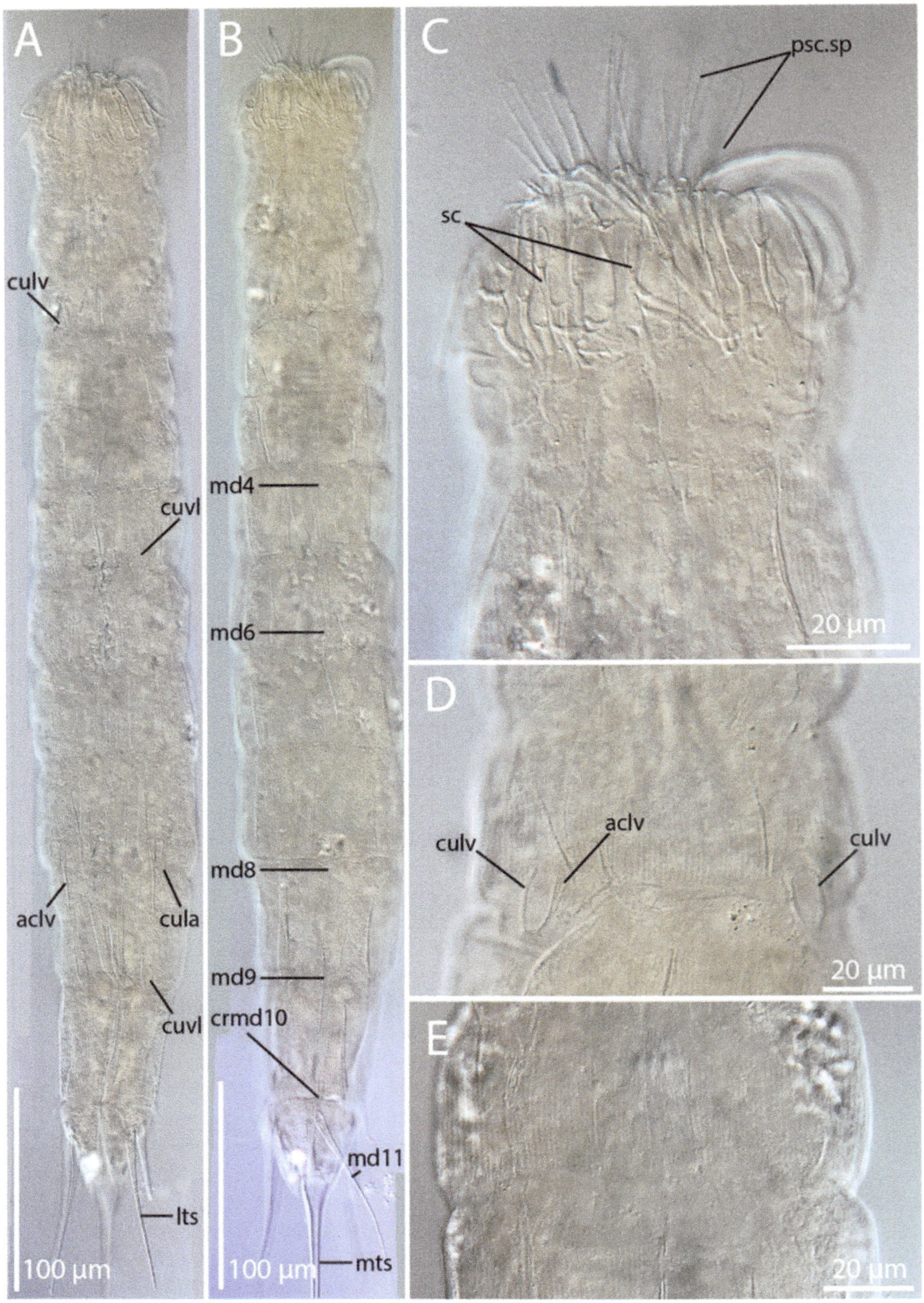

Figure 5 Overviews and details of trunk morphology. Light micrographs showing overviews and details of trunk morphology in *Zelinkaderes yong* sp. nov., holotypic male (KIN-000818). **A**. Ventral view of the entire specimen with incomplete midterminal spine. **B**. Dorsal view of the specimen shown in A. **C**. Partly everted introvert with scalids and scalid spines. **D**. Detail of the second trunk segment with cuspidate spines and the barely visible acicular spines. **E**. Ventral view of the third trunk segment showing the characteristic cuticular ornamentation with a longitudinal striation formed by the cuticular hairs and the pectinate fringe on the posterior border of the segment. Abbreviations: *aclv* acicular lateroventral spine, *crmd10* crenulated middorsal spine on segment 10, *cula* cuspidate lateral accessory spine, *culv* cuspidate lateroventral spine, *cuvl* cuspidate ventrolateral spine, *lts* lateral terminal spine, *md4-9 + 11* acicular middorsal spines on segments 4 to 9 and 11, *mts* midterminal spine, *psc.sp* primary spinoscalid spines, *sc* spinoscalid.

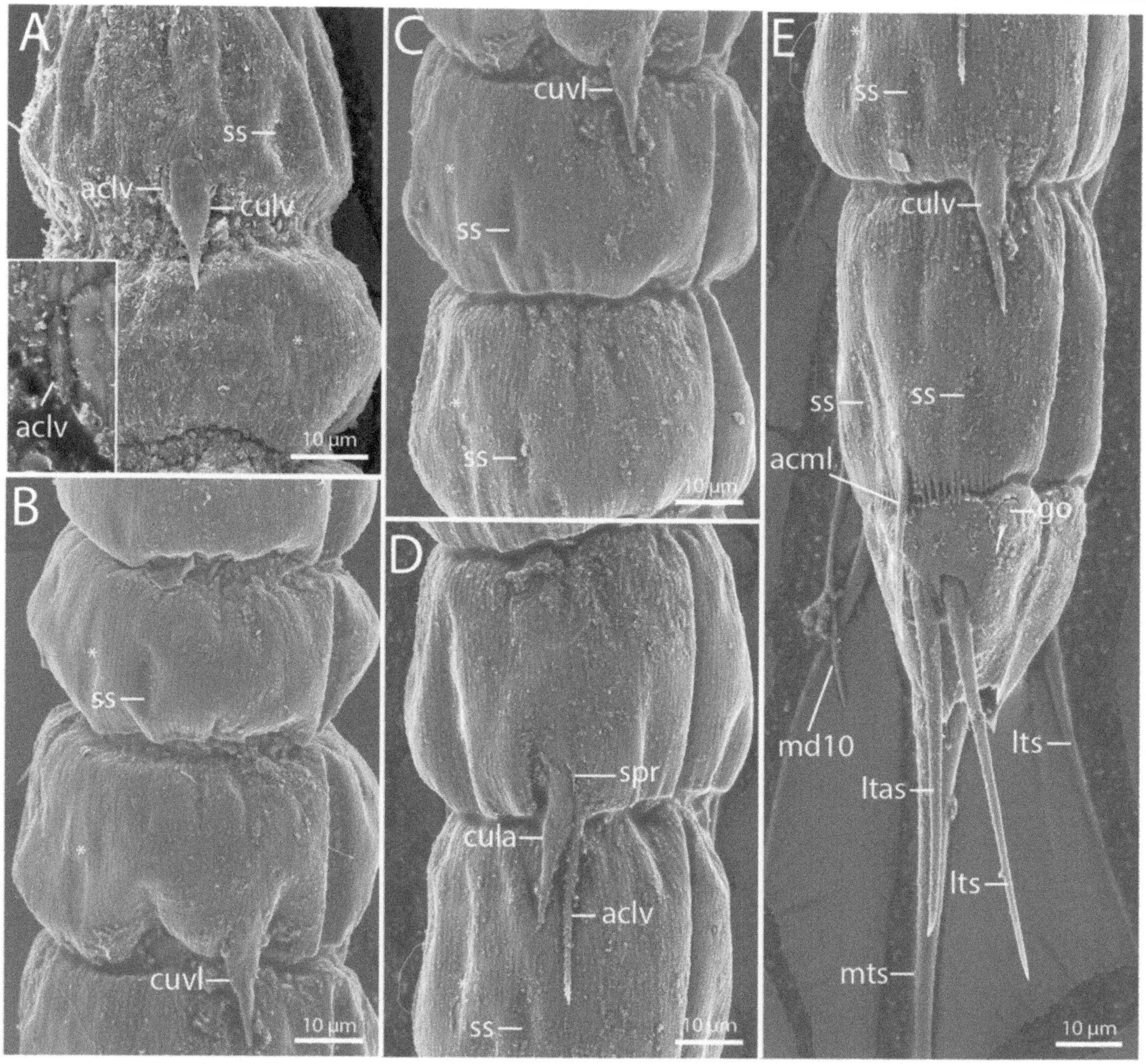

Figure 6 Details of female *Zelinkaderes yong* sp. nov. paratype (KIN-000825). A. Ventrolateral view on trunk segments 2 and 3 with lateroventral cuspidate spine and minute acicular spine. The inset shows a close-up of the minute acicular spine. **B**. Ventrolateral view of segments 4 and 5 with sublateral sensory spot on segment 4 and midlateral hairless patch (asterisk) as well as ventrolateral cuspidate spine on segment 5. **C**. Ventrolateral view of segments 6 and 7 with midlateral sensory spots and midlateral hairless patches (asterisk) on both segments. **D**. Ventrolateral view of segments 8 and 9, with lateral accessory cuspidate spine and lateroventral acicular spine on segment 8, and midlateral hairless patch (asterisk) and sublateral sensory spot on segment 9. **E**. Ventrolateral view of segments 9 to 11 of female specimen. Abbreviations: *aclv* acicular lateroventral spine, *acml* acicular midlateral spine, *cula* cuspidate lateral accessory spine, *culv* cuspidate lateroventral spine, *cuvl* cuspidate ventrolateral spine, *go* gonopore, *ltas* lateral terminal accessory spine, *lts* lateral terminal spine, *md10* acicular middorsal spine on segment 10, *mts* midterminal spine, *spr* spiny protrusion, *ss* sensory spot. Hairless patches are marked with asterisk '*'.

diagnostic trait that separates *Zelinkaderes yong* sp. nov. from other species of *Zelinkaderes* is its absence of lateroventral or lateral accessory acicular spines on segment 9. Such a spine is present in all four previously described species. Otherwise, it differs most from *Z. submersus* that shows cuspidate spines in the lateral series on all segments from segments 4 to 9 (Higgins 1990) and from *Z. brightae* that has ventrolateral cuspidate spines on segments 4 to 6 (Sørensen et al. 2007). In this part of the trunk (segments 4 to 6), *Z. yong* sp. nov. only possesses cuspidate spines on segment 5. *Z. floridensis* shows a closer resemblance to the new species, but besides the differences on segment 9, *Z. floridensis* is unique among its congeners by the absence of cuspidate spines on segment 2 (Higgins 1990). The cuspidate spines on segment 5 in *Z. floridensis* are furthermore located more laterally

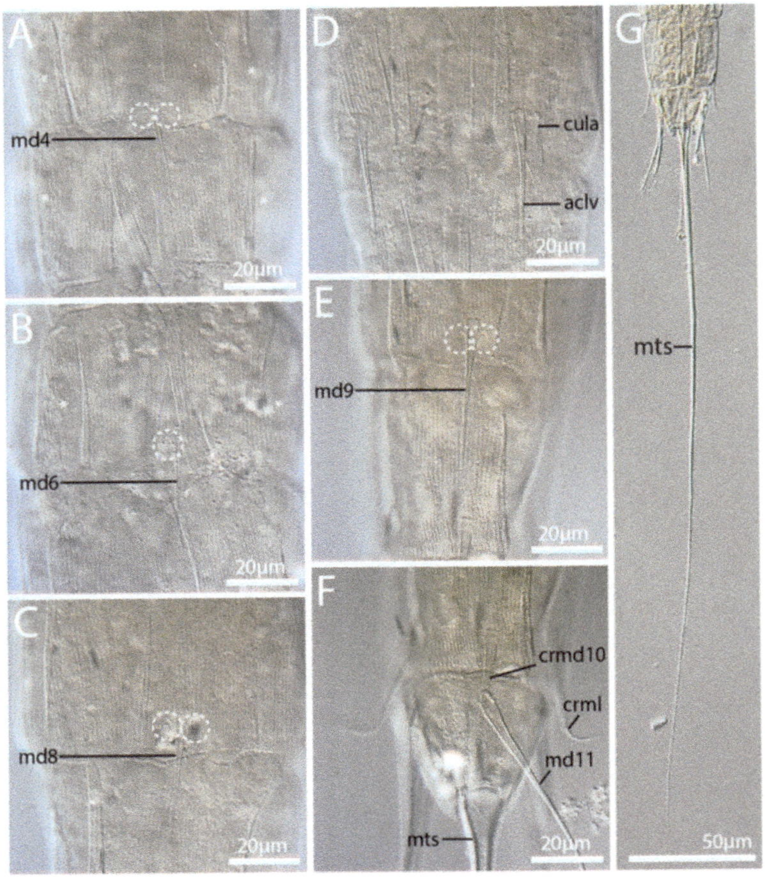

Figure 7 Light micrographs with details of *Zelinkaderes yong* sp. nov., holotypic male (KIN-000818). **A**. Dorsal view of segments 4 and 5 with weakly developed pectinate fringe and middorsal spine flanked by paradorsal sensory spots (dashed circles) on segment 4 as well as hairless patches on segments 4 and 5 (asterisk). **B**. Dorsal view of segment 6 with middorsal spine, hairless patches (asterisk), and paradorsal sensory spot (dashed circle). **C**. Dorsal view of segment 8 with middorsal spine flanked by paradorsal sensory spots (dashed circles). **D**. Ventral view of segment 8 with acicular and cuspidate spines. **E**. Dorsal view of segments 9 and 10 with middorsal spine of segment 9. Presence of paradorsal sensory spots (dashed circles) is only evident on segment 9. The hairless patches on segment 9 are out of focus and not indicated. **F**. Dorsal view of segments 10 and 11 with middorsal spine and lateral acicular spines on segment 10 and middorsal spine and base of the midterminal spine on segment 11. **G**. Posterior part of the specimen showing the long midterminal spine (mt) that can reach the same length as the trunk of the animal. Abbreviations: *aclv* acicular lateroventral spine, *crmd10* crenulated middorsal spine on segment 10, *crml* crenulated midlateral spine, *cula* cuspidate lateral accessory spine, *md4-9 + 11* acicular middorsal spines on segments 4 to 9 and 11, *mts* midterminal spine. Hairless patches are marked with asterisk '*'.

compared to the spines in *Zelinkaderes yong* sp. nov. The new species shows most resemblance with *Zelinkaderes klepali*. Except for the lateral accessory acicular spine on segment 9, present in *Z. klepali* and absent in *Z. yong* sp. nov., the spine patterns of the two species are identical. Also, the distribution of large sensory spots is identical for the two species. Bauer-Nebelsick (1995) reports the positions of some of them as dorsolateral, but according to the provided illustrations, these sensory spots are pretty much located in midlateral positions as in *Z. yong* sp. nov. The midlateral position of the sensory spots on segments 2, 6, and 7 in *Z. klepali* was also confirmed by personal examinations of the type specimens, carried out by the last author during a visit at the Natural History Museum of Vienna some years ago. Unfortunately, these examinations could not clarify whether small sensory spots are located near the attachment points of the middorsal spines, but at least Figure twenty three in Bauer-Nebelsick (1995) shows indications of paradorsal sensory spots around the middorsal spine of segment 4. Confirmation of the existence of additional paradorsal sensory spots would probably require SEM examinations, but the SEM specimens used in the description of *Z. klepali* are unfortunately no longer available (pers. comm. M. Bright). Based on studies of LM material, paradorsal sensory spots similar or identical to the spots of *Z. yong* are present on segments 4, 6, 8, and 9 of *Z. klepali* from the Red Sea (Neuhaus pers. comm.) One remarkable difference regarding sensory spots is the apparent absence of type 3 sensory spots around the base of the midterminal spine in *Z. yong* sp. nov. Such sensory spots have been documented from both *Z. klepali* and *Z. brightae* (Bauer-Nebelsick 1995; Sørensen et al. 2007), and they even appear to be quite common in species of other genera with midterminal spine, e.g., *Triodontoderes, Antygomonas, Semnoderes, Sphenoderes, Tubulideres*, and *Wollunquaderes* (see Sørensen 2007; Sørensen et al. 2007, Sørensen et al. 2009, Sørensen et al. 2010c; Sørensen and Rho 2009, Sørensen and Thormar 2010).

Additional systematic and morphological remarks

Zelinkaderes yong sp. nov. fits well into the genus *Zelinkaderes*, with oral styles consisting of two units, segments 1 and 2 consisting of one closed cuticular ring, segments 3 to 11 with midventral articulation of the tergal plate, middorsal spines on segments 4, 6, and 8 to 11, cuspidate spines on segments 5 and 9, cuspidate spines and acicular spines on segment 8, and a conspicuous longitudinal striation pattern formed by rows of cuticular hairs on segments 1 to 10. However, it differs slightly in the composition of its neck region. When Sørensen and Rho (2009) described *Triodontoderes anulap* and assigned it to Zelinkaderidae, they also proposed an emended diagnosis for the family. In accordance with observations from *T. anulap, Z. brightae, Z. klepali*, and perhaps also *Z. floridensis*, Sørensen and Rho (2009) suggested that the presence of distally tripartite placids in the neck could be a common trait for all species of the family. However, this tripartition could not be confirmed for *Z. yong* sp. nov. because its placids are so inconspicuous that they can hardly be identified. We see this as a reduction that results from the extremely thin cuticle that is found in *Z. yong* sp. nov. Strongly reduced (or perhaps absence of) placids in species with thin cuticle are also known from other species, e.g., *Cateria* sp. (see Figure '5.1.9.' in Neuhaus 2013). Zelinkaderids might be either a highly derived taxon or, as suggested by Higgins 1990, a primitive form, closely resembling the common ancestor of the whole phylum.

The introvert in species of *Zelinkaderes* differs from most other kinorhynchs, due to its reduced number of spinoscalids. This reduction was already noticed in some of the earliest descriptions (see Higgins 1990; Bauer-Nebelsick 1995). In the description of *Z. brightae*, Sørensen et al. (2007) added more details to the *Zelinkaderes* introvert morphology by describing other structures, termed 'fringes' and 'spines', that apparently were replacing spinoscalids and suggested that these structures marked the positions of spinoscalids. We argue that the fringes and spines are reduced spinoscalids, but they might also represent anlagen of not fully developed spinoscalids. The scalid patterns observed in *Z. brightae* are very similar with those in *Z. yong* sp. nov. Actually, it appears that the section-wise arrangement of not only spinoscalids but also these peculiar fringes that could mark the position of reduced spinoscalids is identical in the two species. Furthermore, if we expect that Bauer-Nebelsick (1995) reported only the fully developed spinoscalids, leaving out the fringes, the introvert of *Z. klepali* would be identical with *Z. brightae* and *Z. yong* sp. nov. as well.

Conclusions

The finding of a new species of *Zelinkaderes* in East Asia extends the distributional range of the genus considerably. Species of the genus are now known from European, African, North American, and East Asian waters, which suggests that the genus basically could be present anywhere in the world and could be considered as cosmopolitan. The new species shows the closest resemblance with *Z. brightae*, but it would be too premature to suggest an actual sister-taxon relationship.

Competing interests

The authors declare that they have no competing interests.

Authors' contributions

AA investigated the specimens, made the figures, and drafted the manuscript. HSR and CYC discovered the new species, collected it at the type locality, and contributed to writing of the manuscript. MVS designed and coordinated the study and helped to draft the manuscript. All authors read and approved on the final manuscript.

Acknowledgements

We wish to thank Wong-Gi Min and Thomas M. Jespersen for their help collecting the new species and processing the samples. This study was supported by a postdoctoral fellowship awarded to AA by the Carlsberg Foundation grant no. 2011_01_0536, by the project of the Survey of Indigenous Biological Resources of Korea (NIBR no. 2014-02-001) to CYC, and the research program of KIOST with contract number PE99204 to HSR.

Author details

[1]Natural History Museum of Denmark, University of Copenhagen, Universitetsparken 15, 2100 Copenhagen, Denmark. [2]Dokdo Research Center, Korean Institute of Ocean Science and Technology, 767-813 Uljin, South Korea. [3]Department of Biological Science, College of Natural Sciences, Daegu University, 712-714 Gyeongsan, South Korea.

References

Bauer-Nebelsick M (1995) *Zelinkaderes klepali* sp.n., from shallow water sands of the Red Sea. Ann Naturhist Mus Wien Ser B 97:57–74

Dujardin F (1851) Observations zoologiques I. Sur un petit animal marin, l'Echinodère, formant un type intermédiaire entre les Crustacés et les Vers. Ann Sci Naturelles 3 Ser Tome 15:158–160

Gerlach SA (1969) *Cateria submersa* sp. n., ein cryptorhager Kinorhynch aus dem sublitoralen Mesopsammal der Nordsee. Ver Inst Meeresforsch Bremerh 12:161–168

Herranz M, Pardos F, Boyle MJ (2013) Comparative morphology of serotonergic-like immunoreactive elements in the central nervous system of kinorhynchs (Kinorhyncha, Cyclorhagida). J Morphol 274:258–274, doi:10.1002/jmor.20089

Higgins RP (1964) Three new Kinorhynchs from the North Carolina coast. Bull Mar Sci Gulf Caribbean 14:479–493

Higgins RP (1990) Zelinkaderidae, a new family of cyclorhagid Kinorhyncha. Smithson Contr Zool 500:1–26

Lundbye H, Rho HS, Sørensen MV (2011) *Echinoderes rex* n. sp. (Kinorhyncha: Cyclorhagida), the largest *Echinoderes* species found so far. Sci Mar 75:41–51, doi:10.3989/scimar.2011.75n1041

Neuhaus B (2013) Kinorhyncha (=Echinodera). In: Schmidt-Rhaesa A (ed) Gastrotricha. Cycloneuralia and Gnathifera. Handbook of zoology. De Gruyter, Berlin, pp 181–348, doi:10.1515/9783110272536.181

Ortmann AE (1891) Die Decapoden-Krebse des Straßburger Museums, mit besonderer Berücksichtigung der von Herrn Dr. Döderlein bei Japan und den Liu-Kiu-Inseln gesammelten und z. Z. im Straßburger Museum aufbewahrten Formen. III. Die Abteilungen der Reptantia Boas: Homaridea, Loricata und Thalassinidea. Zool Jb (Systematik, Geographie und Biologie der Thiere) 6:1–53

Park J-K, Rho HS, Kristensen RM, Kim W, Giribet G (2006) First molecular data on the phylum Loricifera - an investigation into the phylogeny of Ecdysozoa with emphasis on the positions of Loricifera and Priapulida. Zool Sci 23:943–954, doi:10.2108/zsj.23.943

Sørensen MV (2007) A new species of *Antygomonas* (Kinorhyncha: Cyclorhagida) from the Atlantic coast of Florida, USA. Cah Biol Mar 48:155–168

Sørensen MV (2013) Phylum Kinorhynch. Zootaxa 3703:63–66, doi:10.11646/zootaxa.3703.1.13

Sørensen MV, Rho HS (2009) *Triodontoderes anulap* gen. et sp. nov. - a new cyclorhagid kinorhynch genus and species from Micronesia. J Mar Biol Assoc UK 89:1269–1279, doi:10.1017/s0025315409000526

Sørensen MV, Thormar J (2010) *Wollunquaderes majkenae* gen. et sp. nov. - a new cyclorhagid kinorhynch genus and species from the Coral Sea, Australia. Mar Biodiv 40:261–275, 10.1007/s12526-010-0048-x

Sørensen MV, Heiner I, Hansen JG (2009) A comparative morphological study of the kinorhynch genera *Antygomonas* and *Semnoderes* (Kinorhyncha: Cyclorhagida). Helgoland Mar Res 63:129–147, doi:10.1007/s10152-008-0132-9

Sørensen MV, Heiner I, Ziemer O, Neuhaus B (2007) *Tubulideres seminoli* gen. et sp. nov. and *Zelinkaderes brightae* sp. nov. (Kinorhyncha, Cyclorhagida) from Florida. Helgoland Mar Res 61:247–265, doi:10.1007/s10152-007-0073-8

Sørensen MV, Pardos F, Herranz M, Rho HS (2010a) New data on the genus *Paracentrophyes* (Homalorhagida, Kinorhyncha), with the description of a new species from the West Pacific. Open Zool J 3:42–59

Sørensen MV, Rho HS, Kim D (2010b) A new species of *Condyloderes* (Cyclorhagida, Kinorhyncha) from Korea. Zool Sci 27:234–242, doi:10.2108/zsj.27.234

Sørensen MV, Rho HS, Kim D (2010c) A new species of the rare genus *Sphenoderes* (Cyclorhagida, Kinorhyncha), with differential notes on *S. indicus* Higgins, 1969. Mar Biol Res 6:472–484, doi:10.1080/17451000903334702

Sørensen MV, Rho HS, Min WG, Kim D, Chang CY (2012) An exploration of *Echinoderes* (Kinorhyncha: Cyclorhagida) in Korean and neighboring waters, with the description of four new species and a redescription of *E. tchefouensis* Lou, 1934. Zootaxa 3368:161–196

Sørensen MV, Rho H, Min W-G, Kim D, Chang C (2013) Occurrence of the newly described kinorhynch genus *Meristoderes* (Cyclorhagida: Echinoderidae) in Korea, with the description of four new species. Helgoland Mar Res 67:291–319, doi:10.1007/s10152-012-0323-2

Zelinka C (1896) Demonstration von Tafeln der *Echinoderes*-Monographie. Verh Dtsch Zool Ges 6:197–199

26

Meiofaunal communities in a tropical seagrass bed and adjacent unvegetated sediments with note on sufficient sample size for determining local diversity indices

Jian-Xiang Liao[1*], Hsin-Ming Yeh[2] and Hin-Kiu Mok[1]

Abstract

Background: Seagrass beds are highly diverse and productive marine habitats for many associated organisms in nearshore coastal waters. The differences in abundance, diversity, and community structure of benthic invertebrates between seagrass beds and adjacent unvegetated sediments have been stated, whereas most studies are primarily focused on macrofauna or based on a comparatively long distance, i.e., more than 10 m. The present study is designed to test if the community structures of meiofauna, especially the free-living nematodes, differ between seagrass beds and adjacent unvegetated sediments on a meter scale.

Results: There are 21 meiofaunal taxa and 63 nematode genera that have been identified from a tropical seagrass bed of *Thalassia hemprichii* in Ludao, Taiwan. Although the compositions of higher meiofaunal taxa are undistinguished, according to correspondence analysis, the assemblages of nematode genera differ substantially between the seagrass bed and unvegetated sediments. Regarding the nematodes, approximately 50% of genera are restricted to the seagrass bed whereas 6% are restricted to unvegetated sediments, which indicate both habitats possessing distinct infaunas. The number of replicates for reasonable estimation of the local diversity index is calculated by the randomization technique. For local seagrass beds, only a single core is sufficient for reliably estimating meiofaunal diversity, but at least three cores or a sample size of 300 individuals is needed for the nematode community.

Conclusions: Nematode assemblages provide more particular differences between seagrass and unvegetated habitats than meiofaunal communities on small spatial scales. Both seagrass beds and adjacent unvegetated sediments harbor specific meiofaunal communities, and hence, the conservation strategy for seagrass should also consider the peripheral bare area of seagrass beds.

Keywords: Meiobenthos; *Thalassia hemprichii*; Community composition; Marine nematode; Randomization; Diversity index

Background

Seagrass beds, which distribute widely in tropical and temperate coastal regions, are one of the most productive marine ecosystems (Duarte and Chiscano 1999). Thalli, roots, and rhizomes of seagrasses offer associated animals microhabitats and shelters (Osenga and Coull 1983; Orth et al. 1984; Castel et al. 1989; Atilla et al. 2005). In addition, various detritus and organic particles in seagrass beds enhance the density and heterogeneity of infauna (Hall and Bell 1993; Danovaro et al. 2002). Meiofauna, benthic organisms between 1 mm and 42 μm, is considered a trophic level between microfauna and macrofauna (animals larger than 1 mm). However, most studies dealing with seagrass beds merely focus on macrofauna, even though meiofauna can consume from 10% to more than 50% of the total primary production in a seagrass system (Danovaro et al. 2002). Moreover, most members of meiofauna are less influenced by hypoxia and predation than

* Correspondence: jianxiangliao@gmail.com
[1]Department of Oceanography, National Sun Yat-sen University, Lienhai Road 70, Kaohsiung 80424, Taiwan
Full list of author information is available at the end of the article

those of macrofauna and have distinct responses to environmental changes (Josefson and Widbom 1988; Castel et al. 1989; Giere 2009). Among meiofaunal communities, free-living nematodes usually constitute the most abundant taxa, albeit seagrass studies of meiofauna have primarily focused on epifauna, i.e., harpacticoid copepods. Nematode assemblages shift consistently with ambient sediments, and they are considered an ecological indicator (Bongers and Ferris 1999).

The unvegetated area around seagrass patches facilitates infauna dispersal and can be a matrix habitat, but few seagrass studies have looked into this bare region (Boström et al. 2006). The abundance and community structure of nematodes show substantial differences between vegetated and peripheral unvegetated sites (Fisher and Sheaves 2003; Hourston et al. 2005; Monthum and Aryuthaka 2006; Fonseca et al. 2011; Leduc and Probert 2011), although no difference between both habitats has also been reported (Ndaro and Ólafsson 1999). The variability in nematode populations on a small spatial scale (centimeter scale) accounts for 30% and 43% in intertidal and shallow subtidal sediments, respectively (Hodda 1990; Li et al. 1997). However, the distances between samples or quadrats in most seagrass studies are separated by more than 10 m or unmentioned (except Hourston et al. 2005; 2 m), whereas the other unanticipated local environmental factors (e.g., water depth) exceed the effect of seagrass (Mills and Berkenbusch 2009). Comparisons between seagrass beds and peripheral unvegetated sediments at small-scale distances elucidate the seagrass effect on meiofaunal communities.

Diversity indices are widely used in ecological research, especially for comparative purposes. The Shannon-Wiener index (*H'*) is the most popular one for measuring diversity in a community. Because of the sample size dependence, a pertinent sampling effort is required for comparison between various communities. The randomization technique has been used to estimate the sufficient sample size for obtaining an approximate value of the diversity index (Soetaert and Heip 1990). However, Hurlbert's modified rarefaction curve (Hurlbert 1971) or the expected species richness, ES(*n*), a sample size-independent index of species richness (Sanders 1968), is much more often used to estimate the diversity of meiofauna (e.g., Shimanaga et al. 2004; Leduc et al. 2010). In addition, most relevant investigations are performed in temperate regions or the deep sea, but the same procedure may not be appropriate for tropical habitats. To our knowledge, there is an absent or scarce report on the marine meiofaunal communities in Taiwan, and a relevant method for determining local diversity indices might improve the future comparison with the other regions.

The present study aims to inspect the community structure of meiofauna, particularly nematodes, which are used to examine the difference between seagrass beds and adjacent unvegetated sediments on a meter scale. Meiofaunal samples were collected quantitatively from sediments below the sickle seagrass *Thalassia hemprichii* and peripheral unvegetated substrates in Ludao, Taiwan. The sampling effort (i.e., the number of replicates) required for estimating the local diversity index in a seagrass habitat is determined using a randomization test to provide a basis for future experimental design of comparative researches.

Methods

Study site and sampling methods

Shihlang is a recreational diving site in Ludao, also known as Green Island, situated off the southeastern coast of Taiwan (Figure 1). The shore is composed of a sandy platform extending 50 m seaward and a *T. hemprichii* meadow (22°39′08″ N, 121°28′25″ E) in the littoral zone. Patches of seagrass distribute at a region 150 m in length and 12 m in width along the coast. The distance between the fringes of seagrass beds to the mean high water line is about 18 m. Because of the freshwater input, the salinity reaches approximately 20 to 25 practical salinity units (PSU) during low tide. Megafaunal bioturbators are frequently present, mainly composed of crustaceans and echinoderms.

Regarding the narrow distribution of the seagrass patches and avoiding intense environmental changes, four 5 × 5 m side-by-side sampling quadrats were selected: three seagrass patches and an adjacent unvegetated control quadrat. The vegetated quadrats densely grew *T. hemprichii* and patches with sparse region, where taking cores was avoided. Within each quadrat, five replicate cores for meiofauna were sampled by a lower end cutoff syringe (diameter = 2 cm), which was pushed into the sediment to a depth of 5 cm. On each sampling occasion, the replicates were at least 1 m apart from each other. All samples collected from the area were 20 m long and 5 m wide at approximately 30-cm water depth during low tide on 17 March 2012. Samples of meiofauna were anesthetized using 7% $MgCl_2$ solution to promote tissue relaxation and then preserved with formaldehyde to a final concentration of 4%. The sediment grain size profile was very similar to the sandy platform, and five additional replicate cores were taken for sediment granulometry as a reference.

Laboratory procedures

Meiofaunal samples were stained with 0.5 g/L Rose Bengal for 1 h and washed through sieves of 1,000- and 42-μm mesh size. All the meiofauna retained on the 42-μm sieve were extracted, counted, and identified to the major taxon level under a stereomicroscope. From each replicate, 100 nematodes (or all individuals if fewer) were picked out, transferred to pure glycerol, mounted onto permanent slides, and identified to the genus level

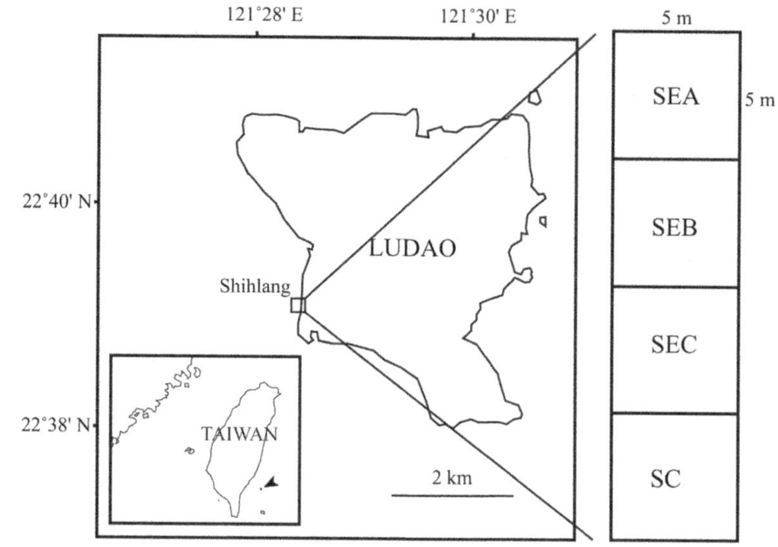

Figure 1 Map of the sampling site in Shihlang (Ludao) with indication of the quadrats. SC, unvegetated quadrat; SEA, SEB, and SEC, seagrass quadrats.

using the pictorial keys (Platt and Warwick 1988; Warwick et al. 1998). The trophic types of nematodes were classified by the morphology of buccal cavities (Wieser 1953). For granulometric analysis, sediment samples were first washed with freshwater to remove salt, dried at 60°C for 48 h, and then separated out on mesh sizes of 1,000, 500, 250, 125, and 62 μm.

Statistical analyses

Multivariate and univariate statistical methods were used to discriminate meiofaunal communities and nematode assemblages between seagrass beds and unvegetated sediments. Community structures were examined using classification and ordination methods to test the consistency of the core replicates. Cluster analysis based on the similarity of the Morisita-Horn index (Horn 1966), which was independent of sample size and diversity, was conducted to group the subsamples. The unweighted pair group method with arithmetic mean (UPGMA) was used to construct the dendrograms. Correspondence analysis (CA) was selected as the ordination method for appropriating to species abundance data. Scaling type 1 (Legendre and Legendre 1998) was chosen for preserving the chi-square distances between sites. For meiofauna and nematode assemblages, the abundance data were square root transformed to reduce the skewness before the multivariate analysis. The square root transformation was used to balance between retaining quantitative information and downweighting the dominant taxa.

The randomization technique was used to calculate a sufficient number of replicates for achieving a reasonable estimate of the diversity index in the sampling site (Soetaert and Heip 1990). According to the results of cluster analysis and CA, if the replicates from identical quadrats were not clustered an obvious group, the abundance data of these replicates were pooled. The numbers of individuals among the core replicates were highly variable, so the mean abundance of meiofauna was chosen as a representative sample size for calculating efficiency. Regarding nematodes, 100 individuals, the most commonly selected number for taxonomic identification, were used to simulate a given sample. All abundance data of seagrass beds were pooled for the randomization simulation by Yeh's computer program (Yeh and Ohta 2002). Pooled data were randomly selected to achieve the predetermined number of sample sizes. The significant differences of ten replicates of the Shannon-Wiener index (H') among every dissimilar sample sizes were examined using the Kruskal-Wallis test. Dunn's test for multiple comparisons was used to determine which values differed significantly if the Kruskal-Wallis test revealed a significant difference.

Results

Environmental and biological variables

The quadrats covered an area 20 m long and 5 m wide, where salinity was approximately 20 PSU during the sampling occurrence at low tide. Sediment grain sizes ($n = 5$) showed peaks around coarse (500 to 1,000 μm) or very coarse (1,000 to 2,000 μm) sand, with the average median grain size 537 ± 51 μm and very few silt (<62 μm) contents (<1%).

A total of 21 meiofaunal taxa were identified in the present study (Table 1). Meiofaunal density in the top 5 cm of sediments varied between 188 and 2,978 individuals 10 cm^{-2}, whereas mean densities showed 1,172 and 1,345 individuals 10 cm^{-2} for vegetated and unvegetated

Table 1 Abundance of higher meiofaunal taxa from seagrass beds and adjacent unvegetated sediments at Shihlang, Ludao

Higher meiofaunal taxa	Seagrass ($n = 15$)		Unvegetated ($n = 5$)	
	Mean	SD	Mean	SD
Acari	1.5	2.0	1.3	0.6
Amphipoda	3.2	4.8	0.6	0.3
Bivalvia	0.2	0.8	0.6	0.3
Ciliophora	11.0	11.2	3.8	1.0
Copepoda	160.7	201.6	607.6	82.3
Cnidaria	0.8	1.9	1.3	0.6
Foraminifera			0.6	0.3
Gastropoda			1.3	0.3
Gastrotricha	3.6	4.0	0.6	0.3
Hirudinea	0.2	0.8		
Isopoda	7.4	15.8		
Nauplius	16.6	45.1	138.9	30.6
Nematoda	853.3	731.1	475.8	126.0
Nemertea	1.1	2.0		
Oligochaeta	5.9	8.2		
Ostracoda	5.7	7.1	22.3	4.7
Polychaeta	71.5	97.6	69.4	12.6
Rotifera	0.2	0.8		
Sipuncula	0.4	1.1	0.6	0.3
Turbellaria	28.7	20.5	15.3	4.3
Undetermined egg			4.5	1.0

The mean and standard deviation (SD) of each meiofaunal taxon are shown in individuals per 10 cm^2.

sites, respectively. Nematoda was the most abundant taxon (72.8%) in the seagrass quadrats, followed by Copepoda (13.7%) and Polychaeta (6.1%). However, Copepoda accounted for 45.2% and dominated the unvegetated site, followed by Nematoda (35.4%), nauplius (10.3%), and Polychaeta (5.2%). The remaining groups represented less than 6% of total meiofauna.

Regarding nematodes, densities within each sample ranged from 51 to 2,682 individuals 10 cm^{-2}. A total of 63 nematode genera belonging to 20 families were identified from 1,680 individuals (Table 2). Ten genera (6.3% of total) were restricted to unvegetated sediments, and 30 genera (47.6%) solely occurred in seagrass beds. *Daptonema* was the most dominant genus and accounted for similar percentages in both habitats, whereas most genera (44 genera) made poor contributions (<1% of total abundance). Four genera, i.e., *Daptonema*, *Perspiria*, *Robbea*, and *Paralinhomoeus*, contributed 50.2% to the total nematode abundance at seagrass sites, whereas three genera, i.e., *Daptonema*, *Longicyatholaimus*, and *Spirinia*, made a larger contribution to the abundance (60.1%) at the adjacent unvegetated site.

Consistency of core replicates

The univariate analyses cannot detect any significant differences in meiofaunal abundance, nematode abundance, and nematode diversity between seagrass and unvegetated habitats. According to the multivariate analysis, most subsamples were clustered at a high similarity level (>0.80) for higher meiofaunal taxa, whereas replicates of various habitats were undetermined and combined randomly (Figure 2a). For nematode genera, all replicates from the seagrass quadrats were clustered to a single group at a similarity level of 0.51, whereas one unvegetated replicate was highly dissimilar from the others (Figure 2b).

The results of CA were similar to what cluster analysis displayed. For higher meiofaunal taxa, the eigenvalues of the first three axes were 0.207, 0.145, and 0.078 and accounted for 27.7%, 19.5%, and 10.4% of the variation, respectively (Figure 3a). Except for one seagrass replicate (SEC3 in Figure 3a, which possessed rare taxa, i.e., the only rotifer in this study and a sipunculan) in the second quadrant, the other replicates were positioned near the origin and distributed along the first axis. A seagrass replicate (SEB5 in Figure 3a, which possessed the fewest nematodes showing a low nematode/copepod ratio as in the unvegetated samples) and the unvegetated samples were aggregated together. For nematode genera, the eigenvalues of the first three axes were 0.582, 0.287, and 0.254 and accounted for 21.5%, 10.6%, and 9.4% of the variation, respectively (Figure 3b). In addition to one unvegetated replicate (SC4 in Figure 3b, which revealed an extremely distinct generic composition from the other replicates: 71 individuals of *Longicyatholaimus*, 10 of *Mesacanthion*, and 2 rare taxa, *Oxystomina* and *Thalassironus*, only appearing in this sample) in the fourth quadrant, the other replicates were distributed along the second axis. All seagrass replicates were located near the origin and well separated from the replicates of unvegetated sediments. The non-metric multidimensional scaling (MDS) ordination for meiofauna and nematodes (data not shown) both represented a considerable degree of similarity to the results of CA but with a high stress value (0.16).

Cluster analysis and CA revealed that the nematodes showed dissimilar community structures between seagrass beds and unvegetated sediments. Therefore, the randomization test was constructed afterward only according to the seagrass samples.

Randomization test

All of the 15 cores from seagrass quadrats were combined in this analysis. For meiofauna, the mean abundance of the seagrass samples (i.e., 368 individuals) was used to simulate a sample size for the randomization test. The cumulative diversity curve (Figure 4a) flattened from the first sample size ($n = 368$) to the total abundance ($n = 5,521$; $H' = 1.0$), and the Kruskal-Wallis test showed no significant

Table 2 Percentages of nematode genera from seagrass beds and adjacent unvegetated sediments at Shihlang, Ludao

Nematode genus	Family	Trophic type	Seagrass	Unvegetated
Acanthonchus	Cyatholaimidae	2A	0.2	0.0
Alaimella	Leptolaimidae	1A	0.1	0.0
Anticoma	Anticomidae	1A	1.3	0.3
Bathyeurystomina	Enchelidiidae	2B	0.2	0.0
Calyptronema	Enchelidiidae	2B	0.2	0.0
Camacolaimus	Leptolaimidae	2A	0.5	0.0
Chromadorina	Chromadoridae	2A	1.1	0.0
Chromadorita	Chromadoridae	2A	*6.1*	*3.4*
Daptonema	Xyalidae	1B	*21.2*	*23.3*
Demonema	Selachnematidae	2B	0.3	0.0
Desmodora	Desmodoridae	2A	0.1	0.6
Desmodorella	Desmodoridae	2A	0.0	0.3
Desmolaimus	Linhomoeidae	1B	1.2	0.3
Dichromadora	Chromadoridae	2A	0.0	0.6
Diodontolaimus	Leptolaimidae	2A	1.1	0.0
Elzalia	Xyalidae	1B	0.8	0.3
Epsilonema	Epsilonematidae	1A	1.7	*2.0*
Eubostrichus	Desmodoridae	1A	0.0	1.1
Eurystomina	Enchelidiidae	2B	0.2	0.3
Filoncholaimus	Oncholaimidae	2B	0.1	0.0
Gammanema	Selachnematidae	2B	0.2	0.3
Graphonema	Chromadoridae	2A	0.0	*6.0*
Halalaimus	Oxystominidae	1A	0.5	0.0
Halaphanolaimus	Leptolaimidae	1A	0.2	0.0
Halichoanolaimus	Selachnematidae	2B	0.4	0.9
Leptolaimus	Leptolaimidae	1A	0.5	0.0
Longicyatholaimus	Cyatholaimidae	2A	0.0	*22.1*
Megadesmolaimus	Linhomoeidae	1B	0.5	0.0
Mesacanthion	Thoracostomopsidae	2B	0.1	*3.2*
Metachromadora	Desmodoridae	2A	*4.9*	0.9
Metalinhomoeus	Linhomoeidae	1B	*3.0*	0.0
Meyersia	Oncholaimidae	2B	0.5	0.3
Microlaimus	Microlaimidae	2A	0.2	0.0
Molgolaimus	Desmodoridae	1A	0.9	0.0
Neochromadora	Chromadoridae	2A	2.5	0.0
Odontanticoma	Anticomidae	2A	0.1	0.0
Oncholaimus	Oncholaimidae	2B	2.0	1.4
Oxystomina	Oxystominidae	1A	0.0	0.6
Paracyatholaimus	Cyatholaimidae	2A	0.9	0.0
Paralinhomoeus	Linhomoeidae	1B	*7.4*	*3.4*
Perspiria	Desmodoridae	2A	*13.4*	0.6
Phanoderma	Phanodermatidae	2A	0.2	0.0
Phanodermella	Phanodermatidae	1A	0.2	0.0
Pomponema	Cyatholaimidae	2A	0.6	*2.0*

Table 2 Percentages of nematode genera from seagrass beds and adjacent unvegetated sediments at Shihlang, Ludao *(Continued)*

Prochromadorella	Chromadoridae	2A	0.4	0.6
Promonhystera	Xyalidae	1B	0.1	0.0
Pseudocella	Leptsomaticae	2A	0.3	0.0
Pseudochromadora	Desmodoridae	2A	*5.9*	1.7
Pseudonchus	Desmodoridae	2B	0.0	1.7
Rhinema	Monoposthidae	2A	0.9	0.0
Robbea	Desmodoridae	1A	*8.1*	0.6
Spilophorella	Chromadoridae	2A	*2.6*	*4.3*
Spirinia	Desmodoridae	2A	0.5	*14.7*
Steineria	Xyalidae	1B	0.9	0.0
Symplocostoma	Enchelidiidae	2B	0.1	0.0
Symplocostomella	Enchelidiidae	2B	0.1	0.0
Synonchus	Leptsomatidae	2A	0.1	0.0
Thalassironus	Ironidae	2B	0.0	0.9
Thalassomonhystera	Monhysteridae	1A	0.1	0.0
Theristus	Xyalidae	1B	*3.2*	0.0
Trichotheristus	Xyalidae	1B	0.0	0.3
Vasostoma	Comesomatidae	2A	0.0	0.3
Viscosia	Oncholaimidae	2B	1.7	0.9

The percentages of the ten most dominant nematode genera for both habitats are shown in italics. Trophic type: 1A, selective deposit feeder; 1B, non-selective deposit feeder; 2A, epistrate feeder; 2B, predator or omnivore.

difference among the cumulative sample sizes. For nematodes, 100 individuals, the most commonly selected number for taxonomic identification, were used to simulate a given sample. After the two replicates ($n = 200$) were pooled, the cumulative diversity curve (Figure 4b) flattened to the total number of individuals ($n = 1,332$; $H' = 2.9$). The Kruskal-Wallis test showed a significant difference among various sample sizes ($p < 0.001$). Only 100 and 200 individuals revealed significant differences from the other cumulative sample sizes. The cumulative sample size of 200 nematodes merely shows a significant difference with 1,300 individuals but is indistinguishable from the ultimate sample size ($n = 1,332$) by Dunn's test (Figure 4b). For reliably estimating the local diversity index of nematode genera, a cumulative sample size of at least 300 individuals is recommended.

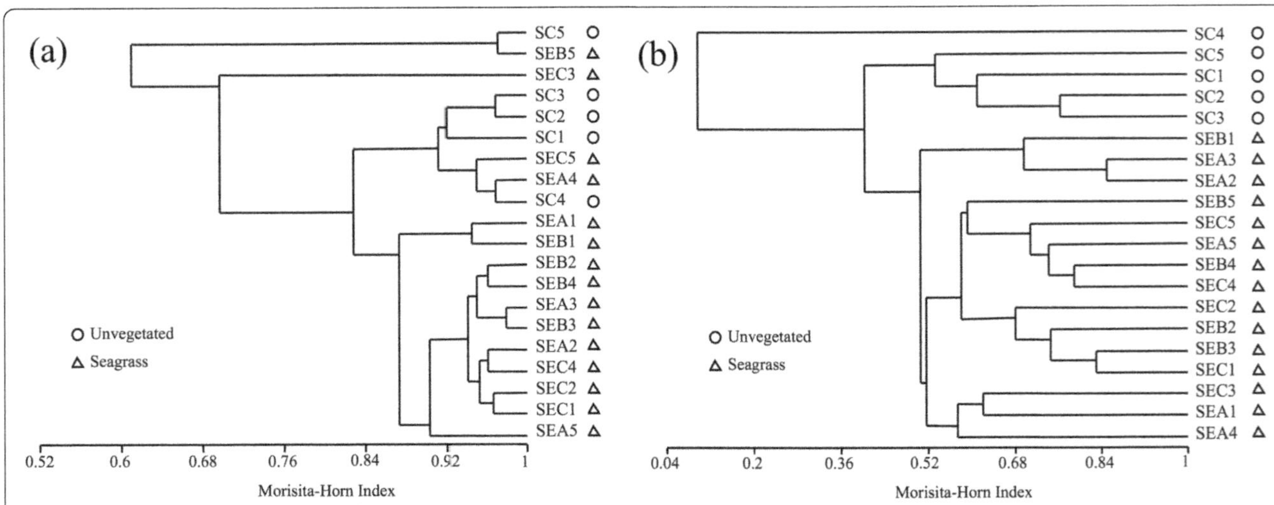

Figure 2 Dendrograms based on the similarity of the Morisita-Horn index for (a) higher meiofaunal taxa and (b) nematode genera. SC, replicates from the unvegetated quadrat; SEA, SEB, and SEC, replicates from each seagrass quadrat. Numbers indicate the number of replicates.

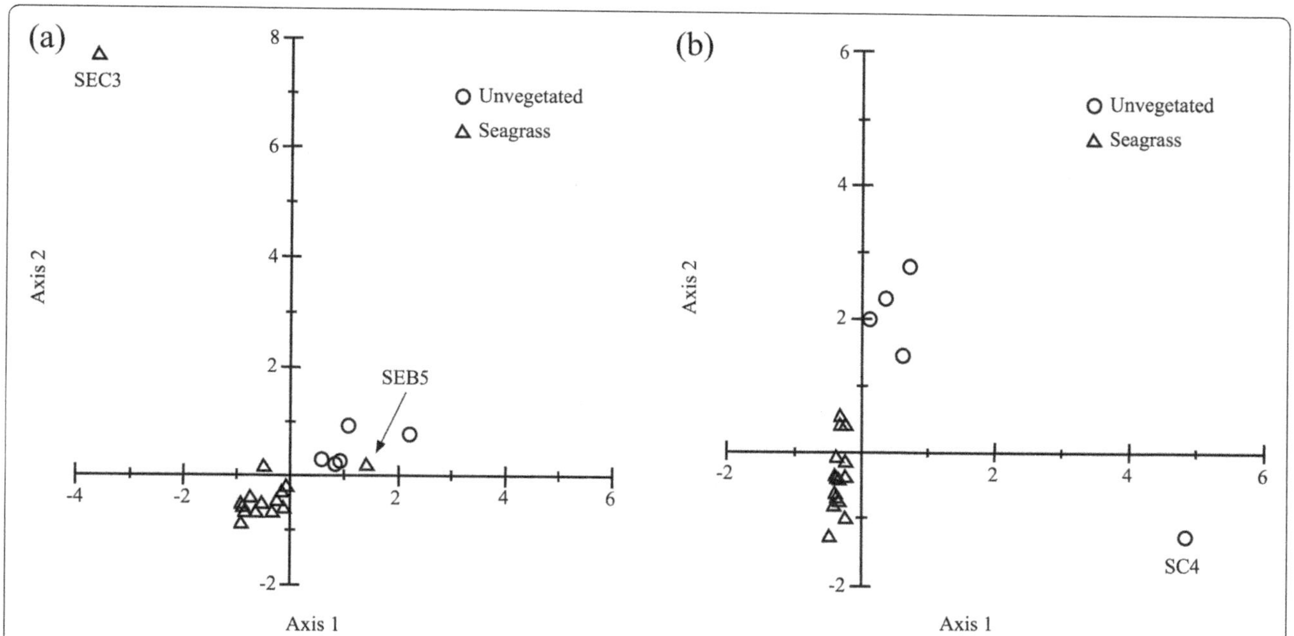

Figure 3 Scatterplots by correspondence analysis of (a) higher meiofaunal taxa and (b) nematode genera. The abundance data have been square root transformed.

Discussion

According to the multivariate analyses, the nematode abundance and diversity recorded in the seagrass bed being higher than those in the adjacent unvegetated sediments are consistent with previous studies (Castel et al. 1989; Danovaro et al. 2002; Hourston et al. 2005; Monthum and Aryuthaka 2006; Leduc and Probert 2011). However, an unexpected pattern was observed in this study where copepods display a reverse trend to nematodes, i.e., the abundance of copepods inside the seagrass bed is lower than that in the unvegetated area. Most surveys have mentioned that copepod densities are higher in seagrass beds (e.g., Hicks 1986; Ansari and Parulekar 1994; Ndaro and Ólafsson 1999; De Troch et al. 2001), but a contrary result has also been reported (Leduc and Probert 2011). Shallow seagrass beds may not substantially influence the abundance of copepods as the deeper ones do (De Troch et al. 2001). The abundance of copepods is positively related to the coarser sediments (Coull 1985), whereas sediments underneath seagrass are commonly finer than those of unvegetated areas. In addition, the dense canopies of seagrasses prevent the growth of microphytobenthos, which are the main diet of copepods (Leduc et al. 2009; Leduc and Probert 2011). The epibenthic and interstitial

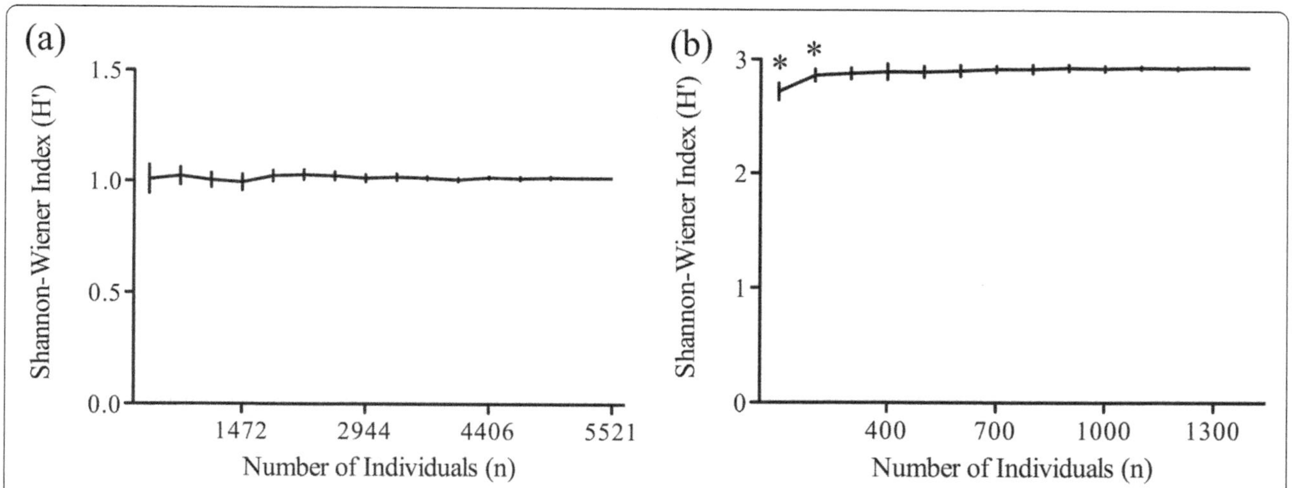

Figure 4 Results of the randomization test for (a) higher meiofaunal taxa and (b) nematode genera. Standard deviations are shown as vertical bars. The sample sizes with an asterisk reveal the significant differences of the Shannon-Wiener index from the other sample sizes.

copepod taxa have distinct environmental demands and stress endurance (Hicks 1986; Giere 2009), so classifying subgroups of copepods may clarify their preference for the microhabitats in seagrass beds.

The seagrass samples bear a rough resemblance in meiofauna as revealed by multivariate analyses but with exceptions. The differences in meiofaunal communities between seagrass beds and adjacent unvegetated sediments may be hardly detectable at a meter scale, and most higher meiofaunal taxa, e.g., copepods and nauplii, have substantially higher dispersal rates than nematodes, which affect their temporary local distribution (Commito and Tita 2002). The active seagrass-associated taxa could forage for foods in peripheral unvegetated sediments but keep close to seagrass beds (Orth et al. 1984). The wave disturbance and salinity changes in the coastal region tend to mitigate small-scale patchiness and local diversity of nematodes (Lambshead and Boucher 2003). However, the compositions of nematode genera have various arrangements in this study, even at a scale within several meters.

Regarding the trophic types of nematodes, epistrate feeders (2A; trophic type by Wieser 1953) are generally dominant in sandy sediments and non-selective deposit feeders (1B) are highly abundant in muddy regions, whereas a recent study has pointed out that the trophic structure is coupling with the potential food sources rather than the other environmental factors (Danovaro and Gambi 2002). *Daptonema*, which possesses the highest percentage (>20%) in both habitats, dominates wide seagrass habitats, especially in sandy sediments (Ndaro and Ólafsson 1999; Hourston et al. 2005; Fonseca et al. 2011). Although *Daptonema* has been treated as a non-selective deposit feeder, many trophic studies have found that they also feed on diatoms at the surface of sand grains and can be regarded as epistrate feeders (Heip et al. 1985; Moens and Vincx 1997). Comparing with the uniform distribution of *Daptonema*, most epistrate feeders (most members of Chromadoridae and Desmodoridae), which have the highest relative dispersal rate of nematodes (Commito and Tita 2002), show a substantial patchy colonization in this study. Many studies have also stated the dominance of epistrate feeders in seagrass sediments (e.g., Hopper and Meyers 1967; Ndaro and Ólafsson 1999; Danovaro and Gambi 2002; Fonseca et al. 2011). The seagrass samples consisted of 8.1% of *Robbea*, a selective deposit feeder (1A), which is coated with particular ectosymbiotic bacteria and occurred extremely rare (0.6%) in unvegetated sediments. A similar deep-dwelling genus, *Catanema*, also dominates Australian seagrass beds (Fisher 2003; Fisher and Sheaves 2003).

Freshwater input in the sampling site (20 PSU during sampling occurrence) allows salinity-tolerant nematodes, including *Chromadorita*, *Daptonema*, *Metachromadora*, *Metalinhomoeus*, *Paralinhomoeus*, *Pomponema*, *Spilophorella*, *Spirinia*, and *Theristus* (Heip et al. 1985), to occur as the dominant genera. The salinity stress drives the abundance of few euryhaline species while rich nutrient supply by seagrass creates a luxuriant microbial food resource, which attracts the rare deposit-feeding nematodes from peripheral sediments (Fisher and Sheaves 2003). Definite differences are present in nematode assemblages between both habitats, and shifts in nematode generic composition have occurred on a meter scale. For Chromadoridae, *Chromadorina* and *Neochromadora* are restricted to seagrass beds, whereas *Dichromadora* and *Graphonema* are restricted to unvegetated sediments. For Desmodoridae, *Robbea*, *Metachromadora*, *Perspiria*, and *Pseudochromadora* are dominant in seagrass beds, whereas *Eubostrichus*, *Pseudonchus*, and *Spirinia* are restricted to or dominant in unvegetated sediments. Various genera of the same family may have distinct preferences for vegetated and unvegetated habitats, or the distribution pattern could be the competitive result in which similar species compete for the plentiful food sources in seagrass sediments. In addition to the chromadorids, desmodorids, and xyalids common to both habitats, linhomoeids (e.g., *Paralinhomoeus* and *Metalinhomoeus*) are obviously occurring in seagrass beds, whereas cyatholaimids (e.g., *Longicyatholaimus*) dominate in unvegetated sediments. Linhomoeids are also restricted to the seagrass beds in Australia, and it has been suggested that they are tolerant to high organic matter content and hypoxia in the sediments underneath seagrasses (Fonseca et al. 2011). *Paralongicyatholaimus*, which resembles *Longicyatholaimus*, is the dominant genus in seagrass beds and adjacent unvegetated area in Thailand (Monthum and Aryuthaka 2006).

Regarding local seagrass beds, only a single core or a cumulative sample size of 368 individuals is sufficient for estimating the diversity of meiofauna, but at least three cores or a cumulative sample size of 300 individuals is needed for nematode diversity. The taxonomic categories (family, genus, or species) obviously influence the requirements of appropriate sample sizes. The Shannon-Wiener index (H'), used as a criterion for the randomization test, is less sample size dependent than Hill's N_1 index (Hill 1973; Soetaert and Heip 1990), but H' has still been applied extensively to compare community diversity and provides an appropriate estimate for a community structure. The diversity of nematodes ($H' = 2.9$), albeit only for genera, is lower than the species diversity in Thailand (3.4 to 3.7; Monthum and Aryuthaka 2006) but resembles the Australian values (2.66 to 3.5; Fisher and Sheaves 2003). The nematode assemblages dominate by a small number of taxa but still show moderate diversity in tropical seagrass beds (Fisher and Sheaves 2003).

At least 300 individuals are recommended for steady estimation of the local diversity of nematode genera.

However, 6 out of 15 of the seagrass samples and 3 out of 5 of the unvegetated samples possess less than 100 individual nematodes in this study. The use of cores with fairly wide diameters (e.g., 3.6 cm) is suggested for achieving reliable quantitative data. A similar result showed that two 10-cm^2 subsampling cores are required to monitor nematode genus diversity in subtidal stations of the North Sea (Li et al. 1997). Water flows or bioturbated activities improve oxygen penetration in sandy habitats and then induce vertical migration of nematodes to a deeper (beneath 5 cm) sediment layer (Steyaert et al. 2003; Giere 2009). The seagrass bed in Shihlang possesses a large amount of coarse sands, where nematodes tend to distribute more deeply (McLachlan et al. 1977; Heip et al. 1985), and thus, the vertical profile of the nematode community can be clarified by sampling deeper than 5- or 10-cm depths (Steyaert et al. 2003; Leduc and Probert 2011).

Conclusions

The community structure of marine nematode genera distinguishes the seagrass habitats from adjacent unvegetated sediments at a scale within several meters, whereas higher meiofaunal taxa are undetermined according to multivariate analysis. The copepod abundance is unexpectedly higher in peripheral unvegetated habitats than in seagrass beds in the present study. Classifying copepods into epibenthic or interstitial subgroups may elucidate their preference for microhabitats. Regarding nematodes, *Daptonema*, a non-selective deposit feeder, dominates both habitats. *Longicyatholaimus* is abundant and restricted to unvegetated sediments, whereas some genera, e.g., *Perspiria* and *Robbea*, are restricted to dominating in sediments under seagrass. According to randomization analysis, only a single core can reliably estimate meiofaunal diversity, whereas three cores are sufficient to estimate the local diversity of nematodes. Albeit the unvegetated areas may not contain as diverse infauna as vegetated sediments, they could be a matrix habitat and provide corridors between fragmented seagrass patches (Boström et al. 2006). In addition, both habitats compose specific infauna, and the unvegetated site should also be considered in the conservation strategy for seagrass.

Competing interests
The authors declare that they have no competing interests.

Authors' contributions
JXL designed and carried out the experiment. JXL and HMY analyzed the data. JXL, HMY, and HKM finalized the manuscript. All authors read and approved the final manuscript.

Acknowledgements
We thank Shih-Wei Su and Yen-Wei Chang for their help in the fieldwork. We are grateful to two anonymous reviewers for their constructive suggestion.

Author details
[1]Department of Oceanography, National Sun Yat-sen University, Lienhai Road 70, Kaohsiung 80424, Taiwan. [2]Coastal and Offshore Resources Research Center, Fisheries Research Institute, Yugang North 3rd Road 6, Kaohsiung 80672, Taiwan.

References
Ansari ZA, Parulekar AH (1994) Meiobenthos in the sediment of seagrass meadows of Lakshadweep atolls, Arabian Sea. Vie Milieu 44:185–190
Atilla N, Fleeger JW, Finelli CM (2005) Effects of habitat complexity and hydrodynamics on the abundance and diversity of small invertebrates colonizing artificial substrates. J Mar Res 63:1151–1172
Bongers T, Ferris H (1999) Nematode community structures as a bioindicator in environmental monitoring. Trends Ecol Evol 14:224–228
Boström C, Jackson EL, Simenstad CA (2006) Seagrass landscapes and their effects on associated fauna: a review. Estuar Coast Shelf Sci 68:383–403
Castel J, Labourg PJ, Escaravage V, Auby I, Garcia ME (1989) Influence of seagrass beds and oyster parks on the abundance and biomass patterns of meio- and macrobenthos in tidal flats. Estuar Coast Shelf Sci 28:71–85
Commito JA, Tita G (2002) Differential dispersal rates in an intertidal meiofauna assemblage. J Exp Mar Biol Ecol 268:237–256
Coull BC (1985) Long-term variability of estuarine meiobenthos: an 11 year study. Mar Ecol Prog Ser 24:205–218
Danovaro R, Gambi C (2002) Biodiversity and trophic structure of nematode assemblages in seagrass system: evidence for a coupling with changes in food availability. Mar Biol 141:667–677
Danovaro R, Gambi C, Mirto S (2002) Meiofaunal production and energy transfer efficiency in a seagrass *Posidonia oceanica* bed in the western Mediterranean. Mar Ecol Prog Ser 234:95–104
De Troch M, Gurdebeke S, Fiers F, Vincx M (2001) Zonation and structuring factors of meiofauna communities in a tropical seagrass bed (Gazi Bay, Kenya). J Sea Res 45:45–61
Duarte CM, Chiscano CL (1999) Seagrass biomass and production: a reassessment. Aquat Bot 65:159–174
Fisher R (2003) Spatial and temporal variations in nematode assemblages in tropical seagrass sediments. Hydrobiologia 493:43–63
Fisher R, Sheaves MJ (2003) Community structure and spatial variability of marine nematodes in tropical Australian pioneer seagrass meadows. Hydrobiologia 495:143–158
Fonseca G, Hutchings P, Gallucci F (2011) Meiobenthic communities of seagrass beds (*Zostera capricorni*) and unvegetated sediments along the coast of New South Wales, Australia. Estuar Coast Shelf Sci 91:69–77
Giere O (2009) Meiobenthology: the microscopic motile fauna of aquatic sediments, 2nd edn. Springer-Verlag, Berlin
Hall MO, Bell SS (1993) Meiofauna on the seagrass *Thalassia testudium*—population characteristics of harpacticoid copepods and associations with algal epiphytes. Mar Biol 116:137–146
Heip C, Vincx M, Vranken G (1985) The ecology of marine nematodes. Oceanogr Mar Biol Ann Rev 23:399–489
Hicks GRF (1986) Distribution and behaviour of meiofaunal copepods inside and outside seagrass beds. Mar Ecol Prog Ser 31:159–170
Hill MO (1973) Diversity and evenness: a unifying notation and its consequences. Ecology 54:427–432
Hodda M (1990) Variation in estuarine littoral nematode populations over three spatial scales. Estuar Coast Shelf Sci 30:325–340
Hopper BE, Meyers SP (1967) Population studies on benthic nematodes within a subtropical seagrass community. Mar Biol 1:85–96
Horn HS (1966) Measurement of "overlap" in comparative ecological studies. Am Nat 100:419–424
Hourston M, Warwick RM, Valesini FJ, Potter IC (2005) To what extent are the characteristics of nematode assemblages in nearshore sediments on the west Australian coast related to habitat type, season and zone? Estuar Coast Shelf Sci 64:601–612
Hurlbert SH (1971) The non-concept of species diversity: a critique and alternative parameters. Ecology 52:577–586
Josefson AB, Widbom B (1988) Differential response of benthic macrofauna and meiofauna to hypoxia in the Gullmar Fjord basin. Mar Biol 100:31–40

Lambshead PJD, Boucher G (2003) Marine nematode deep-sea biodiversity – hyperdiverse or hype? J Biogeogr 30:475–485
Leduc D, Probert PK (2011) Small-scale effect of intertidal seagrass (*Zostera muelleri*) on meiofaunal abundance, biomass, and nematode community structure. J Mar Biol Ass UK 91:579–591
Leduc D, Probert PK, Duncan A (2009) A multi-method approach for identifying meiofaunal trophic connections. Mar Ecol Prog Ser 383:95–111
Leduc D, Probert PK, Nodder SD (2010) Influence of mesh size and core penetration on estimates of deep-sea nematode abundance, biomass, and diversity. Deep-Sea Res I 57:1354–1362
Legendre P, Legendre L (1998) Numerical ecology, 2nd English edition. Elsevier Science B. V, Amsterdam
Li J, Vincx M, Herman PMJ, Heip C (1997) Monitoring meiobenthos using cm-, m- and km-scales in the Southern Bight of the North Sea. Mar Env Res 43:265–278
McLachlan A, Erasmus T, Furstenberg JP (1977) Migrations of sandy beach meiofauna. Zool Afr 12:257–277
Mills VS, Berkenbusch K (2009) Seagrass (*Zostera muelleri*) patch size and spatial location influence infaunal macroinvertebrate assemblages. Estuar Coast Shelf Sci 81:123–129
Moens T, Vincx M (1997) Observations on the feeding ecology of estuarine nematodes. J Mar Biol Ass UK 77:211–227
Monthum Y, Aryuthaka C (2006) Spatial distribution of meiobenthic community in Tha Len seagrass bed, Krabi Province, Thailand. Coast Mar Sci 30:146–153
Ndaro S, Ólafsson E (1999) Soft-bottom fauna with emphasis on nematode assemblage structure in a tropical intertidal lagoon in Zanzibar, eastern Africa: I. Spatial variability. Hydrobiologia 405:133–148
Orth RJ, Heck KL Jr, van Montfrans J (1984) Faunal communities in seagrass beds: a review of the influence of plant structure and prey characteristics on predator–prey relationships. Estuaries 7:339–350
Osenga GA, Coull BC (1983) *Spartina alterniflora* Loisel root structure and meiofaunal abundance. J Exp Mar Biol Ecol 67:221–225
Platt HM, Warwick RM (1988) Free-living marine nematodes. Part II. British chromadorids, Brill/Backhuys, Leiden
Sanders HL (1968) Marine benthic diversity: a comparative study. Am Nat 102:243–282
Shimanaga M, Kitazato H, Shirayama Y (2004) Temporal patterns in diversity and species composition of deep-sea benthic copepods in bathyal Sagami Bay, central Japan. Mar Biol 144:1097–1110
Soetaert K, Heip C (1990) Sample-size dependence of diversity indices and the determination of sufficient sample size in a high-diversity deep-sea environment. Mar Ecol Prog Ser 59:305–307
Steyaert M, Vanaverbeke J, Vanreusel A, Barranguet C, Lucas C, Vincx M (2003) The importance of fine-scale, vertical profiles in characterising nematode community structure. Estuar Coast Shelf Sci 58:353–366
Warwick RM, Platt HM, Somerfield PJ (1998) Free-living marine nematodes. Part III. Monhysterids, Field Studies Council, Shrewsbury
Wieser W (1953) Die Beziehung zwischen Mundhöhlengestalt, Ernährungsweise und Vorkommen bei freilebenden marinen Nematoden. Arkiv Zoologi 4:439–484
Yeh HM, Ohta S (2002) Influence of velocity and types of beam trawl towing on deep-sea demersal fish and decapod crustacean samples. J Oceanogr 58:505–517

27

Current and potential geographical distribution of *Platymeris biguttatus* (Linnaeus, 1767) with description of nymphs

Dominik Chłond, Agnieszka Bugaj-Nawrocka* and Łukasz Junkiert

Abstract

Background: The description of *Platymeris biguttatus* (Linnaeus 1767) nymphal instars as well as the prediction of the potentially suitable ecological niche was the main goal of this study. Our research was based on 258 specimens of *P. biguttatus* species of museum collections. A set of 23 environmental predictor variables covering Africa was used at ecological niche modeling - a method performed using the Maxent software to prepare potential distribution maps for this species.

Results: The results suggested the most suitable areas seen as potentially suitable ecological niche for *P. biguttatus* in Africa. A jackknife test showed that temperature seasonality and percentage of tree cover were among the most important environmental variables affecting the distribution of the species. The analysis of climate preferences shows that most of the potentially suitable niches for this species were located in the area of tropical savanna climate, with a small participation of tree vegetation.

Conclusions: *P. biguttatus* was only known to be widely distributed in the tropical part of continental Africa. Thanks to the ecological niche modeling methods and the museum data on the occurrence of the species, we introduced new information about potentially suitable ecological niches and the possible range of distribution.

Keywords: Africa; Comparison of niches; Ecological niche modeling; Habitat suitability; Species distribution

Background

Platymeris biguttatus (Linnaeus 1767) is one of thirteen species of the genus *Platymeris* Laporte 1833 described so far. It is a medium-sized genus belonging to the subfamily Reduviinae, and all known species of these assassin bugs are large-sized insects, distributed exclusively in Africa (Maldonado Capriles 1990). *P. biguttatus* (L.) is widely distributed in the tropical part of continental Africa, and it can be easily recognized amongst the other representatives of this genus by the white spots on the hemelytra, yellow bands on the legs and black connexiva. Redescription of the species as well as notes on its biology in captivity were published by Li et al. (2010). Although the article contains pictures and presents the biology of nymphs of this species, it does not provide a description of each nymphal instars. Due to the availability of a large number of individuals of this species deposited in European museum collections as well as numerous live specimens breed in captivity in our laboratory, examination and description of morphology of each nymphal instars of this species was possible, and it is provided in the present paper.

Ecological niche modeling was introduced into this study to learn more about some ecological aspects of the mentioned species. This method is widely used in the faunal and floral studies, and it is very useful for biogeography, e.g., to identify potential migration routes of species (mostly invasive ones), to plan the establishment of protected areas, or even to predict the effects of climate change (Beck 2012; Berzitis et al. 2014; Ochoa-Ochoa et al. 2009; Wieczorek and Bugaj-Nawrocka 2014).

As many factors, both biotic and abiotic, affect the distribution of species or plant communities, it is important to choose the right method and environment variables. Our study was aimed at determining which niches might be potentially suitable for *P. biguttatus*, interpreting the

* Correspondence: abugaj-nawrocka@us.edu.pl
Department of Zoology, Faculty of Biology and Environmental Protection, University of Silesia, Bankowa 9, 40-007 Katowice, Poland

influence of predictors on the distribution and comparison with the results obtained for *Platymeris rhadamanthus* (Chłond and Bugaj-Nawrocka 2014). The maximum entropy model (Maxent) is based on a machine learning method and has many advantages (e.g., it avoids commission errors, incorporates interactions between variables, utilizes continuous and categorical data) that make it a well-suited method for species distribution modeling (Phillips et al. 2006). It is also worth to note that Maxent seems to work better than other software and has a better performance in predicting the distribution of species, especially when a limited number of sample localities is available (Elith et al. 2006; Hernandez et al. 2006; Papeş and Gaubert 2007).

Methods

Taxonomy

Numerous live specimens breed in captivity in our laboratory were available at various stages of nymphs. External structures of nymph specimens were examined using a stereoscopic microscope Olympus SZH10 (Olympus Corporation, Shinjuku-ku, Japan). All drawings were made using a camera lucida. Measurements are given in millimeters. Measurements were performed on 30 individuals of each nymphal stage - a total of 150 individuals were measured. Photos were taken with a Sony SLT-A37 camera with additional 13-mm and 31-mm extension tubes.

Material

See Additional file 1 for a detailed list of all examined individuals of *P. biguttatus*. This list, as well as the list in Additional file 2, provides the information about occurrence localities.

Occurrence data

A total of 50 occurrence localities were assembled for the species *P. biguttatus*. It has been localized in Senegal, Gambia, Guinea, Mali, Ivory Coast, Togo, Benin, Niger, Nigeria, Chad, Sudan, Ethiopia, Somalia, the Democratic Republic of Congo, Uganda, Kenya, Tanzania, Zambia, Zimbabwe, and Mozambique (Figure 1). All occurrence data was based on the detailed review of 258 specimens studied in museum collections (see abbreviations for depositories in Additional file 1). Records with unspecified

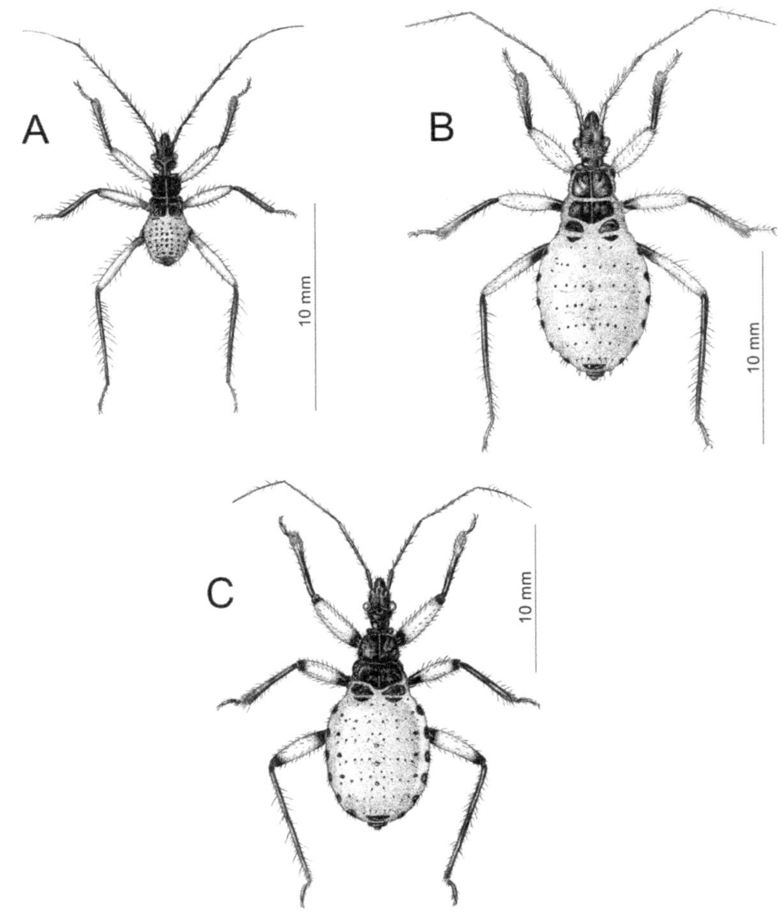

Figure 1 Dorsal habitus of the first three instar nymphs of *P. biguttatus*. (A) first instar nymph; **(B)** second instar nymph; **(C)** third instar nymph.

or unknown localities were not used. All localities were georeferenced using Google Earth 7.1.2.2041 (Google Inc. 2013, Mountain View, CA, USA) (geographical projection, decimal degrees, datum: WGS84).

Environmental predictors

As potential predictors of habitat distribution for *P. biguttatus*, environmental variables were used. The study was based on bioclimatic variables which were downloaded from WorldClim 1.4 dataset (Hijmans et al. 2005; http://www.worldclim.org/bioclim). The Global Land One-km Base Elevation Project (GLOBE) (GLOBE Task Team and others 1999; http://www.ngdc.noaa.gov/mgg/topo/globe.html) provided a digital elevation model (DEM). Also a layer of the tree cover continuous fields was used. It shows vegetation as continuous fields of land cover, where every pixel has a percentage value for tree cover (DeFries et al. 2000; http://glcf.umd.edu/data/treecover). All variables were cut to the boundaries of Africa and a spatial resolution of 30 arc-seconds (approximately 1 km) was chosen.

To determine the possible climatic preferences, results for potential ecological niche were compared to the Köppen-Geiger climate classification system updated by Peel et al. (2007).

Ecological niche modeling

Maxent software (version 3.3.3 k), based on maximum entropy algorithm, was used to develop a species distribution model. The following settings were chosen: maximum number of background points = 10,000, convergence threshold = 0.00001, regularization multiplier = 1, maximum number of iterations = 5,000. These settings allowed for adequate time for model convergence (Stohlgren et al. 2011; Elmore et al. 2013; Solhjouy-Fard and Sarafrazi 2013). A 50-fold cross-validation was run using the full occurrence dataset randomly split into a 25% test and 75% training dataset (Phillips et al. 2006; Phillips and Dudík 2008). The logistic output of Maxent was used, which gives the results for the prediction values ranging from 0 (unsuitable habitat) to 1 (optimal habitat). A jackknife test was selected to show relative importance of each predictor by comparing models with all environmental variable combinations by measuring variable importance. A receiver operating characteristic (ROC) analysis was used to assess the quality of the model (Fielding and Bell 1997). A measure of the area under the ROC curve (AUC) was also calculated. It showed the performance of the model and the weight of both the omission error and the commission error. The AUC assumes a range of values from 0 to 1, but the useful models produce AUC values from 0.7 to 0.9, and AUC values above 0.9 characterize models with almost perfect discrimination (Fielding and Bell 1997; Pearce and Ferrier 2000). It is worth noting that recent studies indicate that the model accuracy (AUC values) depends on the number of records, the used algorithm, and the type of data used (Raes and ter Steege 2007; Wisz et al. 2008; Aguirre-Gutiérrez et al. 2013), so it should not be considered to be the most important determinant of the accuracy of the model.

Raw environmental data was extracted from all raster layers at species occurrence records using SAGA GIS 2.1.0 (SAGA Development Team 2013). A Spearman rank correlation was performed in the R software (version 3.0.2) (R Development Core Team 2011) using Rattle package (version 3.0.2 r169) (Williams 2011) to minimize the number of variables by discarding these which were highly correlated ($r > 0.75$). The variables that did not have any significant contribution to the model were removed, and six uncorrelated environmental factors were selected (Table 1).

As species occurrence points came from the museum data and were not collected randomly, a bias file was provided during Maxent modeling. The bias grid file was generated in SAGA GIS, weighted by a Gaussian kernel with a standard deviation (SD) of 200 km (see Elith et al. 2010). To avoid extreme values, the resulting grid was scaled to have a maximum of 20 and a minimum of 1.

Moreover, to identify niche overlap from Maxent model predications for *P. biguttatus* and *P. rhadamanthus* (Chłond and Bugaj-Nawrocka 2014), SAGA GIS 2.1.0 software was used. The ASCII median output files of Maxent for both species were reclassified and then added together to obtain a raster that contained the suitable niche classes for both species.

Results

Taxonomy

Description of nymphs

First instar (Figure 1A) Color: Head and thorax black with dark brown ventral parts and red ecdysial lines. Antennae with black scapus and pedicellus, both segments covered by black setae. Basiflagellomere with black basal

Table 1 Variables selected for the modeling

Variables	Description	Occurrences			
		Min.	Median	Mean	Max.
Bio02	Mean diurnal range	7.1	11.6	11.7	15.8
Bio04	Temperature seasonality	0.47	1.24	1.47	3.3
Bio06	Minimal temperature of coldest month	7.3	16.1	16.1	22.8
Bio14	Precipitation of driest month	0	5	11.5	59
Bio15	Precipitation seasonality	32	77.5	88.6	157
Tree	Tree vegetation cover	0	13	14	35

Variables with explanations and altitude ranges for the localities with occurrence of *P. biguttatus*. Temperatures given in °C; precipitations given in mm; vegetation cover (Tree) given in %.

part and gradually paler into the apex. Medial part of basiflagellomere greyish, apical part white; both parts with distinctly visible black rings in the location of black setae. Distiflagellomere greyish with black basal part, covered by black and white setae. Legs bicolorous with dark brown coxa and trochanters. Femurs yellow with dark brown basal part of hind femur. Tibiae and tarsi dark brown. Abdomen red with black sclerotized points and plates on dorsal part and dark brown ostioles of dorsal abdominal glands. Tip of abdomen black. Setae on head, thorax, and abdomen black. Legs with pale and dark brown setae.

Structure: Body dull with shiny head and thorax, covered by various-sized setae. Head with medium-sized and long setae. Ecdysial line visible in lateral parts of transversal suture of the head. In the middle part ecdysial line is u-shaped and connected with longitudinal ecdysial line placed in 1/2 of postocular part of the head and on the neck. Clypeus covered by long setae, mandibular, and maxillary plates covered by medium-sized setae. Labrum visible in all length of first visible labial segment. Scapus short and wide with distinct enlargement in basal part. Pedicellus slightly thinner than scapus, both segments covered by long, erected, and robust setae. Basiflagellomere the longest, distiflagellomere slightly shorter than basiflagellomere, both segments covered by long, thin, erected, and regularly placed setae. Distiflegallomere covered also by pale short, adherent setae (except 1/5 of basal part). Tergites of thorax covered by a few various-sized setae, two of them are visible outside the area of tergites. Posterior margin of thorax in dorsal view with two elongated sclerotizations, placed laterally. Pro-, meso-, and metaepleura with numerous, short, and robust spine-like processes. Thorax beneath covered by three transversal rows of few various-sized setae. Legs covered by long, regularly arranged setae. Abdomen dorsally covered by very dense and very short setae. On each abdominal segment visible long setae placed on distinct sclerotized plates and arranged in transversal rows - six setae on each segment, except for the row on second abdominal segment, which contains only four setae. A pair of long setae visible on lateral margins of each abdominal segment. On segments IV to VI visible large ostioles of dorsal abdominal glands. Segments IX and X with visible sclerotized plates of tergites. Ventral side of abdomen covered by transversal rows of long and thin setae (one row on each segment) without any sclerotized elements.

Measurements (in mm): Body length: 7.2 to 7.5; head length: 1.55 to 1.61; head width: 1.03 to 1.05; length of anteocular part: 0.88 to 0.9; length of postocular part: 0.4 to 0.41; length of synthlipsis: 0.54 to 0.55; lengths of antennal segments I:II:III:IV: 0.65 to 0.67: 1.7 to 1.75: 2.6 to 2.64: 2.54 to 2.58; lengths of rostral segments I:II:III: 0.55 to 0.57: 0.9 to 0.92: 0.43. to 0.46; maximum length of thorax: 1.65 to 1.7 maximum width of thorax: 1.88 to 1.91; maximum width of abdomen: 3.1 to 3.19; maximum length of abdomen: 3.78 to 3.86.

Second instar (Figure 1B) Color: Head and thorax black with dark brown ventral parts and red ecdysial lines. Antennae with dark brown scapus and pedicellus, both segments covered by black setae. Apex of pedicellus pale. Basiflagellomere greyish with black spots and dark brown setae. Distiflagellomere dark brown with dark setae on 1/4 of basal part and pale setae on 3/4 of apical part. Legs bicolorous with brown coxa and trochanters. Femurs yellow with brown basal and apical part of femurs; 1/4 of basal part of hind femur brown. Tibiae dark brown with slightly paler apex. Tarsi brown. Legs covered by dark setae. Abdomen red with black sclerotized parts. Ostioles of dorsal abdominal glands light brown. Tip of abdomen black. Setae on head, thorax, and abdomen black. Legs with pale and dark brown setae.

Structure: Body dull with shiny head and thorax, covered by medium-sized setae. Head and thorax with dense and very small granulations as well as several large granules. Ecdysial line visible in lateral parts of transversal suture of the head and on the anterior margin of the eyes. In the middle part, the ecdysial line is u-shaped and connected with longitudinal ecdysial line placed in 1/2 of postocular part of the head and on the neck. Clypeus covered by long setae, mandibular, and maxillary plates covered by medium-sized setae. Labrum visible in 1/2 of basal part of first visible labial segment. Scapus short and wide with distinct enlargement in basal part. Pedicellus slightly thinner than scapus, both segments covered by long, erected, and robust setae. Basiflagellomere the longest, distiflagellomere slightly shorter than basiflagellomere. Basiflagellomere covered by long, thin, erected, and regularly placed setae, setae placed on the shorter apex. Distiflegallomere covered in 1/3 of basal part by regularly arranged and long setae (dark with pale apices) and in 2/3 of apical part by few long, erected setae, and very dense, short, and adherent setae (pale). Collar of pronotum with distinctly developed lateral calli. Calli with one short and robust setae. Tergites of thorax covered by a few various-sized setae, long, and robust setae placed on distinct granulations. Lateral margins of tergites of thorax covered by medium-sized and short setae. Posterior margin of tergite of prothorax flattened. Tergite of mesothorax with two symmetrical arranged setae placed on large granulations. Tergites of metathorax with five medium-sized setae placed on granulations. Pro-, meso-, and metaepleura with numerous, short, and robust spine-like processes. Ventral part of pronotum with well-formed stridulitrum. On the sides of stridulitrum visible conical spines with one setae on the apex. Anterior margin of mesothorax with double ridge visible ventrally. Mesothorax beneath covered by one transversal row of four medium-sized setae. Unsclerotized parts of metathorax as well as posterior part of mesothorax covered by very short and dense setae (absent in the other parts of thorax). Legs covered by various-sized, regularly arranged setae. Abdomen dorsally covered by very dense and very short setae. II abdominal segment with two crescent-shaped sclerotizations. On each abdominal segment visible long setae placed on distinct sclerotized plates and arranged in transversal rows. Muscle attachments are visible in lateral parts of II to VIII abdominal segment in dorsal and ventral view. Dorsolateral sclerotizations covered by medium-sized setae visible on abdominal segments III to IX. On segments IV to VI visible a medium-sized ostioles of dorsal abdominal glands. Segments IX and X with visible sclerotized plates of tergites. Ventral side of abdomen covered by transversal rows of long and thin setae. A sclerotization of sternites in middle part visible on abdominal segments VIII and IX.

Measurements (in mm): Body length: 10.8 to 11.5; head length: 2.17 to 2.3; head width: 1.55 to 1.59; length of anteocular part: 1.21 to 1.25; length of postocular part: 0.69 to 0.72; length of synthlipsis: 0.89 to 0.93; lengths of antennal segments I:II:III:IV: 0.96 to 1.02: 2.51 to 2.55: 3.33 to 3.37: 2.95 to 2.99; lengths of rostral segments I:II:III: 0.89 to 0.92: 1.04 to 1.07: 0.52 to 054; maximum length of thorax: 2.91 to 2.95; maximum width of thorax: 2.74 to 2.83; maximum width of abdomen: 4.48 to 4.55; maximum length of abdomen: 5.43 to 5.54.

Third instar (Figure 1C) Color: Head and thorax black with dark brown ventral parts and red ecdysial lines. Antennae with black scapus and pedicellus, both segments covered by black setae. Basiflagellomere greyish with black spots and dark brown setae. Distiflagellomere greyish with dark setae on 1/3 of basal part and pale setae on 2/3 of apical part. Legs bicolorous with dark brown to black coxa and trochanters. Femurs yellow with black basal and apical part of femurs; 1/4 of basal part of hind femur black. Tibiae black with slightly paler apex and fossula spongiosa. Tarsi brown with paler ventral side. Legs covered by brown and black setae. Abdomen dark brown with black sclerotized parts. Ostioles of dorsal abdominal glands brown. Tip of abdomen black. Setae on head, thorax, and abdomen black. Legs with pale, brown, and black setae.

Structure: Body dull with shiny parts of head and thorax, covered by medium-sized setae. Head and thorax as well as sclerotized parts of the abdomen with dense and various-sized granulations. Ecdysial line visible in lateral parts of transversal suture of the head and on the anterior margin of the eyes. In the middle part the ecdysial line is reverse omega shaped and connected with longitudinal ecdysial line placed in 1/2 of postocular part of the head and on the neck. Clypeus, mandibular, and maxillary plates covered by medium-sized setae, Labrum wide, visible in 1/2 of basal part of first visible labial segment. Scapus short and wide with distinct enlargement in basal part. Pedicellus distinctly thinner than scapus, both segments covered by long, erected, and robust setae. Basiflagellomere the longest, distiflagellomere slightly shorter than basiflagellomere. Basiflagellomere covered by medium-sized, thin, suberected, and regularly placed setae. Distiflegallomere covered in 1/3 of basal part by regularly arranged and long setae (brown) and in 2/3 of apical part by few medium-sized, erected setae, and very dense, short, and suberected setae (pale). Collar of pronotum with distinctly developed lateral calli and small, regularly arranged granulations. On anterior margin of collar a line of medium-sized and dense setae placed parallel to the neck. Calli with one long and robust and a few short setae. Tergites of thorax covered by various-sized (mostly short) setae, placed on large granulations. Lateral margins of tergites of thorax covered by medium-sized setae. Tergite of prothorax with diagonal short sulci placed between lines of granulations. Tergite of mesothorax with medium-sized setae placed on large granulations in the middle and on lateral margins. Tergites of metathorax with medium-sized setae placed on granulations. Pro-, meso-, and metaepleura with numerous, small, regularly arranged granulations. Ventral part of pronotum with well-formed stridulitrum. On the sides of stridulitrum, visible conical spines with four setae on the apex. Pro- and mesocoxal cavities partially sclerotized; mesocoxal cavity with row of three long and robust erected setae placed postolaterally on inner margin of sclerotized part of cavity. Mesothorax beneath covered by one transversal row of medium-sized setae and with partially sclerotized metacoxal cavity. Unsclerotized parts of metathorax as well as posterior part of mesothorax with dense punctuation. Legs covered by various-sized, regularly arranged setae. Abdomen dorsally covered by short setae. First abdominal spiracle distinctly visible, placed near erected, medium-sized setae. II abdominal segment with two crescent-shaped sclerotizations covered by small, regularly arranged granulations. On each abdominal segment, visible medium-sized setae placed on distinct sclerotized plates and arranged in transversal rows. Muscle attachments visible in lateral parts of II to VII abdominal segment in dorsal and ventral view. Dorsolateral sclerotizations covered by medium-sized setae visible on abdominal segments III to IX. On segments IV to VI visible medium-sized ostioles of dorsal abdominal glands (ostiole on segment VI smaller). Segments IX and X with visible sclerotized plates of tergites. Ventral side of abdomen wedge shaped, covered by transversal rows of long and thin setae and with distinctly visible dark spiracles and muscle attachments. A sclerotization of sternites in middle part visible on abdominal segments

VIII and IX; sclerotizations covered by long and very short setae.

Measurements (in mm): Body length: 13.7 to 14.8; head length: 2.88 to 2.95; head width: 2.07 to 2.11; length of anteocular part: 1.5 to 1.59; length of postocular part: 0.59 to 0.64; length of synthlipsis: 0.99-1.06; lengths of antennal segments I:II:III:IV: 1.1 to 1.12: 3.2 to 3.23: 3.7 to 3.74: 3.24 to 3.26; lengths of rostral segments I:II:III: 1.29 to 1.32: 1.24 to 1.25: 0.7 to 0.8; maximum length of thorax: 4.01 to 4.06; maximum width of thorax: 3.2 to 3.32; maximum width of abdomen: 5.56 to 5.62; maximum length of abdomen: 6.91 to 7.12.

Fourth instar (Figure 2A)

Color: Head and thorax black with dark brown ventral parts and red ecdysial lines. Antennae with black scapus and pedicellus, both segments covered by black setae. Basiflagellomere dark brown with pale spots and dark brown setae. Distiflagellomere brown with paler spots, covered by dark and pale setae. Legs bicolorous. Coxa dark brown with black apical margin. Trochanters black. Femurs yellow with black basal and apical part of femurs; 1/3 of basal part of hind femur black. Tibiae black with slightly paler apex and fossula spongiosa. Fore and middle tarsi brown with paler ventral side, hind tarsi black. Legs covered by brown and black setae. Abdomen dark brown with black sclerotized parts. Ostioles of dorsal abdominal glands dark brown. Tip of abdomen black. Setae on head, thorax, and abdomen black. Legs with brown and black setae.

Structure: Body dull with shiny parts of head and thorax, covered by medium-sized setae. Head and thorax as well as sclerotized parts of the abdomen with various-sized granulations. Ecdysial line visible in lateral parts of transversal suture of the head and on the anterior margin of the eyes. In the middle part, ecdysial line w-shaped and connected with longitudinal ecdysial line placed in 1/2 of postocular part of the head and on the neck. Clypeus, mandibular, and maxillary plates covered by medium-sized setae. Labrum wide, visible almost 1/2 of basal part of first visible labial segment. Scapus short and wide, slightly curved, club-shaped with enlargement in basal part. Pedicellus distinctly thinner than scapus. Scapus covered by long, suberected, and robust setae. Pedicellus covered by suberected and adherent, long, robust setae. Basiflagellomere the longest, covered by medium-sized, thin, suberected, and regularly placed setae. Distiflegallomere covered by regularly arranged and long setae (brown and pale) and by very dense, short, and suberected, pale setae (except basal part). Collar of pronotum distinctly separate with large lateral calli and small, regularly arranged granulations. On anterior margin of collar, a line of medium-sized setae placed subparallel to the neck. Calli with few robust and long setae. Tergites of thorax covered

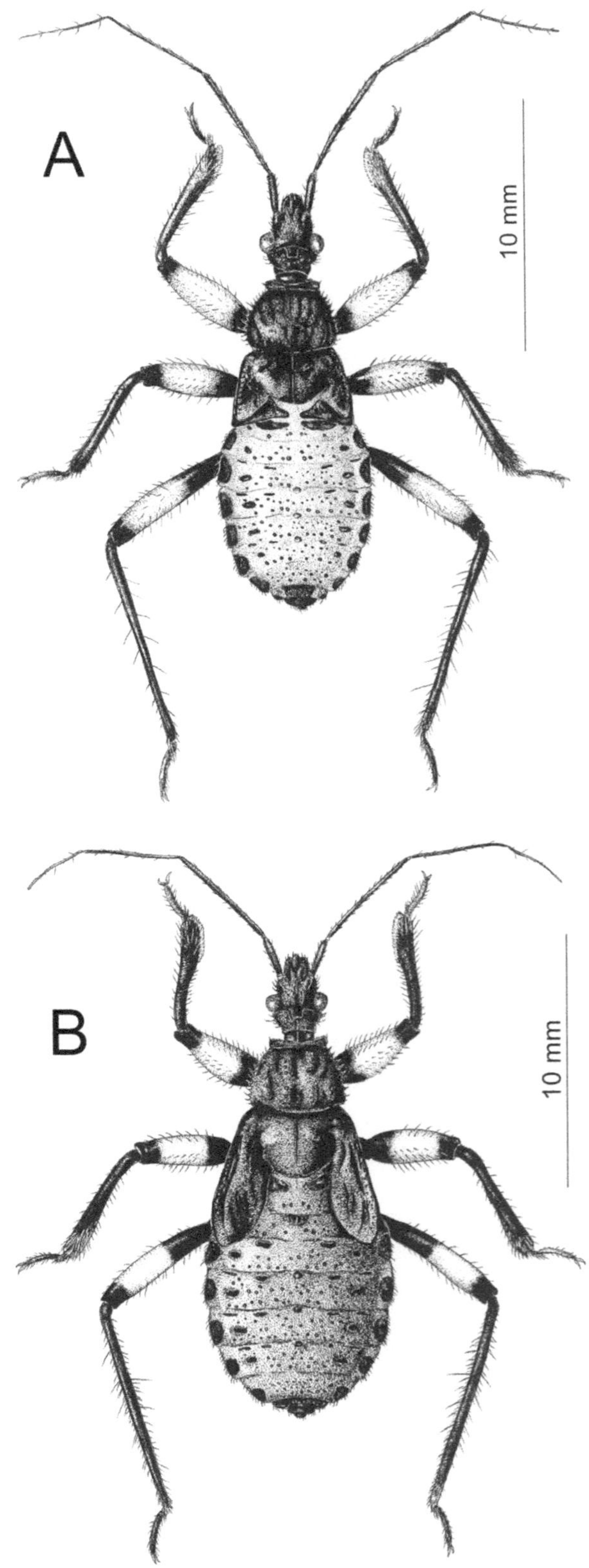

Figure 2 Dorsal habitus of the next two instar nymphs of *P. biguttatus*. **(A)** fourth instar nymph; **(B)** fifth instar nymph.

by various-sized (mostly short) setae, placed on large granulations. Lateral margins of tergites of thorax covered by medium-sized and very dense setae. Tergite of prothorax with diagonal short sulci placed between lines of granulations and with two distinct concaves in anterior part. Tergite of mesothorax with distinctly visible wing pads, covered by small granulations and medium-sized setae placed on large granulations. Tergites of metathorax with medium-sized setae placed on granulations. Tergites of meso- and metathorax with a few large, conical granulations with long setae. Pro-, meso-, and metaepleura with numerous, small, regularly arranged granulations. Ventral part of pronotum with well-formed stridulitrum. On the sides of stridulitrum, visible conical spines with four setae on the apex. Pro- and mesocoxal cavities almost completely sclerotized. Unsclerotized parts of metathorax as well as posterior part of mesothorax with slightly visible punctuation. Legs covered by various-sized setae, visibly much denser on ventral side. Abdomen dorsally covered by short setae. First abdominal spiracle large. II abdominal segment with two crescent-shaped, elongated sclerotizations covered by small, regularly arranged granulations. On each abdominal segment, visible medium-sized setae placed on distinct sclerotized plates and arranged in transversal rows. Muscle attachments very distinctly visible in lateral parts of II to VIII abdominal segment in dorsal and ventral view. Dorsolateral sclerotizations covered by medium-sized setae visible on abdominal segments III to IX. On segments IV to VI visible medium-sized ostioles of dorsal abdominal glands (ostiole on segment VI very small). Segments IX and X with visible sclerotized plates of tergites. Ventral side of abdomen flattened, covered by transversal rows of long and thin setae and with distinctly visible dark spiracles and two rows of muscle attachments. A sclerotization of sternites in middle part visible on abdominal segments VIII and IX (both connected); sclerotizations covered by long and very short setae.

Measurements (in mm): Body length: 19.4 to 20.5; head length: 3.75 to 3.8; head width: 2.8 to 2.84; length of anteocular part: 2.01 to .2 to 04; length of postocular part: 0.89 to 0.93; length of synthlipsis: 1.9 to 2.02; lengths of antennal segments I:II:III:IV: 1.83 to 1.86: 4.7 to 4.76: 4.8 to 4.9: 3.55 to 3.6; lengths of rostral segments I:II:III: 1.89 to 1.93: 1.75 to 1.8: 0.91 to 0.95; maximum length of thorax: 6.2 to 6.3; maximum width of thorax: 5.1 to 5.18; maximum width of abdomen: 8.82 to 8.95; maximum length of abdomen: 10.5 to 11.1.

Fifth instar (Figure 2B) Color: Head and thorax black with dark brown ventral parts and red ecdysial lines. Antennae with black scapus and pedicellus, both segments covered by black setae. Basiflagellomere and distiflagellomere brown with paler spots, covered by dark and pale setae. Legs bicolorous. Coxa dark brown with black basal part and apical margin. Trochanters black. Femurs yellow with black basal and apical part - 1/3 of basal part of fore and middle femurs black, 1/2 of basal part of hind femur black. Tibiae black with pale fossula spongiosa. Tarsi black with paler ventral side. Legs covered by brown and black setae. Abdomen dark brown with black sclerotized parts. Ostioles of dorsal abdominal glands dark brown. Tip of abdomen black. Setae on head, thorax, and abdomen black. Legs with brown and black setae.

Structure: Body dull with shiny parts of head and thorax, covered by short setae. Head and thorax as well as sclerotized parts of the abdomen with large and very small granulations. Ecdysial line in the middle part w-shaped and connected with longitudinal ecdysial line placed in over 1/2 of postocular part of the head and on the neck. Postocular part of the head distinctly gibbous in the area of ocellar tubercles. Clypeus, mandibular, and maxillary plates covered by medium-sized setae. Mandibular plates distinctly visible, covering the basal part of clypeus in lateral view. Labrum wide, visible in over 1/2 of basal part of first visible labial segment. Labium covered by dense, medium-sized, regularly arranged setae. Scapus short, wide, and slightly curved, covered by short, suberected, and robust setae. Pedicellus covered by suberected, medium-sized, short setae. Basiflagellomere the longest, covered by short, thin, adherent, and regularly placed setae. Distiflegallomere covered by regularly arranged and long setae (brown and pale) and by short, suberected, pale setae. Collar of pronotum with large lateral calli and small, regularly arranged granulations. On anterior margin of collar, two lines of medium-sized setae placed parallel and subparallel to the neck. Each calli with one conical granule. Tergite of prothorax with diagonal sulci placed between lines of large granulations and with two distinct concaves in anterior part. Pterothorax covered by short setae and granulations (distinctly visible areas without granulations). Lateral margins of tergites of thorax covered by medium-sized and very dense setae and with visible pale spots in apical part of each margin. Pterothorax with distinctly visible wing pads, covered by small various-sized granulations and medium-sized setae placed on lateral margin. Pro-, meso-, and metaepleura with numerous, small, regularly arranged granulations. Conical spines on the sides of stridulitrum with one long setae placed apically. Pro- and mesocoxal cavities almost completely sclerotized. Unsclerotized parts of metathorax as well as posterior part of mesothorax with slightly visible, very scarce punctuation. Legs covered by various-sized setae, visibly much denser on ventral side. Abdomen dorsally covered by short setae. Evaporatory area of first abdominal gland distinctly visible. On each abdominal segment, visible medium-sized setae placed on distinct sclerotized plates and arranged in transversal rows. Muscle attachments very distinctly visible in lateral parts of III to VIII abdominal segment in dorsal and ventral view. Dorsolateral sclerotizations covered by medium-sized setae visible on abdominal segments III to IX. Ostioles of dorsal abdominal glands visible on segments IV to VI (ostiole on segment VI smaller). Segments IX and X with visible sclerotized plates of tergites. Ventral side of abdomen flattened, covered by transversal rows of long and thin setae and with distinctly visible dark spiracles and two rows of muscle attachments. A sclerotization of sternites in middle part visible on abdominal segments VIII and IX (both connected); sclerotizations covered by long and very short setae.

Measurements (in mm): Body length: 26.5 to 28.2; head length: 4.1 to 4.2; head width: 3.9 to 3.95; length of anteocular part: 2.1 to 2.15; length of postocular part: 1.25 to 1.3; length of synthlipsis: 1.8 to 1.83; lengths of antennal segments I:II:III:IV: 2.28 to 2.31: 6.68 to 6.72: 5.1 to 5.22: 4.31 to .4.34; lengths of rostral segments I:II: III: 2.1 to 2.2: 2.25 to 2.31: 1.05 to 1.1; maximum length of thorax: 6.96 to 7.05; maximum width of thorax: 5.8 to 5.9; maximum width of abdomen: 10.1 to 10.5; maximum length of abdomen: 10.22 to 10.35.

For detailed photo of nymphs stages, see Additional file 3.

Evaluation of model and the importance of environmental predictors

The mean AUC was 0.834, indicating a good discriminatory capacity of the model. The jackknife test (see Additional file 2) showed that the environmental variable with highest gain when used in isolation was temperature seasonality (Bio04) followed by tree cover (Tree_cover), and the variable that decreased the gain the most when it was omitted was also tree cover and Bio04. Thus, these variables appear to have the most useful information by themselves. They also contain a significant portion of information that is not contained in the other variables.

Other variables used in the model were mean diurnal range (Bio02), minimal temperature of coldest month (Bio06), precipitation of driest month (Bio14), and precipitation seasonality (Bio15). Temperature variables (Bio02 and Bio06) were of greater importance as variables that carried the most useful information by themselves, while the precipitation variables (Bio14 and Bio15) decreased the gain when they were omitted.

For Maxent model outputs, see Additional file 2.

The potential species distribution

The resulting map shows the median of the 50 model replicates output grid. The median was used because it is much more resistant to outliers as an average than the arithmetic mean. As the logistic output was used and the result in values ranging from 0 to 1 was given, an output map with four ranges of presence probability of suitable habitat was created: 0% to 25% as unsuitable habitat, 25% to 50% as acceptable habitat, 50% to 75% as quite suitable and over 75% as suitable habitat (Figure 3) (Additional file 3).

See Additional file 2 for plots that show how the distribution of occurrence records of *P. biguttatus* refers to the used predictors.

P. biguttatus can be quite widespread in Africa. According to the model, the most suitable areas for this species were mainly restricted to the western part of Senegal, Gambia and Guinea Bissau, the south-western part of Guinea, Sierra Leone and Liberia, the southern part of Ivory Coast, Ghana, Togo, Benin and Nigeria, the south-western part of Cameroon, and the north-western part of Equatorial Guinea, as well as the western part of Angola and the north-western part of Namibia. There were also potentially suitable areas nearby Lake Mai-Ndombe and Lake Tumba in the Democratic Republic of Congo, and at the north of these lakes in the basin of the river Congo. In the eastern part of the African continent, suitable areas were found in the central part of Somalia, the eastern part of Kenya, the south-eastern part of Tanzania and the north-eastern part of Mozambique, as well as in the vicinity of the African Great Lakes, in Zanzibar, Comoros, Mayotte, and the northern and eastern parts of Madagascar.

The niche overlap between *P. biguttatus* and *P. rhadamanthus*

By comparing the results for overlap between the potential niches of *P. biguttatus* and *P. rhadamanthus*, it was found that at a threshold of 0.25 (which means that the area of more than 25% was considered to be potentially useful for the species and any data values below the threshold was classified as unsuitable habitat), 14% of the area considered to be potentially suitable for both species may be a common niche (on a continental scale). However, with a threshold of 0.5 (50%), this value decreases to 4%. At the 0.25 threshold, common areas are mainly located in the south of Ivory Coast, Ghana, Togo and Benin, Central African Republic, in the center of Ethiopia, in the south of Congo, and in the Democratic Republic of the Congo outside the territory of Congo Basin, as well as in the area of Angola, Uganda, Rwanda, Burundi, Kenya, Tanzania, Zambia, Malawi, Mozambique, South Africa, and Madagascar. At the 0.5 threshold, common areas are restricted to the areas in Central African Republic, Congo, Democratic Republic of the Congo, Uganda, Kenya, Tanzania, and Mozambique (for the results see the Additional file 3).

The climatic preferences

The most important factors affecting species distributions, because of the continental range used in the

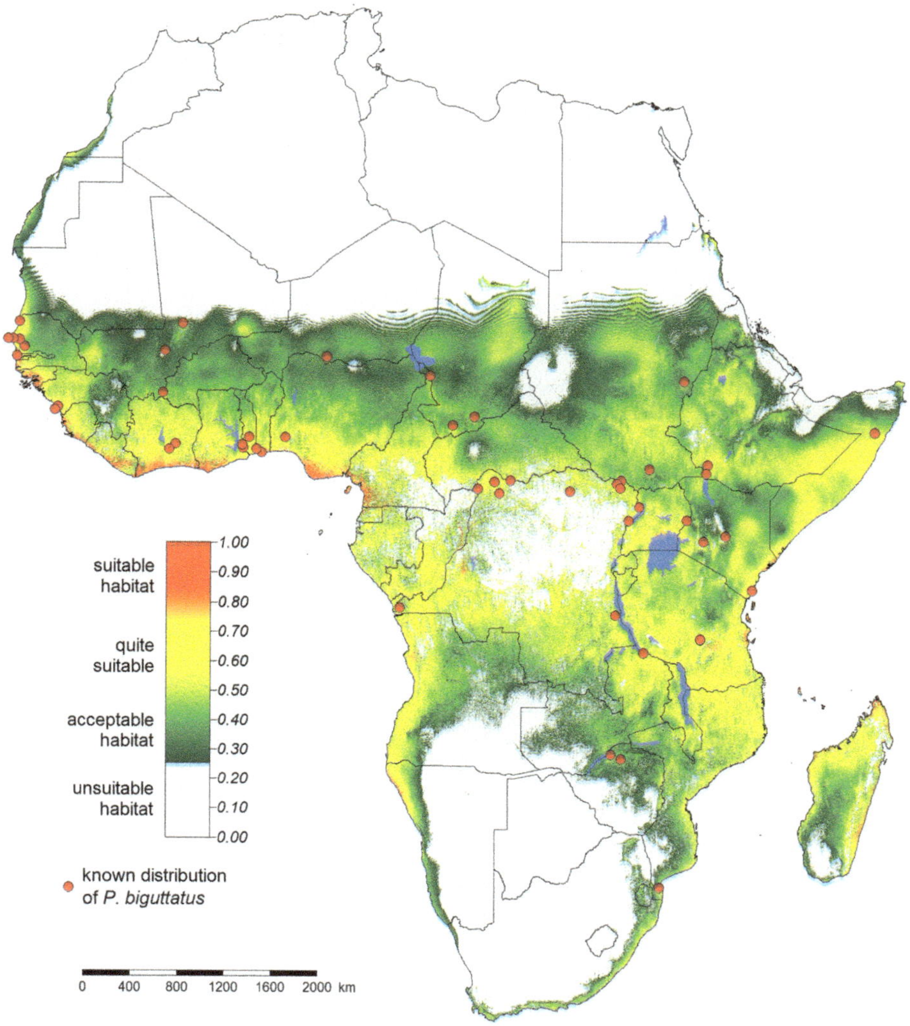

Figure 3 Map of potentially suitable niches for *P. biguttatus*. Predicted probability of presence of a suitable niche in Africa according to selected variables; the legend indicates the presence probability in percentage. Inland continuous lines represent country boundaries.

model, were climate variables. Possible climatic preferences of this species were inferred by comparing potential ecological niches to the Köppen-Geiger climate classification.

The analysis of climate data from known localities of *P. biguttatus* shows that it prefers places with an average annual temperature of about 24.8°C (76.6°F) and a maximum up to 28.6°C (82.5°F). The precipitation in the driest month (Bio14) oscillates about 11 mm/m^2, while an average of annual precipitation is equal to 1,164 mm/m^2, and values above 2,000 mm/m^2 in individual cases.

By assigning to the locations of the occurrence of *P. biguttatus* the mean temperature and mean total precipitation for each month, a climatic diagram was created (Figure 4). The diagram shows a clear correlation between the increase of precipitation and the decrease of temperature. Tropical savanna climate is a grassland biome, located in regions of subtropical and tropical latitudes. It is characterized by periods of low precipitation and higher thermal amplitudes in the annual and diurnal cycle than in typical tropical climates.

Discussion

To provide some new information about niche preferences as well as potential distributions of *P. biguttatus*, the ecological niche modeling method was used. Because many specimens obtained from museums were collected from the same location, in order to avoid model disturbances by strong intercorrelations, duplicate records were removed to prevent unnecessary strengthening of points (Veloz 2009). A bias grid was also used to upweight records with few neighbors in geographic space (Elith et al. 2010).

Because the presented model was built with climatic, tree cover, and elevation data layers, and it employed the continental range, it mainly allowed for drawing conclusions about the possible climatic preferences of the

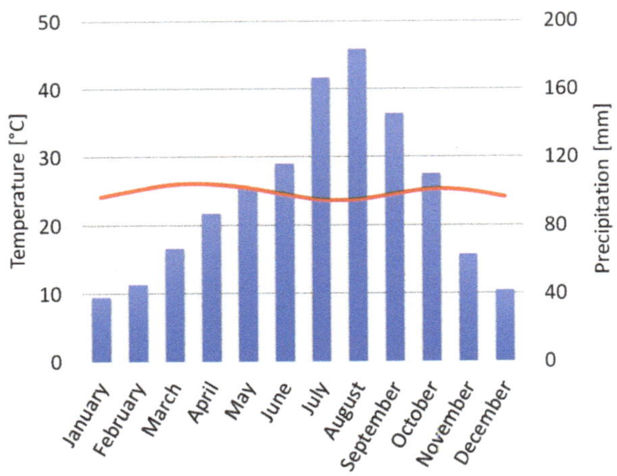

Figure 4 Climate diagram of the distribution of air temperature and precipitation. Line shows the temperature distribution during the year and bars represent the distribution of precipitation throughout the year. For each month, the average value from all known locations of this species is given.

species (see Dungan et al. 2002). Elevation data proved unimportant to the model, but the climate variables showed the possible tolerance ranges for temperatures and precipitations.

The analysis of climate classification shows that most of the potentially suitable niches suggested by the model for *P. biguttatus* were in the area of tropical savanna climate - with a pronounced dry season and monthly mean temperature above 18°C (64°F) in every month of the year. Some of potentially suitable niche areas were also present in the area of hot semi-arid climate, mainly in Ethiopia, Somalia, Kenya, and Tanzania. This kind of climate is characterized by hot summers and mild to warm winters. Areas with this type of climate are usually dominated by grasses and/or shrubs, because of the low precipitation during the year (but the short wet season is well-defined).

As well as other representatives of Reduviidae, *P. biguttatus* is predatory, preying on several different species of prey. For this reason, it is difficult to use biotic interactions in the model, but the analysis of the results from tree cover layer gives a picture of the habitat preferences of the species - open areas with small participation of tree vegetation but also away from the typical desert areas. We believe that this is a characteristic of the food preferences of its prey (grassy vegetation), so the tolerance to climatic factors of this species may be much wider.

Some of the data from unidentified materials (marked with * in the Additional file 1) supports the results of modeling by confirmation of presence of the suitable niche for example in Sierra Leone, Cameroun, or Ghana. Those materials have not been used in the modeling process because of the fact that the description of these specimens contained only the country in which they were collected, without the exact location.

It is worth noting that the model suggests suitable conditions for *P. biguttatus* in the area of Madagascar. There is a geographical barrier which does not allow it to penetrate to the island; however, its introduction could threaten the local fauna.

It should be remembered that even if the model suggests some locations as potentially suitable, the species could be absent there (Anderson 2003; Martínez-Meyer 2005). It is not possible to take into account all the potential environmental factors, like the presence of another predator or lack of prey resources. Therefore, it would be very worthwhile to carry out some field studies to test the hypothesis of a potential distribution and also to obtain data on the ecology of this species and to improve knowledge about its prey in a natural environment.

Conclusions

A description of each nymphal instars of *P. biguttatus* (L.) has been carried out. So far it was only known that this species occurs in widely understood tropical part of continental Africa. The ecological niche modeling methods could be performed thanks to the museum data on the occurrence of the species. It resulted in new information about potentially suitable ecological niches and the possible range of distribution. Moreover, it provides valuable information about climatic preferences of the species and allows for comparison of niche overlap.

Competing interests

The authors declare that they have no competing interests.

Authors' contributions

DC performed taxonomic analysis and collected data on the occurrence of specimens; drafted the manuscript. AB-N performed ecological niche modeling; analyzed the data on the occurrence; analyzed climatic preferences of the species; prepared photos, maps, and diagrams; drafted the manuscript. ŁJ prepared drawings. Authors have read and approved the final manuscript.

Acknowledgements

We are very grateful to E. De Coninck (MRAC), J. Constant (IRSNBG), J. Deckert (ZMB), E. Guilbert (MNHN), G. Lindberg (NHRS), A. Orosz and D. Rédei (HMNH), M. Webb (BMNH), and H. Zettel (NMW) for all their kind help and hospitality during the first author's visit in the collections.

References

Aguirre-Gutiérrez J, Carvalheiro LG, Polce C, van Loon EE, Raes N, Reemer M et al (2013) Fit-for-purpose: species distribution model performance depends on evaluation criteria – Dutch hoverflies as a case study. PLoS One 8(5):406, e63708

Anderson RP (2003) Real vs. artefactual absences in species distributions: tests for *Oryzomys albigularis* (Rodentia: Muridae) in Venezuela. J Biogeogr 30:591–605

Beck J (2012) Predicting climate change effects on agriculture from ecological niche modeling: who profits, who loses? Clim Chang 116:177–189

Berzitis EA, Minigan JN, Hallett RH, Newman JA (2014) Climate and host plant availability impact the future distribution of the bean leaf beetle (*Cerotoma trifurcata*). Glob Chang Biol. doi:10.1111/gcb.12557

Chłond D, Bugaj-Nawrocka A (2014) Model of potential distribution of *Platymeris rhadamanthus* Gerstaecker, 1873 with redescription of species. Zool Stud 53:8

DeFries R, Hansen M, Townshend JRG, Janetos AC, Loveland TR (2000) A new global 1 km data set of percent tree cover derived from remote sensing. Glob Chang Biol 6:247–254

Dungan JL, Perry JN, Dale MRT, Legendre P, Citron-Pousty S, Fortin M-J et al (2002) A balanced view of scale in spatial statistical analysis. Ecography 25:626–640

Elith J, Graham CH, Anderson RP, Dudik M, Ferrier S, Guisan A et al (2006) Novel methods improve prediction of species' distributions from occurrence data. Ecography 29:129–151

Elith J, Kearney M, Phillips S (2010) The art of modelling range-shifting species. Methods Ecol Evol 1:330–342

Elmore AJ, Julian JP, Guinn SM, Fitzpatrick MC (2013) Potential stream density in Mid-Atlantic U.S. watersheds. PLoS One 8(8):e74819

Fielding AH, Bell JF (1997) A review of methods for the assessment of prediction errors in conservation presence/absence models. Environ Conserv 24:38–49

GLOBE Task Team and others (1999) In: Hastings DA, Dunbar PK, Elphingstone GM, Bootz, M, Murakami H, Maruyama H, Masaharu H, Holland P, Payne J, Bryant NA, Logan TL, Muller J-P, Schreier G, MacDonald JS (eds) The Global Land One-kilometer Base Elevation (GLOBE) Digital Elevation Model, Version 1.0. National Oceanic and Atmospheric Administration, National Geophysical Data Center, 325 Broadway, Boulder, Colorado 80305-3328, USA. Digital Database on http://www.ngdc.noaa.gov/mgg/topo/globe.html and CD-ROMs (accessed 16 November 2014)

Google Inc (2013) Google Earth, version 7.2.1.2041 [Software]. Available from https://www.google.com/earth/

Hernandez PA, Graham CH, Master LL, Albert DL (2006) The effect of sample size and species characteristics on performance of different species distribution modeling methods. Ecography 29:773–785

Hijmans RJ, Cameron SE, Parra JL, Jones PG, Jarvis A (2005) Very high resolution interpolated climate surfaces for global land areas. Int J Climatol 25:1965–1978

Laporte FL (1833) Essai d'une classification systematique de l'ordre des Hémiptères (Hémiptéres Hétéroptères, Latr.). Mag Zool (ser 2) 1:1–88, plus supplement

Li H, Zhao G, Cao L, Xu K, Cai W (2010) Taxonomic and bionomic notes on the white spot assassin bug *Platymeris biguttatus* (Linnaeus) (Hemiptera: Reduviidae: Reduviinae). Zootaxa 2644:47–56

Linnaeus C (1767) Systema Naturæ per regna tria naturæ, secundum classes, ordines, genera, species, cum characteribus, differentiis, synonymis, locis. Editio duodecima, reformata. Tomus II. Laurentius Salvius, Holmiae

Maldonado Capriles J (1990) Systematic catalogue of the Reduviidae of the world (Insecta: Heteroptera). (Special edition of the Caribbean Journal of Science). University of Puerto Rico, Mayagüez, Puerto Rico

Martínez-Meyer E (2005) Climate change and biodiversity: some considerations in forecasting shifts on species' potential distributions. Biodivers Inform 2:42–55

Ochoa-Ochoa L, Urbina-Cardona JN, Vázquez L-B, Flores-Villela O, Bezaury-Creel J (2009) The effects of governmental protected areas and social initiatives for land protection on the conservation of Mexican amphibians. PLoS One 4(9):e6878

Papeş M, Gaubert P (2007) Modelling ecological niches from low numbers of occurrences: assessment of the conservation status of poorly known viverrids (Mammalia, Carnivora) across two continents. Divers Distrib 13:890–902

Pearce J, Ferrier S (2000) Evaluating the predictive performance of habitat models developed using logistic regression. Ecol Model 133(3):225–245

Peel MC, Finlayson BL, McMahon TA (2007) Updated world map of the Köppen-Geiger climate classification. Hydrol Earth Syst Sci 11:1633–1644

Phillips SJ, Dudík M (2008) Modelling of species distributions with Maxent: new extensions and a comprehensive evaluation. Ecography 31:161–175

Phillips SJ, Anderson RP, Schapire RE (2006) Maximum entropy modeling of species geographic distributions. Ecol Model 190:231–259

Raes N, ter Steege H (2007) A null-model for significance testing of presence-only species distribution models. Ecography 30:727–736

R Development Core Team (2011) R: a language and environment for statistical computing (version 3.0.2). R Foundation for Statistical Computing, Vienna, Austria

SAGA Development Team (2013) System for Automated Geoscientific Analyses (SAGA) (version 2.1.0). Institute of Geography at the University of Hamburg, Germany

Solhjouy-Fard S, Sarafrazi A (2013) Potential impacts of climate change on distribution range of *Nabis pseudoferus* and *N. palifer* (Hemiptera: Nabidae) in Iran. Ent Sci 17(3):283–292

Stohlgren TJ, Jarnevich CS, Esaias WE, Morisette JT (2011) Bounding species distribution models. Curr Zool 57(5):642–647

Veloz SD (2009) Spatially autocorrelated sampling falsely inflates measures of accuracy for presence-only niche models. J Biogeogr 36:2290–2299

Wieczorek K, Bugaj-Nawrocka A (2014) Invasive Aphids of the Tribe Siphini: Model of Potentially Suitable Ecological Niche. Agric For Entomol 16:434–443

Williams GJ (2011) Data mining with Rattle and R: the art of excavating data for knowledge discovery. Use R! series. Springer–Verlag, Heidelberg

Wisz MS, Hijmans RJ, Li J, Peterson AT, Graham CH, Guisan A et al (2008) Effects of sample size on the performance of species distribution models. Divers Distrib 14:763–773

Permissions

All chapters in this book were first published in ZS, by Springer; hereby published with permission under the Creative Commons Attribution License or equivalent. Every chapter published in this book has been scrutinized by our experts. Their significance has been extensively debated. The topics covered herein carry significant findings which will fuel the growth of the discipline. They may even be implemented as practical applications or may be referred to as a beginning point for another development.

The contributors of this book come from diverse backgrounds, making this book a truly international effort. This book will bring forth new frontiers with its revolutionizing research information and detailed analysis of the nascent developments around the world.

We would like to thank all the contributing authors for lending their expertise to make the book truly unique. They have played a crucial role in the development of this book. Without their invaluable contributions this book wouldn't have been possible. They have made vital efforts to compile up to date information on the varied aspects of this subject to make this book a valuable addition to the collection of many professionals and students.

This book was conceptualized with the vision of imparting up-to-date information and advanced data in this field. To ensure the same, a matchless editorial board was set up. Every individual on the board went through rigorous rounds of assessment to prove their worth. After which they invested a large part of their time researching and compiling the most relevant data for our readers.

The editorial board has been involved in producing this book since its inception. They have spent rigorous hours researching and exploring the diverse topics which have resulted in the successful publishing of this book. They have passed on their knowledge of decades through this book. To expedite this challenging task, the publisher supported the team at every step. A small team of assistant editors was also appointed to further simplify the editing procedure and attain best results for the readers.

Apart from the editorial board, the designing team has also invested a significant amount of their time in understanding the subject and creating the most relevant covers. They scrutinized every image to scout for the most suitable representation of the subject and create an appropriate cover for the book.

The publishing team has been an ardent support to the editorial, designing and production team. Their endless efforts to recruit the best for this project, has resulted in the accomplishment of this book. They are a veteran in the field of academics and their pool of knowledge is as vast as their experience in printing. Their expertise and guidance has proved useful at every step. Their uncompromising quality standards have made this book an exceptional effort. Their encouragement from time to time has been an inspiration for everyone.

The publisher and the editorial board hope that this book will prove to be a valuable piece of knowledge for researchers, students, practitioners and scholars across the globe.

List of Contributors

Hsi-Te Shih and Yu-Ching Lai
Department of Life Science, National Chung Hsing University, Taichung 40227, Taiwan

Noor Us Saher
University of Karachi, Karachi 75270, Pakistan

Ehsan Kamrani
Marine Biology Department, Hormozgan University, Bandar Abbas 79145, Iran

Peter KL Ng
Department of Biological Sciences, National University of Singapore, Singapore 119260, Republic of Singapore

Min-Yun Liu
Taiwan Ocean Research Institute, National Applied Research Laboratories, Qieding, Kaohsiung City 85243, Taiwan

Lukasz Wejnerowski, Slawek Cerbin and Marcin Krzysztof Dziuba
Department of Hydrobiology, Faculty of Biology, Adam Mickiewicz University, Umultowska 89, 61-614 Poznań, Poland

Mei-Ling Bai, Fan-Sian Lin, Yu-Ching Lee and Chih-hao Hsieh
Institute of Oceanography, National Taiwan University, Taipei 10617, Taiwan

Mei-Ling Bai and Chih-hao Hsieh
Ocean Data Bank, Ministry of Science and Technology, Taipei 10617, Taiwan

Gwo-Ching Gong
Center of Excellence for the Oceans, National Taiwan Ocean University, Keelung 20224, Taiwan

Chih-hao Hsieh
Institute of Ecology and Evolutionary Biology, National Taiwan University, Taipei 10617, Taiwan

Nelson Ferretti
Centro de Estudios Parasitológicos y de Vectores CEPAVE (CCT-CONICET-La Plata), Calle 2 N°584, La Plata B1902CHX, Argentina

Kaede Sato, Daichi Arima, Yoshiyuki Abe and Atsushi Yamaguchi
Laboratory of Marine Biology, Graduate School of Fisheries Science, Hokkaido University, 3-1-1 Minatomachi, Hakodate, Hokkaido 041-8611, Japan

Kohei Matsuno
Arctic Environmental Research Center, National Institute of Polar Research, 10-3 Midori-cho, Tachikawa, Tokyo 190-8518, Japan

Raja Jelassi, Hajer Khemaissia and Karima Nasri-Ammar
Université de Tunis El Manar II, Faculté des Sciences de Tunis, Unité de recherche Bio-Écologie et Systématique Évolutive, Tunis, Tunisie

Martin Zimmer
FB Organismische Biologie, AG Ökologie, Biodiversität & Evolution der Tiere, Universität Salzburg, Salzburg, Austria

Dieter Garbe-Schönberg
Institut für Geowissenschaften, ICP-MS Labor, Universität zu Kiel, Kiel, Germany

Atsushi Yamaguchi and Tomoe Homma
Laboratory of Marine Biology, Graduate School of Fisheries Science, Hokkaido University, 3-1-1 Minatomachi, Hakodate, Hokkaido 041-8611, Japan

Kohei Matsuno
Arctic Environmental Research Centre, National Institute of Polar Research, 10-3 Midori-cho, Tachikawa, Tokyo 190-8518, Japan

Cheol Min Lee
Division of Forest Ecology, Korea Forest Research Institute, 57 Hoegi-ro, Dongdaemun-gu, Seoul 130-712, Republic of Korea

Jin Wook Park
Department of Landscape Architecture, Catholic University of Daegu, 13-13 Hayang-ro, Hayang-eup, Gyeongsan-si, Daegu 712-702, Republic of Korea

Tae-Sung Kwon
Division of Forest Insect Pests and Diseases, Korea Forest Research Institute, 57 Hoegi-ro, Dongdaemun-gu, Seoul 130-712, Republic of Korea

Sung-Soo Kim
Research Institute for East Asian Environment and Biology, 293-27 Amsa 3 dong, Gangdong-gu, Seoul 143-203, Republic of Korea

Jae Won Ryu
Ewha Womans University Natural History Museum, 52 Ewhayeodae-gil, Seodaemun-gu, Seoul 120-750, Republic of Korea

Seung Jae Jung
Department of Biology, Kyung Hee University, 26 Kyungheedae-ro, Dongdaemun-gu, Seoul 130-701, Republic of Korea

Sun Kyung Lee
Entomology program, Department of Agricultural Biotechnology, Seoul National University, Seoul, Republic of Korea

Soledad Ibáñez and Marcela A Vidal
Departamento de Ciencias Básicas, Facultad de Ciencias, Universidad del Bío-Bío, Casilla 447, 4051381 Chillán, Chile

Juan Carlos Ortiz
Departamento de Zoología, Facultad de Ciencias naturales y Oceanográficas, Universidad de Concepción, Casilla 160-C, Concepción, Chile

Fernando Torres-Pérez
Instituto de Biología, P. Universidad Católica de Valparaíso, Av. Universidad 330, Campus Curauma, Valparaíso, Chile

Alejandro Travaini and Julieta Pedrana
Centro de Investigación de Puerto Deseado, Universidad Nacional de la Patagonia Austral, CONICET, Avenida Prefectura Naval s/n, 9050 Puerto Deseado, Santa Cruz, Argentina

Sonia C Zapata and Juan I Zanón
Centro de Investigación de Puerto Deseado, Universidad Nacional de la Patagonia Austral, Avenida Prefectura Naval s/n, 9050 Puerto Deseado, Santa Cruz, Argentina

Javier Bustamante
Department of Wetland Ecology, Remote Sensing and GIS Lab (LAST-EBD), Estación Biológica de Doñana, CSIC, Américo Vespucio s/n, 41092 Sevilla, Spain

Julieta Pedrana
Recursos Naturales y Gestión Ambiental, Instituto Nacional de Tecnología Agropecuaria, EEA Balcarce, CC 276, 7620 Balcarce, Buenos Aires, Argentina

Juan I Zanón
Centro para el Estudio y Conservación de las Aves Rapaces en Argentina, UNLPam, INCITAP, CONICET, Avda. Uruguay 151, 6300 Santa Rosa, La Pampa, Argentina

Alejandro Rodríguez
Department of Conservation Biology, Estación Biológica de Doñana, CSIC, Américo Vespucio s/n, 41092 Sevilla, Spain

Atilla Arslan
Department of Biology, Faculty of Science, Selçuk University, 42031 Konya, TR, Turkey

Jan Zima
Institute of Vertebrate Biology, Academy of Sciences of the Czech Republic, Kvetna 8, 603 65 Brno, CZ, Czech Republic

Javier Sánchez-Hernández and Fernando Cobo
Department of Zoology and Physical Anthropology, Faculty of Biology, University of Santiago de Compostela, Campus Sur s/n, 15782 Santiago de Compostela, Spain

Javier Sánchez-Hernández
Department of Arctic and Marine Biology, Faculty for Biosciences, Fisheries and Economics, UiT The Arctic University of Norway, N-9037 Tromsø, Norway

Fernando Cobo
Station of Hydrobiology 'Encoro do Con', Castroagudín s/n, 36617 Vilagarcía de Arousa, Pontevedra, Spain

Brian J Stucky
Department of Ecology and Evolutionary Biology, University of Colorado, Campus Box 334 UCB, Boulder, CO 80309, USA

Maria Carolina de Almeida Castilho
Departamento de Zoologia, Instituto de Biociências, Universidade Estadual Paulista, CP. 510, CEP 18618-000, Botucatu, SP, Brazil

Maria José dos Santos Wisniewski, Cínthia Bruno de Abreu and Tereza Cristina Orlando
Instituto de Ciências da Natureza, Universidade Federal de Alfenas-MG, CEP 37130-000, Alfenas, MG, Brazil

Carlos Lara, Berenice Pérez and Citlalli Castillo-Guevara
Centro de Investigación en Ciencias Biológicas, Universidad Autónoma de Tlaxcala, Km. 10.5 Autopista Tlaxcala-San Martín Texmelucan, San Felipe Ixtacuixtla, Tlaxcala 90120, Mexico

Martín Alejandro Serrano-Meneses
Laboratorio de Biología Evolutiva, Centro Tlaxcala de Biología de la Conducta, Universidad Autónoma de Tlaxcala, Carretera Tlaxcala-Puebla, Km. 1.5, 90062 Tlaxcala, Mexico

Milena Roszkowska, Marta Ostrowska and Łukasz Kaczmarek
Department of Animal Taxonomy and Ecology, Faculty of Biology, Adam Mickiewicz University in Poznań, Umultowska 89, 61-614, Poznań, Poland

Milena Roszkowska and Łukasz Kaczmarek
Laboratorio de Ecología Natural y Aplicada de Invertebrados, Universidad Estatal Amazónica, Campus Principal Km 2.1/2 via a Napo (Paso Lateral) Puyo, Pastaza, Ecuador

Chang-En Chou and Hsueh-Wen Chang
Department of Biological Sciences, National Sun Yat-sen University, 70, Lien-hai Road, Kaohsiung 804, Taiwan

Te-Yu Liao
Department of Oceanography, National Sun Yat-sen University, 70, Lien-hai Road, Kaohsiung 804, Taiwan

Shui-Kai Chang
Institute of Marine Affairs, National Sun Yat-sen University, 70, Lien-hai Road, Kaohsiung 804, Taiwan

Rick A Adams and Emily R Snode
School of Biological Sciences, Ross Hall, University of Northern Colorado, 501 20 Street, Greeley, CO 80639, USA

Jessica S Fortin, André D Dallaire and Marie-Odile Benoit-Biancamano
Département de pathologie et de microbiologie, Faculté de medicine vétérinaire, Université de Montréal, 3200 Sicotte, Saint-Hyacinthe, QC J2S 2M2, Canada

Ariane Santamaria-Bouvier and Stéphane Lair
Département des sciences cliniques, Faculté de medicine vétérinaire, Université de Montréal, 3200 Sicotte, Saint-Hyacinthe, QC J2S 2M2, Canada

Jose E F Alfaya and Gregorio Bigatti
LARBIM, IBIOMAR- Centro Nacional Patagónico (CENPAT), CONICET, Bvd. Brown 2915, U9120ACV Puerto Madryn, Chubut, Argentina
Facultad de Ciencias Naturales, Universidad Nacional de la Patagonia San Juan Bosco (UNPSJB), Bvd. Brown S/N, U9120ACV Puerto Madryn, Chubut, Argentina

Hiroshi Kajihara
Faculty of Science, Hokkaido University, Kita 10, Nishi 8, Kita-ku, Sapporo 060-0810, Japan

Malin Strand
The Swedish Species Information Centre, SLU, Bäcklösavägen 10, Box 7007, SE-750 07 Uppsala, Sweden

Per Sundberg
Department of Biological and Environmental Sciences, University of Gothenburg, Medicinaregatan 18, PO Box 463, SE-40530 Gothenburg, Sweden

Annie Machordom
Museo Nacional de Ciencias Naturales (MNCN), CSIC, José Gutiérrez Abascal, 2, 28006 Madrid, Spain

Yuriana Gómez-Ortiz and Octavio Monroy-Vilchis
Estación Biológica Sierra Nanchititla, Facultad de Ciencias, Universidad Autónoma del Estado de México, Instituto Literario 100, Colonia Centro, Toluca, CP 50000, México

Germán D Mendoza-Martínez
Departamento de Producción Agrícola y Animal, Universidad Autónoma Metropolitana, Xochimilco, Calzada del Hueso 1100, Distrito Federal 04960, México

Bai-Shi Wang, Zhen-Long Wang, Jun-Dong Tian, Zhen-Wei Cui and Ji-Qi Lu
Institute of Biodiversity and Ecology, Zhengzhou University, Kexue Dadao 100, Zhengzhou 450001, PR China

Jun-Dong Tian
Current address: Department of Wildlife Diseases, Leibniz Institute for Zoo and Wildlife Research, Alfred-Kowalke-Straße 17, Berlin 10315, Germany

Christos Sokos, Konstantinos Andreadis and Nikolaos Papageorgiou
Laboratory of Wildlife and Freshwater Fisheries, School of Forestry and Natural Environment, Aristotle University of Thessaloniki, Thessaloniki GR-54124, Hellas (Greece)

Christos Sokos
Research Division, Hunting Federation of Macedonia and Thrace, Ethnikis, Antistasis 173-175, GR-55134 Thessaloniki, Hellas

Hugues-Alexandre Blain and Iván Lózano-Fernández
IPHES, Institut Català de Paleoecologia Humana i Evolució Social, c/ Marcel • lí Domingo s/n (Edifici W3), Campus Sescelades, 43007 Tarragona, Spain
Area de Prehistoria, Universitat Rovira i Virgili (URV), Avinguda de Catalunya 35, 43002 Tarragona, Spain

Gottfried Böhme
Museum für Naturkunde, Leibniz-Institut für Evolutions- und Biodiversitätsforschung an der Humboldt-Universität zu Berlin, Invalidenstrasse 43, 10115 Berlin, Germany

Andreas Altenburger and Martin Vinther Sørensen
Natural History Museum of Denmark, University of Copenhagen, Universitetsparken 15, 2100 Copenhagen, Denmark

Hyun Soo Rho
Dokdo Research Center, Korean Institute of Ocean Science and Technology, 767-813 Uljin, South Korea

Cheon Young Chang
Department of Biological Science, College of Natural Sciences, Daegu University, 712-714 Gyeongsan, South Korea

Jian-Xiang Liao and Hin-Kiu Mok
Department of Oceanography, National Sun Yat-sen University, Lienhai Road 70, Kaohsiung 80424, Taiwan

Hsin-Ming Yeh
Coastal and Offshore Resources Research Center, Fisheries Research Institute, Yugang North 3rd Road 6, Kaohsiung 80672, Taiwan

Dominik Chłond, Agnieszka Bugaj-Nawrocka and Łukasz Junkiert
Department of Zoology, Faculty of Biology and Environmental Protection, University of Silesia, Bankowa 9, 40-007 Katowice, Poland